Fundamental Astronomy

Springer
*Berlin
Heidelberg
New York
Barcelona
Budapest
Hong Kong
London
Milan
Paris
Santa Clara
Singapore
Tokyo*

H. Karttunen P. Kröger H. Oja
M. Poutanen K. J. Donner (Eds.)

Fundamental Astronomy

Third Revised and Enlarged Edition
With 402 Illustrations Including 36 Colour Plates

 Springer

Dr. *Hannu Karttunen*
Tuorla Observatory, University of Turku,
FIN-21500 Piikkö, Finland

Dr. *Markku Poutanen*
Finnish Geodetic Institute, Geodeetinrinne 2,
FIN-02430 Masala, Finland

Dr. *Pekka Kröger*
Isonniitynkatu 9 C9,
FIN-00520 Helsinki, Finland

Dr. *Karl Johan Donner*
Observatory, P.O. Box 14,
FIN-00014 University of Helsinki, Finland

Dr. *Heikki Oja*
Observatory, P.O. Box 14,
FIN-00014 University of Helsinki, Finland

Cover photograph: The 15-m SEST (Swedish-ESO Submillimetre Telescope) at La Silla, Chile. (Photograph European Southern Observatory)

Frontispiece: The η Carinae nebula, NGC3372, is a giant HII region in the Carina spiral arm of our galaxy at a distance of 8000 light-years. (Photograph European Southern Observatory)

Title of the original Finnish edition: *Tähtitieteen perusteet* (Ursan julkaisuja 56)
© Tähtitieteellinen yhdistys Ursa, Helsinki 1984, 1995

Sources for the illustrations are given in the captions and more fully at the end of the book. Most of the uncredited illustrations are
© Ursa Astronomical Association, Raatimiehenkatu 3A2, FIN-00140 Helsinki, Finland

ISBN 3-540-60936-9   3rd Edition
Springer-Verlag Berlin Heidelberg New York

ISBN 3-540-57203-1   2nd Edition
Springer-Verlag Berlin Heidelberg New York

Library of Congress Cataloging-in-Publication Data. Tähtitieteen perusteet. English. Fundamental astronomy / H. Karttunen ... [et al.] eds. – 3rd rev. and enl. ed. p. cm. Includes bibliographical references and index. ISBN 3-540-60936-9 (softcover :alk. paper) 1. Astronomy. I. Karttunen, Hannu.   II. Title   QB43.2. T2613   1996   96-18691

Production editor: A. Kübler, Heidelberg
Cover design: E. Kirchner, Heidelberg
Typesetting: K + V Fotosatz GmbH, D-64743 Beerfelden
SPIN: 10533322   55/3144-543210 – Printed on acid-free paper

Preface to the Third Edition

In addition to small corrections and notes on new results this third edition includes two major revisions. Development of telescopes and instruments has been fast, and the chapter on observations and instruments was getting out of date. Material on the CCD camera has been added; the text in the section on future telescopes has been updated and included in other sections, since most of the telescopes have already been built. The chapter on galaxies has been reorganized to reflect the physical properties instead of morphological classification.

We are grateful to our keen readers who have sent us comments even on minute typographical errors found in the previous edition.

Helsinki, June 1996 *The Editors*

Preface to the Second Edition

For this second English edition, the text was brought up to date to correspond to the situation as of April 1993. Small revisions were made throughout the book, and some sections in Chapters 3 (Observations and Instruments), 8 (The Solar System), 13 (The Sun), and 18 (The Milky Way) were rewritten. Many new photographs were added, some of them as replacements for older ones.

In many reviews of this book's English as well as German first editions, more exercises were expected at the ends of the chapters. We have added worked-out examples to nearly all chapters that didn't have them before, and we added some homework problems to the chapters too. The answers to these problems are to be found at the end of the Appendices. We hope the new examples and problems facilitate the work of the instructor who may choose to use this book.

We want to thank Dr. Christian Naundorf of Bonn, Drs. Erik Heyn Olsen and J.O. Petersen of Copenhagen, Dr. Björn Sundelius of Göteborg, and an anonymous German book buyer, who pointed out many typographical errors and inaccuracies in our book. For financial support in modifying the second English edition, we thank Suomalaisen kirjallisuuden edistämisvarojen valtuuskunta.

Helsinki, October 1993   *The Editors*

Preface to the First Edition

The main purpose of this book is to serve as a university textbook for a first course in astronomy. However, we believe that the audience will also include many serious amateurs, who often find the popular texts too trivial. The lack of a good handbook for amateurs has become a problem lately, as more and more people are buying personal computers and need exact, but comprehensible, mathematical formalism for their programs. The reader of this book is assumed to have only a standard high-school knowledge of mathematics and physics (as they are taught in Finland); everything more advanced is usually derived step by step from simple basic principles. The mathematical background needed includes plane trigonometry, basic differential and integral calculus, and (only in the chapter dealing with celestial mechanics) some vector calculus. Some mathematical concepts the reader may not be familiar with are briefly explained in the appendices or can be understood by studying the numerous exercises and examples. However, most of the book can be read with very little knowledge of mathematics, and even if the reader skips the mathematically more involved sections, (s)he should get a good overview of the field of astronomy.

This book has evolved in the course of many years and through the work of several authors and editors. The first version consisted of lecture notes by one of the editors (Oja). These were later modified and augmented by the other editors and authors. Hannu Karttunen wrote the chapters on spherical astronomy and celestial mechanics; Vilppu Piirola added parts to the chapter on observational instruments, and Göran Sandell wrote the part about radio astronomy; chapters on magnitudes, radiation mechanisms and temperature were rewritten by the editors; Markku Poutanen wrote the chapter on the solar system; Juhani Kyröläinen expanded the chapter on stellar spectra; Timo Rahunen rewrote most of the chapters on stellar structure and evolution; Ilkka Tuominen revised the chapter on the Sun; Kalevi Mattila wrote the chapter on interstellar matter; Tapio Markkanen wrote the chapters on star clusters and the Milky Way; Karl Johan Donner wrote the major part of the chapter on galaxies; Mauri Valtonen wrote parts of the galaxy chapter, and, in collaboration with Pekka Teerikorpi, the chapter on cosmology. Finally, the resulting, somewhat inhomogeneous, material was made consistent by the editors.

The English text was written by the editors, who translated parts of the original Finnish text, and rewrote other parts, updating the text and correcting errors found in the original edition. The parts of text set in smaller print are less important material that may still be of interest to the reader.

For the illustrations, we received help from Veikko Sinkkonen, Mirva Vuori and several observatories and individuals mentioned in the figure captions. In the practical work, we were assisted by Arja Kyröläinen and Merja Karsma. A part of the translation was read and corrected by Brian Skiff. We want to express our warmest thanks to all of them.

Financial support was given by the Finnish Ministry of Education and Suomalaisen kirjallisuuden edistämisvarojen valtuuskunta (a foundation promoting Finnish literature), to whom we express our gratitude.

Helsinki, June 1987   *The Editors*

Contents

The Virgo cluster of galaxies. Each galaxy contains hundreds of billions of stars
(Photograph National Optical Astronomy Observatories)

Chapter 1 **Introduction**

1.1 The Role of Astronomy

On a dark, cloudless night, at a distant location far away from the city lights, the starry sky can be seen in all its splendour (Fig. 1.1). It is easy to understand how these thousands of lights in the sky have affected people throughout the ages. After the Sun, necessary to all life, the Moon, governing the night sky and continuously changing its phases, is the most conspicuous object in the sky. The stars seem to stay fixed. Only some relatively bright objects, the planets, move with respect to the stars.

The phenomena of the sky aroused people's interest a long time ago. The Cro Magnon people made bone engravings 30,000 years ago, which may depict the phases of the Moon. These calendars are the oldest astronomical documents, 25,000 years older than writing.

Agriculture required a good knowledge of the seasons. Religious rituals and prognostication were based on the locations of the celestial bodies. Thus time reckoning be-

Fig. 1.1. The North America nebula in the constellation of Cygnus. The brightest star on the right is α Cygni or Deneb. (Photo M. Poutanen and H. Virtanen)

came more and more accurate, and people learned to calculate the movements of celestial bodies in advance.

During the rapid development of seafaring, when voyages extended farther and farther from home ports, position determination presented a problem for which astronomy offered a practical solution. Solving these problems of navigation were the most important tasks of astronomy in the 17th and 18th centuries, when the first precise tables on the movements of the planets and on other celestial phenomena were published. The basis for these developments was the discovery of the laws governing the motions of the planets by Copernicus, Tycho Brahe, Kepler, Galilei and Newton.

Astronomical research has changed man's view of the world from geocentric, anthropocentric conceptions to the modern view of a vast universe where man and the Earth play an insignificant role. Astronomy has taught us the real scale of the nature surrounding us.

Modern astronomy is fundamental science, motivated mainly by man's curiosity, his wish to know more about Nature and the Universe. Astronomy has a central role in forming a scientific view of the world. "A scientific view of the world" means a model of the universe based on observations, thoroughly tested theories and logical reasoning. Observations are always the ultimate test of a model: if the model does not fit the observations, it has to be changed, and this process must not be limited by any philosophical, political or religious conceptions or beliefs.

1.2 Astronomical Objects of Research

Modern astronomy explores the whole Universe and its different forms of matter and energy. Astronomers study the contents of the Universe from the level of elementary particles and molecules (with masses of 10^{-30} kg) to the largest superclusters of galaxies (with masses of 10^{50} kg).

Astronomy can be divided into different branches in several ways. The division can be made according to either *the methods* or *the objects of research*.

The Earth (Fig. 1.2) is of interest to astronomy for many reasons. Nearly all observations must be made through *the atmosphere*, and the phenomena of the upper atmosphere and magnetosphere reflect the state of interplanetary space. The Earth is also the most important object of comparison for planetologists.

The Moon is still studied by astronomical methods, although spacecraft and astronauts have visited its surface and brought samples back to the Earth. To amateur astronomers, the Moon is an interesting and easy object for observations.

In the study of *the planets* of the solar system, the situation in the 1980's was the same as in lunar exploration 20 years earlier: the surfaces of the planets and their moons have been mapped by fly-bys of spacecraft or by orbiters, and spacecraft have softlanded on Mars and Venus. This kind of exploration has tremendously added to our knowledge of the conditions on the planets. Continuous monitoring of the planets, however, can still only be made from the Earth, and many bodies in the solar system still await their spacecraft.

The Solar System is governed by *the Sun*, which produces energy in its centre by nuclear fusion. The Sun is our nearest *star*, and its study lends insight into conditions on other stars.

Fig. 1.2. The Earth as seen from the Moon. The picture was taken on the last Apollo flight in December, 1972. (Photo NASA)

Some thousands of stars can be seen with the naked eye, but even a small telescope reveals millions of them. Stars can be classified according to their observed characteristics. A majority are like the Sun; we call them *main sequence stars*. However, some stars are much larger, *giants* or *supergiants*, and some are much smaller, *white dwarfs*. Different types of stars represent different stages of stellar evolution. Most stars are components of *binary* or *multiple systems*; many are *variable*: their brightness is not constant.

Among the newest objects studied by astronomers are the *compact stars: neutron stars* and *black holes.* In them, matter has been so greatly compressed and the gravitational field is so strong that Einstein's general theory of relativity must be used to describe matter and space.

Stars are points of light in an otherwise seemingly empty space. Yet *interstellar space* is not empty, but contains large clouds of *atoms, molecules, elementary particles* and *dust*. New matter is injected into interstellar space by erupting and exploding stars; at other places, new stars are formed from contracting interstellar clouds.

Stars are not evenly distributed in space, but form concentrations, *clusters of stars*. These consist of stars born near each other, and in some cases, remaining together for billions of years.

The largest concentration of stars in the sky is *the Milky Way*. It is a massive stellar system, a *galaxy*, consisting of over 200 billion stars. All the stars visible to the naked eye belong to the Milky Way. Light travels across our galaxy in 100,000 years.

The Milky Way is not the only galaxy, but one of almost innumerable others. Galaxies often form *clusters of galaxies,* and these clusters can be clumped together into *superclusters*. Galaxies are seen at all distances as far away as our observations reach. Still further out we see *quasars* − the light of the most distant quasars we see now was emitted when the Universe was one-tenth of its present age.

The largest object studied by astronomers is the whole *Universe. Cosmology,* once the domain of theologicians and philosophers, has become the subject of physical theories and concrete astronomical observations.

Table 1.1. The share of different branches of astronomy in *Astronomy and Astrophysics Abstracts* for the second half of 1991. This index service contains short abstracts of all astronomical articles published during the half year covered by the book

Branch	Pages	Percent
Astronomical instruments and techniques	95	8
Positional astronomy, celestial mechanics	27	2
Space research	14	1
Theoretical astrophysics	177	15
Sun	112	10
Earth	35	3
Planetary system	120	10
Stars	201	17
Interstellar matter, nebulae	72	6
Radio sources, x-ray sources, cosmic rays	39	3
Stellar system, galaxy, extragalactic objects, cosmology	284	24

Among the different branches of research, *spherical*, or *positional, astronomy* studies the coordinate systems on the celestial sphere, their changes and the apparent places of celestial bodies in the sky. *Celestial mechanics* studies the movements of bodies in the solar system, in stellar systems and among the galaxies and clusters of galaxies. *Astrophysics* studies celestial objects, using methods of modern physics. It thus has a central position in almost all branches of astronomy (Table 1.1).

Astronomy can be divided into different areas according to *the wavelength* used in observations. We can speak of *radio, infrared, optical, ultraviolet, x-ray* or *gamma astronomy,* depending on which wavelengths of the electromagnetic spectrum are used. In the future, *neutrinos* and *gravitational waves* may also be observed.

1.3 The Scale of the Universe

The masses and sizes of astronomical objects are usually enormously large. But to understand their properties, the smallest parts of matter, molecules, atoms and elementary particles, must be studied. The densities, temperatures and magnetic fields in the Universe vary within much larger limits than can be reached in laboratories on the Earth.

The greatest natural density met on the Earth is 22,500 kg/m^3 (osmium), while in neutron stars, densities of the order of 10^{18} kg/m^3 are possible. The density in the best vacuum achieved on the Earth is only 10^{-9} kg/m^3, but in interstellar space, the density of the gas may be 10^{-21} kg/m^3 or still less. Modern accelerators can give particles energies of the order of 10^{12} electron volts (eV). Cosmic rays coming from the sky may have energies of over 10^{20} eV.

It has taken man a long time to grasp the vast dimensions of space. The scale of the solar system was known relatively well already in the 17th century, the first measurements of stellar distances were made in the 1830's, and the distances to the galaxies were established only in the 1920's.

We can get some kind of picture of the distances involved (Fig. 1.3) by considering the time required for light to travel from a source to the retina of the human eye. It takes 8 minutes for light to travel from the Sun, 5 1/2 hours from Pluto and 4 years

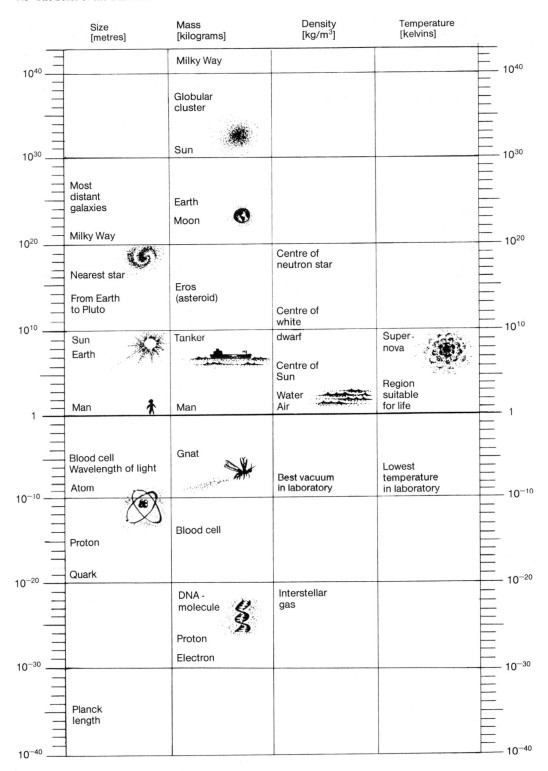

Fig. 1.3. The dimensions of the Universe

Fig. 1.4. The globular cluster M13. There are over a million stars in the cluster. (Photo Palomar Observatory)

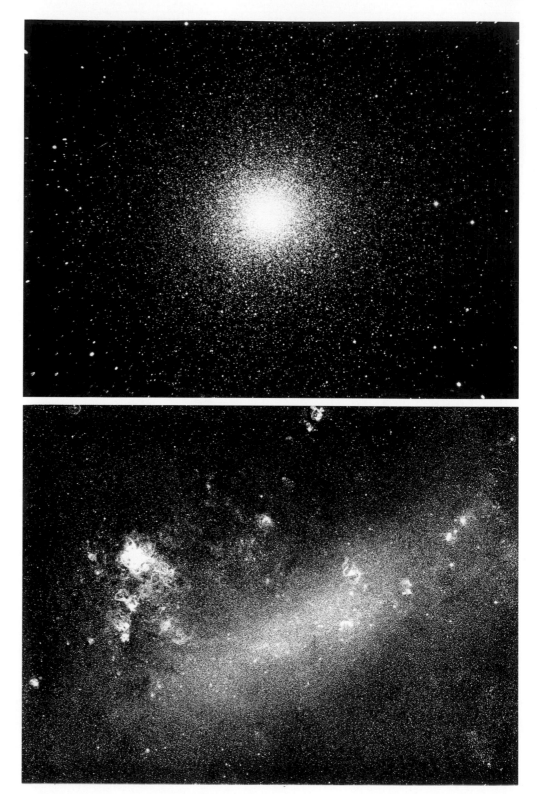

Fig. 1.5. The Large Magellanic Cloud, our nearest neighbour galaxy. (Photo National Optical Astronomy Observatories, Cerro Tololo Inter-American Observatory)

from the nearest star. We cannot see the centre of the Milky Way, but the many globular clusters around the Milky Way are at approximately similar distances. It takes about 20,000 years for the light from the globular cluster of Fig. 1.4 to reach the Earth. It takes 150,000 years to travel the distance from the nearest galaxy, the Magellanic Cloud seen on the southern sky (Fig. 1.5). The photons that we see now started their voyage when Neanderthal Man lived on the Earth. The light coming from the Andromeda Galaxy in the northern sky originated 2 million years ago. Around the same time, the first actual human using tools, Homo habilis, appeared.

The most distant objects known, the quasars, are so far away that their radiation, seen on the Earth now, was emitted long before the Sun or the Earth were born.

Chapter 2 Spherical Astronomy

Spherical astronomy is a science studying astronomical coordinate frames, directions and apparent motions of celestial objects, determination of position from astronomical observations, observational errors, etc. We shall concentrate mainly on astronomical coordinates, apparent motions of stars and time reckoning. Also, some of the most important star catalogues will be introduced.

For simplicity we will assume that the observer is always on the Northern hemisphere. Although all definitions and equations are easily generalized for both hemispheres, this might be unnecessarily confusing. In spherical astronomy, all angles are usually expressed in degrees; we will also use degrees unless otherwise mentioned.

2.1 Spherical Trigonometry

For the coordinate transformations of spherical astronomy, we need some mathematical tools, which we present now.

If a plane passes through the centre of a sphere, it will split the sphere into two identical hemispheres along a circle called a *great circle* (Fig. 2.1). A line perpendicular to the plane and passing through the centre of the sphere intersects the sphere at the *poles* P and P'. If a sphere is intersected by a plane not containing the centre, the intersection curve is a *small circle*. There is exactly one great circle passing through two given points Q and Q' on a sphere (unless these points are antipodal, in which case all circles passing through both of them are great circles). An arc QQ' of this great circle is the shortest path on the surface of the sphere between these points.

A *spherical triangle* is not just any three-cornered figure lying on a sphere; its sides must be arcs of great circles. The spherical triangle in Fig. 2.2 has the arcs AB, BC and AC as its sides. If the radius of the sphere is r, the length of the arc AB is

Fig. 2.1

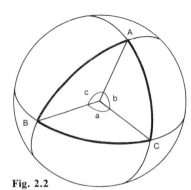

Fig. 2.2

Fig. 2.1. A great circle is the intersection of a sphere and a plane passing through its centre. P and P' are the poles of the great circle. The shortest path from Q to Q' follows the great circle

Fig. 2.2. A spherical triangle is bounded by three arcs of great circles, AB, BC and CA. The corresponding central angles are c, a and b

$$|AB| = rc, \quad [c] = \text{rad},$$

where c is the angle subtended by the arc AB as seen from the centre. This angle is called the *central angle* of the side AB. Because lengths of sides and central angles correspond to each other in a unique way, it is customary to give the central angles instead of the sides. In this way, the radius of the sphere does not enter into the equations of spherical trigonometry. An angle of a spherical triangle can be defined as the angle between the tangents of the two sides meeting at a vertex, or as the dihedral angle between the planes intersecting the sphere along these two sides. We denote the angles of a spherical triangle by capital letters (A, B, C) and the opposing sides, or, more correctly, the corresponding central angles, by lowercase letters (a, b, c).

The sum of the angles of a spherical triangle is always greater than 180 degrees; the excess

$$E = A + B + C - 180° \tag{2.1}$$

is called the *spherical excess*. It is not a constant, but depends on the triangle. Unlike in plane geometry, it is not enough to know two of the angles to determine the third one. The area of a spherical triangle is related to the spherical excess in a very simple way:

$$\text{Area} = Er^2, \quad [E] = \text{rad}. \tag{2.2}$$

This shows that the spherical excess equals the solid angle in steradians (see Appendix A.1), subtended by the triangle as seen from the centre.

To prove (2.2), we extend all sides of the triangle $\triangle$ to great circles. These great circles will form another triangle $\triangle'$, antipodal to $\triangle$. The area of the slice $S(A)$ bounded by the two sides of the angle A (shaded area in Fig. 2.3) is obviously $2A/2\pi$ times the area of the sphere: $S(A) = 2A/2\pi \times 4\pi = 4A$. Similarly, the slices $S(B)$ and $S(C)$ cover areas $S(B) = 4B$ and $S(C) = 4C$. Together, the three slices cover the whole surface of the sphere, the equal triangles $\triangle$ and $\triangle'$ belonging to every slice, and each point outside the triangles, to exactly one slice. Thus the area $S(A) + S(B) + S(C)$ equals the area of the sphere plus four times the area of $\triangle$:

$$S(A) + S(B) + S(C) = 4A + 4B + 4C = 4\pi + 4\triangle, \quad \text{whence}$$

$$\triangle = A + B + C - \pi = E.$$

As in the case of plane triangles, we can derive relationships between the sides and angles of spherical triangles. The easiest way to do this is by inspecting certain coordinate transformations. Suppose we have two rectangular coordinate frames $Oxyz$ and $Ox'y'z'$ (Fig. 2.4), such that the $x'y'z'$-frame is obtained from the xyz-frame by rotating it around the x-axis by an angle χ.

The position of a point P on a unit sphere is uniquely determined by giving two angles. The angle ψ is measured counterclockwise from the positive x-axis along the xy-plane; the other angle θ tells the angular distance from the xy-plane. In an analogous way, we can define the angles ψ' and θ', which give the position of the point P in the $x'y'z'$-frame. The rectangular coordinates of the point P as functions of these angles are:

$$
\begin{aligned}
x &= \cos\psi\cos\theta, & x' &= \cos\psi'\cos\theta', \\
y &= \sin\psi\cos\theta, & y' &= \sin\psi'\cos\theta', \\
z &= \sin\theta, & z' &= \sin\theta'.
\end{aligned}
\tag{2.3}
$$

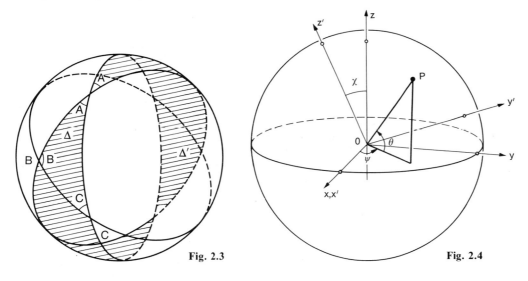

Fig. 2.3

Fig. 2.4

Fig. 2.3. If the sides of a spherical triangle are extended all the way around the sphere, they form another triangle $\triangle'$, antipodal and equal to the original triangle $\triangle$. The shaded area is the slice $S(A)$

Fig. 2.4. The location of a point P on the surface of a unit sphere can be expressed by rectangular xyz-coordinates or by two angles, ψ and θ. The $x'y'z'$-frame is obtained by rotating the xyz-frame around its x-axis by an angle χ

We also know that the dashed coordinates are obtained from the undashed ones by a rotation in the yz-plane (Fig. 2.5):

$$x' = x \; ,$$
$$y' = y \cos \chi + z \sin \chi \; , \tag{2.4}$$
$$z' = - y \sin \chi + z \cos \chi \; .$$

By substituting the expressions of the rectangular coordinates (2.3) into (2.4), we have

$$\cos \psi' \cos \theta' = \cos \psi \cos \theta \; ,$$
$$\sin \psi' \cos \theta' = \sin \psi \cos \theta \cos \chi + \sin \theta \sin \chi \; , \tag{2.5}$$
$$\sin \theta' = - \sin \psi \cos \theta \sin \chi + \sin \theta \cos \chi \; .$$

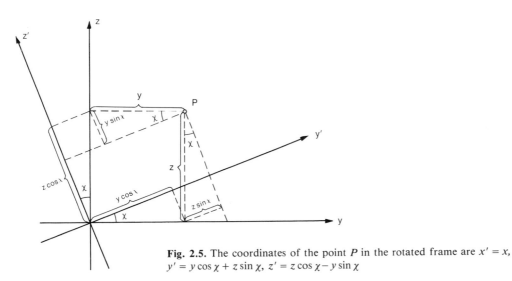

Fig. 2.5. The coordinates of the point P in the rotated frame are $x' = x$, $y' = y \cos \chi + z \sin \chi$, $z' = z \cos \chi - y \sin \chi$

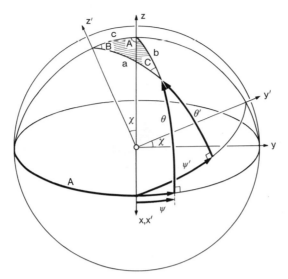

Fig. 2.6. To derive triangulation formulae for the spherical triangle ABC, the spherical coordinates ψ, θ, ψ' and θ' of the vertex C are expressed in terms of the sides and angles of the triangle

In fact, these equations are quite sufficient for all coordinate transformations we may encounter. However, we shall also derive the usual equations for spherical triangles. To do this, we set up the coordinate frames in a suitable way (Fig. 2.6). The z-axis points towards the vertex A and the z'-axis, towards B. Now the vertex C corresponds to the point P in Fig. 2.4. The angles ψ, θ, ψ', θ' and χ can be expressed in terms of the angles and sides of the spherical triangle:

$$\psi = A - 90° \ ,$$

$$\theta = 90° - b \ ,$$

$$\psi' = 90° - B \ , \tag{2.6}$$

$$\theta' = 90° - a \ ,$$

$$\chi = c \ .$$

Substitution into (2.5) gives

$$\cos(90° - B)\cos(90° - a) = \cos(A - 90°)\cos(90° - b) \ ,$$

$$\sin(90° - B)\cos(90° - a) = \sin(A - 90°)\cos(90° - b)\cos c + \sin(90° - b)\sin c \ ,$$

$$\sin(90° - a) = -\sin(A - 90°)\cos(90° - b)\sin c + \sin(90° - b)\cos c \ ,$$

or

$$\sin B \sin a = \sin A \sin b \ ,$$

$$\cos B \sin a = -\cos A \sin b \cos c + \cos b \sin c \ , \tag{2.7}$$

$$\cos a = \cos A \sin b \sin c + \cos b \cos c \ .$$

Equations for other sides and angles are obtained by cyclic permutations of the sides a, b, c and the angles A, B, C. For instance, the first equation also yields

$$\sin C \sin b = \sin B \sin c \ ,$$

$$\sin A \, \sin c = \sin C \sin a \ .$$

All these variations of the *sine formula* can be written in an easily remembered form:

$$\frac{\sin a}{\sin A} = \frac{\sin b}{\sin B} = \frac{\sin c}{\sin C} \; . \qquad (2.8)$$

If we take the limit, letting the sides *a*, *b* and *c* shrink to zero, the spherical triangle becomes a plane triangle. If all angles are expressed in radians, we have approximately

$$\sin a \approx a \; , \quad \cos a \approx 1 - \tfrac{1}{2} a^2 \; .$$

Substituting these approximations into the sine formula, we get the familiar sine formula of plane geometry:

$$\frac{a}{\sin A} = \frac{b}{\sin B} = \frac{c}{\sin C} \; .$$

The second equation in (2.7) is the *sine-cosine formula,* and the corresponding plane formula is a trivial one:

$$c = b \cos A + a \cos B \; .$$

This is obtained by substituting the approximations of sine and cosine into the sine-cosine formula and ignoring all quadratic and higher-order terms. In the same way we can use the third equation in (2.7), the *cosine formula,* to derive the planar cosine formula:

$$a^2 = b^2 + c^2 - 2bc \cos A \; .$$

2.2 The Earth

A position on the Earth is usually given by two spherical coordinates. If necessary, also a third coordinate, e.g. the distance from the centre, can be used. The reference plane is the *equatorial plane,* perpendicular to the rotation axis and intersecting the surface of the Earth along the *equator.* Small circles parallel to the equator are called *parallels of latitude.* Semicircles from pole to pole are *meridians.* The geographical *longitude* is the angle between the meridian and the zero meridian passing through Greenwich Observatory. We shall use positive values for longitudes west of Greenwich and negative values east of Greenwich. Sign convention, however, varies; so it is usually better to say explicitly whether the longitude is east or west of Greenwich. Sometimes the longitude is expressed by giving the time difference between local time and Greenwich time (time systems will be discussed in Sect. 2.12). A complete revolution corresponds to 24 hours; hence 1 hour equals 15 degrees.

The *latitude* is usually supposed to mean the *geographical latitude,* which is the angle between the plumb line and the equatorial plane. The latitude is positive in the Northern hemisphere and negative in the Southern one. The geographical latitude can be determined by astronomical observations (Fig. 2.7): the altitude of the celestial pole measured from the horizon equals the geographical latitude. (The celestial pole is the intersection of the rotation axis of the Earth and the infinitely distant celestial sphere; we shall return to these concepts a little later.)

Because the Earth is rotating, it is slightly flattened. The shape defined by the surface of the oceans, called the *geoid,* is very close to an oblate spheroid, the short axis of

Fig. 2.7. The latitude ϕ is obtained by measuring the altitude of the celestial pole. The celestial pole can be imagined as a point at an infinite distance in the direction of the Earth's rotation axis

Fig. 2.8. Due to the flattening of the Earth, the geographic latitude ϕ and geocentric latitude ϕ' are different

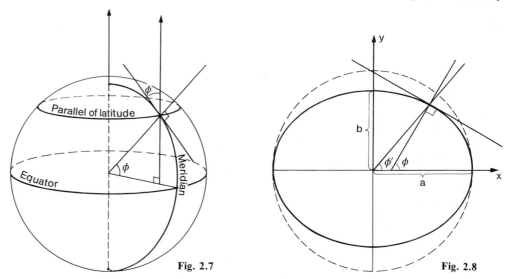

Fig. 2.7

Fig. 2.8

which coincides with the rotation axis. In 1979, the International Astronomical Union (IAU) adopted the following dimensions for the meridional ellipse:

equatorial radius $a = 6{,}378{,}140 \, \text{m}$,

polar radius $b = 6{,}356{,}755 \, \text{m}$,

flattening $f = (a-b)/a = 1/298.257$.

The angle between the equator and the normal to the ellipsoid approximating the true Earth is called the *geodetic latitude*. Because the surface of a liquid (like an ocean) is perpendicular to the plumb line, the geodetic and geographical latitudes are practically the same.

Because of the flattening, the plumb line does not point to the centre of the Earth except at the poles and on the equator. An angle corresponding to the ordinary spherical coordinate (the angle between the equator and the line from the centre to a point on the surface), the *geocentric latitude* is therefore a little smaller than the geographic latitude (Fig. 2.8).

We now derive an equation between the geographic latitude ϕ and geocentric latitude ϕ', assuming the Earth is an oblate spheroid and the geographic and geodesic latitudes are equal. The equation of the meridional ellipse is

$$\frac{x^2}{a^2} + \frac{y^2}{b^2} = 1 \ .$$

The direction of the normal to the ellipse at a point (x, y) is given by

$$\tan \phi = -\frac{dx}{dy} = \frac{a^2}{b^2}\frac{y}{x} \ .$$

The geocentric latitude is obtained from

$$\tan \phi' = y/x \ .$$

Hence

$$\tan \phi' = \frac{b^2}{a^2} \tan \phi = (1 - e^2) \tan \phi , \quad \text{where}$$

$$e = \sqrt{1 - b^2/a^2}$$

is the eccentricity of the ellipse. The difference $\Delta \phi = \phi - \phi'$ has a maximum 11.5′ at the latitude 45°.

Since the coordinates of celestial bodies in astronomical almanacs are given with respect to the centre of the Earth, the coordinates of nearby objects must be corrected for the difference in the position of the observer, if high accuracy is required. The easiest way to do this is to use the geocentric coordinates of the observer.

If a certain longitude difference corresponds to a distance l_0 on the equator, the arc subtending the same longitude difference along a parallel of latitude ϕ has a length $l = l_0 \cos \phi' \approx l_0 \cos \phi$. One arc minute along a meridian is called a *nautical mile*. Since the radius of curvature varies with latitude, the length of the nautical mile also depends on the latitude. Therefore one nautical mile has been defined to be equal to one minute of arc at $\phi = 45°$, whence 1 nm = 1852 m.

2.3 The Celestial Sphere

The ancient universe was confined within a finite spherical shell. The stars were fixed to this shell and thus were all equidistant from the Earth, which was at the centre of the spherical universe. This simple model is still in many ways as useful as it was in antiquity: it helps us to easily understand the diurnal and annual motions of stars, and, more important, to predict these motions in a relatively simple way. Therefore we will assume for the time being that all the stars are located on the surface of an enormous sphere and that we are at its centre. Because the radius of this celestial sphere is practically infinite, we can neglect the effects due to the changing position of the observer, caused by the rotation and orbital motion of the Earth. These effects will be considered in Sects. 2.8 and 2.11.

Since the distances of the stars are ignored, we need only two coordinates to specify their directions. Each coordinate frame has some fixed reference plane passing through the centre of the celestial sphere and dividing the sphere into two hemispheres along a great circle. One of the coordinates indicates the angular distance from this reference plane. There is exactly one great circle going through the object and intersecting this plane perpendicularly; the second coordinate gives the angle between that point of intersection and some fixed direction.

2.4 The Horizontal System

The most natural coordinate frame from the observer's point of view is the *horizontal frame* (Fig. 2.9). Its reference plane is the tangent plane of the Earth passing through the observer; this horizontal plane intersects the celestial sphere along the *horizon*. The

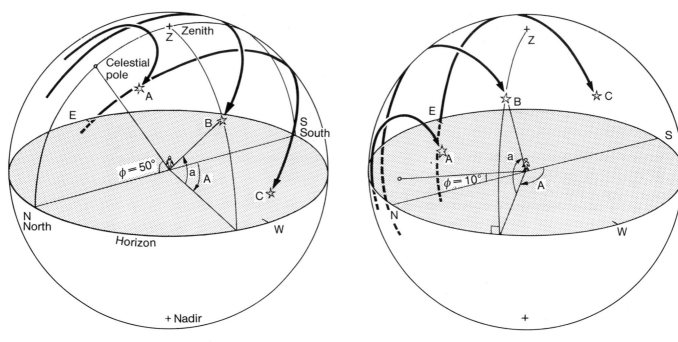

Fig. 2.9. The apparent motions of stars during a night, as seen from latitude $\phi = 50°$

Fig. 2.10. The same stars as in Fig. 2.9, as seen from latitude $\phi = 10°$

point just above the observer is called the *zenith* and the antipodal point below the observer is the *nadir*. (These two points are the poles corresponding to the horizon.) Great circles through the zenith are called *verticals*. All verticals intersect the horizon perpendicularly.

By observing the motion of a star over the course of a night, an observer finds out that it follows a track like one of those in Fig. 2.9. Stars rise in the east, reach their highest point, or *culminate*, on the vertical NZS, and set in the west. The vertical NZS is called the *meridian*. North and south directions are defined as the intersections of the meridian and the horizon.

One of the horizontal coordinates is the *altitude* or *elevation, a*, which is measured from the horizon along the vertical passing through the object. The altitude lies in the range $[-90°, +90°]$; it is positive for objects above the horizon and negative for the objects below the horizon. The *zenith distance*, or the angle between the object and the zenith, is obviously

$$z = 90° - a \ . \tag{2.9}$$

The second coordinate is the *azimuth, A*; it is the angular distance of the vertical of the object from some fixed direction. Unfortunately, in different contexts, different fixed directions are used; thus it is always advisable to check which definition is employed. The azimuth is usually measured from the north or south, and though clockwise is the preferred direction, counterclockwise measurements are also occasionally made. In this book, we have adopted a fairly common astronomical convention, measuring the azimuth *clockwise* from the *south*. Its values are usually normalized between 0° and 360°.

In Fig. 2.9, we can see the altitude and azimuth of a star *B* at some instant. As the star moves along its daily track, both of its coordinates will change. Another difficulty with this coordinate frame is its local character. In Fig. 2.10, we have the same stars as in Fig. 2.9, but the observer is now further south. We can see that the coordinates of the same star at the same moment are different for different observers. Since the horizontal coordinates are time and position dependent, they cannot be used, for instance, in star catalogues.

2.5 The Equatorial System

The direction of the rotation axis of the Earth remains almost constant and so does the equatorial plane perpendicular to this axis. Therefore the equatorial plane is a suitable reference plane for a coordinate frame that has to be independent of time and the position of the observer.

The intersection of the celestial sphere and the equatorial plane is a great circle, which is called the *equator of the celestial sphere*. The *north pole* of the celestial sphere is one of the poles corresponding to this great circle. It is also the point in the northern sky where the extension of the Earth's rotational axis meets the celestial sphere. The celestial north pole is at a distance of about one degree (which is equivalent to two full moons) from the moderately bright star Polaris. The meridian always passes through the north pole; it is divided by the pole into north and south meridians.

The angular separation of a star from the equatorial plane is not affected by the rotation of the Earth. This angle is called the *declination* δ. Stars seem to revolve around the pole once every day (Fig. 2.11). To define the second coordinate, we must again

Fig. 2.11. At night, stars seem to revolve around the celestial pole. The altitude of the pole from the horizon equals the latitude of the observer. (Photo Lick Observatory)

agree on a fixed direction, unaffected by the Earth's rotation. From a mathematical point of view, it does not matter which point on the equator is selected. However, for later purposes, it is more appropriate to employ a certain point with some valuable properties, which will be explained in the next chapter. This point is called the *vernal equinox*. Because it used to be in the constellation Aries (the Ram), it is also called the first point of Aries and denoted by the sign of Aries, ♈. Now we can define the second coordinate as the angle from the vernal equinox measured along the equator. This angle is the *right ascension* α (or R.A.) of the object, measured counterclockwise from ♈.

Since declination and right ascension are independent of the position of the observer and the motions of the Earth, they can be used in star maps and catalogues.

As will be explained later, one of the axes of a telescope (its hour axis) is usually parallel to the rotation axis of the Earth. The other axis (declination axis) is perpendicular to the hour axis. Declinations can be read immediately on the declination dial of the telescope. But the zero point of the right ascension seems to move in the sky, due to the diurnal rotation of the Earth. So we cannot use the right ascension to find an object unless we know the direction of the vernal equinox. Since the south meridian is a well-defined line in the sky, we use it to establish a local coordinate corresponding to right ascension. The *hour angle* is measured clockwise from the meridian. The hour angle of a object is not a constant, but grows at a steady rate, due to the Earth's rotation. The hour angle of the vernal equinox is called the *sidereal time* Θ. Figure 2.12 shows that for any object,

$$\Theta = h + \alpha \, , \tag{2.10}$$

where h is the object's hour angle and α, its right ascension.

Since hour angle and sidereal time change with time at a constant rate, it is practical to express them in units of time. Also the closely related right ascension is customarily given in time units. Thus 24 hours equals 360 degrees, 1 hour = 15 degrees, 1 minute of time = 15 minutes of arc, and so on. All these quantities are in the range [0h, 24h].

In practice, the sidereal time can be readily determined by pointing a telescope to an easily recognizable star and reading its hour angle on the hour angle dial of the tele-

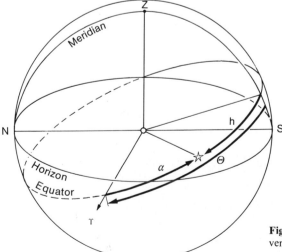

Fig. 2.12. The sidereal time Θ (the hour angle of the vernal equinox) equals the hour angle plus right ascension of any object

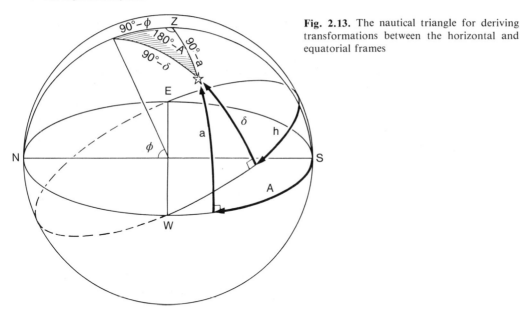

Fig. 2.13. The nautical triangle for deriving transformations between the horizontal and equatorial frames

scope. The right ascension found in a catalogue is then added to the hour angle, giving the sidereal time at the moment of observation. For any other time, the sidereal time can be evaluated by adding the time elapsed since the observation. If we want to be accurate, we have to use a sidereal clock to measure time intervals. A sidereal clock runs 3 min 56.56 s fast a day as compared with an ordinary solar time clock:

$$24\,\text{h solar time} = 24\,\text{h}\ 3\,\text{min}\ 56.56\,\text{s sidereal time} \ . \tag{2.11}$$

The reason for this is the orbital motion of the Earth: stars seem to move faster than the Sun across the sky; hence, a sidereal clock must run faster. (This is further discussed in Sect. 2.12.)

Transformations between the horizontal and equatorial frames are easily obtained from spherical trigonometry. Comparing Figs. 2.6 and 2.13, we find that we must make the following substitutions into (2.5):

$$\psi = 90° - A \ ,$$
$$\theta = a \ ,$$
$$\psi' = 90° - h \ , \tag{2.12}$$
$$\theta' = \delta \ ,$$
$$\chi = 90° - \phi \ .$$

The angle ϕ in the last equation is the altitude of the celestial pole, or the latitude of the observer. Making the substitutions, we get

$$\sin h \cos \delta = \sin A \cos a \ ,$$
$$\cos h \cos \delta = \cos A \cos a \sin \phi + \sin a \cos \phi \ , \tag{2.13}$$
$$\sin \delta = -\cos A \cos a \cos \phi + \sin a \sin \phi \ .$$

The inverse transformation is obtained by substituting

$$\psi = 90° - h \; ,$$
$$\theta = \delta \; ,$$
$$\psi' = 90° - A \; , \tag{2.14}$$
$$\theta' = a \; ,$$
$$\chi = -(90° - \phi) \; ,$$

whence

$$\sin A \cos a = \sin h \cos \delta \; ,$$
$$\cos A \cos a = \cos h \cos \delta \sin \phi - \sin \delta \cos \phi \; , \tag{2.15}$$
$$\sin a = \cos h \cos \delta \cos \phi + \sin \delta \sin \phi \; .$$

Since altitude and declination are in the range $[-90°, +90°]$, it suffices to know the sine of either of these angles to determine the angle unambiguously. Azimuth and right ascension, however, can have any value from $0°$ to $360°$ ($[0h, 24h]$), and to solve for them, we have to know both the sine and cosine to choose the correct quadrant.

The altitude of an object is greatest when it is on the south meridian (the great circle arc between the celestial poles containing the zenith). At that moment (called *upper culmination*, or *transit*), its hour angle is $0h$. At the *lower culmination*, the hour angle is $h = 12h$. When $h = 0h$, we get from the last equation in (2.15)

$$\sin a = \cos \delta \cos \phi + \sin \delta \sin \phi = \cos(\phi - \delta) = \sin(90° - \phi + \delta) \; .$$

The altitude at the upper culmination is

$$a_{\text{max}} = \begin{cases} 90° - \phi + \delta \; , & \text{if the object culminates south of zenith} \; , \\ 90° + \phi - \delta \; , & \text{if the object culminates north of zenith} \; . \end{cases} \tag{2.16}$$

The altitude is positive for objects with $\delta > \phi - 90°$. Objects with declinations less than $\phi - 90°$ can never be seen at the latitude ϕ. On the other hand, when $h = 12h$ we have

$$\sin a = -\cos \delta \cos \phi + \sin \delta \sin \phi = -\cos(\delta + \phi) = \sin(\delta + \phi - 90°) \; ,$$

and the altitude at the lower culmination is

$$a_{\text{min}} = \delta + \phi - 90° \; . \tag{2.17}$$

Stars with $\delta > 90° - \phi$ will never set. For example, in Helsinki ($\phi \approx 60°$), all stars with a declination higher than $30°$ are such *circumpolar* stars. And stars with a declination less than $-30°$ can never be observed there.

We shall now study briefly how the (α, δ)-frame can be established by observations. Suppose we observe a circumpolar star at its upper and lower culmination (Fig. 2.14). At the upper transit, its altitude is $a_{\text{max}} = 90° - \phi + \delta$ and at the lower transit, $a_{\text{min}} = \delta + \phi - 90°$. Eliminating the latitude, we get

$$\delta = \tfrac{1}{2}(a_{\text{min}} + a_{\text{max}}) \; . \tag{2.18}$$

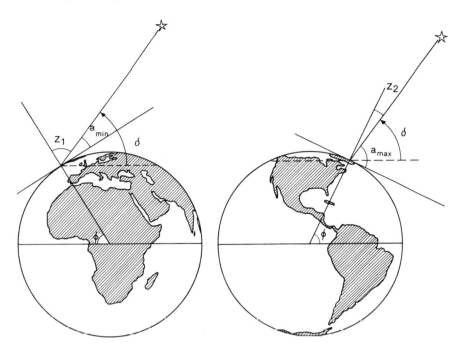

Fig. 2.14. The altitude of a circumpolar star at upper and lower culmination

Thus we get the same value for the declination, independent of the observer's location. Therefore we can use it as one of the absolute coordinates. From the same observations, we can also determine the direction of the celestial pole as well as the latitude of the observer. After these preparations, we can find the declination of any object by measuring its distance from the pole.

The equator can be now defined as the great circle, all of whose points are at a distance of 90 degrees from the pole. The zero point of the second coordinate (right ascension) can then be defined as the point where the Sun seems to cross the equator from south to north.

From the last equation in (2.15), we find the hour angle h of an object at the moment its altitude is a:

$$\cos h - -\tan \delta \tan \phi + \frac{\sin a}{\cos \delta \cos \phi} \ . \tag{2.19}$$

This equation can be used for computing rising and setting times. Then $a = 0$ and the hour angles corresponding to rising and setting times are obtained from

$$\cos h = -\tan \delta \tan \phi \ . \tag{2.20}$$

If the right ascension α is known, we can use (2.10) to compute the sidereal time Θ. (Later, in Sect. 2.12, we shall study how to transform the sidereal time into ordinary time.)

If higher accuracy is needed, we have to correct for the refraction of light caused by the atmosphere of the Earth (see Sect. 2.8). In that case, we must use a small negative value for a in (2.19). This value, the *horizontal refraction,* is about $-35'$. And to find the rising and setting times of the upper limb of the Sun, we must set $a = -51'$ $(= -35' - 16')$.

2.6 The Ecliptic System

The orbital plane of the Earth has another important coordinate frame, the *ecliptic*, as its reference plane. The ecliptic can also be defined as the great circle on the celestial sphere described by the Sun in the course of one year. This frame is used mainly for planets and other bodies of the solar system. The orientation of the Earth's equatorial plane remains invariant, unaffected by annual motion. In spring, the Sun appears to move from the Southern hemisphere to the Northern one (Fig. 2.15). The time of this remarkable event as well as the direction to the Sun at that moment are called the *vernal equinox*. At the vernal equinox, the Sun's right ascension and declination are zero. The equatorial and ecliptic planes intersect along a straight line directed towards the vernal equinox. Thus we can use this direction as the zero point for both the equatorial and ecliptic coordinate frames. The point opposite the vernal equinox is *the autumnal equinox*; it is the point at which the Sun crosses the equator from north to south.

The *ecliptic latitude* β is the angular distance from the ecliptic; it is in the range $[-90°, +90°]$. The other coordinate is the *ecliptic longitude* λ, measured counterclockwise from the vernal equinox.

Transformation equations between the equatorial and ecliptic frames can be derived analogously to as (2.13) and (2.15):

$$\sin \lambda \cos \beta = \sin \delta \sin \varepsilon + \cos \delta \cos \varepsilon \sin \alpha \; ,$$
$$\cos \lambda \cos \beta = \cos \delta \cos \alpha \; , \tag{2.21}$$
$$\sin \beta = \sin \delta \cos \varepsilon - \cos \delta \sin \varepsilon \sin \alpha \; .$$

$$\sin \alpha \cos \delta = - \sin \beta \sin \varepsilon + \cos \beta \cos \varepsilon \sin \lambda \; ,$$
$$\cos \alpha \cos \delta = \cos \lambda \cos \beta \; , \tag{2.22}$$
$$\sin \delta = \sin \beta \cos \varepsilon + \cos \beta \sin \varepsilon \sin \lambda \; .$$

The angle ε appearing in these equations is the *obliquity of the ecliptic*, or the angle between the equatorial and ecliptic planes. Its value is roughly $23°26'$ (but also see Sect. 8.3).

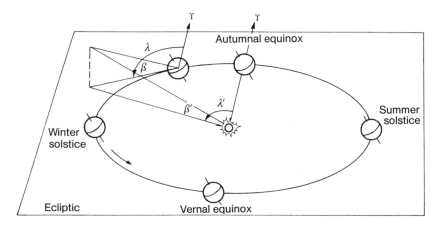

Fig. 2.15. The ecliptic geocentric (λ, β) and heliocentric (λ', β') coordinates are equal only if the object is very far away. The geocentric coordinates depend also on the Earth's position in its orbit

Fig. 2.16. The galactic coordinates *l* and *b*

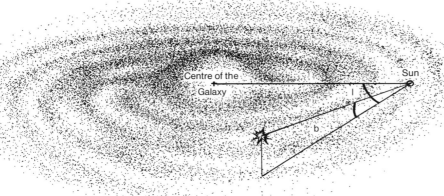

Depending on the problem to be solved, we may encounter heliocentric (origin at the Sun), geocentric (origin at the centre of the Earth) or topocentric (origin at the observer) coordinates. For very distant objects the differences are negligible, but not for bodies of the solar system. To transform heliocentric coordinates to geocentric coordinates or vice versa, we must also know the distance of the object. This transformation is most easily accomplished by computing the rectangular coordinates of the object and the new origin, then changing the origin and finally evaluating the new latitude and longitude from the rectangular coordinates.

2.7 The Galactic Coordinates (Fig. 2.16)

For studies of the Milky Way Galaxy, the most natural reference plane is the plane of the Milky Way. Since the Sun lies very close to that plane, we can put the origin at the Sun. The *galactic longitude l* is measured counterclockwise (like right ascension) from the direction of the centre of the Milky Way (in Sagittarius, $\alpha = 17$h 45.7 min, $\delta = -29°00'$). The *galactic latitude b* is measured from the galactic plane, positive northwards and negative southwards. This definition was officially adopted only in 1959, when the direction of the galactic centre was determined from radio observations accurately enough. The old galactic coordinates l^I and b^I had the intersection of the equator and the galactic plane as their zero point.

2.8 Perturbations of Coordinates

Even if a star remains fixed with respect to the Sun, its coordinates can change, due to several disturbing effects. Naturally its altitude and azimuth change constantly because of the rotation of the Earth, but even its right ascension and declination are not quite free from perturbations.

Precession. Since most of the members of the solar system orbit close to the ecliptic, they tend to pull the equatorial bulge of the Earth towards it. Most of this "flattening" torque is caused by the Moon and the Sun. But the Earth is rotating and therefore the torque cannot change the inclination of the equator relative to the ecliptic. Instead, the rotation axis turns in a direction perpendicular to the axis and the torque, thus describing a cone once in roughly 26,000 years. This slow turning of the rotation axis is called *precession*. Because of precession, the vernal equinox moves along the ecliptic clockwise about 50 seconds of arc every year, thus increasing the ecliptic longitudes of all objects at the same rate. At present the rotation axis points about one degree away from Polaris, but after 12,000 years, the celestial pole will be roughly in the direction of Vega. The changing ecliptic longitudes also affect right ascension and declination. Thus we have to know the instant of time, or *epoch*, for which the coordinates are given.

Let us now derive expressions for the changes in right ascension and declination. Taking the last transformation equation in (2.22),

$$\sin \delta = \cos \varepsilon \sin \beta + \sin \varepsilon \cos \beta \sin \lambda \; ,$$

and differentiating, we get

$$\cos \delta \, d\delta = \sin \varepsilon \cos \beta \cos \lambda \, d\lambda \; .$$

Applying the second equation in (2.21) to the right-hand side, we have, for the change in declination,

$$d\delta = d\lambda \sin \varepsilon \cos \alpha \; . \tag{2.23}$$

By differentiating the equation

$$\cos \alpha \cos \delta = \cos \beta \cos \lambda \; ,$$

we get

$$- \sin \alpha \cos \delta \, d\alpha - \cos \alpha \sin \delta \, d\delta = - \cos \beta \sin \lambda \, d\lambda \; ;$$

and, by substituting the previously obtained expression for $d\delta$ and applying the first equation (2.21), we have

$$\sin \alpha \cos \delta \, d\alpha = d\lambda (\cos \beta \sin \lambda - \sin \varepsilon \cos^2 \alpha \sin \delta)$$
$$= d\lambda (\sin \delta \sin \varepsilon + \cos \delta \cos \varepsilon \sin \alpha - \sin \varepsilon \cos^2 \alpha \sin \delta) \; .$$

Simplifying this, we get

$$d\alpha = d\lambda (\sin \alpha \sin \varepsilon \tan \delta + \cos \varepsilon) \; . \tag{2.24}$$

If $d\lambda$ is the annual increment of the ecliptic longitude (about 50''), the precessional changes in right ascension and declination in one year are thus

$$d\delta = d\lambda \sin \varepsilon \cos \alpha \; ,$$
$$d\alpha = d\lambda (\sin \varepsilon \sin \alpha \tan \delta + \cos \varepsilon) \; . \tag{2.25}$$

These expressions are usually written in the form

$$d\delta = n \cos \alpha , \quad d\alpha = m + n \sin \alpha \tan \delta , \qquad (2.26)$$

where

$$m = d\lambda \cos \varepsilon , \quad n = d\lambda \sin \varepsilon \qquad (2.27)$$

are the *precessional constants*. Since the obliquity of the ecliptic is not exactly a constant but changes with time, m and n also vary slowly with time. However, this variation is so slow that usually we can regard m and n as constants unless the time interval is very long. The values of these constants for some epochs are given in Table 2.1.

Table 2.1. Precession constants m and n. Here, "a" means a tropical year

Epoch	m	n	
1800	3.07048 s/a	1.33703 s/a = 20.0554''/a	
1850	3.07141	1.33674	20.0511
1900	3.07234	1.33646	20.0468
1950	3.07327	1.33617	20.0426
2000	3.07419	1.33589	20.0383

Nutation. The Moon's orbit is inclined with respect to the ecliptic, resulting in precession of its orbital plane. One revolution takes 18.6 years, producing perturbations with the same period in the Earth's precession. This effect, *nutation*, changes ecliptic longitudes as well as the obliquity of the ecliptic (Fig. 2.17). Calculations are now much more complicated, but fortunately nutational perturbations are relatively small, only fractions of an arc minute.

Parallax. If we observe an object from different points, we see it in different directions. The difference of the observed directions is called the *parallax*. Since the amount of parallax depends on the distance of the observer from the object, we can utilize the parallax to measure distances. Human stereoscopic vision is based (at least to some extent) on this effect. For astronomical purposes we need much longer baselines than the distance between our eyes (about 7 cm). Appropriately large and convenient baselines are the radius of the Earth and the radius of its orbit.

Distances to the nearest stars can be determined from the *annual parallax*, which is the angle subtended by the radius of the Earth's orbit (called the *astronomical unit*, AU) as seen from the star. (We shall discuss this further in Sect. 2.11.)

By *diurnal parallax* we mean the change of direction due to the daily rotation of the Earth. In addition to the distance of the object, the diurnal parallax also depends on the latitude of the observer. If we talk about the parallax of a body in our solar system, we always mean the angle subtended by the Earth's equatorial radius (6378 km) as seen from the object (Fig. 2.18). This equals the apparent shift of the object with respect to background stars seen by an observer at the equator if (s)he observes the object moving from the horizon to the zenith. The parallax of the Moon, for example, is about 57′, and that of the Sun, 8.79″.

Fig. 2.17. The effect of precession and nutation on the direction of the rotation axis of the Earth

Fig. 2.18. The horizontal parallax π of an object is the angle subtended by the Earth's equatorial radius as seen from the object

Aberration. Because of the finite speed of light, an observer in motion sees an object shifted in the direction of her/his motion (Fig. 2.19). This change of apparent direction is called *aberration*. The amount of the aberration is

$$a = \frac{v}{c}\sin\theta \; , \tag{2.28}$$

where v is the velocity of the observer, c is the speed of light and θ is the angle between the true direction of the object and the velocity vector of the observer. The greatest possible value of the aberration due to the orbital motion of the Earth, v/c, called the *aberration constant*, is 21″. The maximal shift due to the Earth's rotation, the diurnal aberration constant, is much smaller, about 0.3″ (Fig. 2.20).

Refraction. Since light is refracted by the atmosphere, the direction of an object differs from the true direction by an amount depending on the atmospheric conditions along the line of sight. Since this refraction varies with atmospheric pressure and temperature, it is very difficult to predict it accurately. However, an approximation good enough for most practical purposes is easily derived. If the object is not too far from the zenith, the atmosphere between the object and the observer can be approximated by a stack of parallel planar layers, each of which has a certain index of refraction n_i (Fig. 2.21). Outside the atmosphere, we have $n = 1$.

Let the true zenith distance be z and the apparent one, ζ. Using the notations of Fig. 2.21, we obtain the following equations for the boundaries of successive layers:

Fig. 2.19a, b. The effect of aberration on the apparent direction of an object. (**a**) Observer at rest. (**b**) Observer in motion

Fig. 2.21. Refraction of a light ray travelling through the atmosphere

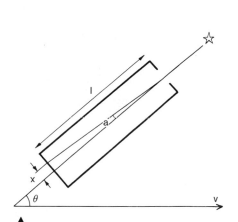

Fig. 2.20. A telescope is pointed in the true direction of a star. It takes a time $t = l/c$ for the light to travel the length of the telescope. The telescope is moving with velocity v, which has a component $v \sin \theta$, perpendicular to the direction of the light beam. The beam will hit the bottom of the telescope displaced from the optical axis by a distance $x = tv \sin \theta = l(v/c) \sin \theta$. Thus the change of direction in radians is $a = x/l = (v/c) \sin \theta$

$$\sin z \quad = n_k \sin z_k \ ,$$
$$\ldots$$
$$n_2 \sin z_2 = n_1 \sin z_1 \ ,$$
$$n_1 \sin z_1 = n_0 \sin \zeta \ ,$$

or

$$\sin z = n_0 \sin \zeta \ . \tag{2.30}$$

$$\tag{2.29}$$

When the *refraction angle* $R = z - \zeta$ is small and is expressed in radians, we have

$$n_0 \sin \zeta - \sin z = \sin(R + \zeta) = \sin R \, \cos \zeta + \cos R \, \sin \zeta$$
$$\approx R \, \cos \zeta + \sin \zeta \ .$$

Thus we get

$$R = (n_0 - 1) \tan \zeta , \quad [R] = \text{rad} \ .$$

Observations give an average value for this expression of the refraction angle

$$R = 58.2'' \tan \zeta \ . \tag{2.31}$$

Light coming from the zenith is not refracted at all if the boundaries between the layers are horizontal. Under some climatic conditions, a boundary (e.g. between cold and warm layers) can be slanted, and in this case, there can be a small zenith refraction, which is of the order of a few arc seconds.

If the object is far from the zenith, the curvature of the atmosphere can no longer be ignored and (2.31) should not be used. The graph in Fig. 2.22 gives the refraction as a function of the true altitude. Altitude is always increased by refraction. On the horizon the change is about 35′, which is slightly more than the diameter of the Sun. When the lower limb of the Sun just touches the horizon, the Sun has in reality already set.

Fig. 2.22. Refraction at different altitudes. The refraction angle R tells how much higher the object seems to be compared with its true altitude a

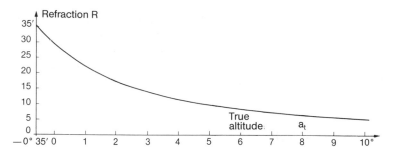

Star catalogues give the *mean places*; i.e. the effects of parallax, aberration and nutation have been removed. Therefore the mean place can differ from the observed or *apparent place*. There is a catalogue published annually that gives the apparent places of certain references stars at intervals of a few days. These positions have been corrected for precession, nutation, parallax and annual aberration. The effects of diurnal aberration and refraction are not included because they depend on the location of the observer.

2.9 Constellations

At any one time, about 1000–1500 stars can be seen in the sky (above the horizon). Under ideal conditions, the number of stars visible to the naked eye can be as high as 5000 on a hemisphere, or 10,000 altogether. Some stars seem to form figures vaguely resembling something; they have been ascribed to various mythological and other animals. This grouping of stars into constellations is a product of human imagination without any physical basis. Different cultures have different constellations, depending on their mythology, history and environment.

The shapes and names of the constellations we are familiar with date back to Mediterranean antiquity. But the names and boundaries were far from unambiguous as late as the 19th century. Therefore the International Astronomical Union (IAU) confirmed fixed boundaries at its 1928 meeting.

The official boundaries of the constellations were established along lines of constant right ascension and declination for the epoch 1875. During the century which has elapsed since then, precession has had time to turn the equatorial frame. However, the boundaries remain fixed with respect to the stars. So a star belonging to a constellation will belong to it forever (unless it is moved across a boundary by its proper motion).

The names of the 88 constellations confirmed by the IAU are given in Table E.21 at the end of the book. The table also gives the abbreviation of the Latin name, its genitive (needed for names of stars) and the English name.

The brightest stars of each constellation are usually denoted by Greek letters. The brightest star is α (alpha), e.g. Deneb in the constellation Cygnus is α Cygni, which is abbreviated as α Cyg. The second brightest star is β (beta), the next one γ (gamma) and so on. After the Greek alphabet has been exhausted, numbers can be employed; e.g. 30 Tau is a bright binary star in the constellation Taurus. In fact, there are quite a few exceptions to this rule; among the brightest stars (which should be α) we find, for example, β Orionis, ε Sagittarii, γ Sagittae and even 46 Leonis Minoris!

As telescopes evolved, more and more stars were seen and catalogued. It soon became impractical to continue this method of naming. Thus most of the stars are known only by their catalogue index numbers. One star may have many different numbers; e.g. the bright Capella (α Aur) is number BD + 45 1077 in the Bonner Durchmusterung and HD 34029 in the Henry Draper catalogue.

2.10 Star Catalogues and Maps (Fig. 2.23)

The first actual star catalogue was published by *Ptolemy* in the second century; this catalogue appeared in the book to be known later as Almagest (which is a Latin corruption of the name of the Arabic translation, Al-mijisti). It had 1025 entries; the positions of these bright stars had been measured by *Hipparchos* 250 years earlier. Ptolemy's catalogue was the only widely used one prior to the 17th century.

The first catalogues still being used by astronomers were prepared under the direction of *Friedrich Wilhelm August Argelander* (1799 – 1875). Argelander served as professor of astronomy in Turku and later in Helsinki, but he made his major contributions in Bonn. Using a 72-mm telescope, he measured the positions and estimated the magnitudes of 320,000 stars. The catalogue, *Bonner Durchmusterung*, contains nearly all stars brighter than magnitude 9.5 between the north pole and declination − 2. (Magnitudes are further discussed in Chap. 4.) Argelander's work was later used as a model for two other large catalogues covering the whole sky. The total number of stars in catalogues is close to one million.

The purpose of these Durchmusterungen or general catalogues was to systematically list a great number of stars. In the so-called zone catalogues, the main goal is to give the positions of stars as exactly as possible. A typical zone catalogue is the German *Katalog der Astronomischen Gesellschaft* (AGK). Twelve observatories, each measuring a certain region in the sky, contributed to this catalogue. The work was begun in the 1870's and completed at the turn of the century.

General and zone catalogues were based on visual observations with a telescope. The evolution of photography made this kind of work unnecessary at the end of the 19th century. Photographic plates could be stored for future purposes, and measuring the positions of stars became easier and faster, making it possible to measure many more stars.

A great international program was started at the end of the 19th century in order to photograph the entire sky. Eighteen observatories participated in this *Carte du Ciel* project, all using similar instruments and plates. The positions of stars were first measured with respect to a rectangular grid exposed on each plate (Fig. 2.24). These coordinates could then be converted into declination and right ascension.

Positions of stars in catalogues are measured with respect to certain comparison stars, the coordinates of which are known with high accuracy. The coordinates of these reference stars are published in fundamental catalogues. The first such catalogue was needed for the AGK catalogue; it was published in Germany in 1879. This *Fundamental Katalog* (FK 1) gives the positions of over 500 stars.

The fundamental catalogues are usually revised every few decades. The latest edition, the FK 5, appeared in 1984. At the same time, a new system of astronomical constants was adopted. The apparent places of the stars in the FK catalogues at 10-day intervals are published annually in the *Apparent Places of Fundamental Stars*.

Fig. 2.23a – d. The representations in four atlases of the Hyades cluster in the constellation Taurus. (**a**) Heis: Atlas Coelestis, published in 1872. (**b**) Bonner Durchmusterung. (**c**) SAO. (**d**) Palomar Sky Atlas, red plate. The big blob is the brightest star of Taurus, α Tauri alias Aldebaran

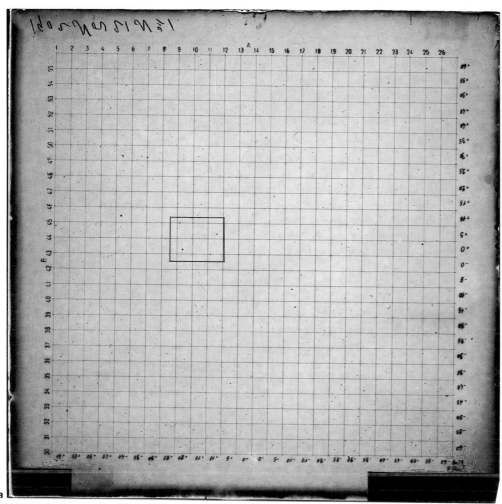

Fig. 2.24. (a) A plate photographed for the Carte du Ciel project in Helsinki on November 21, 1902. The centre of the field is at $\alpha = 18\,h\,40\,min$, $\delta = 46°$, and the area is $2° \times 2°$. Distance between coordinate lines (exposed separately on the plate) is 5 minutes of arc. **(b)** The framed region on the same plate. **(c)** The same area on a plate taken on November 7, 1948. The bright star in the lower right corner (SAO 47747) has moved about 12 seconds of arc. The brighter, slightly drop-shaped star to the left is a binary star (SAO 47767); the separation between its components is 8″

In addition to the FK catalogues there are other fundamental catalogues, for example, *Benjamin Boss's General Catalogue of 33,342 stars.*

One of the most widely used catalogues is the SAO catalogue, published by the Smithsonian Astrophysical Observatory in the 1960's. It contains the exact positions, magnitudes, proper motions, spectral classifications, etc. of 258,997 stars brighter than magnitude 9. The catalogue is also available on magnetic tape.

Star maps have been published since ancient times, but the earlierst maps were globes showing the celestial sphere as seen from the outside. At the beginning of the 17th century, a German, *Johannes Bayer*, published the first map showing the stars as seen from inside the celestial sphere, as we see them in the sky. Constellations were usually decorated with drawings of mythological figures.

The above-mentioned SAO catalogue is accompanied by a star map containing all the stars in the catalogue. The most important map is a photographic atlas the full name of which is *The National Geographic Society – Palomar Observatory Sky Survey*. The plates for this atlas were taken with the 1.2-m Schmidt camera on Mount Palomar. The Palomar Sky Survey was completed in the 1950's. It consists of 935 pairs of photographs: each region has been photographed in red and blue light. The size of each plate is about 35 cm × 35 cm, covering an area of 6.6° × 6.6°. The prints are negatives (black stars on a light background), because in this way, fainter objects are visible. The limiting magnitude is about 19 in blue and 20 in red.

The Palomar atlas covers the sky down to declination −20°. Work to map the rest of the sky was carried out later at two observatories in the Southern hemisphere, at Siding Spring Observatory in Australia, and at the European Southern Observatory (ESO) in Chile. The instruments and the scale on the plates are similar to those used earlier for the Palomar Sky Survey, but the atlas is distributed on film transparencies instead of paper prints.

New steps in mapping the sky were taken in the beginning of the 1990's. At the Palomar Observatory, a new photographic atlas of the northern sky is being prepared using modern techniques in photographing. ESO is undertaking responsibility of producing copies of the plates.

The first astrometric satellite Hipparchos started observing the stars at the end of 1989. Although Hipparchos didn't reach the planned geosynchronous orbit, it has given exact positions of hundreds of thousands of stars.

2.11 Positional Astronomy

The position of a star can be measured either with respect to some reference stars or with respect to a fixed coordinate frame. Absolute coordinates are usually determined using a *meridian circle*, which is a telescope that can be turned only in the meridional plane (Fig. 2.25). It has only one axis, which is aligned exactly in the east-west direction. Since all stars cross the meridian in the course of day, they all come to the field of the meridian circle at some time or other. When a star culminates, its altitude and the time of transit are recorded. If the time is determined with a sidereal clock, the sidereal time immediately gives the right ascension of the star, since the hour angle is $h = 0\,\mathrm{h}$. The other coordinate, declination, is obtained from the altitude:

$$\delta = a - (90° - \phi) \,,$$

where a is the observed altitude and ϕ is the geographic latitude of the observatory.

Relative coordinates are measured on photographic plates containing some known reference stars. The scale of the plate as well as the orientation of the coordinate frame can be determined from the reference stars. After this has been done, the right ascension and declination of any object on the plate can be calculated if its coordinates on the plate are measured.

The shift in the direction of a star with respect to distant background stars due to the annual motion of the Earth is called the *trigonometric parallax* of the star. It gives the distance of a star: the smaller the parallax, the farther away the star is. Trigonometric parallax is, in fact, the only direct method we currently have of measuring distances to stars. Later we shall be introduced to some other, indirect methods, which require certain assumptions on the motions or structure of stars.

The same method of triangulation is employed to measure distances of earthly objects. To measure distances to stars, we have to use the longest baseline available, the diameter of the orbit of the Earth. During the course of one year, a star will appear to describe a circle if it is at the pole of the celestial sphere, a segment of line if it is in the ecliptic, or an ellipse otherwise. The semi-major axis of this ellipse is called the *parallax* of the star is usually denoted by π. It equals the angle subtended by the radius of the Earth's orbit as seen from the star (Fig. 2.26). The unit of distance used in astronomy is *parsec* (pc). At a distance of one parsec, one astronomical unit subtends an angle of one arc second. Since one radian is about 206 265″, 1 pc equals 206 265 AU. Furthermore, because 1 AU = 1.496×10^{11} m, 1 pc $\approx 3.086 \times 10^{16}$ m. If the parallax is given in arc seconds, the distance is simply

$$r = 1/\pi, \quad [r] = \text{pc}, \quad [\pi] = '' \, . \tag{2.32}$$

In popular astronomical texts, distances are usually given in light-years, one light-year being the distance light travels in one year, or 9.5×10^{15} m. Thus one parsec is about 3.26 light-years.

Fig. 2.26. The trigonometric parallax π of a star is the angle subtended by the radius of the orbit of the Earth, or one astronomical unit as seen from the star

The first parallax measurement was accomplished by *Friedrich Wilhelm Bessel* (1784 – 1846) in 1838. He found the parallax of 61 Cygni to be 0.3″. The nearest star Proxima Centauri has a parallax of 0.762″ and thus a distance of 1.31 pc.

In addition to the motion due to the annual parallax, many stars seem to move slowly in a direction that does not change with time. This effect is caused by the relative motion of the Sun and the stars through space; it is called the *proper motion*. The appearance of the sky and the shapes of the constellations are constantly, though extremely slowly, changed by the proper motions of the stars.

The velocity of a star with respect to the Sun can be divided into two components (Fig. 2.27), one of which is directed along the line of sight (the radial component or the *radial velocity*), and the other perpendicular to it (the tangential component). The tangential velocity results in the proper motion, which can be measured by taking plates at intervals of several years or decades. The proper motion μ has two components, one giving the change in declination μ_δ and the other, in right ascension, $\mu_\alpha \cos \delta$. The coefficient $\cos \delta$ is used to correct the scale of right ascension: hour circles (the great circles with $\alpha =$ constant) approach each other towards the poles, so the coordinate difference must be multiplied by $\cos \delta$ to obtain the true angular separation. The total proper motion is

$$\mu = \sqrt{\mu_\alpha^2 \cos^2 \delta + \mu_\delta^2} \, . \tag{2.33}$$

The greatest known proper motion belongs to Barnard's Star, which moves across the sky at the enormous speed of 10.3 arc seconds per year. It needs less than 200 years to travel the diameter of a full moon.

In order to measure proper motions, we must observe stars for decades. The radial component, on the other hand, is readily obtained from a single observation, thanks to the *Doppler effect*. By the Doppler effect we mean the change in frequency and wavelength of radiation due to the radial velocity of the radiation source. The same effect can be observed, for example, in the sound of an ambulance, the pitch being higher when the ambulance is approaching and lower when it is receding.

The formula for the Doppler effect for small velocities can be derived as in Fig. 2.28. The source of radiation transmits electromagnetic waves, the period of one cycle being T. In time T, the radiation approaches the observer by a distance $s = cT$, where c is the speed of propagation. During the same time, the source moves with respect to the observer a distance $s' = vT$, where v is the speed of the source, positive

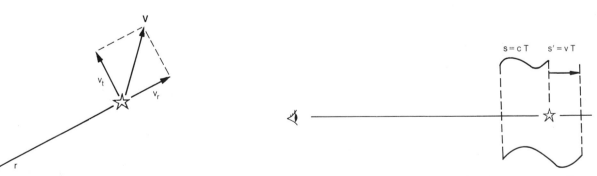

Fig. 2.27. The radial and tangential components v_r and v_t of the velocity v of a star. The latter component is observed as proper motion

Fig. 2.28. The wavelength of radiation increases if the source is receding for a receding source and negative for an approaching one. We find that the length of one cycle, the wavelength λ, equals

$$\lambda = s + s' \ .$$

If the source were at rest, the wavelength of its radiation would be $\lambda_0 = cT$. The motion of the source changes the wavelength by an amount

$$\Delta\lambda = \lambda - \lambda_0 = cT + vT - cT = vT \ ,$$

and the relative change in wavelength is

$$\frac{\Delta\lambda}{\lambda_0} = \frac{v}{c} \ . \tag{2.34}$$

This is valid only when $v \ll c$. For very high velocities, we must use the relativistic formula

$$\frac{\Delta\lambda}{\lambda_0} = \sqrt{\frac{1 + v/c}{1 - v/c}} - 1 \ . \tag{2.35}$$

In astronomy, the Doppler effect can be seen in stellar spectra, in which the spectral lines are often displaced towards the blue (shorter wavelengths) or red (longer wavelengths) end of the spectrum. A *blueshift* means that the star is approaching, while a *redshift* indicates that it is receding.

The displacements due to the Doppler effect are usually very small. In order to measure them, a reference spectrum is exposed on the plate next to the stellar spectrum. The lines in the reference spectrum are produced by a light source at rest in the laboratory. If the reference spectrum contains some lines found also in the stellar spectrum, the displacements can be measured.

Displacements of spectral lines give the radial velocity v_r of the star, and the proper motion μ can be measured from photographic plates. To find the tangential velocity v_t, we have to know the distance r, obtainable from e.g. parallax measurements. Tangential velocity and proper motion are related by

$$v_t = \mu r \ . \tag{2.36}$$

The total velocity v of the star is now

$$v = \sqrt{v_r^2 + v_t^2} \ .$$

If μ is given in arc seconds per year and r in pc we have to make the following unit transformations to get v_t in km/s:

$$1 \text{ rad} = 206\,265'' \ ,$$
$$1 \text{ year} = 3.156 \times 10^7 \text{ s} \ ,$$
$$1 \text{ pc} = 3.086 \times 10^{13} \text{ km} \ .$$

Hence

$$v_t = 4.74\,\mu r \ . \tag{2.37}$$

2.12 Time Reckoning

Measuring time was once one of the most important duties of astronomy. The angular velocity of the Earth around its axis is relatively constant, and therefore one *day* was often taken as the basic unit of time measurements. The orbital motion of the Earth around the Sun is also very accurately periodic; thus another convenient basic unit is one *year*. Nowadays we use various gadgets (like atomic clocks), based on some physical phenomena, to measure time with higher precision than obtainable by any astronomical means, and one *second* is used as the basic unit.

A fairly smoothly flowing astronomical time is the sidereal time, which we defined as the hour angle of the vernal equinox. A good basic unit is a *sidereal day*, which is the time between two successive upper culminations of the vernal equinox. After one sidereal day the celestial sphere with all its stars has returned to its original position with respect to the observer. The flow of sidereal time is as constant as the rotation of the Earth; only during the last few decades has it been possible to measure its small irregularities.

Figure 2.29 shows the Sun and the Earth at vernal equinox. When the Earth is at the point *A*, the Sun culminates and, at the same time, a new sidereal day begins in the city with the huge black arrow standing in its central square. After one sidereal day, the Earth has moved along its orbit almost one degree of arc to the point *B*. Therefore the Earth has to turn almost a degree further before the Sun will culminate. The *solar* or *synodic day* is therefore 3 min 56.56 s (sidereal time) longer than the sidereal day. This means that the beginning of the sidereal day will move around the clock during the course of one year. After one year, sidereal and solar time will again be in phase. The number of sidereal days in one year is one higher than the number of solar days.

When we talk about rotational periods of planets, we usually mean sidereal periods. The length of day, on the other hand, means the rotation period with respect to the Sun. If the orbital period around the Sun is P, sidereal rotation period τ_* and synodic day τ, we now know that the number of sidereal days in time P, P/τ_*, is one higher than the number of synodic days, P/τ:

$$\frac{P}{\tau_*} - \frac{P}{\tau} = 1 \quad \text{or}$$

$$\frac{1}{\tau} = \frac{1}{\tau_*} - \frac{1}{P}. \tag{2.38}$$

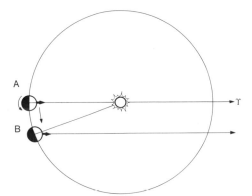

Fig. 2.29. One sidereal day is the time between two successive transits or upper culminations of the vernal equinox. By the time the Earth has moved from *A* to *B*, one sidereal day has elapsed

This holds for a planet rotating in the direction of its orbital motion (counterclockwise). If the sense of rotation is opposite, or *retrograde*, the number of sidereal days in one orbital period is one less than the number of synodic days, and the equation becomes

$$\frac{1}{\tau} = \frac{1}{\tau_*} + \frac{1}{P}. \tag{2.39}$$

For the Earth, we have $P = 365.2564$ d, and $\tau = 1$ d, whence (2.38) gives $\tau_* = 0.99727$ d $= 23$ h 56 min 4 s, solar time.

Since our everyday life follows the alternation of day and night, it is more convenient to base our timekeeping on the motion of the Sun rather than that of the stars. Unfortunately, solar time does not flow at a constant rate. There are two reasons for this. First, the orbit of the Earth is not exactly circular, but an ellipse, which means that the velocity of the Earth along its orbit is not constant. Second, the Sun moves along the ecliptic, not the equator. Thus its right ascension does not increase at a constant rate. The change is fastest at the end of December (4 min 27 s per day) and slowest in mid-September (3 min 35 s per day). As a consequence, the hour angle of the Sun (which determines the solar time) also grows at an uneven rate.

To find a solar time flowing at a constant rate, we define a fictitious *mean sun*, which moves along the celestial equator with constant angular velocity, making a complete revolution in one year. By year we mean here the *tropical year*, which is the time it takes for the Sun to move from one vernal equinox to the next. In one tropical year, the right ascension of the Sun increases exactly 24 hours. The length of the tropical year is 365 d 5 h 48 min 46 s $= 365.2422$ d. Since the direction of the vernal equinox moves due to precession, the tropical year differs from the sidereal year, during which the Sun makes one revolution with respect to the background stars. One sidereal year is 365.2564 d.

Using our artificial mean sun, we now define an evenly flowing solar time, the *mean solar time* T_M, which is equal to the hour angle h_M of the centre of the mean sun plus 12 hours (so that each day will begin at midnight, to annoy astronomers):

$$T_M = h_M + 12\,\text{h}. \tag{2.40}$$

This is essentially the time our life as well as our clocks are based on.

The difference between the true solar time T and the mean time T_M is called the *equation of time*:

$$\text{E.T.} = T - T_M. \tag{2.41}$$

(In spite of the identical abbreviation, this has nothing to do with a certain species of little green men.) The greatest positive value of E.T. is about 16 minutes and the greatest negative value about -14 minutes (see Fig. 2.30). This is also the difference between the true noon (the meridian transit of the Sun) and the mean noon.

Both true solar time and mean time are *local times*, depending on the hour angle of the sun, real or artificial. If one observes the true solar time by direct measurement and computes the mean time from (2.41), a digital watch will probably be found to disagree with both of them. The reason for this is that we do not use local time in our everday life; instead we use the *zonal time* of the nearest *time zone*.

Fig. 2.30. Equation of time. A sundial always shows (if correctly installed) true local solar time. To find the local mean time the equation of time must be subtracted from the local solar time

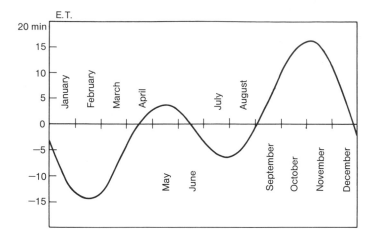

In the past, each city had its own local time. When travelling became faster and more popular, the great variety of local times became an inconvenience. At the end of the 19th century, the Earth was divided into 24 zones, the time of each zone differing from the neighboring ones by one hour. On the surface of the Earth, one hour in time corresponds to 15° in longitude; the time of each zone is determined by the local mean time at one of the longitudes 0°, 15°, . . . , 345°.

The time of the zero meridian going through Greenwich is used as an international reference, Universal Time. In most European countries, time is one hour ahead of this (Fig. 2.31).

In summer, many countries switch to *daylight savings time*, during which time is one hour ahead of the ordinary time. The purpose of this is to make the time when people are awake coincide with daytime in order to save electricity, particularly in the evening, when people go to bed one hour earlier. During daylight savings time, the difference between the true solar time and the official time can grow even larger.

Fig. 2.31. The time zones. The map gives the difference of the local zonal time from the Greenwich mean time (UT). During daylight savings time, one hour must be added to the given figures. When travelling across the date line westward, the date must be incremented by one day, and decremented if going eastward. For example, a traveller taking a flight from Honolulu to Tokyo on Monday morning will arrive on Tuesday, even though (s)he does not see a single night en route

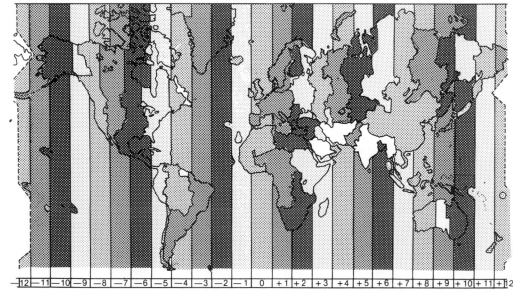

2.13 Astronomical Time Systems

The basis of astronomical time systems is the *Greenwich mean solar time*, denoted by UT (Universal Time), sometimes also by GMT. The universal time obtained directly from observations is called UT0. If we correct this for changes of the meridian due to motions of the Earth's rotation axis (precession, nutation, motion of the pole), we get UT1. Even this is not accurate enough for modern time measurements. UT1 suffers from tiny irregularities caused by variations in the Earth's rotation rate. These variations have at least two periods, one year and half a year. Removing these effects we get UT2, which has relative errors of the order 10^{-7}.

Even UT2 is not exact because of long-term changes in the Earth's rotation, such as slowing down due to tidal friction. Therefore astronomers have been willing to abandon timekeeping based on the Earth's rotation.

First the astronomers used *ephemeris time* for tabulating ephemerides. It was based on the Earth's movement around the Sun, which is much more accurate than the Earth's rotation. In the 1980's *dynamical time* was taken into use. For geocentric phenomena, the *terrestrial dynamical time* (TDT) was used. According to the theory of relativity, each observer has her/his own proper time. TDT corresponds to the proper time of the Earth after the effect of rotation has been removed. The *barycentric dynamical time* (TDB) was used for ephemerides related to the centre of gravity of the solar system. The dynamical time scales were specified so that their difference does not keep growing, but varies only periodically.

In the beginning of 1990's the International Astronomical Union adopted a new time system which agrees with the international atomic time and the SI second, as defined below, more accurately than the dynamical time. The new time reference for apparent geocentric ephemerides is called *terrestrial time* (TT), which can be considered as equivalent to TDT.

Nowadays it is possible to use certain physical processes to measure time more accurately than by any astronomical methods. In 1967, one second was defined as 9,192,631,770 times the period of the light emitted by cesium 133 isotope in its ground state, transiting from hyperfine level $F = 4$ to $F = 3$. Later, this definition was revised to include small relativistic effects produced by gravitational fields. The accuracy of this atomic time is about 10^{-12}.

The currently used *international atomic time*, TAI, is an average of several accurate atomic clocks. The relative errors in TAI are about 4×10^{-13}. The atomic and terrestrial times are related approximately by

$$TT \approx TAI + 32.184 \, s \; . \tag{2.42}$$

These extremely accurate time measurement methods have made it possible to control universal time related to the Earth's rotation with high accuracy. *Coordinated universal time*, UTC, was adopted in 1972. All currently used time signals are synchronized with UTC; UTC, modified with the time zone difference, is the time our clocks should show, if they are running correctly. As a new feature, UTC involves *leap seconds* to correct for variations in the Earth's rotation. UTC follows UT1 in such a way that the difference never exceeds 0.9 seconds. Should the error grow larger, a leap second is added (or subtracted) at the end of June or December. The Earth's rotation rate has been slowing down so that since 1972, a leap second has been added nearly every year.

2.14 Calendars

Our calendar is a result of long evolution. The main problem it must contend with is the incommensurability of the basic units, day, month and year: the numbers of days and months in a year are not integers. This makes it rather complicated to develop a calendar that takes correctly into account the alternation of seasons, day and night, and perhaps also the lunar phases.

Our calendar has its origin in the Roman calendar, which, in its earliest form, was based on the phases of the Moon. From around 700 B.C. on, the length of the year has followed the apparent motion of the Sun; thus originated the division of the year into twelve months. One month, however, still had a length roughly equal to the lunar cycle. Hence one year was only 354 days long. To keep the year synchronized with the seasons, a leap month had to be added to every other year.

Eventually the Roman calendar got mixed up. The mess was cleared by Julius Caesar in about 46 B.C., when the *Julian calendar* was developed upon his orders. The year had 365 days and a leap day was added to every fourth year.

In the Julian calendar, the average length of one year is 365 d 6 h, but the tropical year is 11 min 14 s shorter. After 128 years, the Julian year begins almost one day too late. The difference was already 10 days in 1582, when a calendar reform was carried out by Pope Gregorius XIII. In the *Gregorian calendar*, every fourth year is a leap year, the years divisible by 100 being exceptions. Of these, only the years divisible by 400 are leap years. Thus 1900 was not a leap year, but 2000 will be. The Gregorian calendar was adopted slowly, at different times in different countries. The transition period did not end before the 20th century.

Even the Gregorian calendar is not perfect. The differences from the tropical year will accumulate to one day in about 3300 years.

Since years and months of variable length make it difficult to compute time differences, especially astronomers have employed various methods to give each day a running number. The most widely used numbers are *Julian dates*. In spite of their name, they are not related to the Julian calendar. The only connection is the length of a *Julian century* of 36,525 days, a quantity appearing in many formulas involving Julian dates. The Julian day number 0 dawned about 4700 B.C. The day number changes always at 12:00 UT.

* Julian Date and Sidereal Time

There are several methods for finding the Julian date. The following one, presented by Meeus, is well adapted for pocket calculators. The functions $\text{INT}(x)$ and $\text{FRAC}(x)$ mean the integral and decimal parts of x, e.g. $\text{INT}(4.7) = 4$ and $\text{FRAC}(4.7) = 0.7$.

Let y be the year (with all four digits), m the month and d the day, and

$$f = \begin{cases} y\,, & \text{if } m \geq 3 \\ y-1\,, & \text{if } m = 1 \text{ or } 2 \end{cases}$$

$$g = \begin{cases} m\,, & \text{if } m \geq 3 \\ m+12\,, & \text{if } m = 1 \text{ or } 2 \end{cases}$$

$$A = 2 - \text{INT}(f/100) + \text{INT}(f/400)\,.$$

The Julian date JD at 0 hours UT is then

$$\text{JD} = \text{INT}(365.25f) + \text{INT}(30.6001(g+1)) + d + A + 1720994.5\,.$$

Example. Find JD at 0 UT on April 15, 1982.

Since $m = 4$, f is 1982 and $g = 4$. Hence

$$A = 2 - \text{INT}(1982/100) + \text{INT}(1982/400) = 2 - 19 + 4 = -13 \quad \text{and}$$

$$\text{JD} = \text{INT}(365.25 \times 1982) + \text{INT}(30.6001 \times 5) + 15 - 13 + 1720994.5 = 2445074.5 \;.$$

The inverse procedure is a little more complicated. The following will work for all positive JD's: Let $I = \text{INT}(\text{JD} + 0.5)$ and $F = \text{FRAC}(\text{JD} + 0.5)$. Now, if $I < 2,299,161$ (Julian calendar), let $A = I$; otherwise (Gregorian calendar), compute A as follows:

$$X = \text{INT}\left(\frac{I - 1867216.25}{36524.25}\right),$$

$$A = I + 1 + x - \text{INT}(x/4) \;.$$

Then compute the following quantities:

$$B = A + 1524 \;,$$

$$C = \text{INT}\left(\frac{B - 122.1}{365.25}\right),$$

$$D = \text{INT}(365.25\,C) \;,$$

$$E = \text{INT}\left(\frac{B - D}{30.6001}\right) .$$

Now the calendar date is obtained from

$$\text{day} \quad = B - D - \text{INT}(30.6001\,E) + F \;,$$

$$\text{month} = \begin{cases} E - 1 \;, & \text{if } E < 13.5 \\ E - 13 \;, & \text{if } E > 13.5 \end{cases}$$

$$\text{year} \quad = \begin{cases} C - 4716 \;, & \text{if month} > 2.5 \\ C - 4715 \;, & \text{if month} < 2.5 \;. \end{cases}$$

Example. What is the date 1000 days after April 15, 1982?

We found that the Julian date corresponding to April 15, 1982 is 2,445,074.5. 1000 days later, we have JD $= 2,446,074.5$.

$$I \;= \text{INT}(2,446,074.5 + 0.5) = 2,446,075 \;,$$

$$F = 0 \;,$$

$$x = \text{INT}\left(\frac{2,446,075 - 1,867,216.25}{36,524.25}\right) = \text{INT}(15.848) = 15 \;,$$

$$A = 2,446,075 + 1 + 15 - \text{INT}(15/4) = 2,446,088 \;,$$

$$B = 2,447,612 \;,$$

$$C = 6700 \;,$$

$$D = 2,447,175 \;,$$

$$E = 14 \;,$$

$$\text{day} \quad = 2,447,612 - 2,447,175 - 428 + 0 = 9 \;,$$

$$\text{month} = 14 - 13 = 1 \;,$$

$$\text{year} \quad = 6700 - 4715 = 1985 \;.$$

So the date is January 9, 1985.

Also the sidereal time can be calculated more accurately than in Exercise 2.15.9 by the following formula:

$$\Theta = 24\,\mathrm{h} \times \mathrm{FRAC}\,(0.27691398 + 100.00219359\,T + 0.000001075\,T^2)\,,$$

where T is the number of Julian centuries since 1900:

$$T = \frac{\mathrm{JD}\,(\mathrm{at}\ 0\ \mathrm{UT}) - 2{,}415{,}020}{36{,}525}\,.$$

This gives the sidereal time in Greenwich at 0 UT.

Example. What is the sidereal time in Greenwich at 20:00 UT on April 15, 1982?

First we have to find the sidereal time at 0 UT:

$$\mathrm{JD} = 2{,}445{,}074.5\,,$$

$$T \;= \frac{2{,}445{,}074.5 - 2{,}415{,}020}{36{,}525} = 0.822847365\,,$$

$$\begin{aligned}
\Theta \;&= 24\,\mathrm{h} \times \mathrm{FRAC}\,(0.276919398 + 82.286494 + 7.279 \times 10^{-7}) \\
&= 24\,\mathrm{h} \times 0.563414 = 13.5219\,\mathrm{h} = 13\,\mathrm{h}\ 31\,\mathrm{min}\ 19\,\mathrm{s}\,.
\end{aligned}$$

Since a sidereal clock runs about 3 min 57 s fast a day as compared with an ordinary clock, the difference accumulating in 20 hours is

$$\frac{20}{24} \times 3\,\mathrm{min}\ 57\,\mathrm{s} = 3\,\mathrm{min}\ 17\,\mathrm{s}\,.$$

Thus the sidereal time at 20:00 UT is 13 h 31 min 19 s + 20 h + 3 min 17 s = 33 h 34 min 46 s = 9 h 34 min 36 s, or 9:34:36.

At the same moment, the sidereal time in Berlin (where the time is 21:00 or 22:00 daylight savings time) would be 13.3°, or 53 min 12 s more, or 10:27:48.

2.15 Examples

2.15.1 *Trigonometric Functions in a Rectangular Spherical Triangle*

Let the angle A be a right angle. When the figure a plane triangle, the trigonometric functions of the angle B would be:

$$\sin B = b/a\,, \qquad \cos B = c/a\,, \qquad \tan B = b/c\,.$$

For the spherical triangle we have to use the equations in (2.7), which are now simply:

$$\sin B \sin a \;= \sin b\,,$$

$$\cos B \sin a = \cos b \sin c\,,$$

$$\cos a = \cos b \cos c\,.$$

The first equation gives the sine of B:

$$\sin B = \frac{\sin b}{\sin a}\,.$$

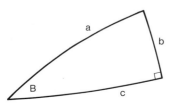

Dividing the second equation by the third one, we get the cosine of B:

$$\cos B = \frac{\tan c}{\tan a} \; .$$

And the tangent is obtained by dividing the first equation by the second one:

$$\tan B = \frac{\tan b}{\sin c} \; .$$

The third equation is the equivalent of the Pythagorean theorem for rectangular triangles.

2.15.2 The Coordinates of New York City

The geographic coordinates are 41° north and 74° west of Greenwich, or $\phi = +41°$, $\lambda = +74°$. In time units, the longitude would be $74/15\,\mathrm{h} = 4\,\mathrm{h}\,56\,\mathrm{min}$ west of Greenwich. The geocentric latitude is obtained from

$$\tan \phi' = \frac{b^2}{a^2} \tan \phi = \left(\frac{6,356,755}{6,378,140}\right)^2 \tan 41° = 0.86347 \Rightarrow \phi' = 40°48'34'' \; .$$

The geocentric latitude is $11'26''$ less than the geographic latitude.

2.15.3 The angular separation of two objects in the sky is quite different from their coordinate difference.

Suppose the coordinates of a star A are $\alpha_1 = 10\,\mathrm{h}$, $\delta_1 = 70°$ and those of another star B, $\alpha_2 = 11\,\mathrm{h}$, $\delta_2 = 80°$. Using the Pythagorean theorem for plane triangles, we would get

$$d = \sqrt{(15°)^2 + (10°)^2} = 18° \; .$$

But if we use the third equation in (2.7), we get

$$\cos d = \cos (\alpha_1 - \alpha_2) \sin (90° - \delta_1) \sin (90° - \delta_2) + \cos (90° - \delta_1) \cos (90° - \delta_2)$$

$$= \cos (\alpha_1 - \alpha_2) \cos \delta_1 \cos \delta_2 + \sin \delta_1 \sin \delta_2$$

$$= \cos 15° \cos 70° \cos 80° + \sin 70° \sin 80°$$

$$= 0.983 \; ,$$

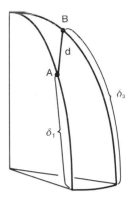

which yields $d = 10.6°$. The figure shows why the result obtained from the Pythagorean theorem is so far from being correct: hour circles (circles with $\alpha = $ const) approach each other towards the poles and their angular separation becomes smaller, though the coordinate difference remains the same.

2.15.4 Find the altitude and azimuth for an object with right ascension $\alpha = 10\,\mathrm{h}$ and declination $\delta = 70°$ if the sidereal time is $\Theta = 18\,\mathrm{h}$ and latitude $\phi = 60°$.

The hour angle is $h = \Theta - \alpha = 8\,\mathrm{h} = 120°$. Next we apply the equations in (2.15):

$$\sin A \cos a = \sin 120° \cos 70° \; ,$$

$$\cos A \, \cos a = \cos 120° \, \cos 70° \, \sin 60° - \sin 70° \, \cos 60° \ ,$$

$$\sin a = \cos 120° \, \cos 70° \, \cos 60° + \sin 70° \, \sin 60° \ .$$

Thus $\sin a = 0.728$ and the altitude is $a = 47°$. To find the azimuth we have to compute its sine and cosine:

$$\sin A = 0.432 \ , \quad \cos A = -0.902 \ .$$

Hence the azimuth is $A = 154°$. The object is in the north-northwest, 47 degrees above the horizon.

2.15.5 The coordinates of Arcturus are $\alpha = 14\,h\,15.7\,min$, $\delta = 19°11'$. Find the sidereal time at the moment Arcturus rises or sets in Boston ($\phi = 42°19'$).

Neglecting refraction, we get

$$\cos h = -\tan 19°11' \, \tan 42°19' = -0.348 \times 0.910 = -0.317 \ .$$

Hence, $h = \pm 108.47° = 7\,h\,14\,min$. The more accurate result is

$$\cos h = -\tan 19°11' \, \tan 42°19' - \frac{\sin 35'}{\cos 19°11' \, \cos 42°19'} = -0.331 \ ,$$

whence $h = \pm 109.35° = 7\,h\,17\,min$. The plus and minus signs correspond to setting and rising, respectively. When Arcturus rises, the sidereal time is

$$\Theta = \alpha + h = 14\,h\,16\,min - 7\,h\,17\,min = 6\,h\,59\,min \ ,$$

and when it sets, the sidereal time is

$$\Theta = 14\,h\,16\,min + 7\,h\,17\,min = 21\,h\,33\,min \ .$$

Note that the result is independent of the date: a star rises and sets at the same sidereal time every day.

2.15.6 The proper motion of Aldebaran is $\mu = 0.20''/a$ and parallax $\pi = 0.048''$. The spectral line of iron at $\lambda = 440.5\,nm$ is displaced $0.079\,nm$ towards the red. What are the radial and tangential velocities and the total velocity?

The radial velocity is found from

$$\frac{\Delta \lambda}{\lambda} = \frac{v_r}{c} \Rightarrow v_r = \frac{0.079}{440.5} \cdot 3 \times 10^8 \, m/s = 5.4 \times 10^4 \, m/s = 54\,km/s \ .$$

Tangential velocity is now given by (2.37), since μ and π are in correct units:

$$v_t = 4.74 \, \mu r = 4.74 \, \mu / \pi = \frac{4.74 \times 0.20}{0.048} = 20\,km/s \ .$$

The total velocity is

$$v = \sqrt{v_r^2 + v_t^2} = \sqrt{54^2 + 20^2} \, km/s = 58\,km/s \ .$$

2.15.7 Find the local time in Paris (longitude $\lambda = -2°$) at 12:00.

Local time coincides with the zonal time along the meridian 15° east of Greenwich. Longitude difference $15° - 2° = 13°$ equals $(13°/15°) \times 60\,\text{min} = 52$ minutes. The local time is 52 minutes less than the official time, or 11:08. This is mean solar time. To find the true solar time, we must add the equation of time. In early February, E.T. $= -14\,\text{min}$ and the true solar time is $11:08 - 14\,\text{min} = 10:54$. At the beginning of November, E.T. $= +16\,\text{min}$ and the solar time would be 11:24. Since $-14\,\text{min}$ and $+16\,\text{min}$ are the extreme values of E.T., the true solar time is in the range $10:54 - 11:24$, the exact time depending on the day of the year. During daylight savings time, we must still subtract one hour from these times.

2.15.8 *Estimating Sidereal Time*

Since the sidereal time is the hour angle of the vernal equinox Υ, it is 0h when Υ culminates or transits the south meridian. At the moment of the vernal equinox, the Sun is in the direction of Υ and thus culminates at the same time as Υ. So the sidereal time at 12:00 local solar time is 0:00, and at the time of the vernal equinox, we have

$$\Theta = T + 12\,\text{h} \ ,$$

where T is the local solar time. This is accurate within a couple of minutes. Since the sidereal time runs about 4 minutes fast a day, the sidereal time, n days after the vernal equinox, is

$$\Theta \approx T + 12\,\text{h} + n \times 4\,\text{min} \ .$$

At autumnal equinox Υ culminates at 0:00 local time, and sidereal and solar times are equal.

Let us try to find the sidereal time in Paris on April 15 at 22:00, European standard time ($= 23:00$ daylight savings time). The vernal equinox occurs on the average on March 21; thus the time elapsed since the equinox is $10 + 15 = 25$ days. Neglecting the equation of time, the local time T is 52 minutes less than the zonal time. Hence

$$\Theta = T + 12\,\text{h} + n \times 4\,\text{min} = 21\,\text{h}\,8\,\text{min} + 12\,\text{h} + 25 \times 4\,\text{min} = 34\,\text{h}\,48\,\text{min}$$

$$= 10\,\text{h}\,48\,\text{min} \ .$$

The time of the vernal equinox can vary about two days in either direction from the average. Therefore the accuracy of the result is roughly 10 min.

2.15.9 *A More Accurate Approximation for Sidereal Time*

As in the previous example, we begin with the equation

$$\Theta = T + 12\,\text{h} \ ,$$

which is accurate only at the exact moment of the vernal equinox. The local time is, according to (2.41),

$$T = T_M + \text{E.T.}$$

At vernal equinox, E.T. $= -7\,\text{min}$; thus

$$\Theta = T_M + 11\,\text{h}\,53\,\text{min}\ .$$

The mean time T_M flows at an even rate and the sidereal time θ runs $3\,\text{min}\,57\,\text{s}$ fast a day. x days after the vernal equinox, we have

$$\Theta = T_M + 11\,\text{h}\,53\,\text{min} + x \times 3\,\text{min}\,57\,\text{s}.$$

We shall now use this to find the sidereal time in Paris on April 15, 1982 at 22:00, European standard time. In 1982, the vernal equinox occurred on March 20, at 23 hours UT, or on March 21, 0 hours European time. Thus

$$x = 25 + 22/24 = 25.9167\ .$$

The local mean time was $T_M = 21\,\text{h}\,8\,\text{min}$; hence the sidereal time is

$$\Theta = 21\,\text{h}\,8\,\text{min} + 11\,\text{h}\,53\,\text{min} + 25.9167 \times (3\,\text{min}\,57\,\text{s}) = 34\,\text{h}\,43\,\text{min} = 10\,\text{h}\,43\,\text{min}\ .$$

2.15.10 Find the rising time of Arcturus in Boston on January 10.

In Exercise 2.15.5, we found the sidereal time of this event, $\Theta = 6\,\text{h}\,59\,\text{min}$. Since we do not know the year, we use the rough method of Exercise 2.15.8. The time between January 1 and vernal equinox (March 21) is about 70 days. Thus the sidereal time on January 1 is

$$\Theta \approx T + 12\,\text{h} - 70 \times 4\,\text{min} = T + 7\,\text{h}\,20\,\text{min}\ ,$$

from which

$$T = \Theta - 7\,\text{h}\,20\,\text{min} = 6\,\text{h}\,59\,\text{min} - 7\,\text{h}\,20\,\text{min} = 30\,\text{h}\,59\,\text{min} - 7\,\text{h}\,20\,\text{min}$$
$$= 23\,\text{h}\,39\,\text{min}\ .$$

The longitude of Boston is 71°W, and the Eastern standard time is (4°/15°) ×60 min = 16 minutes less, or 23:23.

2.16 Exercises

2.16.1 Find the distance between Helsinki and Seattle along the shortest route. Where is the northernmost point of the route, and what is its distance from the North Pole? The longitude of Helsinki is 25 °E and latitude 60°; the longitude of Seattle is 122 °W and latitude 48°. Assume that the radius of the Earth is 6370 km.

2.16.2 A star crosses the South meridian at an altitude of 85°, and the North meridian at 45°. Find the declination of the star and the latitude of the observer.

2.16.3 Where are the following statements true?

a) Castor (α Gem, declination 31°56′) is circumpolar.
b) Betelgeuze (α Ori, $\delta = 7°24′$) culminates at zenith.
c) α Cen ($\delta = -60°46′$) rises to an altitude of 30°.

2.16.4   In his *Old Man and the Sea* Hemingway wrote:

It was dark now as it becomes dark quickly after the Sun sets in September. He lay against the worn wood of the bow and rested all that he could. The first stars were out. He did not know the name of Rigel but he saw it and knew soon they would all be out and he would have all his distant friends.

How was Hemingway's astronomy?

2.16.5   The right ascension of the Sun in June 1st, 1983, was 4 h 35 min and declination 22°00′. Find the ecliptic longitude and latitude of the Sun and the Earth.

2.16.6   Show that the galactic coordinates are obtained from

$$\cos b \sin(l_N - l) = \cos \delta \sin(\alpha - \alpha_P) \ ,$$
$$\cos b \cos(l_N - l) = -\cos \delta \sin \delta_P \cos(\alpha - \alpha_P) + \sin \delta \cos \delta_P \ ,$$
$$\sin b = \cos \delta \cos \delta_P \cos(\alpha - \alpha_P) + \sin \delta \sin \delta_P \ ,$$

where the direction of the galactic north pole is $\alpha_P = 12\,\text{h } 51.4\,\text{min}$, $\delta_P = 27°08′$ and the galactic longitude of the celestial pole is $l_N = 123.0°$.

2.16.7   Show that on the Arctic Circle the Sun

a) rises at the same sidereal time Θ_0 between December 22 and June 22,
b) sets at the same sidereal time Θ_0 between June 22 and December 22.

What is Θ_0?

2.16.8   The coordinates of Sirius for the epoch 1900.0 were $\alpha = 6\,\text{h } 40\,\text{min } 45\,\text{s}$, $\delta = -16°35′$, and the components of its proper motion were $\mu_\alpha = -0.037\,\text{s a}^{-1}$, $\mu_\delta = -1.12''\text{a}^{-1}$. The parallax of Sirius is 0.375″ and radial velocity $-8\,\text{km s}^{-1}$.

a) Find the coordinates of Sirius for 2000.0.
b) What are the tangential and total velocities of Sirius?
c) When will Sirius be closest to the Sun?
d) What will then be its proper motion and parallax?

Chapter 3 **Observations and Instruments**

Up to the end of the Middle Ages, the most important means of observation in astronomy was the human eye. It was aided by various mechanical devices to measure the positions of celestial bodies in the sky. The telescope was invented in Holland at the beginning of the 17th century, and in 1609, Galileo Galilei made his first astronomical observations with this new instrument. Astronomical photography was introduced at the end of the 19th century, and during the last few decades, many kinds of electronic detectors have been adopted for the study of electromagnetic radiation from space. The electromagnetic spectrum from the shortest gamma rays to long radio waves can now be used for astronomical observations.

3.1 Observing Through the Atmosphere

With satellites and spacecraft, observations can be made outside the atmosphere. Yet, the great majority of astronomical observations are carried out from the surface of the Earth. In the preceding chapter, we discussed refraction, which changes the apparent altitudes of objects. The atmosphere affects observations in many other ways as well. The air is never quite steady, and there are layers with different temperature and densities; this causes convection and turbulence. When the light from a star passes through the unsteady air, rapid changes in refraction in different directions result. Thus, the amount of light reaching a detector, e.g. the human eye, constantly varies; the star is said to *scintillate* (Fig. 3.1). The planets shine more steadily, since they are not point sources like the stars.

Fig. 3.1. Scintillation of Sirius during four passes across the field of view. The star was very low on the horizon. (Photo by Pekka Parviainen)

A telescope collects light over a larger area, which evens out rapid changes and diminishes scintillation. Instead, differences in refraction along different paths of light through the atmosphere smear the image and point sources are seen in telescopes as vibrating speckles. This phenomenon is called *seeing*, and the size of the so-called seeing disc may vary from less than an arc second to several tens of arc seconds. If the size of the seeing disc is small, we speak of good seeing. Seeing and scintillation both tend to blot out small details when one looks through a telescope, for example, at a planet.

Some wavelength regions in the electromagnetic spectrum are strongly absorbed by the atmosphere. The most important transparent interval is the so-called *optical window,* from about 300 – 800 nm. This interval coincides with the region of sensitivity of the human eye (about 400 – 700 nm).

At wavelengths under 300 nm, absorption by atmospheric ozone prevents radiation from reaching the ground. The ozone is concentrated in a thin layer at a height of about 20 – 30 km, and this layer protects the Earth from harmful ultraviolet radiation. At still shorter wavelengths, the main absorbers are O_2, N_2 and free atoms. Nearly all of the radiation under 300 nm is absorbed by the upper parts of the atmosphere.

At wavelengths longer than visible light, in the near-infrared region, the atmosphere is fairly transparent up to 1.3 μm. There are some absorption belts caused by water and molecular oxygen, but the atmosphere gets more opaque only at wavelengths of more than 1.3 μm. At these wavelengths, radiation reaches the lower parts of the atmosphere only in a few narrow windows. All wavelengths between 20 μm and 1 mm are totally absorbed. At wavelengths longer than 1 mm, there is the *radio window* extending up to about 20 m. At still longer wavelengths, the ionosphere in the upper parts of the atmosphere absorbs all radiation (Fig. 3.2). The exact upper limit of the radio window depends on the strength of the ionosphere, which varies during the day. (The structure of the atmosphere is described in Chap. 8.)

At optical wavelengths (300 – 800 nm), light is scattered by the molecules and dust in the atmosphere, and the radiation is attenuated. Scattering and absorption together are called *extinction*. Extinction must be taken into account when one measures the brightness of celestial bodies (Chap. 4).

In the 19th century, Lord Rayleigh succeeded in explaining why the sky is blue. Scattering caused by the molecules in the atmosphere is inversely proportional to the fourth power of the wavelength. Thus, blue light is scattered more than red light. The blue light we see all over the sky is scattered sunlight. The same phenomenon colours the setting sun red, because, owing to the long, oblique path through the atmosphere, all the blue light has been scattered away.

Fig. 3.2. The transparency of the atmosphere at different wavelengths. In the white parts, radiation reaches the surface of the Earth; the hatched part indicates absorption. The radiation is also absorbed by interstellar gas, as shown in the lowermost figure. (Modified from Sky and Telescope, June 1975)

Fig. 3.3. Night views from the top of Mount Wilson. The upper photo was taken in 1908, the lower one in 1988. The lights of Los Angeles, Pasadena, Hollywood and more than 40 other towns are reflected in the sky, causing considerable disturbance to astronomical observations. (Photos by Ferdinand Ellerman and International Dark-Sky Association)

In astronomy, one often has to observe very faint objects. Thus, it is important that the background sky be as dark as possible, and the atmosphere as transparent as possible. That is why the large observatories have been built on mountain tops far from the cities. The air above an observatory site must be very dry, the number of cloudy nights few, and the seeing good.

Astronomers have looked all over the Earth for optimal conditions and have found some exceptional sites. In the 1970's, several new major observatories were founded at these sites. Among the best sites in the world are: the extinguished volcano Mauna Kea on Hawaii, rising more than 4000 m above the sea; the dry mountains in northern Chile; the Sonora desert in the U.S., near the border of Mexico; and the mountains on La Palma, in the Canary Islands. Many older observatories are severely plagued by the lights of nearby cities. This was one of the reasons why the old 2.5-m telescope on Mt. Wilson (Fig. 3.3) was retired in 1985.

In radio astronomy, atmospheric conditions are not very critical except when observing at the shortest wavelengths. Constructors of radio telescopes have much greater freedom in choosing their sites than optical astronomers. Still, radio telescopes are also

Fig. 3.4. A lens telescope or refractor and a mirror telescope or reflector

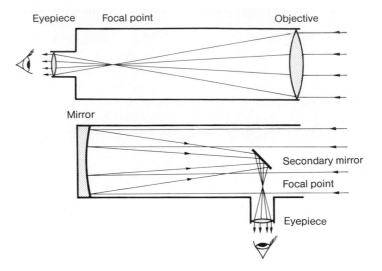

Eyepiece   Focal point   Objective

Mirror

Secondary mirror

Focal point

Eyepiece

often constructed in uninhabited places to isolate them from disturbing radio and television broadcasts.

3.2 Optical Telescopes

The telescope fulfills three major tasks in astronomical observations:

1) It collects the light from objects, making possible the study of very faint sources.
2) It increases the apparent angular diameter of the object and thus improves resolution.
3) It is used to measure the positions of objects.

The light-collecting surface in a telescope is either a lens or a mirror. Thus, optical telescopes are divided into two types, lens telescopes or *refractors* and mirror telescopes or *reflectors* (Fig. 3.4).

Geometrical Optics. Refractors have two lenses, the *objective* which collects the incoming light and forms an image in the focal plane, and the *eyepiece* which is a small magnifying glass for looking at the image (Fig. 3.5). The lenses are at the opposite ends of a tube which can be directed towards any desired point. The distance between the eyepiece and the focal plane can be adjusted to get the image into focus. The image

Fig. 3.5. The scale and magnification of a refractor. The object extends an angle u. The objective forms an image of the object in the focal plane. When the image is viewed through the eyepiece, it is seen in an angle u'

formed by the objective lens can also be registered, e.g. on a photographic plate, as in an ordinary camera.

The diameter of the objective, D, is called the *aperture* of the telescope. The ratio of the aperture D to the focal length, f, $F = D/f$, is called the *aperture ratio*. This quantity is used to characterize the light-gathering power of the telescope. If the aperture ratio is large, near unity, one has a powerful, "fast" telescope; this means that one can take photographs using short exposures, since the image is bright. A small aperture ratio (the focal length much greater than the aperture) means a "slow" telescope.

In astronomy, as in photography, the aperture ratio is often denoted by f/n (e.g. $f/8$), where n is the focal length divided by the aperture. For fast telescopes, this ratio can be $f/1 \ldots f/3$, but usually it is smaller, $f/8 \ldots f/15$.

The *scale* of the image formed in the focal plane of a refractor can be geometrically determined from Fig. 3.5. When the object is seen at the angle u, it forms an image of height s,

$$s = f \tan u \approx fu , \tag{3.1}$$

since u is a very small angle. If a telescope has a focal length of, for instance, 343 cm, one arc minute corresponds to

$$\begin{aligned}
s &= 343 \text{ cm} \times 1' \\
&= 343 \text{ cm} \times (1/60) \times (\pi/180) \\
&= 1 \text{ mm.}
\end{aligned}$$

The *magnification* ω is (from Fig. 3.5)

$$\omega = u'/u \approx f/f' , \tag{3.2}$$

where we have used the equation $s = fu$. Here, f is the focal length of the objective and f' that of the eyepiece. For example, if $f = 100$ cm and we use an eyepiece with $f' = 2$ cm, the magnification is 50-fold. The magnification is not an essential feature of a telescope, since it can be changed simply by changing the eyepiece.

A more important characteristic, which depends on the aperture of the telescope, is the *resolving power,* which determines, for example, the minimum angular separation of the components of a binary star that can be seen as two separate stars. The theoretical limit for the resolution is set by the diffraction of light: The telescope does not form a point image of a star, but rather a small disc, since light "bends around the corner", like all radiation (Fig. 3.6).

The theoretical resolution of a telescope is often given in a form introduced by Rayleigh (*Diffraction by a Circular Aperture, p. 67)

$$\sin \theta \approx \theta = 1.22 \, \lambda/D .$$

As a practical rule, we can say that two objects are seen as separate if the angular distance between them is

$$\theta \gtrsim \lambda/D , \quad [\theta] = \text{rad} . \tag{3.3}$$

This formula can be applied to optical as well as radio telescopes. For example, if one makes observations at a typical yellow wavelength ($\lambda = 550$ nm), the resolving power of a reflector with an aperture of 1 m is about $0.2''$. However, seeing spreads out the

Fig. 3.6a−e. Diffraction and resolution. The image of a single star (**a**) consists of concentric diffraction rings, which can be displayed as a mountain diagram (**b**). Wide pairs of stars can be easily resolved (**c**). For resolving close binaries, different criteria can be used. One is the Rayleigh limit 1.22 λ/D (**d**). In practice, the resolution can be written λ/D, which is near the Dawes limit (**e**). (Photo Sky and Telescope)

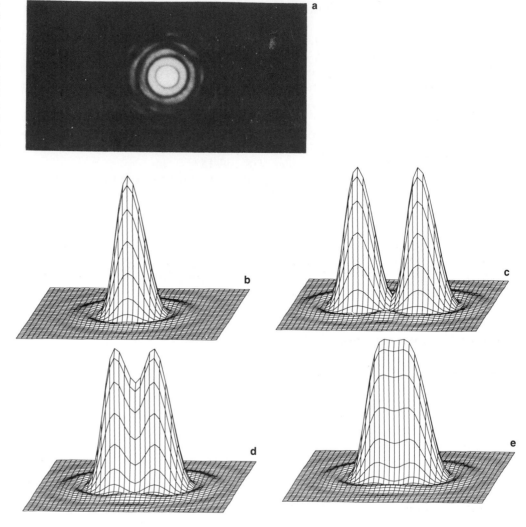

image to a diameter of typically 1 arc second. Thus, the theoretical diffraction limit cannot be reached on the surface of the Earth.

In photography, the image is further spread in the photographic plate, decreasing the resolution as compared with visual observations. The grain size of photographic emulsions is about 0.01 − 0.03 mm, which is also the minimum size of the image. For a focal length of 1 m, the scale is 1 mm = 206″, and, thus, 0.01 mm corresponds to about 2 arc seconds. This is similar to the theoretical resolution of a telescope with an aperture of 7 cm in visual observations.

In practice, the resolution of visual observations is determined by the ability of the eye to see details. In night vision (when the eye is perfectly adapted to darkness), the resolving capability of the human eye is about 2′.

The *maximum magnification* ω_{max} is the largest magnification that is worth using in telescopic observations. Its value is obtained from the ratio of the resolving capability of the eye, $e \approx 2' = 5.8 \times 10^{-4}$ rad, to the resolving power of the telescope, θ,

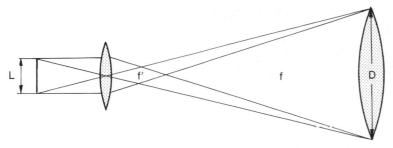

Fig. 3.7. The exit pupil is the image of the objective lens formed by the eyepiece

$$\omega_{\max} = e/\theta \approx eD/\lambda = \frac{5.8 \times 10^{-4} D}{5.5 \times 10^{-7}\,\mathrm{m}} \approx D/1\,\mathrm{mm} \ .$$

If we use for example an objective with a diameter of 100 mm, the maximum magnification is about 100. The eye has no use for larger magnifications.

The *minimum magnification* $\omega_{\min}$ is the smallest magnification that is useful in visual observations. Its value is obtained from the condition that the diameter of the so-called exit pupil L of the telescope must be smaller or equal to the pupil of the eye.

The exit pupil is the image of the objective lens, formed by the eyepiece, through which the light from the objective goes behind the eyepiece. From Fig. 3.7 we obtain

$$L = \frac{f'}{f}D = \frac{D}{\omega} \ .$$

Thus the condition $L \le d$ means that

$$\omega \ge \frac{D}{d} \ .$$

In the night, the diameter of the pupil of the human eye is about 6 mm, and thus the minimum magnification of a 100 mm telescope is about 17.

Refractors. In the first refractors, which had a simple objective lens, the observations were hampered by the so-called *chromatic aberration.* Since glass refracts different colours by different amounts, all colours do not meet at the same focal point (Fig. 3.8),

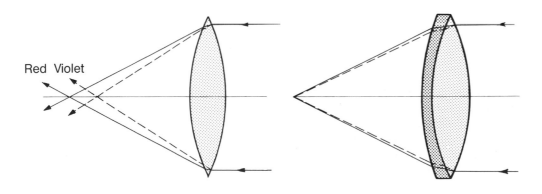

Red Violet

Fig. 3.8. Chromatic aberration. Light rays of different colours are refracted to different focal points (*left*). The aberration can be corrected with an achromatic lens consisting of two parts (*right*)

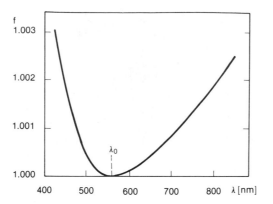

Fig. 3.9. The wavelength dependence of the focal length of a typical achromatic objective for visual observations. The focal length has a minimum near $\lambda = 550$ nm, where the eye is most sensitive. In bluer light ($\lambda \approx 450$ nm) or in redder light ($\lambda \approx 800$ nm), the focal length increases by a factor of about 1.002

but the focal length increases with increasing wavelength. To remove this aberration, *achromatic lenses* consisting of two parts were adopted in the 18th century. The colour dependence of the focal length is much smaller than in single lenses, and at some wavelength, λ_0, the focal length has an extremum (usually a minimum). Near this point the change of focal length with wavelength is very small (Fig. 3.9). If the telescope is intended for visual observations, we choose $\lambda_0 \approx 550$ nm, corresponding to the maximum sensitivity of the eye. Objectives for photographic refractors are usually constructed with $\lambda_0 \approx 425$ nm, since normal photographic plates are most sensitive to the blue part of the spectrum.

By combining three or even more lenses of different glasses in the objective, the chromatic aberration can be corrected still better (as in apochromatic objectives). Also, special glasses have been developed where the wavelength dependences of the refractive index cancel out so well that two lenses already give a very good correction of the chromatic aberration. They have, however, hardly been used in astronomy so far.

The largest refractors in the world have an aperture of about one metre [102 cm in the Yerkes Observatory telescope (Fig. 3.10), finished in 1897, and 91 cm in the Lick Observatory telescope (1888)]. The aperture ratio is typically $f/10\ldots f/20$.

The use of refractors is limited by their small field of view and awkwardly long structure. Refractors are used, e.g. for visual observations of binary stars and in various meridian telescopes for measuring the positions of stars. In photography they can be used for accurate position measurements, for example, to find parallaxes.

A wider field of view is obtained by using more complex lens systems, and telescopes of this kind are called *astrographs*. Astrographs have an objective made up of typically $3-5$ lenses and an aperture ratio of less than 60 cm. The aperture ratio is $f/5\ldots f/7$, and the field of view, about $5°$. Astrographs are used to photograph large areas of the sky, e.g. for proper motion studies and for statistical brightness studies of the stars.

Reflectors. The most common telescope type in astrophysical research is the mirror telescope or reflector. As a light-collecting surface, it employs a mirror coated with a thin layer of aluminium. The form of the mirror is usually parabolic. A parabolic mirror reflects all light rays entering the telescope parallel to the main axis into the same focal point. The image formed at this point is observed through an eyepiece or is registered on a photographic plate. One of the advantages of reflectors is the absence of chromatic aberration, since all wavelengths are reflected to the same point. The largest

Fig. 3.10. The largest refractor in the world is at the Yerkes Observatory, University of Chicago. It has an objective lens with a diameter of 102 cm. (Photo by Yerkes Observatory)

reflectors in the world are the 10-m Keck telescope in Hawaii, the 6-m BTA telescope in the Caucasus and the 5-m Hale telescope on Mt. Palomar.

In the very largest telescopes, the observer can sit with his instruments in a special cage at the *primary focus* (Fig. 3.11) without eclipsing too much of the incoming light. In smaller telescopes, this is not possible, and the image must be inspected from outside the telescope.

Isaac Newton, who invented the reflector, guided the light perpendicularly out from the telescope with a small flat mirror. Therefore the focus of the image in such a system is called the *Newton focus*. A typical aperture ratio of a Newtonian telescope is $f/3 \ldots f/10$. Another possibility is to bore a hole in the centre of the primary mirror and reflect the rays through it with a small hyperbolic secondary mirror in the front end

Fig. 3.11. Different locations of the focus in reflectors: primary focus, Newton focus, Cassegrain focus and Coudé focus. The Coudé system in this figure cannot be used for observations near the zenith. More complex Coudé systems usually have three flat mirrors after the primary and secondary mirrors

of the telescope. In such a design, the rays meet in the *Cassegrain focus*. Cassegrain systems have aperture ratios of $f/8...f/15$.

The effective focal length (f_e) of a Cassegrain telescope is determined by the position and convexity of the secondary mirror. Using the notations of Fig. 3.12, we get

$$f_e = \frac{b}{a} f_p .$$

If we choose $a \ll b$, we have $f_e \gg f_p$. In this way one can construct short telescopes with long focal lengths. Cassegrain systems are especially well suited for spectrographic,

Fig. 3.12. The principle of a Cassegrain-reflector. A concave (paraboloid) primary mirror P_1 reflects the light rays parallel to the main axis towards the primary focus S_1. A convex secondary mirror P_2 (hyperboloid) reflects the rays back through a small hole in the centre of the main mirror, to the secondary focus S_2 outside the telescope

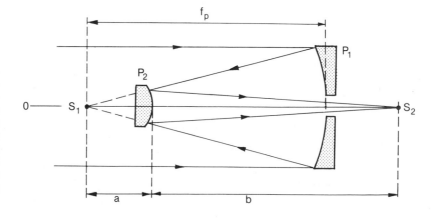

Fig. 3.13. The Coudé system of the Kitt Peak 2.1-m reflector. (Drawing National Optical Astronomy Observatories, Kitt Peak National Observatory)

photometric and other instruments that can be mounted in the secondary focus, easily accessible to the observers.

More complicated arrangements use several mirrors to guide the light through the declination axis of the telescope to a fixed *Coudé focus* (from the French word *couder*, to bend), which can even be situated in a separate room near the telescope (Fig. 3.13). The focal length is thus very long and the aperture ratio $f/30 \ldots f/40$. The Coudé focus is used mainly for accurate spectroscopy, since the large spectrographs can be stationary and their temperature can be held accurately constant. A drawback is that much light is lost in the reflections in the several mirrors of the Coudé system. An

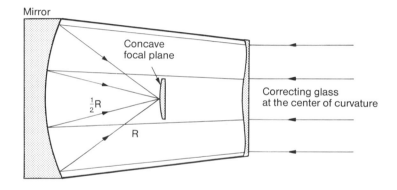

Fig. 3.14. The principle of the Schmidt camera. A correcting glass at the centre of curvature of a concave spherical mirror deviates parallel rays of light and compensates for the spherical aberration of the spherical mirror. (In the figure, the form of the correcting glass and the change of direction of the light rays have been greatly exaggerated.) Since the correcting glass lies at the centre of curvature, the image is practically independent of the incoming angle of the light rays. Thus there is no coma or astigmatism, and the images of stars are points on a spherical surface at a distance of $1/2\,R$, where R is the radius of curvature of the spherical mirror. In photography, the plate must be bent into the form of the focal plane

Fig. 3.15. The large Schmidt telescope of the European Southern Observatory. The diameter of the mirror is 1.62 m and of the free aperture, 1 m. (Photo ESO)

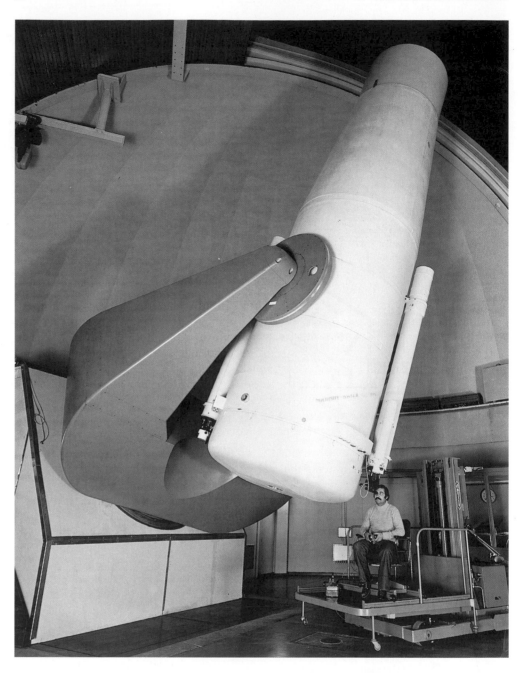

aluminized mirror reflects about four-fifths of the light falling on it, and thus in a Coudé system of, e.g. five mirros (including the primary and secondary mirrors), only $0.8^5 \approx 30\%$ of the light reaches the detector.

The reflector has its own aberration, *coma*. It affects images displaced from the optical axis. Light rays do not converge at one point, but form a figure like a comet. Due to the coma, the classical reflector with a paraboloid mirror has a very small correct field of view. The coma limits the diameter of the useful field to 2 – 20 minutes of arc, depending on the aperture ratio of the telescope. The 5-m Palomar telescope, for instance, has a useful field of view of about 4′, corresponding to about one-eighth of the diameter of the Moon. In practice, the small field of view can be enlarged by various correcting lenses, giving a field of about 1° in the area of a typical photographic plate (e.g. $25 \times 25 \text{ cm}^2$).

If the primary mirror were spherical, there would be no coma. However, this kind of mirror has its own error, *spherical aberration:* light rays from the centre and edges converge at different points. To remove the spherical aberration, the Estonian astronomer *Bernhard Schmidt* developed a thin correcting lens that is placed in the way of the incoming light. Schmidt cameras (Fig. 3.14) have a very wide (about 7°), nearly faultless field of view, and the correcting lens is so thin that it absorbs very little light. The images of the stars are very sharp.

In Schmidt telescopes (Fig. 3.15), the diaphragm with the correcting lens is positioned at the centre of the radius of curvature of the mirror (this radius equals twice the focal length). To collect all the light from the edges of the field of view, the diameter of the mirror must be larger than that of the correcting glass. The Palomar Schmidt camera, for example, has an aperture of 122 cm (correcting lens)/183 cm (mirror), and a focal length of 300 cm. The largest Schmidt telescope in the world is at Tautenburg, GDR, and its corresponding values are 134/203/400 cm.

A disadvantage of the Schmidt telescope is the curved focal plane, consisting of a part of a sphere. When the telescope is used for photography, the plate must be bent along the curved focal plane. Another possibility of correcting the curvature of the field of view is the use of an extra correcting lens near the focal plane. Schmidt cameras have proved to be very effective in mapping the sky. They have been used to photograph the Palomar Sky Atlas mentioned in the previous chapter, and its continuation, the *ESO/SRC Southern Sky Atlas.*

The Schmidt camera is an example of a *catadioptric telescope*, which has both lenses and mirrors. *Schmidt Cassegrain telescopes* used by many amateurs are modifications of the Schmidt camera. They have a secondary mirror mounted at the centre of the correcting lens; the mirror reflects the image through a hole in the primary mirror. Thus the effective focal length can be rather long, although the telescope itself is very short.

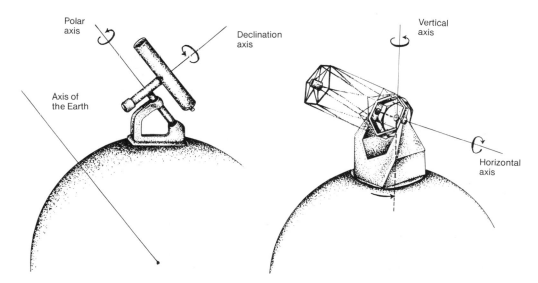

Fig. 3.16. The equatorial mounting (*left*) and the azimuthal mounting (*right*)

Another common catadioptric telescope is the *Maksutov*. Both surfaces of the correcting lens as well as the primary mirror of a Maksutov telescope are concentric spheres.

Another way of removing the coma of the classical reflectors is to use more complicated mirror surfaces. The *Ritchey-Chrétien* system has hyperboloidal primary and secondary mirrors, providing a fairly wide useful field of view. Ritchey-Chrétien optics are used in many large telescopes.

Mountings of Telescopes. A telescope has to be mounted on a steady support to prevent it shaking, and it must be smoothly rotated during observations. There are two principal types of mounting, *equatorial* and *azimuthal* (Fig. 3.16).

a

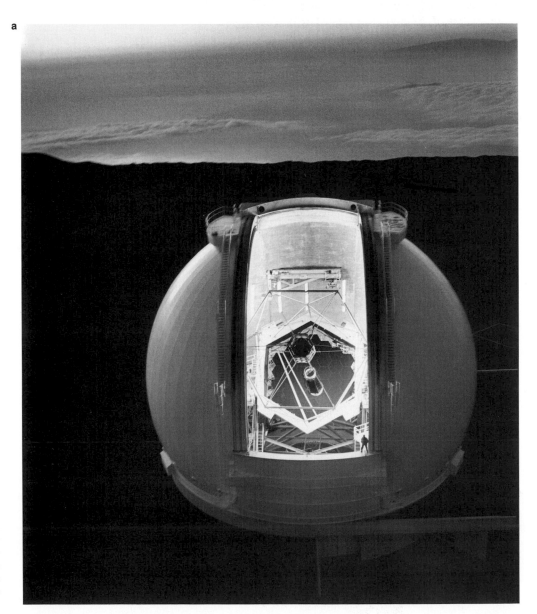

Fig. 3.17 a – c. The largest telescopes in the world. **(a)** The William M. Keck Telescope on the summit of Mauna Kea, Hawaii, was completed in 1992. The 10-m mirror consists of 36 hexagonal segments. **(b)** The BTA, Big Azimuthal Telescope, is situated in the Caucasus in the southern Soviet Union. Its mirror has a diameter of 6 m. It was set in operation at the end of 1975. **(c)** For nearly 30 years, the 5.1-m Hale telescope on Mt. Palomar, California, USA, was the largest telescope in the world. (Photos Roger Ressmeyer-Starlight for the California Association for Research in Astronomy, Spetsialnaya Astrofizitsheskaya Observatorya and Palomar Observatory)

In the equatorial mounting, one of the axes is directed towards the celestial pole. It is called the *polar axis* or *hour axis*. The other one, the *declination axis*, is perpendicular to it. Since the hour axis is parallel to the axis of the Earth, the apparent rotation of the sky can be compensated for by turning the telescope around this axis at a constant rate.

The declination axis is the main technical problem in the equatorial mounting. When the telescope is pointing to the south, its weight causes a force perpendicular to the axis. When the telescope is tracking an object and turns westward, the bearings must take an increasing load parallel with the declination axis.

In the azimuthal mounting, one of the axes is vertical, the other one horizontal. This mounting is easier to construct than the equatorial mounting, and is more stable for very large telescopes. In order to follow the rotation of the sky, the telescope must be turned around both of the axes with changing velocities. The field of view will also rotate; this rotation must be compensated for when the telescope is used for photography.

If an object approaches close to the zenith, its azimuth will change 180° in a very short time. Therefore, around the zenith there is a small region where observations with an azimuthal telescope are not possible.

The largest telescopes in the world were equatorially mounted until the development of computers made possible the more complicated guidance needed for azimuthal mountings. Some of the new large telescopes, like the six-metre BTA, are already azimuthally mounted. (In fact, the abbreviation BTA stands for "Big Azimuthal Telescope".) Azimuthally mounted telescopes have two additional obvious places for foci, the *Nasmyth foci* at both ends of the horizontal axis.

Another type of mounting is the *coelostat*, where rotating mirrors guide the light into a stationary telescope. This system is used especially in solar telescopes.

To measure absolute positions of stars and accurate time, telescopes aligned with the north-south direction are used. They can be rotated around one axis only, the east-west horizontal axis. *Meridian circles* or *transit instruments* with this kind of mounting were widely constructed for different observatories during the 19th century.

The Largest Telescopes. The last two decades have witnessed the birth of many new observatories. Some of them have been founded on the southern hemisphere, which had previously almost completely lacked large telescopes. At this writing (1996), there are a dozen telescopes in the world with mirrors measuring over 3 m in diameter. Some of them are shown in Figs. 3.17 and 3.18.

New Techniques. Detectors are already approaching the theoretical limit of efficiency, where all incident photons are registered. Ultimately, to detect even fainter objects the only solution is to increase the light gathering area; however the mirrors are also getting close to the practical maximum size. Thus, new technical solutions are needed.

One new feature is *active optics*, used, for example, in the ESO 3.5 metre NTT (New Technology Telescope) at La Silla, Chile. The mirror is very thin, but its shape is kept exactly correct by a computer-controlled support mechanism. The weight and production cost of such a mirror are much smaller than those of a conventional thick mirror. Because of the smaller weight the supporting structure can also be made lighter.

Developing the support mechanism further leads to *adaptive optics*. A reference star (or an artificial beam) is monitored constantly in order to obtain the shape of the seeing disk. The shape of the main mirror is adjusted hundreds of times a second to keep the

Fig. 3.18 a – c. Some new large telescopes in the southern hemisphere. (**a**) The 4-m telescope at the Cerro Tololo Inter-American Observatory was set in operation in 1976. It is quite similar to the Mayall 3.9-m telescope at Kitt Peak. (**b**) The first big telescope in the southern hemisphere, the Anglo-Australian telescope, started observations in 1974. (**c**) The European Southern Observatory was founded by Belgium, France, the Netherlands, Sweden and West Germany in 1962. Denmark joining them five years later. The ESO 3.5-m New Technology Telescope at La Silla, Northern Chile, was inaugurated in 1990. (Photos Cerro Tololo Inter-American Observatory, Anglo-Australian Observatory and European Southern Observatory)

image as concentrated as possible. Some experiments have already been made, but the method is not yet commonly used.

The mirrors of large telescopes need not be monolithic, but can be made of smaller pieces that are, e.g., hexagonal. These *mosaic mirrors* are very light and can be used to build up mirrors with diameters of several tens of metres. Using active optics, the hexagons can be accurately focussed. The California Association for Research in Astronomy has constructed the William M. Keck telescope with a 10-m mosaic mirror. It is located on Mauna Kea, and the last segment was installed in 1992. A second, similar telescope, Keck II, is being built near the first telescope. When it is completed in 1996, the pair will form a huge binocular telescope.

The reflecting surface does not have to be continuous, but can consist of several separate mirrors. Such a telescope has been operating on Mt. Hopkins, Arizona, since 1979 (Fig. 3.19). It is the *Multiple-Mirror Telescope* (MMT), with six 1.8-m mirrors, together corresponding to a single mirror having a diameter of 4.5 m. In the MMT, the mirrors can be used singly or their images can be guided to a common focus.

The European Southern Observatory is constructing its own multi-mirror telescope. ESO's Very Large Telescope has four closely located mirrors. The diameter of each mirror is 8 m, and the total area corresponds to one telescope with a 16-m mirror. The resolution is even better, since the "aperture", i.e. the maximum distance between the mirrors, is several tens of meters.

Fig. 3.19. The mirror of a telescope can be made up of several smaller mirrors, which are much easier to manufacture, as in the Multi-Mirror Telescope MMT on Mt. Hopkins in Arizona (Photo Multiple Mirror Telescope Observatory)

Fig. 3.20. The U.S. Hubble Space Telescope after launch in 1990. It is the most effective telescope in the world, since it operates above the atmospheric disturbances. (Photo NASA)

One of the most important astronomical instruments of the century is the U.S. *Hubble Space Telescope*, launched in 1990 (Fig. 3.20). It has a mirror with a diameter of 2.4 m. The resolution of the telescope (after correction of the faulty optics) is near the theoretical diffraction limit since there is no disturbing atmosphere.

The Space Telescope is the first large optical telescope in Earth orbit. In the future, satellites will continue to be mainly used for those wavelength regions where the radiation is absorbed by the atmosphere. Due to budgetary reasons, the majority of astronomical observations will still be carried out on the Earth, and great attention will be given to improving ground-based observatories and detectors.

*Diffraction by a Circular Aperture

Consider a circular hole of radius R in the xy-plane. Coherent light enters the hole from the direction of the negative z-axis (see figure). We consider light rays leaving the hole parallel to the xz-plane forming an angle θ with the x-axis. The light waves interfere on a screen far away. The phase difference between a wave through a point (x, y) and a wave going through the centre of the hole can be calculated from the different path lengths $s = x \sin \theta$:

$$\delta = \frac{s}{\lambda} 2\pi = \frac{2\pi \sin \theta}{\lambda} x \equiv kx .$$

Thus, the phase difference δ depends on the x-coordinate only. The sum of the amplitudes of the waves from a small surface element is proportional to the area of the element $dxdy$. Let the amplitude coming through the centre of the hole be $da_0 = dxdy\hat{i}'$. The amplitude coming from the point (x, y) is then

$$da = dxdy(\cos \delta \hat{i}' + \sin \delta \hat{j}') .$$

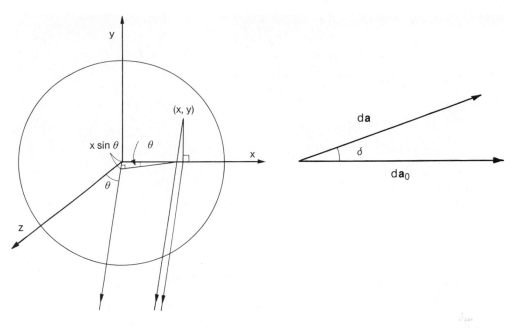

We sum up the amplitudes coming from different points of the hole:

$$a = \int\limits_{\text{aperture}} da = \int\limits_{x=-R}^{R} \int\limits_{y=-\sqrt{R^2-x^2}}^{\sqrt{R^2-x^2}} (\cos kx\,\hat{\boldsymbol{i}}' + \sin kx\,\hat{\boldsymbol{j}}')\,dy\,dx$$

$$= 2 \int\limits_{-R}^{R} \sqrt{R^2-x^2}\,(\cos kx\,\hat{\boldsymbol{i}}' + \sin kx\,\hat{\boldsymbol{j}}')\,dx .$$

Since sine is an odd function $[\sin(-kx) = -\sin(kx)]$, we get zero when we integrate the second term. Cosine is an even function, and so

$$a \propto \int\limits_{0}^{R} \sqrt{R^2-x^2}\,\cos kx\,dx .$$

We substitute $x = Rt$ and define $p = kR = (2\pi R \sin\theta)/\lambda$, thus getting

$$a \propto \int\limits_{0}^{1} \sqrt{1-t^2}\,\cos pt\,dt .$$

The zero points of the intensity observed on the screen are obtained from the zero points of the amplitude,

$$J(p) = \int\limits_{0}^{1} \sqrt{1-t^2}\,\cos pt\,dt = 0 .$$

Inspecting the function $J(p)$, we see that the first zero is at $p = 3.8317$, or

$$\frac{2\pi R \sin\theta}{\lambda} = 3.8317 .$$

The radius of the diffraction disc in angular units can be estimated from the condition

$$\sin\theta = \frac{3.8317}{2\pi R}\,\lambda \approx 1.22\,\frac{\lambda}{D} ,$$

where $D = 2R$ is the diameter of the hole.

3.3 Detectors and Instruments

Only a limited amount of information can be obtained by looking through a telescope with the unaided eye. Up until the end of the 19th century, this was the only way to make observations. The invention of photography in the middle of the last century, brought a revolution in astronomy. The next important step forward in optical astronomy was the development of photoelectric photometry in the 1940's and 1950's. A new revolution, comparable to that caused by the invention of photography, took place in the middle of the 1970's with the introduction of different semiconductor detectors. The sensitivity of detectors has grown so much that today, a 60-cm telescope can be used for observations similar to those made with the Palomar 5-m telescope when it was set in operation in the 1940's.

The Photographic Plate. *Photography* has been and still is one of the most common methods of observation in astronomy. In astronomical photography a glass plate is used, rather than film, since it keeps its shape better. The sensitive layer on the surface of the plate is made up of a silver halide, usually silver bromide, AgBr. A photon absorbed by the halide excites an electron that can move from one atom to another. A silver ion, Ag^+, can catch the electron, becoming a neutral atom. When the necessary amount of silver atoms have been accumulated at one place, they form a latent image. The latent image can be made into a permanent negative by treating the plate after exposure with different chemicals, which transform the silver bromide crystals enclosing the latent image into silver ("development"), and remove the unexposed crystals ("fixing").

The photographic plate has many advantages over the human eye. The plate can register up to millions of stars (picture elements) at one time, while the eye can observe at most one or two objects at a time. The image on a plate is practically permanent – the picture can be studied at any time. In addition, the photographic plate is cheap and easy to use, as compared to many other detectors. The most important feature of a plate is its capability to collect light over an extended time: the longer exposures are used, the more silver atoms are formed on the plate (the plate darkens). By increasing the exposure times, fainter objects can be photographed. The eye has no such capacity: if a faint object does not show through a telescope, it cannot been seen, no matter how long one stares.

One disadvantage of the photographic plate is its low sensitivity. Only one photon in a thousand causes a reaction leading to the formation of a silver grain. Thus, the *quantum efficiency* of the plate is only 0.1%. Several chemical treatments can be used to sensitize the plate before exposure. This brings the quantum efficiency up to a few percent. Another disadvantage is the fact that a silver bromide crystal that has been exposed once does not register anything more, i.e. a saturation point is reached. On the other hand, a certain number of photons are needed to produce an image. Doubling the number of photons does not necessarily double the density (the 'blackness' of the image): the density of the plate depends nonlinearly on the amount of incoming light. The sensitivity of the plate is also strongly dependent on the wavelength of the light. The reasons mentioned above make the maximum accuracy with which brightnesses can be measured on a photographic plate usually less than 5%. Thus the photographic plate makes a poor photometer, but it can be excellently used, e.g. for measuring the positions of stars (positional astronomy) and for mapping the sky.

Photocathodes, Photomultipliers. A *photocathode* is a more effective detector than the photographic plate. It is based on the photoelectric effect. A light quantum, or photon, hits the photocathode and loosens an electron. The electron moves to the positive electrode, or anode, and gives rise to an electric current that can be measured. The quantum efficiency of a photocathode is about $10-20$ times better than that of a photographic plate; optimally, an efficiency of 30% can be reached. A photocathode is also a linear detector: if the number of electrons is doubled, the outcoming current is also doubled.

The photomultiplier is one of the most important applications of the photocathode. In this device, the electrons leaving the photocathode hit a dynode. For each electron hitting the dynode, several others are released. When there are several dynodes in a row, the original weak current can be intensified a millionfold. The photomultiplier measures all the light entering it, but does not form an image. Photomultipliers are mostly used in photometry, and an accuracy of $0.1-1\%$ can be attained.

Photometers, Polarimeters. A detector measuring brightness, a photometer, is usually located behind the telescope in the Cassegrain focus. In the focal plane there is a small hole, the *diaphragm*, which lets through light from the object under observation. In this way, light from other stars in the field of view can be prevented from entering the photometer. A *field lens* behind the diaphragm refracts the light rays onto a photocathode. The outcoming current is intensified further in a *preamplifier*. The photomultiplier needs a voltage of $1000-1500$ volts. The power source, main amplifier and recording apparatus are usually not attached to the telescope. The measurements are nowadays generally recorded in digital form on discs or magnetic tape.

Observations are often made in a certain wavelength interval, instead of measuring all the radiation entering the detector. In this case, a *filter* is used to prevent other wavelengths from reaching the photomultiplier. A photometer can also consist of several

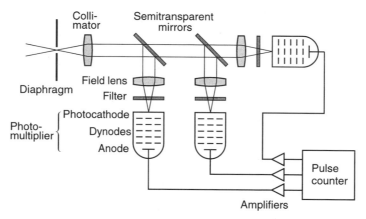

Fig. 3.21. The principle of a photoelectric multicolour photometer. Light collected by a telescope arrives from the left. The light enters the photometer through a small hole in the focal plane, the diaphragm. A lens collimates the light into a parallel beam. Semitransparent mirrors divide the beam and direct it to several photomultipliers. A field lens guides the light through a filter onto the photocathode of the photomultiplier. The quanta of light, photons, release electrons from the cathode. The electrons are accelerated towards the dynodes with a voltage of about 1500 V. The electrons hitting the dynodes release still more electrons, and the current is greatly enhanced. Every electron emitted from the cathode gives rise to a pulse of about 10^9 electrons at the anode; the pulse is amplified and registered by a pulse counter. In this way, the photons from the star are counted

photomultipliers (Fig. 3.21), which simultaneously measure different wavelength bands. In such an instrument beam splitters or semitransparent mirrors split the light beam through fixed filters to the photomultipliers.

In a device called *the photopolarimeter*, a *polarizing filter* is used, either alone or in combination with other filters. The degree and direction of polarization can be found by measuring the intensity of the radiation with different orientations of the polarizers.

In practice, the diaphragm of a photometer will always also let through part of the background sky around the observed object. The measured brightness is in reality the combined brightness of the object and the sky. In order to find the brightness of the object, the background brightness must be measured separately and subtracted from the combined brightness. The accuracy of the measurements is decreased if long observation times are used and the background brightness undergoes fast changes. The problem can be solved by observing the brightness of the background sky and the object simultaneously.

Photometric observations are often relative. If one is observing, for example, a variable star, a *reference star* close to the actual target is observed at regular intervals. Using the observations of this reference star it is possible to derive a model for the slow changes in the atmospheric extinction (see Chap. 4) and remove their effect. The instrument can be calibrated by observing some *standard stars*, whose brightness is known very accurately.

Image Intensifiers. Different *image intensifiers* based on the photocathode have been used since the 1960's. In the intensifier, the information about the starting point of the electron on the photocathode is preserved and the intensified image is formed on a fluorescent screen. The image can then be registered, e.g. by photographing it. One of the advantages of the image intensifier is that even faint objects can be imaged using relatively short exposures, and observations can be made at wavelengths where usual photographic plates are insensitive. The images can also be "exposed" with electrons on special film.

Another common type of detector is based on the TV camera (*Vidicon camera*). The electrons released from the photocathode are accelerated with a voltage of a few kilovolts before they hit the electrode where they form an image in the form of an electric charge distribution. After exposure, the charge at different points of the electrode is read by scanning its surface with an electron beam, row by row. This produces a video signal, which can be transformed into a visible image on a TV tube. The information can also be saved in digital form. In the most advanced systems, the scintillations caused by single electrons on the fluorescent screen of the image intensifier can be registered and stored in the memory of a computer. For each point in the image, there is a memory location, called a picture element or *pixel*.

Since the middle of the 1970's, detectors using *semiconductor techniques* have begun to be used in increasing numbers. With semiconductor detectors, a quantum efficiency factor of about 70 – 80% can be attained; thus, sensitivity cannot be improved much more. The wavelength regions suitable for these new detectors are much wider than in the case of the photographic plate. The detectors are also linear. Computers are used for collecting, saving and analyzing the output data available in digital form.

CCD Camera. The most important new detector is the *CCD camera* (Charge Coupled Device). The detector consists of a surface made up of light-sensitive silicon diodes, arranged in a rectangular array of image elements or *pixels*. The largest cameras can have as many as 4096×4096 pixels, although most are considerably smaller.

A photon hitting the detector can release an electron, which will remain trapped inside a pixel. After exposure, varying potential differences are used to move the accumulated charges row by row to a readout buffer. In the buffer the charges are moved pixel by pixel to an analogue-digital converter, which transmits the digital value to a computer. Reading an image also clears the detector (Fig. 3.22). If the exposures are very short the readout times may take a substantial part of the observing time.

The CCD camera is nearly linear: the number of electrons is directly proportional to the number of photons. Calibration of the data is much easier than with photographic plates.

The quantum efficiency, i.e. the number of electrons per incident photon, is high, and the CCD camera is much more sensitive than a photographic plate. The sensitivity is highest in the red wavelength range, about $600-800$ nm, where the quantum efficiency can be as high as $80-90\%$.

The range of the camera extends far to the infrared. In the ultraviolet the sensitivity drops, due to the absorption of the silicon, very rapidly below about 500 nm. Two methods have been used to avoid this problem. One is to use a coating that absorbs the ultraviolet photons and emits light of a longer wavelength. Another possibility is to make the chip very thin to reduce the absorption.

The thermal noise of the camera generates *dark current* even if the camera is in total darkness. To reduce the noise the camera must be cooled. Astronomical CCD cameras

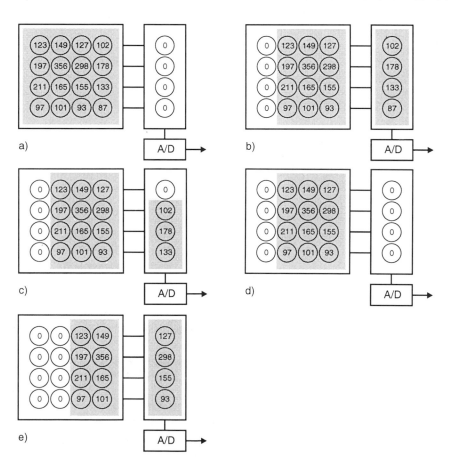

Fig. 3.22 a – e. The principle of reading a CCD camera. (**a**) During an exposure electrons are trapped in potential wells corresponding to pixels of the camera. The number at each pixel shows the number of electrons. (**b**) After the exposure each horizontal line is moved one pixel to the right; the right-most row moves to the readout buffer. (**c**) The contents of the buffer are moved down by one pixel. The lower-most charge moves to the A/D converter, which sends the number of electrons to the computer. (**d**) After moving the buffer down several times one vertical row has been read. (**e**) The image is again shifted right by one pixel. This procedure is repeated till the whole image is read

are usually cooled with liquid nitrogen, which efficiently removes most of the dark current. However, the sensitivity is also reduced when the camera is cooled; so too cold is not good either. The temperature must be kept constant in order to obtain consistent data. For amateurs there are already moderately priced CCD cameras, which are electrically cooled.

The dark current can easily be measured by taking exposures with the shutter closed. Subtracting this from the observed image gives the real number of electrons due to incident light.

The sensitivity of individual pixels may be slightly different. This can be corrected for by taking an image of an evenly illuminated field, like a twilight sky. This image is called a *flat-field*. When observations are divided by the flat-field, the error caused by different pixels is removed.

The CCD camera is very stable. Therefore it is not necessary to repeat the dark current and flat-field observations very frequently. Typically these calibration exposures are taken during evening and morning twilights, just before and after actual observations.

Cosmic rays are charged particles that can produce extraneous bright dots in CCD images. They are usually limited to one or two pixels, and are easily identified. Typically a short exposure of a few minutes contains a few traces of cosmic rays. Instead of a single long exposure it is usually better to take several short ones, clean the images from cosmic rays, and finally add the images on a computer.

A more serious problem is the *readout noise* of the electronics. In the first cameras it could be hundreds of electrons per pixel. In modern cameras it is still several electrons. This gives a limit to the faintest detectable signal: if the signal is weaker than the readout noise, it is indistinguishable from the noise.

Although the CCD camera is a very sensitive detector, even bright light cannot damage it. A photomultiplier, on the other hand, can be easily destroyed by letting in too much light. However, one pixel can only store a certain number of electrons, after which it becomes *saturated*. Excessive saturation can cause the charge to overflow to the neighbouring pixels. If the camera becomes badly saturated it may have to be read several times to completely remove the charges.

The largest CCD cameras are quite expensive, and even they are still rather small compared with photographic plates and films. Therefore photography still has some use in recording extended objects.

Spectrographs. The simplest spectrograph is a prism that is placed in front of a telescope. This kind of device is called an *objective prism spectrograph*. The prism spreads out the different wavelengths of light into a spectrum which can be registered, e.g. on a photographic plate. During the exposure, the telescope is usually slightly moved perpendicularly to the spectrum, in order to increase the width of the spectrum. With an objective prism spectrograph, large numbers of spectra can be photographed, e.g. for spectral classification.

For more accurate information, the *slit spectrograph* must be used (Fig. 3.23). It has a narrow slit in the focal plane of the telescope. The light is guided through the slit to a collimator that reflects or refracts all the light rays into a parallel beam. After this, the light is dispersed into a spectrum by a prism, and focused with a camera onto a photographic plate. A comparison spectrum is exposed on the next side to the stellar spectrum to determine the precise wavelengths. A big slit spectrograph is often placed at a stationary location in a separate room in the Coudé focus of the telescope.

Fig. 3.23. The principle of a slit spectrograph. Light rays entering through a slit are collimated (made parallel to each other), dispersed into a spectrum by a prism and projected onto a photographic plate

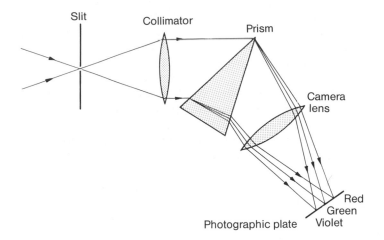

Instead of the prism, a *diffraction grating* can be used to form the spectrum. A grating has narrow grooves, side by side, typically several hundred per millimetre. When light is reflected by the walls of the grooves, the adjoining rays interfere with each other and give rise to spectra of different orders. There are two kinds of gratings: *reflection* and *transmission gratings*. In a reflection grating, no light is absorbed by the glass as in the prism or transmission grating. A grating usually has higher dispersion, or ability to spread the spectrum, than a prism. The dispersion can be increased by increasing the density of the grooves of the grating. In slit spectrographs, the reflection grating is most commonly used.

Interferometers. The resolution of a big telescope is in practice limited by seeing, and thus increasing the aperture does not necessarily improve the resolution. To get nearer to the theoretical resolution limit set by diffraction [c.f. (3.3)], different *interferometers* can be used.

There are two types of optical interferometers. One kind uses an existing large telescope; the other, a system of two or more separate telescopes. In both cases, the light rays are allowed to interfere. By analyzing the outcoming interference pattern, the structures of close binaries can be studied, apparent angular diameters of the stars can be measured, etc.

One of the earliest interferometers was *a Michelson interferometer* that was built shortly before 1920 for the largest telescope of that time. In front of the telescope, at the ends of a six-metre-long beam, there were flat mirrors reflecting the light into the telescope. The form of the interference pattern changed when the separation of the mirrors was varied. In practice, the interference pattern was disturbed by seeing, and only a few positive results were obtained with this instrument.

The diameters of over 30 of the brightest stars have been measured using *intensity interferometers*. Such a device consists of two separate telescopes that can be moved in relation to each other. This method is suitable for the brightest objects only.

In 1970, the Frenchman *Antoine Labeyrie* introduced the principle of *speckle interferometry*. In traditional imaging, the pictures from long exposures consist of a large number of instantaneous images, "speckles", that together form the seeing disc. In speckle interferometry, very short exposures and large magnifications are used and hundreds of pictures are taken. When these pictures are combined and analyzed (usually in digital form), the actual resolution of the telescope can nearly be reached.

3.4 Radio Telescopes

Radio astronomy represents a relatively new branch of astronomy. It covers a frequency range from a few megahertz (100 m) up to frequencies of about 300 GHz (1 mm), thereby extending the observable electromagnetic spectrum by many orders of magnitude. The low-frequency limit of the radio band is determined by the opacity of the ionosphere, while the high-frequency limit is due to strong absorption from oxygen and water bands in the lower atmosphere. Neither of these limits is very strict, and under favourable conditions, radio astronomers can work into the submillimetre region or through ionospheric holes during sunspot minima.

At the beginning of the 20th century, attempts were made to observe radio emission from the Sun. These experiments, however, failed because of the low sensitivity of the antenna-receiver systems, and because of the opaqueness of the ionosphere at the low frequencies at which most of the experiments were carried out. The first observations of cosmic radio emission were later made by the American engineer Karl G. Jansky in 1932, while studying thunderstorm radio disturbances at a frequency of 20.5 MHz (14.6 m). He discovered radio emission of unknown origin, which varied within a 24-hour period. Somewhat later he identified the source of this radiation to be in the direction of the centre of our Galaxy.

The real birth of radio astronomy may perhaps be dated to the late 1930's, when Grote Reber started systematical observations with his homemade 9.5-m paraboloid antenna. Thereafter, radio astronomy developed quite rapidly, and has greatly improved our knowledge of the Universe. Observations are made both in the continuum (broad band) and in spectral lines (radio spectroscopy). Much of our knowledge about the structure of our Milky Way comes from radio observations of the 21-cm line of neutral hydrogen and, more recently, from the 2.6-mm line of the carbon monoxide molecule. Radio astronomy has resulted in many important discoveries; e.g. both pulsars and quasars were first found by radio astronomical observations. The importance of the field can also be seen from the fact that the Nobel prize in physics has recently been awarded twice to radio astronomers.

A radio telescope collects radiation in an aperture or antenna, from which it is transformed to an electric signal by a receiver, called a radiometer. This signal is then amplified, detected and integrated, and the output is registered on some recording device, e.g. a strip chart recorder, or in a modern radio telescope, on a magnetic tape or a computer disk. Because the received signal is very weak, one has to use sensitive receivers. These are often cooled to minimize the noise, which could otherwise mask the signal from the source. Because radio waves are electromagnetic radiation, they are reflected and refracted like ordinary light waves. In radio astronomy, however, mostly reflecting telescopes are used.

At low frequencies, the antennas are usually dipoles (similar to those used for radio or TV), but in order to increase the collecting area and improve the resolution, one uses whole fields of dipoles, dipole arrays, where all dipole elements are connected to each other (Fig. 3.24).

The most common antenna type, however, is a parabolic reflector, which works exactly as an optical mirror telescope. At long wavelengths, the reflecting surface does not need to be solid, because the long wavelength photons cannot see the holes in the reflector and is therefore usually made in the form of a metal mesh. At high frequencies, the surface has to be smooth, and in the millimetre-submillimetre range, radio astronomers even use large optical telescopes, which they equip with their own radio-

Fig. 3.24. The University of Maryland's Clark Lake Radio Telescope located near Borrego Springs, CA. This telescope is an electronically steerable array of 3000 meter aperture and is tunable from 15 to 125 MHz. (Photo University of Maryland)

meters. To ensure a coherent amplification of the signal, the surface irregularities should be less than one-tenth of the wavelength used.

The main difference between a radio telescope and an optical telescope is in the recording of the signal. Radio telescopes are not imaging telescopes (except for synthesis telescopes, which will be described later); but the feed horn, which is located at the antenna focus, transfers the signal to a receiver. The wavelength and phase information is, however, preserved.

The resolution of a radio telescope, θ, can be deduced from the same formula (3.3) as for an optical telescope, i.e. λ/D, where λ is the wavelength used and D is the diameter of the aperture. Since the wavelength ratio between radio and visible light is of the order of 10000, radio antennas with diameters of several kilometres are needed in order to achieve the same resolution as for optical telescopes. In the early days of radio astronomy, poor resolution was the biggest drawback for the development and recognition of radio astronomy. For example, the antenna used by Jansky had a fan beam with a resolution of about $30°$ in the narrower direction. Therefore radio observations could not be compared with optical observations. Neither was it possible to identify the radio sources with optical counterparts.

The world's biggest radio telescope is the Arecibo antenna in Puerto Rico, whose main reflector is fixed and built into a 305-m diameter, natural round valley covered by a metal mesh (Fig. 3.25). In the late 1970's, the antenna surface and receivers were upgraded, enabling the antenna to be used down to wavelengths of 5 cm. The mirror of the Arecibo telescope is not parabolic but spherical, and the antenna is equipped with a movable feed system, which makes observations possible within a 20° radius around zenith.

The biggest completely steerable radio telescope is at Effelsberg, near the city of Bonn in West Germany (Fig. 3.26). This antenna has a parabolic main reflector with a diameter of 100 m. The inner 80 m of the dish is made of solid aluminium panels, while the outmost portion of the disk is a metal mesh structure. By using only the inner portion of the telescope, it has been possible to observe down to wavelengths of 4 mm. The oldest and perhaps best-known big radio telescope is the 76-m antenna at Jodrell Bank in Britain, which was completed in the end of the 1950's.

The biggest telescopes are usually incapable of operating below wavelengths of 1 cm, because the surface cannot be made accurate enough. However, the millimetre range has become more and more important. In this wavelength range, there are many transitions of interstellar molecules, and one can achieve quite high angular resolution even with a single dish telescope. At present, the typical size of a mirror of a millimetre telescope is about 10 m; the best-known one was the former 11-m Kitt Peak telescope in Arizona, USA. Its main mirror was replaced with a more accurate 12-m reflector in 1982. The surface is now so accurate that it is possible to make observations even below 1 mm, if the weather conditions are favourable. The development of this field is, however, rapid, and at present several big millimeter telescopes have started operations (Table E.24). Among them are the 40-m Nobeyama telescope in Japan, which can be

Fig. 3.25. The largest radio telescope in the world is the Arecibo dish in Puerto Rico. It has been constructed over a natural bowl and is 300 m in diameter. (Photo Arecibo Observatory)

Fig. 3.26. The largest fully steerable radio telescope is situated in Effelsberg near Bonn. Its diameter is 100 m. (Photo G. Hutschenreiter, Max-Planck-Institut für Radioastronomie)

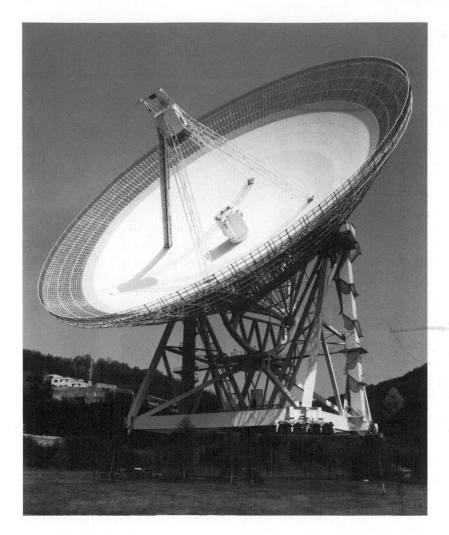

used down to 3 mm, the 30-m IRAM telescope at Pico Veleta in Spain, which is usable down to 1 mm, and the 15-m Swedish-ESO Submillimetre Telescope at La Silla, Chile, operating down to 0.6 mm (Fig. 3.27).

As already mentioned, the resolving power of a radio telescope is far poorer than that of an optical telescope. The biggest radio telescopes can at present reach a resolution of 5 arc sec, and that only at the very highest frequencies. To improve the resolution by increasing the size is difficult, because the present telescopes are already close to the practical upper limit. However, by combining radio telescopes and interferometers, it is possible to achieve even better resolution than with optical telescopes.

As early as 1891, Michelson used an interferometer for astronomical purposes. While the use of interferometers has proved to be quite difficult in the optical wavelength regime, interferometers are extremely useful in the radio region. To form an interferometer, one needs at least two antennas coupled together. The spacing between the antennas, D, is called the *baseline*. Let us first assume that the baseline is perpendicular to the line of sight (Fig. 3.28). Then the radiation arrives at both antennas

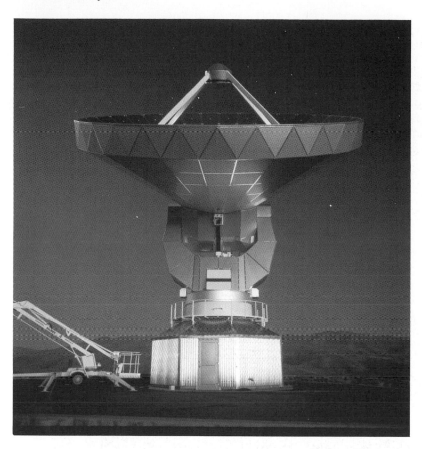

Fig. 3.27. The 15 metre SEST submillimetre telescope at La Silla in Chile is located in a dry climate at an altitude of 2400 m. Observations can be made down to wavelengths of 0.6 mm. (Photo ESO)

with the same phase, and the summed signal shows a maximum. However, due to the rotation of the Earth, the direction of the baseline changes, producing a phase difference between the two signals. The result is a sinusoidal interference pattern, in which minima occur when the phase difference is 180 degrees. The distance between the peaks is given by

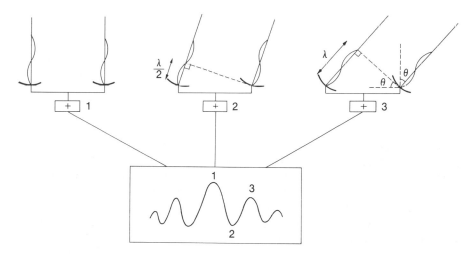

Fig. 3.28. The principle of an interferometer. If the radio radiation reaches the radio telescopes in the same phase, the waves amplify each other and a maximum is obtained in the combined radiation (*cases 1 and 3*). If the incoming waves are in opposite phase, they cancel each other (*case 2*)

$$\theta D = \lambda ,$$

where θ is the angle between the baseline and the source and λ is the wavelength of the received signal. The resolution of an interferometer is thus equal to that of an antenna with a linear size equal to D.

If the source is not a point source, the radiation emitted from different parts of the source will have phase differences when it enters the antennas. In this case, the minima of the interference pattern will not be zero, but will have some positive value P_{min}. If we denote the maximum value of the interference pattern by P_{max}, the ratio

$$\frac{P_{max} - P_{min}}{P_{max} + P_{min}}$$

gives a measure of the source size (fringe visibility).

More accurate information about the source structure can be obtained by changing the spacing between the antennas, i.e. by moving the antennas with respect to each other. If this is done, interferometry is transformed into a technique called *aperture synthesis*.

The theory and techniques of aperture synthesis were developed by the British astronomer Sir Martin Ryle. In Fig. 3.29, the principle of aperture synthesis is illustrated. If the telescopes are located on an east-west track, the spacing between them,

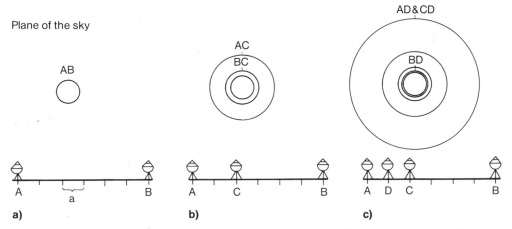

Fig. 3.29 a – c. To illustrate the principle of aperture synthesis, let us consider an east-west oriented interferometer pointed towards the celestial north. Each antenna is identical, has a diameter D and operates at a wavelength λ. The minimum spacing between each antenna element is a, and the maximum spacing is $6a$. In (**a**) there are only two antennas, A and B, displaced by the maximum spacing $6a$. When the earth rotates, antennas A and B will, in the course of 12 hours, track a circle on the plane of the sky with a diameter $\lambda/6a$, the maximum resolution that can be achieved with this interferometer. In (**b**) the antenna C is added to the interferometer, thus providing two more baselines, which track the circles AC and BC with radii of $\lambda/2a$ and $\lambda/4a$, respectively. In (**c**) there is still another antenna D added to the interferometer. In this case, two of the baselines are equal, AD and CD, and therefore only two new circles are covered on the plane of the sky. By adding more interferometer elements, one can fill in the missing parts within the primary beam, i.e. the beam of one single dish, and thus obtain a full coverage of the beam. It is also evident from (**c**) that not all of the antenna positions are needed to provide all the different spacings; some antenna spacings will, in such a case, be equal and therefore provide no additional information.

Obtaining a full aperture synthesis with an east-west interferometer always takes 12 hours, if all spacings are available. Usually, however, several antenna elements are movable, in which case a full aperture synthesis can take a long time before all spacings are filled in projected onto the sky, will describe a circle or an ellipse, depending on the position of the source as the the Earth rotates around its axis. If one varies the distance between the telescopes, one will get a series of circles or ellipses on the sky during a 12-hour interval. As we can see from Fig. 3.29, one does not have to cover all the spacings between the telescopes, because any antenna combination which has the same relative distance will describe the same path on the sky. In this way, one can synthesize an antenna, a filled aperture, with a size equal to the maximum spacing between the telescopes. Interferometers working according to this principle are called *aperture synthesis telescopes*. If one covers all the spacings up to the maximum baseline, the result will be an accurate map of the source over the primary beam of an individual antenna element. Aperture synthesis telescopes therefore produce an image of the sky, i.e. a "radio photograph".

A typical aperture synthesis telescope consists of one fixed telescope and a number of movable telescopes, usually located on an east-west track, although T or Y configurations are also quite common. The number of telescopes used determines how fast one can synthesize a larger disk, because the number of possible antenna combinations increases as $n(n-1)$, where n is the number of telescopes. It is also possible to synthesize a large telescope with only one fixed and one movable telescope, by changing the spacing between the telescopes every 12 hours, but then a full aperture synthesis can require several months of observing time. In order for this technique to work, the source must be constant, i.e. the signal cannot be time variable during the observing session.

The most efficient aperture synthesis telescope at present is the VLA (Very Large Array) in New Mexico, USA (Fig. 3.30). It consists of 27 paraboloid antennas, each with a diameter of 25 m, which are located on a Y-shaped track. The Y-formation was chosen because it provides a full aperture synthesis in 8 hours. Each antenna can be moved be a specially built carrier, and the locations of the telescopes are chosen to give optimal

Fig. 3.30. The VLA at Socorro, New Mexico, is a synthesis telescope consisting of 27 movable antennas

spacings for each configuration. In the largest configuration, each arm is about 21 km long, thereby resulting in an antenna with an effective diameter of 35 km. If the VLA is used in its largest configuration and at its highest frequency, 23 GHz (1.3 cm), the resolution achieved is 0.1 arc second, clearly superior to any optical telescope. A similar resolution can also be obtained with the British MERLIN telescope, where already existing telescopes have been coupled together by radio links. Other well-known synthesis telescopes are the Cambridge 5-km array in Britain and the Westerbork array in the Netherlands, both located on east-west tracks.

Even higher resolution can be obtained with an extension of the aperture synthesis technique, called VLBI (Very Long Baseline Interferometry). With the VLBI technique, the spacing between the antennas is restricted only by the size of the earth. VLBI uses preexisting antennas (often on different continents), which are all pointed towards the same source. In this case, the signal is recorded on magnetic tapes, together with accurate timing signals from atomic clocks. The magnetic tapes are correlated against each other in special processing centres, resulting in maps similar to those obtained with a normal aperture synthesis telescope. With VLBI techniques it is possible to achieve resolutions of 0.0001″. Because interferometry is very sensitive to the distance between the telescopes, the VLBI technique also provides one of the most accurate methods to measure distances. Currently one can measure distances with an accuracy of a few centimetres on intercontinental baselines. This is utilized in geodetic VLBI experiments, which study continental drift and polar motion as a function of time.

In radio astronomy, the maximum size of single antenna has also been reached. The trend is to build synthesis antennas, similar to the VLA in New Mexico. In the 1990's the United States built a chain of antennas extending across the whole continent, and the Australians have constructed a similar, but north-south antenna chain across their country.

More and more observations are being made in the submillimetre region. The disturbing effect of atmospheric water vapour becomes more serious at shorter wavelengths; thus, submillimetre telescopes must be located on mountain tops, as required for optical telescopes. All parts of the mirror are actively controlled in order to accurately maintain the proper form, as are the new optical telescopes. Several new submillimetre telescopes are under construction.

3.5 Other Wavelength Regions

All wavelengths of the electromagnetic spectrum reach the Earth from the sky. However, as mentioned in Sect. 3.1, not all radiation reaches the ground. The wavelength regions absorbed by the atmosphere have been studied more extensively since the 1970's, using Earth-orbiting satellites. Besides the optical and radio regions, there are only some narrow wavelength ranges at the infrared wavelengths that can be observed from high mountain tops.

The first observations in each new wavelength region were usually carried out from balloons, but not until rockets came into use could observations be made from outside the atmosphere. The first actual observations of an x-ray source, for instance, were made on a rocket flight in June 1962, when the detector rose above the atmosphere for about 6 minutes. Satellites have made it possible to map the whole sky in the wavelength regions invisible from the ground.

Gamma Radiation. Gamma ray astronomy studies radiation quanta with energies of $10^5 - 10^{14}$ eV. The boundary between gamma and x-ray astronomy, 10^5 eV, corresponds to a wavelength of 10^{-11} m. The boundary is not fixed; the regions of hard (= high-energy) x-rays and soft gamma rays partly overlap.

While ultraviolet, visible and infrared radiation are all produced by changes in the energy states of the electron envelopes of atoms, gamma rays and hard x-rays are produced by transitions in atomic nuclei or in mutual interactions of elementary particles. Thus observations of the shortest wavelengths give information on processes different from those giving rise to longer wavelengths.

The first observations of gamma sources were obtained at the end of the 1960's, when a device in the OSO 3 satellite (Orbiting Solar Observatory) detected gamma rays from the Milky Way. Later on, some satellites were especially designed for gamma astronomy, notably SAS 2, COS B, HEAO 1 and 3, and Granat. The most effective satellite so far, the Compton Gamma Ray Observatory, was launched in 1991.

The quanta of gamma radiation have energies a million times greater than those of visible light, but they cannot be observed with the same detectors. These observations are made with various *scintillation detectors*, usually composed of several layers of detector plates, where gamma radiation is transformed by the photoelectric effect into visible light, detectable by photomultipliers.

The energy of a gamma quantum can be determined from the depth to which it penetrates the detector. Analyzing the trails left by the quanta gives information on their approximate direction. The field of view is limited by a grating. The directional accuracy is low, and in gamma astronomy, the resolution is far below that in other wavelength regions.

X-rays. The observational domain of x-ray astronomy includes the energies between 10^2 and 10^5 eV, or the wavelengths $10 - 0.01$ nm. The regions $10 - 0.1$ nm and $0.1 - 0.01$ nm are called *soft* and *hard x-rays*, respectively. X-rays were discovered in the late 19th century. Systematic studies of the sky at x-ray wavelengths only became possible in the 1970's with the advent of satellite technology.

The first all-sky mapping was made in the early 1970's by SAS 1 (Small Astronomical Satellite), also called Uhuru. At the end of the 1970's, two High-Energy Astronomy Observatories, HEAO 1 and 2 (the latter called Einstein) mapped the sky with much higher sensitivity than Uhuru.

The Einstein Observatory was able to detect sources about a thousand times fainter than earlier x-ray telescopes. In optical astronomy, this would correspond to a jump from a 15-cm reflector to a 5-m telescope. Thus x-ray astronomy developed in 20 years as much as optical astronomy had in 300 years.

The latest x-ray satellites have been the German-American Rosat and the Japanese Asuka in the beginning of the 1990's.

Besides satellites mapping the whole sky, there have been several satellites observing the x-ray radiation of the Sun. The most effective telescopes so far were installed in the Skylab space station, and they were used to study the Sun in 1973 – 74.

The first x-ray telescopes used detectors similar to those in gamma astronomy. Their directional accuracy was never better than a few arc minutes. The more precise x-ray telescopes utilize the principle of *grazing reflection* (Fig. 3.31). An x-ray hitting a surface perpendicularly is not reflected, but absorbed. If, however, x-rays meet the mirror nearly parallel to its surface, just grazing it, a high quality surface can reflect the ray.

Fig. 3.31. X-rays are not reflected
by an ordinary mirror, and the
principle of grazing reflection
must be used for collecting them.
Radiation meets the paraboloid
mirror at a very small angle, is reflected onto a hyperboloid mirror
and further to a focal point. In
practice, several mirrors are
placed one inside another, collecting radiation in a common
focus. (Drawing by H. Wolter)

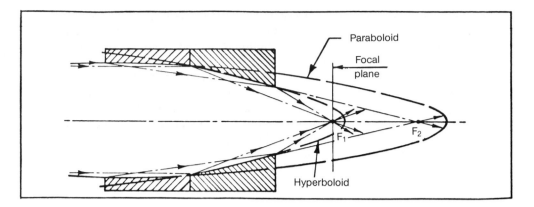

The mirror of an x-ray reflector is on the inner surface of a slowly narrowing cone. The outer part of the surface is a paraboloid, and the inner part, a hyperboloid. The rays are reflected by both surfaces and meet at a focal plane. In practice, several tubes are installed, one within another. For instance, the four cones of the Einstein Observatory had as much polished optical surface as a normal telescope with a diameter of 2.5 m. The resolution in x-ray telescopes is of the order of a few arc seconds, and the field of view, about 1°.

The detectors in x-ray astronomy are usually *Geiger-Müller counters, proportional counters* or scintillation detectors. Geiger-Müller and proportional counters are boxes filled with gas. The walls form a cathode, and an anode wire runs through the middle of the box; in more accurate counters, there are several anode wires. An x-ray quantum entering the box ionizes the gas, and the potential difference between the anode and cathode gives rise to a current of electrons and positive ions.

Ultraviolet Radiation. Between x-rays and the optical region, lies the domain of ultraviolet radiation, with wavelengths between 10 and 400 nm. Most ultraviolet observations have been carried out in the so-called *soft UV region*, i.e. at wavelengths near those of optical light, since most of the UV radiation is absorbed by the atmosphere. The wavelengths below 300 nm are completely blocked out. The short wavelength region from 10 to 91.2 nm is called *extreme ultraviolet* (EUV, XUV).

Extreme ultraviolet was one of the last regions of the electromagnetic radiation to be observed systematically. The reason for this is that the absorption of interstellar hydrogen makes the sky practically opaque at these wavelengths. The visibility in most directions is limited to some hundred light years in the vicinity of the Sun. In some directions however, the density of the interstellar gas is so low that even extragalactic objects can be seen. The first dedicated EUV satellite was the Extreme Ultraviolet Explorer (EUVE), launched in 1992. It has observed hundreds of EUV sources. In EUV, grazing reflection telescopes similar to those used in x-ray astronomy are employed.

In nearly all branches of astronomy, important information is obtained by observations of ultraviolet radiation. Many emission lines from stellar chromospheres or coronae, the Lyman lines of atomic hydrogen, and most of the radiation from hot stars are found in the UV domain. In the *near-ultraviolet*, telescopes can be made similar to optical telescopes and, equipped with a photometer or spectrometer, installed in a satellite orbiting the Earth.

The most effective satellites in the UV have been the European TD-1, the American Orbiting Astronomical Observatories OAO 2 and 3 (Copernicus), the International Ultraviolet Explorer IUE and the Soviet Astron. The instruments of the TD-1 satellite included both a photometer and a spectrometer. The satellite measured the magnitudes of over 30000 stars in four different spectral regions between 135 and 274 nm, and registered UV spectra from over 1000 stars. The OAO satellites were also used to measure magnitudes and spectra, and OAO 3 worked for over eight years. The IUE satellite launched in 1978 has been one of the most successful astronomical satellites (Fig. 3.32). IUE has a 45-cm Ritchey-Chrétien telescope with an aperture ratio of $f/15$ and field of view of 16 arc minutes. The satellite has two spectrographs to measure spectra of higher or lower resolution in wavelength intervals of 115 – 200 nm or 190 – 320 nm. For registration of the spectra, a Vidicon camera is used, and its images are transmitted down to the Earth. Unlike earlier satellites, IUE can be used almost like an Earth-based telescope, with the observer at the control station continually following and, if necessary, changing the observational programme. The Soviet Astron satellite, launched in 1983, had a French-made spectrometer. The aperture of Astron's telescope was 80 cm, similar to that of OAO 3.

Infrared Radiation. Radiation with longer wavelengths than visible light is called infrared radiation. This region extends from about 1 micrometre to 1 millimetre, where the radio region begins. Sometimes the *near-infrared*, at wavelengths below 5 μm, and the *submillimetre domain*, at wavelengths between 0.1 and 1 mm, are considered separate wavelength regions.

In infrared observations, radiation is collected by a telescope, as in the optical region. The incoming radiation consists of radiation from the object, from the background and from the telescope itself. Both the source and the background must be continually measured, the difference giving the radiation from the object. The background

Fig. 3.32. One of the most effective satellites in ultraviolet astronomy has been the International Ultraviolet Explorer, IUE, launched in 1978. (Photo NASA)

Fig. 3.33. (a) The most productive infrared satellite so far has been the Dutch-American IRAS (Infrared Astronomical Satellite). It found over 200 000 new infrared objects while mapping the whole sky in 1983. (Photo Fokker)

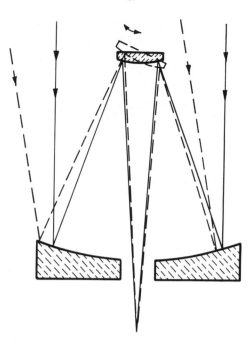

Fig. 3.33. (b) Refractors are not suitable for infrared telescopes, because infrared radiation cannot penetrate glass. The Cassegrain reflectors intended especially for infrared observations have secondary mirrors nodding rapidly back and forth between the object and the background near the object. By subtracting the brightness of the background from the brightness of the object, the background can be eliminated

measurements are usually made with a Cassegrain secondary mirror oscillating between the source and the background at a rate of, say, 100 oscillations per second, and thus the changing background can be eliminated. To register the measurements, semiconductor detectors are used. The detector must always be cooled to minimize its own thermal radiation. Sometimes the whole telescope is cooled.

Infrared observatories have been built on high mountain tops, where most of the atmospheric water vapour remains below. Some favoured sites are, e.g. Mauna Kea on Hawaii, Mt. Lemon in Arizona and Pico del Teide on Tenerife. For observations in the far-infrared, these mountains are not high enough; these observations are carried out e.g. on aeroplanes. One of the best-equipped planes is the Kuiper Airborne Observatory, named after the well-known planetary scientist Gerard Kuiper.

Balloons and satellites are also used for infrared observations. The most successful infrared observatory so far has been the InfraRed Astronomy Satellite IRAS, built in cooperation by the U.S. and the Netherlands (Fig. 3.33). It was launched in the beginning of 1983 and worked for over eight months. It mapped the whole sky at four wavelengths (12, 25, 60 and 100 µm), and found over 200000 new infrared objects.

A very successful satellite was the 1989-launched COBE (Cosmic Background Explorer), which has mapped the background radiation in submillimetre and infrared wavelengths.

3.6 Other Forms of Energy

Besides electromagnetic radiation, energy arrives from space in other forms: particles (*cosmic rays, neutrinos*) and *gravitational radiation*.

Cosmic Rays. Cosmic rays, consisting of electrons and totally ionized nuclei of atoms, are received in equal amounts from all directions. Their incoming directions do not reveal their origin, since cosmic rays are electrically charged; thus their paths are continually changed when they move through the magnetic fields of the Milky Way. The high energies of cosmic rays mean that they have to be produced by high-energy phenomena, like supernova explosions. The majority of cosmic rays are protons (nearly 90%) and helium nuclei (10%), but some are heavier nuclei; their energies lie between 10^8 and 10^{20} eV.

The most energetic cosmic rays give rise to *secondary radiation* when they hit molecules of the atmosphere. This secondary radiation can be observed from the ground, but primary cosmic rays can only be directly observed outside the atmosphere. The detectors used to observe cosmic rays are similar to those used in particle physics. Since Earth-based accelerators reach energies of only about 10^{12} eV, cosmic rays offer an excellent "natural" laboratory for particle physics. Many satellites and spacecraft have detectors for cosmic rays.

Neutrinos. Neutrinos are elementary particles with no electric charge and a mass equal to zero or, at any rate, less than 1/10000 of the mass of the electron. Most neutrinos are produced in nuclear reactions within stars; since they react very weakly with other matter, they escape directly from the stellar interior.

Neutrinos are very difficult to observe; the first method of detection was the radiochemical method. As a reactive agent, e.g. tetrachlorethene (C_2Cl_4) can be used. When a neutrino hits a chlorium atom, the chlorium is transformed into argon, and an electron is freed:

$$^{37}Cl + \nu \rightarrow {}^{37}Ar + e^-.$$

The argon atom is radioactive and can be observed. Instead of chlorium, lithium and gallium might be used to detect neutrinos. The first gallium detectors have been running in Italy and Russia since the end of the 1980's.

Another observation method is based on the Čerenkov radiation produced by neutrinos in extremely pure water. The flashes of light are registered using photomultipliers, and thus it is possible to find out the direction of the radiation. This method is used, for example, in the Japanese Kamiokande detector.

Neutrino detectors must be located deep under the ground to protect them from the secondary radiation caused by cosmic rays.

The detectors have observed neutrinos from the Sun, and Supernova 1987A in the Large Magellanic Cloud was also observed in 1987. Neutrino detectors must be located deep under the ground, e.g. in old mines, to protect them from the secondary radiation caused by cosmic rays.

Gravitational Radiation. Gravitational astronomy is as young as neutrino astronomy. The first attempts to measure gravitational waves were made in the 1960's. Gravitational radiation is emitted by accelerating masses, just as electromagnetic radiation is emitted by electric charges in accelerated motion. Detection of gravitational waves is very difficult, and they have yet to be directly observed.

The first type of gravitational wave antenna was the *Weber cylinder*. It is an aluminium cylinder which starts vibrating at its proper frequency when hit by a gravitational pulse. The distance between the ends of the cylinder changes by about 10^{-17} m, and the changes in the length are studied by strain sensors welded to the side of the cylinder.

Another type of modern gravity radiation detectors measures "spatial strain" induced by gravity waves and consists of two sets of mirrors in directions perpendicular to each other (Michelson interferometer, Fig. 3.34), or one set of parallel mirrors (Fabry-Perot interferometer). The relative distances between the mirrors are monitored by laser interferometers. If a gravity pulse passes the detector, the distances change and the changes can be measured. The longest baselines between the mirrors are some tens of metres, but kilometre long baselines are planned. One of the largest detectors is at the Max-Planck Institut für Quantenoptik in Garching, where an armlength of 30 metres has been run with 110 passes, leading to a light path of 3.3 km.

Fig. 3.34. View of one arm of the 3-m prototype Michelson laser interferometer set up to study large-scale devices of this kind for gravitational wave detection. (Courtesy Max-Planck Institute for Quantum Optics)

3.7 Examples

3.7.1 The distance between the components of the binary star ζ Herculis is 1.38″. What should the diameter of a telescope be to resolve the binary? If the focal length of the objective is 80 cm, what should the focal length of the eyepiece be to resolve the components, when the resolution of the eye is 2′?

In the optical region, we can use the wavelength value of $\lambda = 550$ nm. The diameter of the objective is obtained from the equation for the resolution (3.3),

$$D \approx \frac{\lambda}{\theta} = \frac{550 \times 10^{-9}}{\dfrac{1.38}{3600} \times \dfrac{\pi}{180}} \; m = 0.08 \; m = 8 \; cm.$$

The required magnification is

$$\omega = \frac{2'}{1.38''} = 87.$$

The magnification is given by

$$\omega = \frac{f}{f'}$$

and, thus, the focal length of the eyepiece should be

$$f' = \frac{f}{\omega} = \frac{80}{87} \approx 0.9 \; cm.$$

3.7.2 A telescope has an objective with a diameter of 90 mm and focal length of 1200 mm.

a) What is the focal length of an eyepiece, the exit pupil of which is 6 mm (about the size of the pupil of the eye)?
b) What is the magnification of such an eyepiece?
c) What is the angular diameter of the Moon seen through this telescope and eyepiece?

a) From Fig. 3.7 we get

$$\frac{D}{L} = \frac{f}{f'} \; ,$$

whence

$$f' = f\frac{L}{D} = 1200 \; mm \; \frac{6 \; mm}{90 \; mm} = 80 \; mm \; .$$

b) The magnification is

$$\omega = \frac{f}{f'} = \frac{1200 \; mm}{80 \; mm} = 15 \; .$$

c) Assuming the angular diameter of the Moon is $\alpha = 0.5°$, its diameter through the telescope is $\omega\alpha = 7.5°$.

3.8 Exercises

3.8.1   The Moon was photographed with a telescope, the objective of which had a diameter of 20 cm and focal length of 150 cm. The exposure time was 0.1 s.

a) What should the exposure time be, if the diameter of the objective were 15 cm and focal length 200 cm?
b) What is the size of the image of the Moon in both cases?
c) Both telescopes are used to look at the Moon with an eyepiece the focal length of which is 25 mm. What are the magnifications?

3.8.2   The radio telescopes at Amherst and Råö are used as an interferometer, the baseline being 2900 km.

a) What is the resolution at 22 GHz in the direction of the baseline?
b) What should be the size of an optical telescope with the same resolution?

Chapter 4   Photometric Concepts and Magnitudes

Most astronomical observations utilize electromagnetic radiation in one way or another. We can obtain information on the physical nature of a radiation source by studying the energy distribution of its radiation. We shall now introduce some basic concepts characterizing electromagnetic radiation.

4.1 Intensity, Flux Density and Luminosity

Let us assume we have some radiation passing through a surface element dA (Fig. 4.1). Some of the radiation will leave dA within a solid angle $d\omega$; the angle between $d\omega$ and the normal to the surface is denoted by θ. The amount of energy with frequency in the range $[v, v+dv]$ entering this solid angle in time dt is

$$dE_v = I_v \cos \theta\, dA\, dv\, d\omega\, dt \ . \tag{4.1}$$

Here, the coefficent I_v is the *specific intensity* of radiation at the frequency in the direction of the solid angle $d\omega$. Its dimension is $\mathrm{W\, m^{-2}\, Hz^{-1}\, sterad^{-1}}$.

The projection of the surface element dA as seen from the direction θ is $dA_n = dA \cos \theta$, which explains the factor $\cos \theta$. If the intensity does not depend on direction, the energy dE_v is directly proportional to the surface element perpendicular to the direction of the radiation.

The intensity including all possible frequencies is called the *total intensity I*, and is obtained by integrating I_v over all frequencies:

$$I = \int_0^\infty I_v\, dv \ .$$

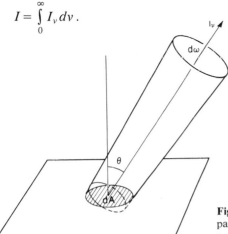

Fig. 4.1. The intensity I_v of radiation is related to the energy passing through a surface element dA into solid angle $d\omega$ in a direction θ

More important quantities from the observational point of view are the *energy flux* (L_ν, L), or, briefly, the *flux*, and the *flux density* (F_ν, F). The flux density gives the power of radiation per unit area; hence its dimension is $\mathrm{W\,m^{-2}\,Hz^{-1}}$ or $\mathrm{W\,m^{-2}}$, depending on whether we are talking about the flux density at a certain frequency or about the total flux density.

Observed flux densities are usually rather small, and $\mathrm{W\,m^{-2}}$ would be an inconveniently large unit. Therefore, especially in radio astronomy, flux densities are often expressed in *Janskys*; one Jansky (1 Jy) equals $10^{-26}\,\mathrm{W\,m^{-2}\,Hz^{-1}}$.

When we are observing a radiation source, we in fact measure the energy collected by the detector during some period of time, which equals the flux density integrated over the radiation-collecting area of the instrument and the time interval.

The flux density F_ν at a frequency ν can be expressed in terms of the intensity as

$$F_\nu = \frac{1}{dA\,d\nu\,dt} \int_S dE_\nu = \int_S I_\nu \cos\theta\,d\omega \ , \tag{4.2}$$

where the integration is extended over all possible directions. Analogously, the total flux density is

$$F = \int_S I \cos\theta\,d\omega \ .$$

For example, if the radiation is *isotropic*, or I is independent of the direction, we get

$$F = \int_S I \cos\theta\,d\omega = I \int_S \cos\theta\,d\omega \ . \tag{4.3}$$

The solid angle element $d\omega$ is equal to a surface element on a unit sphere. In spherical coordinates it is (Fig. 4.2; also cf. Appendix A.5):

$$d\omega = \sin\theta\,d\theta\,d\phi \ .$$

Substitution into (4.3) gives

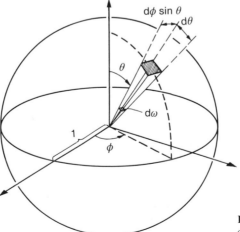

Fig. 4.2. An infinitesimal solid angle $d\omega$ is equal to the corresponding surface element on a unit sphere

$$F = I \int\limits_{\theta=0}^{\pi} \int\limits_{\phi=0}^{2\pi} \cos\theta \sin\theta \, d\theta \, d\phi = 0 \, ,$$

so there is no net flux of radiation. This means that there are equal amounts of radiation entering and leaving the surface. If we want to know the amount of radiation passing through the surface, we can find, for example, the radiation leaving the surface. For isotropic radiation this is

$$F = I \int\limits_{\theta=0}^{\pi/2} \int\limits_{\phi=0}^{2\pi} \cos\theta \sin\theta \, d\theta \, d\phi = \pi I \, . \tag{4.4}$$

In the astronomical literature, terms such as intensity and brightness are used rather vaguely. Flux density is hardly ever called flux density but intensity or (with luck) flux. Therefore the reader should always carefully check the meaning of these terms.

Flux means the power going through some surface, expressed in watts. The flux emitted by a star into a solid angle ω is $L = \omega r^2 F$, where F is the flux density observed at a distance r. *Total flux* is the flux passing through a closed surface encompassing the source. Astronomers usually call the total flux of a star the *luminosity L*. We can also talk about the luminosity L_ν at a frequency ν ($[L_\nu] = \text{W Hz}^{-1}$).

If the source (like a typical star) radiates isotropically, its radiation at a distance r is distributed evenly on a spherical surface whose area is $4\pi r^2$ (Fig. 4.3). If the flux density of the radiation passing through this surface is F, the total flux is

$$L = 4\pi r^2 F \, . \tag{4.5}$$

If we are outside the source, where radiation is not created or destroyed, the luminosity does not depend on distance. The flux density, on the other hand, falls off proportional to $1/r^2$.

For extended objects (as opposed to objects such as stars visible only as points) we can define the *surface brightness* as the flux density per unit solid angle (Fig. 4.4). Now the observer is at the apex of the solid angle. The surface brightness is independent of distance, which can be understood in the following way. The flux density arriving from an area A is inversely proportional to the distance squared. But also the solid angle sub-

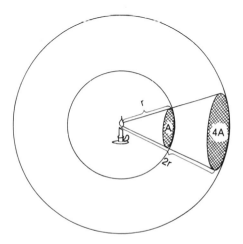

Fig. 4.3. An energy flux which at a distance r from a point source is distributed over an area A is spread over an area $4A$ at a distance $2r$. Thus the flux density decreases inversely proportional to the distance squared

Fig. 4.4. An observer sees radiation coming from a constant solid angle ω. The area giving off radiation into this solid angle increases when the source moves further away ($A \propto r^2$). Therefore the surface brightness or the observed flux density per unit solid angle remains constant

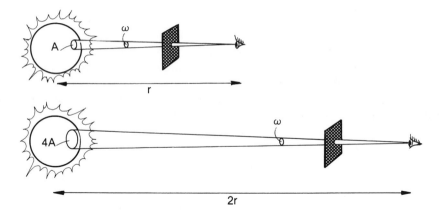

tended by the area A is proportional to $1/r^2 (\omega = A/r^2)$. Thus the surface brightness $B = F/\omega$ remains constant.

The *energy density u* of radiation is the amount of energy per unit volume (Jm^{-3}). Suppose we have radiation with intensity I arriving from a solid angle $d\omega$ perpendicular to the surface dA (Fig. 4.5). In the time dt, the radiation travels a distance $c\,dt$ and fills a volume $dV = c\,dt\,dA$. Thus the energy in the volume dV is (now $\cos\theta = 1$)

$$dE = I\,dA\,d\omega\,dt = \frac{1}{c}I\,d\omega\,dV .$$

Hence the energy density du of the radiation arriving from the solid angle $d\omega$ is

$$du = \frac{dE}{dV} = \frac{1}{c}I\,d\omega ,$$

and the total energy density is obtained by integrating this over all directions:

$$u = \frac{1}{c}\int_S I\,d\omega . \tag{4.6}$$

For isotropic radiation we get

$$u = \frac{4\pi}{c}I . \tag{4.7}$$

Fig. 4.5. In time dt, the radiation fills a volume $dV = c\,dt\,dA$, where dA is the surface element perpendicular to the propagation direction of the radiation

4.2 Apparent Magnitudes

As early as the second century B.C., Hipparchos divided the visible stars into six classes according to their apparent brightness. The first class contained the brightest stars and the sixth the faintest ones still visible to the naked eye.

The response of the human eye to the brightness of light is not linear. If the flux densities of three stars are in the proportion $1:10:100$, the brightness difference of the first and second star seems to be equal to the difference of the second and third star. Equal brightness ratios correspond to equal apparent brightness differences: the human perception of brightness is logarithmic.

The rather vague classification of Hipparchos was replaced in 1856 by Norman R. Pogson. The new, more accurate classification followed the old one as closely as possible, resulting in another of those illogical definitions typical of astronomy. Since a star of the first class is about one hundred times brighter than a star of the sixth class, Pogson defined the ratio of the brightnesses of classes n and $n+1$ as $\sqrt[5]{100} \approx 2.512$.

The brightness class or *magnitude* can be defined accurately in terms of the observed flux density F ($[F] = \mathrm{W\ m^{-2}}$). We decide that the magnitude 0 corresponds to some preselected flux density F_0. All other magnitudes are then defined by the equation

$$m = -2.5 \lg \frac{F}{F_0} . \tag{4.8}$$

Note that the coefficient is exactly 2.5, not 2.512! Magnitudes are dimensionless quantities, but to remind us that a certain value is a magnitude, we can write it, for example, as 5 mag or 5^m.

It is easy to see that (4.8) is equivalent to Pogson's definition. If the magnitudes of two stars are m and $m+1$ and their flux densities F_m and F_{m+1}, respectively, we have

$$m - (m+1) = -2.5 \lg \frac{F_m}{F_0} + 2.5 \lg \frac{F_{m+1}}{F_0}$$

$$= -2.5 \lg \frac{F_m}{F_{m+1}} ,$$

whence

$$\frac{F_m}{F_{m+1}} = \sqrt[5]{100} .$$

In the same way we can show that the magnitudes m_1 and m_2 of two stars and the corresponding flux densities F_1 and F_2 are related by

$$m_1 - m_2 = -2.5 \lg \frac{F_1}{F_2} . \tag{4.9}$$

Magnitudes extend both ways from the original six values. The magnitude of the brightest star, Sirius, is in fact negative, -1.5. The magnitude of the Sun is -26.8, and that of a full moon, -12.5. The faintest observed objects have magnitudes of about 25.

4.3 Magnitude Systems

The *apparent magnitude m*, which we have just defined, depends on the instrument we use to measure it. The sensitivity of the detector is different at different wavelengths. Also, different instruments detect different wavelength ranges. Thus the flux measured by the instrument equals not the total flux, but only a fraction of it.

Depending on the method of observation, we can define various magnitude systems. Different magnitudes have different zero points, i.e. they have different flux densities F_0 corresponding to the magnitude 0. The zero points are usually defined by a few selected standard stars.

In daylight the human eye is most sensitive to radiation with a wavelength of about 550 nm, the sensitivity decreasing towards red (longer wavelengths) and violet (shorter wavelengths). The magnitude corresponding to the sensitivity of the eye is called the *visual magnitude* m_v.

Photographic plates are usually most sensitive at blue and violet wavelengths, but they are also able to register radiation not visible to the human eye. Thus the *photographic magnitude* m_{pg} usually differs from the visual magnitude. The sensitivity of the eye can be simulated by using a yellow filter and plates sensitized to yellow and green light. Magnitudes thus observed are called *photovisual magnitudes* m_{pv}.

If, in ideal case, we were able to measure the radiation at all wavelengths, we would get the *bolometric magnitude* m_{bol}. In practice this is very difficult, since part of the radiation is absorbed by the atmosphere; also, different wavelengths require different detectors. (In fact there is a gadget called the bolometer, which, however, is not a real bolometer but an infrared detector.) The bolometric magnitude can be derived from the visual magnitude if we know the *bolometric correction* BC:

$$m_{bol} = m_v - BC . \tag{4.10}$$

By definition, the bolometric correction is zero for radiation of solar type stars (or, more precisely, stars of the spectral class F5). Although the visual and bolometric magnitudes can be equal, the flux density corresponding to the bolometric magnitude must always be higher. The reason of this apparent contradiction is in the different values of F_0.

The more the radiation distribution differs from that of the Sun, the higher the bolometric correction is. The correction is positive for stars both cooler or hotter than the Sun. Sometimes the correction is defined as $m_{bol} = m_v + BC$, in which case $BC \leq 0$ always. The chance for errors is, however, very small, since we must have $m_{bol} \leq m_v$.

The most accurate magnitude measurements are made using photoelectric photometers. Usually filters are used to allow only a certain wavelength band to enter the detector. One of the multicolour magnitude systems used widely in photoelectric photometry is the UBV system developed in the early 1950's by Harold L. Johnson and William W. Morgan. Magnitudes are measured through three filters, U = ultraviolet, B = blue and V = visual. Figure 4.6 and Table 4.1 give the wavelength bands of these filters. The magnitudes observed through these filters are called U, B and V magnitudes, respectively.

The UBV system was later augmented by adding more bands. One commonly used system is the five colour UBVRI system, which includes R = red and I = infrared filters.

Table 4.1. Wavelength bands of the UBVRI and uvby filters and their effective ($\approx$ average) wavelengths

Magnitude		Passband [nm]	Effective wavelength [nm]
U	ultraviolet	300 – 400	360
B	blue	360 – 550	440
V	visual	480 – 680	550
R	red	530 – 950	700
I	infrared	700 – 1200	880
		Band width [nm]	
u	ultraviolet	30	350
v	violet	19	411
b	blue	18	467
y	yellow	23	547

There are also other broad band systems, but they are not as well standardized as UBV, which has been defined moderately well using a great number of standard stars all over the sky. The magnitude of an object is obtained by comparing it to magnitudes of standard stars.

In Strömgren's four-colour or uvby system, the bands passed by the filters are much narrower than in the UBV system. The uvby system is also well standardized, but is not quite as common as UBV. Other narrow band systems exist as well. By adding more filters, more information on the radiation distribution can be obtained.

In any multicolour system, we can define *colour indices*; a colour index is a difference of two magnitudes. By subtracting the B magnitude from U, we get the colour index U-B, and so on. If the UBV system is used, it is common to give only the V magnitude and the colour indices U-B and B-V.

The constants F_0 in (4.8) for U, B and V magnitudes have been selected in such a way that the colour indices B-V and U-B are zero for stars of spectral type A0 (for spectral types, see Chap. 9). The surface temperature of such a star is about 10 000 K. For example, Vega (α Lyr, spectral class A0V) has V = 0.03, B-V = U-B = 0.00. The Sun has V = $-$ 26.8, B-V $\approx$ 0.66 and U-B $\approx$ 0.10.

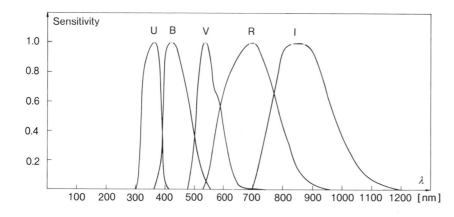

Fig. 4.6. Relative transmission profiles of filters used in the UBVRI magnitude system. The maxima of the bands are normalized to unity. NB: Recently, different definitions of the R and I bands have also been used

Before the UBV system was developed, a colour index C.I., defined as

$$C.I. = m_{pg} - m_v,$$

was used. Since m_{pg} gives the magnitude in blue and m_v in visual, this index is related to B-V. In fact,

$$C.I. = (B-V) - 0.11 .$$

4.4 Absolute Magnitudes

Thus far we have discussed only apparent magnitudes. They do not tell us anything about the true brightness of stars, since the distances differ. A quantity measuring the intrinsic brightness of a star is the *absolute magnitude*. It is defined as the apparent magnitude at a distance of 10 parsecs from the star (Fig. 4.7).

We shall now derive an equation which relates the apparent magnitude m, the absolute magnitude M and the distance r. Because the flux emanating from a star into a solid angle ω has, at a distance r, spread over an area ωr^2, the flux density is inversely proportional to the distance squared. Therefore the ratio of the flux density at a distance r, $F(r)$, to the flux density at a distance of 10 parsecs, $F(10)$, is

$$\frac{F(r)}{F(10)} = \left(\frac{10 \text{ pc}}{r}\right)^2 . \tag{4.11}$$

Thus the difference of magnitudes at r and 10 pc, or the *distance modulus* $m - M$, is

$$m - M = -2.5 \lg \frac{F(r)}{F(10)} = -2.5 \lg \left(\frac{10 \text{ pc}}{r}\right)^2 \quad \text{or}$$

$$m - M = 5 \lg \frac{r}{10 \text{ pc}} . \tag{4.12}$$

Fig. 4.7. The flux density at a distance of 10 parsecs from the star defines its absolute magnitude

For historical reasons, this equation is almost always written as

$$m - M = 5 \lg r - 5 \, , \tag{4.13}$$

which is valid *only* if the distance is expressed in parsecs. (The logarithm of a dimensional quantity is, in fact, physically absurd.) Sometimes the distance is given in kiloparsecs or megaparsecs, which require different constant terms in (4.13). To avoid confusion, we highly recommend the form (4.12).

Absolute magnitudes are usually denoted by capital letters. Note, however, that the U, B and V magnitudes are apparent magnitudes. The corresponding absolute magnitudes are M_U, M_B and M_V.

The absolute bolometric magnitude can be expressed in terms of the luminosity. Let the total flux density at a distance $r = 10$ pc be F and let $F_\odot$ be the equivalent quantity for the Sun. Since the luminosity is $L = 4 \pi r^2 F$, we get

$$M_{\text{bol}} - M_{\text{bol}, \odot} = -2.5 \lg \frac{F}{F_\odot} = -2.5 \lg \frac{L/4 \pi r^2}{L_\odot /4 \pi r^2} \, , \quad \text{or}$$

$$M_{\text{bol}} - M_{\text{bol}, \odot} = -2.5 \lg \frac{L}{L_\odot} \, . \tag{4.14}$$

The absolute bolometric magnitude $M_{\text{bol}} = 0$ corresponds to a luminosity $L_0 = 3.0 \times 10^{28}$ W.

4.5 Extinction and Optical Thickness

Equation (4.12) shows how the apparent magnitude increases (and brightness decreases!) with increasing distance. If the space between the radiation source and the observer is not completely empty, but contains some interstellar medium, (4.12) no longer holds, because part of the radiation is absorbed by the medium (and usually reemitted at a different wavelength, which may be outside the band defining the magnitude), or scattered away from the line of sight. All these radiation losses are called the *extinction*.

Now we want to find out how extinction depends on distance. Assume we have a star radiating a flux L_0 into a solid angle ω in some wavelength range. Since the medium absorbs and scatters radiation, the flux L will now decrease with increasing distance r (Fig. 4.8). In a short distance interval $[r, r+dr]$, the extinction dL is proportional to the flux L and the distance travelled in the medium:

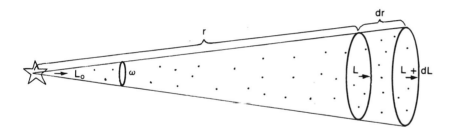

Fig. 4.8. The interstellar medium absorbs and scatters radiation; this usually reduces the energy flux L in the solid angle $\omega (dL \leqq 0)$

$$dL = -\alpha L \, dr \, . \tag{4.15}$$

The factor α tells how effectively the medium can obscure radiation. It is called the *opacity*. From (4.15) we see that its dimension is $[\alpha] = m^{-1}$. The opacity is zero for a perfect vacuum and approaches infinity when the substance becomes really murky. We can now define a dimensionless quantity, the *optical thickness* τ by

$$d\tau = \alpha \, dr \, . \tag{4.16}$$

Substituting this into (4.15) we get

$$dL = -L \, d\tau \, .$$

Next we integrate this from the source (where $L = L_0$ and $\tau = 0$) to the observer:

$$\int_{L_0}^{L} \frac{dL}{L} = -\int_0^{\tau} d\tau \, , \quad \text{which gives}$$

$$L = L_0 e^{-\tau} \, . \tag{4.17}$$

Here, τ is the optical thickness of the material between the source and the observer and L, the observed flux. Now, the flux L falls off exponentially with increasing optical thickness. Empty space is perfectly transparent, i.e. its opacity is $\alpha = 0$; thus the optical thickness does not increase in empty space, and the flux remains constant.

Let F_0 be the flux density on the surface of a star and $F(r)$, the flux density at a distance r. We can express the fluxes as

$$L = \omega r^2 F(r) \, , \qquad L_0 = \omega R^2 F_0 \, ,$$

where R is the radius of the star. Substitution into (4.17) gives

$$F(r) = F_0 \frac{R^2}{r^2} e^{-\tau} \, .$$

For the absolute magnitude we need the flux density at a distance of 10 parsecs, $F(10)$, which is still evaluated without extinction:

$$F(10) = F_0 \frac{R^2}{(10 \text{ pc})^2} \, .$$

The distance modulus $m - M$ is now

$$m - M = -2.5 \lg \frac{F(r)}{F(10)} = 5 \lg \frac{r}{10 \text{ pc}} - 2.5 \lg e^{-\tau}$$

$$= 5 \lg \frac{r}{10 \text{ pc}} + (2.5 \lg e) \tau \quad \text{or}$$

$$m - M = 5 \lg \frac{r}{10 \text{ pc}} + A \, , \tag{4.18}$$

where $A \geq 0$ is the extinction in magnitudes due to the entire medium between the star and the observer. If the opacity is constant along the line of sight, we have

$$\tau = \alpha \int_0^r dr = \alpha r ,$$

and (4.18) becomes

$$m - M = 5 \lg \frac{r}{10 \text{ pc}} + ar , \tag{4.19}$$

where the constant $a = 2.5 \, \alpha \lg e$ gives the extinction in magnitudes per unit distance.

Colour Excess. Another effect caused by the interstellar medium is the reddening of light: blue light is scattered and absorbed more than red. Therefore the colour index B-V increases. The visual magnitude of a star is, from (4.18),

$$V = M_V + 5 \lg \frac{r}{10 \text{ pc}} + A_V , \tag{4.20}$$

where M_V is the absolute visual magnitude and A_V is the extinction in the V passband. Similary, we get for the blue magnitudes

$$B = M_B + 5 \lg \frac{r}{10 \text{ pc}} + A_B .$$

The observed colour index is now

$$B\text{-}V = M_B - M_V + A_B - A_V \quad \text{or}$$
$$B\text{-}V = (B\text{-}V)_0 + E_{B\text{-}V} , \tag{4.21}$$

where $(B\text{-}V)_0 = M_B - M_V$ is the *intrinsic colour* of the star and $E_{B\text{-}V} = (B\text{-}V) - (B\text{-}V)_0$ is the *colour excess*. Studies of the interstellar medium show that the ratio of the visual extinction A_V to the colour excess $E_{B\text{-}V}$ is almost constant for all stars:

$$R = \frac{A_V}{E_{B\text{-}V}} \approx 3.0 .$$

This makes it possible to find the visual extinction if the colour excess is known:

$$A_V \approx 3.0 \, E_{B\text{-}V} . \tag{4.22}$$

When A_V is obtained, the distance can be solved directly from (4.20).

We shall study interstellar extinction in more detail in Sect. 16.1 ("Interstellar Dust").

Atmospheric Extinction. As we mentioned in Sect. 3.1, the Earth's atmosphere also causes extinction. The observed magnitude m depends on the location of the observer and the zenith distance of the object, since these factors determine the distance the light

Fig. 4.9. If the zenith distance of a star is z, the light of the star travels a distance $H/\cos z$ in the atmosphere; H is the height of the atmosphere

has to travel in the atmosphere. To compare different observations, we must first *reduce* them i.e. remove the atmospheric effects somehow. The magnitude m_0 thus obtained can then be compared with other observations.

If the zenith distance z is not too large, we can approximate the atmosphere by a plane layer of constant thickness (Fig. 4.9). If the thickness of the atmosphere is used as a unit, the light must travel a distance

$$X = 1/\cos z = \sec z$$

in the atmosphere. The quantity X is the *air mass*. According to (4.19), the magnitude increases linearly with the distance X:

$$m = m_0 + kX, \tag{4.23}$$

where k is the *extinction coefficient*.

The extinction coefficient can be determined by observing the same source several times during a night with as wide a zenith distance range as possible. The observed magnitudes are plotted in a diagram as a function of the air mass X. The points lie on straight line the slope of which gives the extinction coefficient k. When this line is extrapolated to $X = 0$, we get the magnitude m_0, which is the apparent magnitude outside the atmosphere.

In practice, observations with zenith distances higher than $70°$ (or altitudes less than $20°$) are not used to determine k and m_0, since at low altitudes the curvature of the atmosphere begins to complicate matters. The value of the extinction coefficient k depends on the observation site and time and also on the wavelength, since extinction increases strongly towards short wavelengths.

4.6 Examples

4.6.1 Show that intensity is independent of distance.

Suppose we have some radiation leaving the surface element dA in the direction θ. The energy entering the solid angle $d\omega$ in time dt is

$$dE = I \cos \theta \, dA \, d\omega \, dt \,,$$

where I is the intensity. If we have another surface dA' at a distance r receiving this radiation from direction θ', we have

$$d\omega = dA' \cos \theta'/r^2 \,.$$

The definition of the intensity gives

$$dE = I' \cos \theta' \, dA' \, d\omega' \, dt \,,$$

where I' is the intensity at dA' and

$$d\omega' = dA \cos \theta/r^2 \,.$$

Substitution of $d\omega$ and $d\omega'$ into the expressions of dE gives

$$I \cos \theta \, dA \, \frac{dA' \cos \theta'}{r^2} \, dt = I' \cos \theta' \, dA' \, \frac{dA \cos \theta}{r^2} \, dt$$

$$\Rightarrow I' = I \,.$$

Thus the intensity remains constant in empty space.

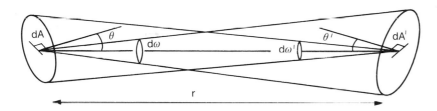

4.6.2 Surface Brightness of the Sun

Assume that the Sun radiates isotropically. Let R be the radius of the Sun, $F_\odot$ the flux density on the surface of the Sun and F the flux density at a distance r. Since the luminosity is constant,

$$L = 4\pi R^2 F_\odot = 4\pi r^2 F \,,$$

the flux density equals

$$F = F_\odot \frac{R^2}{r^2} \,.$$

At a distance $r \gg R$, the Sun subtends a solid angle

$$\omega = \frac{A}{r^2} = \frac{\pi R^2}{r^2} \,,$$

where $A = \pi R^2$ is the cross section of the Sun. The surface brightness B is

$$B = \frac{F}{\omega} = \frac{F_\odot}{\pi} .$$

Applying (4.4) we get

$$B = I_\odot .$$

Thus the surface brightness is independent of distance and equals the intensity. We have found a simple interpretation for the somewhat abstract concept of intensity.

The flux density of the Sun on the Earth, the *solar constant*, is $S_\odot = 1390 \text{ W m}^{-2}$. The angular diameter of the Sun is $\alpha = 32'$, whence

$$\frac{R}{r} = \frac{\alpha}{2} = \frac{1}{2} \times \frac{32}{60} \times \frac{\pi}{180} = 0.00465 \text{ rad} .$$

The solid angle subtended by the Sun is

$$\omega = \pi \left(\frac{R}{r} \right)^2 = \pi \times 0.00465^2 = 6.81 \times 10^{-5} \text{ sterad}$$

and the surface brightness

$$B = \frac{S_\odot}{\omega} = 2.04 \times 10^7 \text{ W m}^{-2} \text{ sterad}^{-1} .$$

4.6.3 *Magnitude of a Binary Star*

Since magnitudes are logarithmic quantities, they can be a little awkward for some purposes. For example, we cannot add magnitudes like flux densities. If the magnitudes of the components of a binary star are 1 and 2, the total magnitude is certainly not 3. To find the total magnitude, we must first solve the flux densities from

$$1 = -2.5 \lg \frac{F_1}{F_0} , \qquad 2 = -2.5 \lg \frac{F_2}{F_0} ,$$

which give

$$F_1 = F_0 \times 10^{-0.4} , \qquad F_2 = F_0 \times 10^{-0.8} .$$

Thus the total flux density is

$$F = F_1 + F_2 = F_0 (10^{-0.4} + 10^{-0.8}) ,$$

and the total magnitude,

$$m = -2.5 \lg \frac{F_0 (10^{-0.4} + 10^{-0.8})}{F_0} = -2.5 \lg 0.55 = 0.64.$$

4.6.4   The distance of a star is $r = 100$ pc and its apparent magnitude, $m = 6$. What is its absolute magnitude?

Substitution into (4.12)

$$m - M = 5 \lg \frac{r}{10 \text{ pc}} \qquad \text{gives} \qquad M = 6 - 5 \lg \frac{100}{10} = 1 \ .$$

4.6.5 The absolute magnitude of a star is $M = -2$ and the apparent magnitude, $m = 8$. What is the distance of the star?

We can solve the distance r from (4.12):

$$r = 10 \text{ pc} \times 10^{(m-M)/5} = 10 \times 10^{10/5} \text{ pc} = 1000 \text{ pc} = 1 \text{ kpc} \ .$$

4.6.6 Although the amount of interstellar extinction varies considerably from place to place, we can use an average value of 2 mag/kpc near the galactic plane. Find the distance of the star in Exercise 4.6.5, assuming such extinction.

Now the distance must be solved from (4.19):

$$8 - (-2) = 5 \lg \frac{r}{10} + 0.002 \, r \ ,$$

where r is in parsecs. This equation cannot be solved analytically, but we can always use a numerical method. We try a simple iteration (Appendix A.6), rewriting the equation as

$$r = 10 \times 10^{2 - 0.0004 \, r} \ .$$

The value $r = 1000$ pc found previously is a good initial guess:

$$
\begin{aligned}
r_0 &= 1000, \\
r_1 &= 10 \times 10^{2 - 0.0004 \times 1000} = 398, \\
r_2 &= 693, \\
&\cdots \\
r_{11} &= 584, \\
r_{12} &= 584.
\end{aligned}
$$

The distance is $r \approx 580$ pc, which is much less than our earlier value, 1000 pc. This should be quite obvious, since due to extinction, radiation is now reduced much faster than in empty space.

4.6.7 What is the optical thickness of a layer of fog, if the Sun seen through the fog seems as bright as a full moon in a cloudless sky?

The apparent magnitudes of the Sun and the Moon are -26.8 and -12.5, respectively. Thus the total extinction in the cloud must be $A = 14.3$. Since

$$A = (2.5 \lg e) \, \tau, \qquad \text{we get}$$

$$\tau = A / (2.5 \lg e) = 14.3/1.08 = 13.2 \ .$$

The optical thickness of the fog is 13.2. In reality, a fraction of the light scatters several times, and a few of the multiply scattered photons leave the cloud along the line of sight, reducing the total extinction. Therefore the optical thickness must be slightly higher than our value.

4.6.8 *Reduction of Observations*

The altitude and magnitude of a star were measured several times during a night. The results are given in the following table.

Altitude	Air mass	Magnitude
50°	1.31	0.90
35°	1.74	0.98
25°	2.37	1.07
20°	2.92	1.17

By plotting the observations as in the following figure, we can determine the extinction coefficient k and the magnitude m_0 outside the atmosphere. This can be done graphically (as here) or using a least-squares fit.

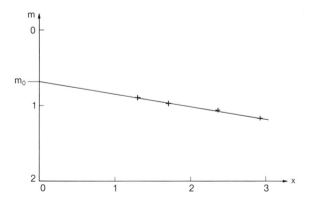

Extrapolation to the air mass $X = 0$ gives $m_0 = 0.68$. The slope of the line gives $k = 0.17$.

4.7 Exercises

4.7.1 The total magnitude of a triple star is 0.0. Two of its components have magnitudes 1.0 and 2.0. What is the magnitude of the third component?

4.7.2 The absolute magnitude of a star in the Andromeda galaxy (distance 690 kpc) is $M = 5$. It explodes as a supernova, becoming one billion (10^9) times brighter. What is its apparent magnitude?

4.7.3 Assume that all stars have the same absolute magnitude and stars are evenly distributed in space. Let $N(m)$ be the number of stars brighter than m magnitudes. Find the ratio $N(m + 1)/N(m)$.

4.7.4 The V magnitude of a star is 15.1, $B - V = 1.6$, and absolute magnitude $M_V = 1.3$. The extinction in the direction of the star in the visual band is $a_V = 1$ mag kpc^{-1}. What is the intrinsic colour of the star?

4.7.5 Stars are observed through a triple window. Each surface reflects away 15% of the incident light.

a) What is the extinction (in magnitudes) caused by the window?
b) What is the optical thickness of the window?

Chapter 5 **Radiation Mechanisms**

In the previous chapters, we have studied the physical properties and detection of electromagnetic radiation. Next we shall briefly discuss concepts related to emission and absorption of radiation. Since we can give here only a summary of some essential results without delving into quantum mechanical explanations, the reader interested in the details is advised to consult any good physics textbook.

5.1 Radiation of Atoms and Molecules

Electromagnetic radiation is emitted or absorbed when an atom or a molecule moves from one energy level to another. If the energy of an atom decreases by an amount ΔE, the atom emits or radiates a quantum of electromagnetic radiation, called a *photon*, whose frequency v is given by the equation

$$\Delta E = h v \quad (h = \text{Planck's constant}) \ . \tag{5.1}$$

Similarly, if the atom receives or absorbs a photon of a frequency v, its energy increases by $\Delta E = h v$.

The classical model describes an atom as a nucleus surrounded by a swarm of electrons. The nucleus consists of Z protons, each having a charge $+ e$ and N electrically neutral neutrons; Z is the charge number of the atom and $A = Z + N$ is its mass number. A neutral atom has as many electrons (charge $- e$) as protons.

An energy level of an atom usually refers to an energy level of its electrons. The energy E of an electron cannot take arbitrary values; only certain energies are allowed: the energy levels are quantized. An atom can emit or absorb radiation only at certain frequencies v_{if} corresponding to energy differences between some initial and final states i and f: $|E_i - E_f| = h v_{if}$. This gives rise to the *line spectrum,* specific for each element (Fig. 5.1). Hot gas under low pressure produces an *emission spectrum* consisting of such discrete lines. If the same gas is cooled down and observed against a source of white light (which has a continuous spectrum), the same lines are seen as dark *absorption lines*.

At low temperatures, most atoms are in their lowest energy state, the *ground state*. Higher energy levels are *excitation states*; a transition from lower to higher state is called *excitation*. Usually the excited atom will return to the lower state, very rapidly radiating a photon (*spontaneous emission*); a typical lifetime of an excited state might be 10^{-8} seconds. The frequency of the emitted photon is given by (5.1). The atom may return to the lower state directly or through some intermediate states, emitting one photon in each transition.

Downward transitions can also be induced by radiation. Suppose our atom has swallowed a photon and become excited. Another photon, whose frequency v corre-

Fig. 5.1a, b. Origin of line spectra. **(a)** Emission spectrum. Atoms of glowing gas returning from excited states to lower states emit photons with frequencies corresponding to the energy difference of the states. Each element emits its own characteristic wavelenghts, which can be measured by spreading the light into a spectrum with a prism or diffraction grating. **(b)** Absorption spectrum. When white light containing all wavelengths travels through gas, the wavelengths characteristic of the gas are absorbed

sponds to some possible downward transition from the excited state, can now irritate the atom, causing it to jump to a lower state, emitting a photon with the same frequency *v*. This is called *induced* or *stimulated emission*. Photons emitted spontaneously leave the atom randomly in all directions with random phases: the radiation is isotropic and incoherent. Induced radiation, on the other hand, is coherent; it propagates in the same direction as, and in phase with, the inducing radiation.

The zero level of the energy states is usually chosen so that a bound electron has negative energy and a free electron, positive energy (cf. the energy integral of planetary orbits, Chap. 7). If an electron with energy $E < 0$ receives more energy than $|E|$, it will leave the atom, which becomes an ion. In astrophysics, ionization is often called a *bound-free* transition (Fig. 5.2). Unlike in excitation, all values of energy ($E > 0$) are now possible. The extraneous part of the absorbed energy goes to the kinetic energy of

Fig. 5.2. Different kinds of transitions between energy levels. Absorption and emission occur between two bound states, whereas ionization and recombination occur between a bound and a free state. Interaction of an atom with an free electron can result in a free-free transition

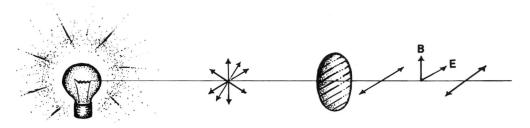

Fig. 5.3. Polarization of light. The light of an incandescent bulb contains all possible directions of vibration, and is therefore unpolarized. Some crystals, for example, pass electric fields oscillating only in certain directions, and the transmitted part of the light becomes linearly polarized. *E* is the electric field and *B* the magnetic field

the liberated electron. The inverse process, in which an atom captures a free electron, is the *recombination* or free-bound transition.

When an electron scatters from a nucleus or an ion without being captured, the electromagnetic interaction can change the kinetic energy of the electron producing *free-free* radiation. An example of this is the *bremsstrahlung* from decelerating electrons (like those hitting the anode of an x-ray tube).

Electromagnetic radiation is transverse wave motion; the electric and magnetic fields oscillate perpendicular to each other and also perpendicular to the direction of propagation. The light of an ordinary incandescent lamp has a random distribution of electric fields vibrating in all directions. If the directions of electric fields in the plane perpendicular to the direction of propagation are not evenly distributed, the radiation is *polarized* (Fig. 5.3). The direction of polarization of *linearly polarized light* means the plane determined by the electric vector and the direction of the light ray. If the electric vector describes a circle, the radiation is *circularly polarized*. If the amplitude of the electric field varies at the same time, the polarization is *elliptic*.

Scattering is an absorption followed by an instantaneous emission at the same wavelength but usually in a new direction. On the macroscopic scale, radiation seems to be reflected by the medium. The light coming from the sky is sunlight scattered from atmospheric molecules. Scattered light is always polarized, the degree of polarization being highest in the direction perpendicular to the direction of the original radiation.

5.2 The Hydrogen Atom

The hydrogen atom is the simplest atom, consisting of a proton and an electron. According to the Bohr model, the electron orbits the proton in a circular orbit. (In spite of the fact that this model has very little to do with reality, it can be successfully used to predict some properties of the hydrogen atom.) Bohr's first postulate says that the angular momentum of the electron must be a multiple of $\hbar$:

$$mvr = n\hbar , \quad \text{where} \tag{5.2}$$

$\hbar = h/2\pi ,$

$m =$ mass of the electron ,

$v =$ speed of the electron ,

$r =$ radius of the orbit ,

$n =$ the principal quantum number, $n = 1, 2, 3, \ldots$.

The quantum mechanical interpretation of Bohr's first postulate is obvious: the electron is represented as a standing wave, and the "length of the orbit" must be a multiple of the de Broglie wavelength, $\lambda = h/p = h/mv$.

A charged particle in a circular orbit (and thus in accelerated motion) should emit electromagnetic radiation, losing energy, were it to obey the rules of classical electrodynamics. Therefore our electron should spiral down towards the nucleus. But obviously, Nature does not behave this way, and we have to accept Bohr's second postulate, which says that an electron moving in an allowed orbit around a nucleus does not radiate. Radiation is emitted only when the electron jumps from a higher energy state to a lower one. The emitted quantum has an energy $h\nu$, equal to the energy difference of these states:

$$h\nu = E_{n_2} - E_{n_1} \ . \tag{5.3}$$

We shall now try to find the energy of an electron in the state E_n. Coulomb's law gives the force pulling the electron towards the proton:

$$F = \frac{1}{4\pi\varepsilon_0} \frac{e^2}{r_n^2} \ , \qquad \text{where} \tag{5.4}$$

ε_0 = the vacuum permittivity = $8.85 \times 10^{-12} \mathrm{N^{-1} m^{-2} C^2}$,

e = the charge of the electron = $1.6 \times 10^{-19} \mathrm{C}$,

r_n = the distance between the electron and the proton .

The acceleration of a particle moving in a circular orbit of radius r_n is

$$a = \frac{v_n^2}{r_n} \ ,$$

and applying Newton's second law ($F = ma$), we get

$$\frac{mv_n^2}{r_n} = \frac{1}{4\pi\varepsilon_0} \frac{e^2}{r_n^2} \ . \tag{5.5}$$

From (5.2) and (5.5) it follows that

$$v_n = \frac{e^2}{4\pi\varepsilon_0 \hbar} \frac{1}{n} \ , \qquad r_n = \frac{4\pi\varepsilon_0 \hbar^2}{me^2} n^2 \ .$$

The total energy of an electron in the orbit n is now

$$E_n = T + V = \frac{1}{2} mv_n^2 - \frac{1}{4\pi\varepsilon_0} \frac{e^2}{r_n}$$

$$= -\frac{me^4}{32\pi^2\varepsilon_0^2\hbar^2} \frac{1}{n^2} = -C\frac{1}{n^2} \ , \tag{5.6}$$

where C is a constant. For the ground state ($n = 1$), we get from (5.6)

Table 5.1. Wavelengths of hydrogen lines (in nm). For historical reasons, the Balmer lines are usually denoted by H_α, H_β, H_γ etc.

n_2	n_1	Lyman 1	Balmer 2	Paschen 3	Brackett 4	Pfund 5
2		121.6	–	–	–	–
3		102.6	656.3	–	–	–
4		97.2	486.1	1875	–	–
5		95.0	434.1	1282	4050	–
6		93.8	410.2	1094	2630	7460
⋮		⋮	⋮	⋮	⋮	⋮
∞		91.2	364.7	821	1460	2280

$$E_1 = -2.18 \times 10^{-18}\,\text{J} = -13.6\,\text{eV} \ .$$

From (5.3) and (5.6) we get the energy of the quantum emitted in the transition $E_{n_2} \to E_{n_1}$:

$$h\nu = E_{n_2} - E_{n_1} = C\left(\frac{1}{n_1^2} - \frac{1}{n_2^2}\right) \ . \tag{5.7}$$

In terms of the wavelength λ, this can be expressed as

$$\frac{1}{\lambda} = \frac{\nu}{c} = \frac{C}{hc}\left(\frac{1}{n_1^2} - \frac{1}{n_2^2}\right) = R\left(\frac{1}{n_1^2} - \frac{1}{n_2^2}\right) \ , \tag{5.8}$$

where R is the *Rydberg constant*, $R = 1.097 \times 10^7\,\text{m}^{-1}$.

Equation (5.8) was derived experimentally for $n_1 = 2$ by Johann Jakob Balmer as early as 1885. That is why we call the set of lines produced by transitions $E_n \to E_2$ the *Balmer series* (Table 5.1). These lines are in the visible part of the spectrum. If the electron returns to its ground state ($E_n \to E_1$), we get the *Lyman series*, which is in ultraviolet. The other series with specific names are the *Paschen series* ($n_1 = 3$), *Brackett series* ($n_1 = 4$) and *Pfund series* ($n_1 = 5$) (see Fig. 5.4).

Fig. 5.4. Transitions of a hydrogen atom. The lower picture shows a part of the spectrum of the star HD193182. On both sides of the stellar spectrum we see an emission spectrum of iron. The wavelengths of these reference lines are known, and they can be used to calibrate the wavelengths of the observed stellar spectrum. The hydrogen Balmer lines are seen as dark absorption lines converging towards the Balmer ionization limit (also called the Balmer discontinuity) at $\lambda = 364.7$ nm to the left. The numbers (15, ..., 40) refer to the quantum number n of the higher energy level. (Photo by Mt. Wilson Observatory)

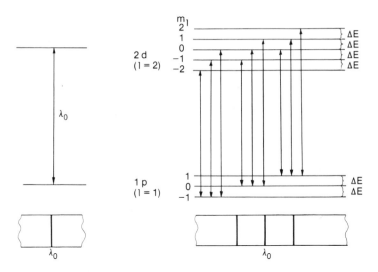

Fig. 5.5. The Zeeman effect. In strong magnetic fields, each energy level of a hydrogen atom splits into $(2\,l+1)$ separate levels, which correspond to different values of the magnetic quantum number $m_l = l,\ l-1,\ \ldots,\ -l$. The energy difference of successive levels is the constant ΔE. For example the p-state $(l = 1)$ splits into three and the d-state $(l = 2)$ into five sublevels. The selection rules require that in electric dipole transitions, Δm_l equals 0 or ± 1, and only nine different transitions between p and d states are possible. Moreover, the transitions with the same Δm_l have the same energy difference. Thus the spectrum has only three separate lines

5.3 Quantum Numbers, Selection Rules, Population Numbers

Quantum Numbers (Fig. 5.5). The Bohr model needs only one quantum number, n, to describe all the energy levels of the electron. This can explain only the coarse features of an atom with a single electron. The quantum mechanical description involves four quantum numbers, one of which is our n, the *principal quantum number*. The other three are the *total angular momentum quantum number l*, the *magnetic quantum number m_l* (which gives the direction of the angular momentum), and the *spin quantum number m_s* (giving the direction of the spin of the electron; the classical analogy of the spin is the rotation of the electron). To understand the fine structure of the spectrum of an atom, we must also include the spin of the nucleus. Quantum mechanically it is also possible to describe atoms with several electrons, although the calculations become very involved.

Selection Rules. The state of an electron cannot change arbitrarily; transitions are restricted by selection rules, which follow from certain conservation laws. The selection rules express how the quantum numbers must change in a transition. Most probable are the electric dipole transitions, which make the atom behave like an oscillating dipole. The probabilities of all other transitions are much smaller, and they are called *forbidden transitions*; examples are magnetic dipole transitions and all quadrupole and higher multipole transitions. Spectral lines originating in forbidden transitions are called *forbidden lines*. The probability of such a transition is so low that under normal circumstances, the transition cannot take place before collisions force the electron to change state. Forbidden lines are possible only if the gas is extremely rarified (like in aurorae and planetary nebulae).

The spins of an electron and nucleus of a hydrogen atom can be either parallel or antiparallel. The energy of the former state is 0.0000059 eV higher. But the selection rules make an electric dipole transition between these states impossible. The transition, which is a magnetic dipole transition, has a very low probability, $A = 2.8 \times 10^{-15}\,\mathrm{s}^{-1}$. This means that the average lifetime of the higher state is $T = 1/A = 11 \times 10^6$ years. Usually collisions change the state of the electron well before this period of time has elapsed. But in interstellar space, the density of hydrogen is so low and the total amount of hydrogen so great that a considerable number of these transitions can take place. The wavelength of the radiation emitted by this transition is 21 cm, which is in the radio band of the spectrum. Extinction at radio wavelengths is very small, and we can observe more distant objects than by using optical wavelengths. The 21-cm radiation has been of crucial importance for surveys of interstellar hydrogen.

Population Numbers. The population number n_i of an energy state i means the number of atoms in that state per unit volume. In thermal equilibrium, the population numbers obey the *Boltzmann distribution*:

$$\frac{n_i}{n_0} = \frac{g_i}{g_0} e^{-\Delta E/kT} , \tag{5.9}$$

where T is the temperature, $\Delta E = E_i - E_0 = h\nu$ is the energy difference between the excited and ground state, and g_i is the statistical weight of the level i (it is the number of different states with the same energy E_i). The subscript 0 always refers to the ground state. Often the population numbers differ from the values given by (5.9), but still we can define an *excitation temperature* T_{exc} in such a way that (5.9) gives correct population numbers, when T is replaced by T_{exc}. The excitation temperature may be different for different energy levels.

5.4 Molecular Spectra

The energy levels of an atom are determined by its electrons. In the case of a molecule, there are many more possibilities: atoms can vibrate around their equilibria and the molecule can rotate around some axis. Both vibrational and rotational states are quantized. Transitions between successive vibrational states typically involve photons in the infrared band, while transitions between rotational states involve photons in the microwave band. These combined with transitions of electrons produce a band spectrum, characteristic for molecules. The spectrum has several narrow bands composed of a great number of lines.

5.5 Continuous Spectra

Due to the Heisenberg uncertainty principle, each spectral line has a natural width. Atoms also have thermal motions, and the Doppler shifts produced by these random motions broaden line profiles.

Each photon has some definite wavelength, and so, basically every spectrum is composed of separate lines. Sometimes, however, the lines are so closely packed and so broad that the spectrum looks continuous (Fig. 5.6). We have already mentioned some processes that produce continuous spectra. Continuous emission spectra can originate in recombinations and free-free transitions. In recombination, an atom captures a free electron whose energy is not quantized; in free-free transitions, both initial and final states are unquantized. Thus the emission line can have any frequency whatsoever.

450   500   550   600   650 [nm]

CO

Fig. 5.6. Spectrum of carbon monoxide CO from 430 nm to 670 nm. The various bands correspond to different vibrational transitions. Each band is composed of numerous rotational lines. Near the right edge of each band the lines are so closely packed that they overlap and at this resolution, the spectrum looks continuous. [R. W. B. Pearse, A. G. Gaydon: *The Identification of Molecular Spectra* (Chapman & Hall Ltd., London 1976), p. 394]

Similarly, ionizations and free-free transitions can give rise to a continuous absorption spectrum.

When the pressure of hot gas is increased, the spectral lines begin to broaden. At high pressure, atoms bump into each other more frequently, and the close neighbors disturb the energy levels. When the pressure is high enough, the lines begin to overlap. Thus the spectrum of hot gas at high pressure is continuous. Electric fields also broaden spectral lines (the Stark effect).

In liquids and solids, the atoms are more densely packed than in gaseous substances. Their mutual perturbations broaden the energy levels, producing a continuous spectrum.

5.6 Blackbody Radiation

A *blackbody* is defined as an object that does not reflect or scatter radiation shining upon it, but absorbs and re-emits the radiation completely. A blackbody is a kind of ideal radiator, which cannot exist in the real world. Yet many objects behave very much as if they were blackbodies.

The radiation of a blackbody depends only on its temperature, being perfectly independent of its shape, material and internal constitution. The wavelength distribution of the radiation follows Planck's law, which is a function of temperature only. The intensity at a frequency v of a blackbody at temperature T is

$$B_v(T) = \frac{2hv^3}{c^2} \frac{1}{\exp(hv/kT) - 1} \, , \qquad \text{where} \qquad (5.10)$$

h = the Planck constant = 6.63×10^{-34} Js ,

c = the speed of light $\cong 3 \times 10^8$ m s^{-1} ,

k = the Boltzmann constant = 1.38×10^{-23} J K^{-1} .

By definition of the intensity, the dimension of B_v is W m^{-2} Hz^{-1} sterad^{-1}.

Blackbody radiation can be produced in a closed cavity whose walls absorb all radiation incident upon them (and coming from inside the cavity). The walls and the radiation in the cavity are in equilibrium; both are at the same temperature, and the walls emit all the energy they receive. Since radiation energy is constantly transformed into thermal energy of the atoms of the walls and back to radiation, the blackbody radiation is also called *thermal radiation*.

The spectrum of a blackbody given by Planck's law (5.10) is continuous. This is true if the size of the radiator is very large compared with the dominant wavelengths. In the case of the cavity, this can be understood by considering the radiation as standing waves trapped in the cavity. The number of different wavelengths is larger, the shorter the wavelengths are compared with the size of the cavity. We already mentioned that spectra of solid bodies are continuous; very often such spectra can be quite well approximated by Planck's law.

We can also write Planck's law as a function of the wavelength. We require that $B_\nu \, d\nu = -B_\lambda \, d\lambda$. The wavelength decreases with increasing frequency; hence the minus sign. Since $\nu = c/\lambda$, we have

$$\frac{d\nu}{d\lambda} = -\frac{c}{\lambda^2} \, , \quad \text{whence} \tag{5.11}$$

$$B_\lambda = -B_\nu \frac{d\nu}{d\lambda} = B_\nu \frac{c}{\lambda^2} \, , \quad \text{or} \tag{5.12}$$

$$B_\lambda(T) = \frac{2hc^2}{\lambda^5} \frac{1}{\exp(hc/\lambda kT) - 1} \, , \quad [B_\lambda] = \mathrm{W\,m^{-2}\,m^{-1}\,sterad^{-1}} \, . \tag{5.13}$$

The functions B_ν and B_λ are defined in such a way that the total intensity can be obtained easily using either of them:

$$B(T) = \int_0^\infty B_\nu \, d\nu = \int_0^\infty B_\lambda \, d\lambda \ .$$

Let us now try to find the total intensity using the first of these integrals:

$$B(T) = \int_0^\infty B_\nu \, d\nu = \frac{2h}{c^2} \int_0^\infty \frac{\nu^3 \, d\nu}{\exp(h\nu/kT) - 1} \ .$$

We now change the integration variable to $x = h\nu/kT$, whence $d\nu = (kT/h) \, dx$:

$$B(T) = \frac{2h}{c^2} \frac{k^4}{h^4} T^4 \int_0^\infty \frac{x^3 \, dx}{e^x - 1} \ .$$

The definite integral in this expression is just a real number[1], independent of the temperature. Thus we find that

$$B(T) = A T^4, \tag{5.14}$$

where the constant A has the value

---

[1] There is no elementary way to evaluate the integral. We can tell those who are familiar with all the exotic functions so beloved by theoretical physicists that the integral can rather easily be expressed as $\Gamma(4)\,\zeta(4)$, where ζ is the Riemann zeta function and Γ is the gamma function. For integral values, $\Gamma(n)$ is simply the factorial $(n-1)!$. The difficult part is showing that $\zeta(4) = \pi^4/90$. This can be done by expanding $x^4 - x^2$ as a Fourier-series and evaluating the series at $x = \pi$.

$$A = \frac{2k^4}{c^2 h^3} \frac{\pi^4}{15} .$$

The flux density F for isotropic radiation of intensity B is (Sect. 4.1):

$$F = \pi B , \quad \text{or}$$

$$F = \sigma T^4 . \tag{5.15}$$

This is the *Stefan-Boltzmann law*, and the constant $\sigma\,(=\pi A)$ is the Stefan-Boltzmann constant,

$$\sigma = 5.67 \times 10^{-8} \mathrm{W\,m^{-2}\,K^{-4}} .$$

From the Stefan-Boltzmann law we get a relation between the luminosity and temperature of a star. If the radius of the star is R, its surface area is $4\pi R^2$, and if the flux density on the surface is F, we have

$$L = 4\pi R^2 F .$$

If the star is assumed to radiate like a blackbody, we have $F = \sigma T^4$, which gives

$$L = 4\pi\sigma R^2 T^4 . \tag{5.16}$$

In fact this defines the *effective temperature* of the star, discussed in more detail in the next chapter.

The luminosity, radius and temperature of a star are interdependent quantities, as we can see from (5.16). They are also related to the absolute bolometric magnitude of the star. Equation (4.14) gives the difference of the absolute bolometric magnitude of the star and the Sun:

$$M_{\mathrm{bol}} - M_{\mathrm{bol},\odot} = -2.5 \lg \frac{L}{L_\odot} . \tag{5.17}$$

But we can now use (5.16) to express the luminosities in terms of radii and temperatures:

$$M_{\mathrm{bol}} - M_{\mathrm{bol},\odot} = -5 \lg \frac{R}{R_\odot} - 10 \lg \frac{T}{T_\odot} . \tag{5.18}$$

As we can see in Fig. 5.7, the wavelength of the maximum intensity decreases with increasing total intensity (equal to the area below the curve). We can find the wavelength $\lambda_{\max}$ corresponding to the maximum intensity by differentiating Planck's function $B_\lambda(T)$ with respect to λ and finding zero of the derivative. The result is *Wien's displacement law*:

$$\lambda_{\max} T = b = \text{const} , \tag{5.19}$$

where the *Wien displacement constant b* is

$$b = 0.0028978 \,\mathrm{K\,m} .$$

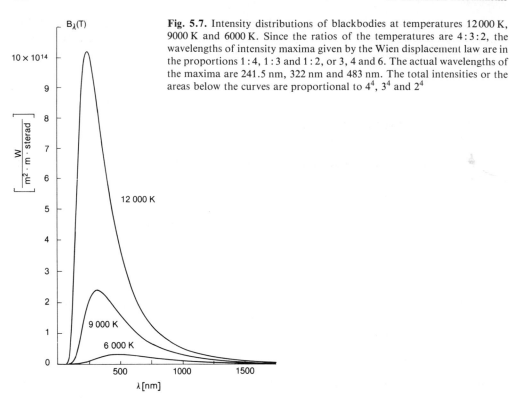

Fig. 5.7. Intensity distributions of blackbodies at temperatures 12 000 K, 9000 K and 6000 K. Since the ratios of the temperatures are 4 : 3 : 2, the wavelengths of intensity maxima given by the Wien displacement law are in the proportions 1 : 4, 1 : 3 and 1 : 2, or 3, 4 and 6. The actual wavelengths of the maxima are 241.5 nm, 322 nm and 483 nm. The total intensities or the areas below the curves are proportional to 4^4, 3^4 and 2^4

We can use the same procedure to find the maximum of B_ν. But the frequency ν_{max} thus obtained is different from $\nu_{max} = c/\lambda_{max}$ given by (5.19). The reason for this is the fact that the intensities are given per unit frequency or unit wavelength, and the dependence of frequency on wavelength is nonlinear.

When the wavelength is near the maximum or much longer than λ_{max}, Planck's function can be approximated by simpler expressions. When $\lambda \approx \lambda_{max}$ (or $hc/\lambda kT \gg 1$), we have

$$e^{hc/\lambda kT} \gg 1 .$$

In this case, we get the *Wien approximation*

$$B_\lambda(T) \approx \frac{2hc^2}{\lambda^5} e^{-hc/\lambda kT} . \tag{5.20}$$

When $hc/\lambda kT \ll 1$ ($\lambda \gg \lambda_{max}$), we have

$$e^{hc/\lambda kT} \approx 1 + hc/\lambda kT ,$$

which gives the *Rayleigh-Jeans approximation*

$$B_\lambda(T) = \frac{2hc^2}{\lambda^5} \frac{\lambda kT}{hc} = \frac{2ckT}{\lambda^4} . \tag{5.21}$$

This is particularly useful in radio astronomy.

Classical physics predicted only the Rayleigh-Jeans approximation. Were (5.21) true for all wavelengths, the intensity would grow beyond all limits when the wavelength approaches zero, contrary to observations. This behaviour was known as the ultraviolet catastrophe.

5.7 Other Radiation Mechanisms

Maser and Laser (Fig. 5.8). The Boltzmann distribution (5.9) shows that usually there are fewer atoms in excited states than in the ground state. There are, however, means to produce a *population inversion*, an excited state containing more atoms than the ground state. This inversion is essential for both the maser and the laser (Microwave/ Light Amplification by Stimulated Emission of Radiation). If the excited atoms are now illuminated with photons having energies equal to the excitation energy, the radiation will induce downward transitions. The number of photons emitted greatly exceeds the number of absorbed photons, and radiation is amplified. Typically, the excited state is a *metastable state*, a state with a very long average lifetime, which means that the contribution of spontaneous emission is negligible. Therefore the resulting radiation is coherent and monochromatic. Several maser sources have been found in interstellar molecular clouds and dust envelopes around stars.

Synchrotron Radiation. A free charge in accelerated motion will emit electromagnetic radiation. Charged particles moving in a magnetic field follow helices around the field lines. As seen from the direction of the field, the motion is circular and therefore accelerated. The moving charge will radiate in the direction of its velocity vector. Such radiation is called *synchrotron radiation*. It will be further discussed in Chap. 16, "The Interstellar Medium".

Maser and synchrotron radiation are examples of non-thermal radiation, which is not related to thermal motions of atoms. Bremsstrahlung is also nonthermal.

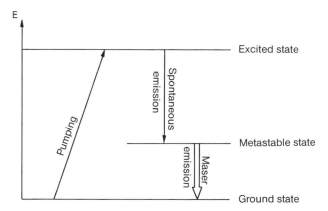

Fig. 5.8. The operational principle of the maser and the laser. A metastable state (a state with a relatively long average lifetime) stores atoms where they accumulate; there are more atoms in the metastable state than in the ground state. This population inversion is maintained by radiatively exciting atoms to a higher excitation state ("pumping"), from which they spontaneously jump down to the metastable state. When the atoms are illuminated by photons with energies equal to the excitation energy of the metastable state, the photons will induce more radiation of the same wavelength, and the radiation is amplified in geometric progression

5.8 Radiative Transfer

Propagation of radiation in a medium, also called radiative transfer, is one of the basic problems of astrophysics. The subject is too complicated to be discussed here in any detail. The fundamental equation of radiative transfer is, however, easily derived.

Assume we have a small cylinder, the bottom of which has an area dA and the length of which is dr. Let I_ν be the intensity of radiation perpendicular to the bottom surfaces going into solid angle $d\omega$ ($[I] = \mathrm{W\,m^{-2}Hz^{-1}sterad^{-1}}$). If the intensity changes by an amount dI_ν in the distance dr, the energy changes by

$$dE = dI_\nu \, dA \, d\nu \, d\omega \, dt$$

in the cylinder in time dt. This equals the emission minus absorption in the cylinder. The absorbed energy is [cf. (4.15)]

$$dE_{\mathrm{abs}} = \alpha_\nu I_\nu \, dr \, dA \, d\nu \, d\omega \, dt \ , \tag{5.22}$$

where α_ν is the opacity of the medium at frequency ν. Let the amount of energy emitted per Hertz at frequency ν into unit solid angle from unit volume and per unit time be j_ν ($[j_\nu] = \mathrm{W\,m^{-3}\,Hz^{-1}\,sterad^{-1}}$). This is called the *emission coefficient* of the medium. The energy emitted into solid angle $d\omega$ from the cylinder is then

$$dE_{\mathrm{em}} = j_\nu \, dr \, dA \, d\nu \, d\omega \, dt \ . \tag{5.23}$$

The equation

$$dE = -dE_{\mathrm{abs}} + dE_{\mathrm{em}}$$

gives then

$$dI_\nu \ = -\alpha_\nu I_\nu \, dr + j_\nu \, dr \ , \qquad \text{or}$$

$$\frac{dI_\nu}{\alpha_\nu \, dr} = -I_\nu + \frac{j_\nu}{\alpha_\nu} \ . \tag{5.24}$$

We shall denote the ratio of the emission coefficient j_ν to the absorption coefficient or opacity α_ν by S_ν:

$$S_\nu = j_\nu / \alpha_\nu . \tag{5.25}$$

S_ν is called the *source function*. Because $\alpha_\nu \, dr = d\tau_\nu$, where τ_ν is the optical thickness at frequency ν, (5.24) can be written as

$$\frac{dI_\nu}{d\tau_\nu} = -I_\nu + S_\nu \ . \tag{5.26}$$

Equation (5.26) is the basic equation of radiative transfer. Without solving the equation, we see that if $I_\nu < S_\nu$, then $dI_\nu / d\tau_\nu > 0$, and the intensity tends to increase in the direction of propagation. And, if $I_\nu > S_\nu$, then $dI_\nu / d\tau_\nu < 0$, and I_ν will decrease. In an equilibrium, the emitted and absorbed energies are equal, in which case we find from (5.22) and (5.23)

$$I_\nu = j_\nu / \alpha_\nu = S_\nu .$$

Substituting this into (5.26), we see that $dI_\nu / d\tau_\nu = 0$. In thermodynamic equilibrium, the radiation of the medium is blackbody radiation, and the source function is given by Planck's law:

$$S_\nu = B_\nu(T) = \frac{2h\nu^3}{c^2} \frac{1}{\exp(h\nu/kT) - 1} . \tag{5.27}$$

Even if the system is not in thermodynamic equilibrium, it may be possible to find an excitation temperature T_{exc} such that $B_\nu(T_{exc}) = S_\nu$. This temperature may depend on frequency.

A formal solution of (5.26) is

$$I_\nu(\tau_\nu) = I_\nu(0)e^{-\tau_\nu} + \int_0^{\tau_\nu} e^{-(\tau_\nu - t)} S_\nu(t)\, dt . \tag{5.28}$$

Here, $I_\nu(0)$ is the intensity of the background radiation, coming through the medium (e.g. an interstellar cloud) and decaying exponentially in the medium. The second term gives the emission in the medium. The solution is only formal, since in general, the source function S_ν is unknown and must be solved simultaneously with the intensity. If $S_\nu(\tau_\nu)$ is constant in the cloud and the background radiation is ignored, we get

$$I_\nu(\tau_\nu) = S_\nu \int_0^{\tau_\nu} e^{-(\tau_\nu - t)}\, dt = S_\nu(1 - e^{-\tau_\nu}) . \tag{5.29}$$

If the cloud is optically thick ($\tau_\nu \gg 1$), we have

$$I_\nu = S_\nu , \tag{5.30}$$

i.e. the intensity equals the source function, and emission and absorption processes are in equilibrium.

*The Equation of Radiative Transfer in an Atmosphere

An important field of application of the theory of radiative transfer is in the study of planetary and stellar atmospheres. In this case, to a good approximation, the properties of the medium only vary in one direction, say along the z-axis. The intensity will then be a function only of z and of θ, where θ is the angle between the z-axis and the direction of propagation of the radiation.

In applications to atmospheres, it is customary to define the optical depth τ_ν in the vertical direction as

$$d\tau_\nu = -\alpha_\nu dz .$$

(Conventionally z increases upwards and the optical depth inwards in the atmosphere.) The vertical line element dz is related to that along the light ray, dr, according to

$$dz = dr \cos\theta .$$

With these notational conventions, (5.24, 25) now yield

$$\cos\theta \frac{dI_\nu(z,\theta)}{d\tau_\nu} = I_\nu - S_\nu \; . \tag{1}$$

This is the form of the equation of radiative transfer usually encountered in the study of stellar and planetary atmospheres.

A formal expression for the intensity emerging from an atmosphere, corresponding to (5.28) can be obtained by integrating Eq. (1) from $\tau_\nu = \infty$ (we assume that the bottom of the atmosphere is at infinite optical depth) to $\tau_\nu = 0$ (corresponding to the top of the atmosphere). This yields

$$I_\nu(0,\theta) = \int_0^\infty S_\nu \exp(-\tau_\nu \sec\theta) \sec\theta \, d\tau_\nu \; . \tag{2}$$

This expression will be used in Sect. 9.6 on the interpretation of stellar spectra.

5.9 Examples

5.9.1 Find the wavelength of the photon emitted in the transition of a hydrogen atom from $n_2 = 110$ to $n_1 = 109$.

Equation (5.8) gives

$$\frac{1}{\lambda} = 1.097 \times 10^7 \, \text{m}^{-1} \left(\frac{1}{109^2} - \frac{1}{110^2} \right) = 16.71 \, \text{m}^{-1} \; ,$$

whence

$$\lambda = 0.060 \, \text{m} \; .$$

(Such radiation was observed for the first time in 1965 by an NRAO radio telescope.)

5.9.2 Derive the Wien displacement laws.

Let us denote $x = hc/\lambda kT$. Planck's law then becomes

$$B_\lambda(T) = \frac{2k^5 T^5}{h^4 c^3} \frac{x^5}{e^x - 1} \; .$$

For a given temperature, the first factor is constant. Thus, it is sufficient to find the maximum of the function $f(x) = x^5/(e^x - 1)$.

First, we must evaluate the derivative of f:

$$f'(x) = \frac{5x^4(e^x - 1) - x^5 e^x}{(e^x - 1)^2} = \frac{x^4 e^x}{(e^x - 1)^2} (5 - 5e^{-x} - x) \; .$$

By definition, x is always strictly positive. Hence $f'(x)$ can be zero only if the factor $5 - 5e^{-x} - x$ is zero. This equation can be solved numerically, and the result is $x = 4.965$. The Wien displacement law is then

$$\lambda_{\max} T = \frac{hc}{xk} = 0.002898 \, \text{K m} \; .$$

In terms of frequency Planck's law is

$$B_v(T) = \frac{2hv^3}{c^2} \frac{1}{e^{hv/kT} - 1} \ .$$

Substituting $x = hv/kT$ we get

$$B_v(T) = \frac{2k^3 T^3}{h^2 c^2} \frac{x^3}{e^x - 1} \ .$$

Now we study the function $f(x) = x^3/(e^x - 1)$:

$$f'(x) = \frac{3x^2(e^x - 1) - x^3 e^x}{(e^x - 1)^2} = \frac{x^3 e^x}{(e^x - 1)^2} (3 - 3e^{-x} - x) \ .$$

This vanishes, when $3 - 3e^{-x} - x = 0$. The solution of this equation is $x = 2.821$. Hence

$$\frac{T}{v_{max}} = 1.701 \times 10^{-11} \ \mathrm{K\,s}$$

or

$$\frac{cT}{v_{max}} = \frac{hc}{kx} \mp 0.00510 \ \mathrm{K\,m} \ .$$

Note that the wavelength corresponding to v_{max} is different from λ_{max}. The reason is that we have used two different forms of Planck's function, one giving the intensity per unit wavelength, the other per unit frequency.

5.9.3 a) Find the fraction of radiation that a blackbody emits in the range $[\lambda_1, \lambda_2]$, where λ_1 and $\lambda_2 \gg \lambda_{max}$. b) How much energy does a 100 W incandescent light bulb radiate in the radio wavelengths, $\lambda \geq 1$ cm? Assume the temperature is 2500 K.

Since the wavelengths are much longer than λ_{max}, we can use the Rayleigh-Jeans approximation. Then

$$B = \int_{\lambda_1}^{\lambda_2} B_\lambda(T) d\lambda$$

$$\approx 2ckT \int_{\lambda_1}^{\lambda_2} \frac{d\lambda}{\lambda^4}$$

$$= \frac{2ckT}{3} \left(\frac{1}{\lambda_1^3} - \frac{1}{\lambda_2^3} \right) \ .$$

The ratio of this to the total intensity is

$$\frac{B}{B_{\text{tot}}} = \frac{5c^3 h^3}{k^3 \pi^4} \frac{1}{T^3} \left(\frac{1}{\lambda_1^3} - \frac{1}{\lambda_2^3} \right) .$$

b) Now $\lambda_1 = 0.01$ m and $\lambda_2 = \infty$. Then

$$B = 100 \text{ W} \times 1.529 \times 10^{-7} \frac{1}{2500^3} \frac{1}{0.01^3} = 9.8 \times 10^{-10} \text{ W} .$$

It is quite difficult to listen to the radio emission of a light bulb with an ordinary radio receiver.

5.10 Exercises

5.10.1   Show that in the Wien approximation the relative error of B_λ is

$$\frac{\Delta B_\lambda}{B_\lambda} = -e^{-hc/\lambda kT} .$$

5.10.2   If the transition of the hydrogen atom $n+1 \to n$ were to correspond to the wavelength 21.05 cm, what would the quantum number n be? The interstellar medium emits strong radiation at this wavelength. Can this radiation be due to such transitions?

5.10.3   The space is filled with background radiation, remnant of the early age of the universe. Currently the distribution of this radiation is similar to the radiation of a blackbody at the temperature of 2.7 K. What is the wavelength of the maximum of this radiation? What is its total intensity? Compare the intensity of the background radiation to the intensity of the Sun at the visual wavelengths.

5.10.4   The temperature of a red giant is $T = 2500$ K and radius 100 times the solar radius. The distance of the star is 10 pc.

a) Find the total luminosity of the star, and the luminosity in the visual band 400 nm $\leq \lambda \leq$ 700 nm.
b) Compare the star with a 100 W lamp that radiates 5% of its energy in the visual band. What is the distance of the lamp if it looks as bright as the star?

Chapter 6   **Temperatures**

Temperatures of astronomical objects range from almost absolute zero to millions of degrees. Temperature can be defined in a variety of ways, and its numerical value depends on the specific definition used. The temperature is welldefined only in a state of thermodynamic equilibrium. Since most astrophysical objects are not in equilibrium, they cannot be assigned a unique temperature. For many purposes it is still useful to describe phenomena in terms of a temperature, but its value will then depend on how it is defined.

Often the temperature is determined by comparing the object, a star for instance, with a blackbody. Although real stars do not radiate exactly like blackbodies, their spectra can usually be approximated by blackbody spectra after the effect of spectral lines has been eliminated. The resulting temperature depends on the exact criterion used to fit Planck's function to observations.

The most important quantity describing the surface temperature of a star is the *effective temperature T_e*. It is defined as the temperature of a blackbody which radiates with the same total flux density as the star. Since the effective temperature depends only on the total radiation power integrated over all frequencies, it is welldefined for all energy distributions even if they deviate far from Planck's law.

In the previous chapter we derived the Stefan-Boltzmann law, which gives the total flux density as a function of temperature. If we now find a value T_e of the temperature such that the Stefan-Boltzmann law gives the correct flux density F on the surface of the star, we have found the effective temperature. The flux density on the surface is

$$F = \sigma T_e^4 . \tag{6.1}$$

The total flux is $L = 4\pi R^2 F$, where R is the radius of the star, and the flux density at a distance r is

$$F' = \frac{L}{4\pi r^2} = \frac{R^2}{r^2} F = \left(\frac{\alpha}{2}\right)^2 \sigma T_e^4 , \tag{6.2}$$

where $\alpha = 2R/r$ is the observed angular diameter of the star. For direct determination of the effective temperature, we have to measure the total flux density and the angular diameter of the star. This is possible only in the few cases in which the diameter has been found by interferometry.

If we assume that at some wavelength λ the flux density F_λ on the surface of the star is obtained from Planck's law, we get the *brightness temperature T_b*. In the isotropic case we have then, $F_\lambda = \pi B_\lambda(T_b)$. If the radius of the star is R and distance from the Earth r, the observed flux density is

$$F'_\lambda = \frac{R^2}{r^2} F_\lambda .$$

Again, F'_λ can be determined only if the angular diameter α is known. The brightness temperature T_b can then be solved from

$$F'_\lambda = \left(\frac{\alpha}{2}\right)^2 \pi B_\lambda(T_b) \ . \tag{6.3}$$

Since the star does not radiate like a blackbody, its brightness temperature depends on the particular wavelength used in (6.3).

In radio astronomy, brightness temperature is used to express the intensity (or surface brightness) of the source. If the intensity at frequency v is I_v, the brightness temperature is obtained from

$$I_v = B_v(T_b) \ .$$

T_b gives the temperature of a blackbody with the same surface brightness as the observed source.

Since radio wavelengths are very long, the condition $h v \ll k T$ of the Rayleigh-Jeans approximation is usually satisfied (except for millimetre and submillimetre bands), and we can write Planck's law as

$$B_v(T_b) = \frac{2 h v^3}{c^2} \frac{1}{\exp(h v / k T_b) - 1} = \frac{2 h v^3}{c^2} \frac{1}{1 + (h v / k T_b) + \ldots - 1}$$
$$\approx \frac{2 k v^2}{c^2} T_b \ . \tag{6.4}$$

Thus we get the following expression for the radio astronomical brightness temperature:

$$T_b = \frac{c^2}{2 k v^2} I_v = \frac{\lambda^2}{2 k} I_v \ . \tag{6.5}$$

A measure of the signal registered by a radio telescope is the *antenna temperature* T_A. After the antenna temperature is measured, we get the brightness temperature from

$$T_A = \eta T_b \ , \tag{6.6}$$

where η is the beam efficiency of the antenna (typically $0.4 \leq \eta \leq 0.8$). Equation (6.6) holds if the source is wide enough to cover the whole beam, i.e. the solid angle Ω_A from which the antenna receives radiation. If the solid angle subtended by the source, Ω_S, is smaller than Ω_A, the observed antenna temperature is

$$T_A = \eta \frac{\Omega_S}{\Omega_A} T_b \ , \quad (\Omega_S < \Omega_A) \ . \tag{6.7}$$

The *colour temperature* T_c can be determined even if the angular diameter of the source is unknown (Fig. 6.1). We only have to know the relative energy distribution in some wavelength range $[\lambda_1, \lambda_2]$; the absolute value of the flux is not needed. The observed flux density as a function of wavelength is compared with Planck's function at different temperatures. The temperature giving the best fit is the colour temperature in

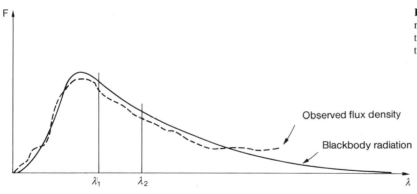

Fig. 6.1. Determination of the colour temperature. The ratio of the flux densities at wavelengths λ_1 and λ_2 gives the temperature of a blackbody with the same ratio. In general, the result depends on the wavelengths chosen

the interval $[\lambda_1, \lambda_2]$. The colour temperature is usually different for different wavelength intervals, since the shape of the observed energy distribution may be quite different from the blackbody spectrum.

A simple method for finding a colour temperature is the following. We measure the flux density F'_λ at two wavelengths λ_1 and λ_2. If we assume that the intensity distribution follows Planck's law, the ratio of these flux densities must be the same as the ratio obtained from Planck's law:

$$\frac{F'_{\lambda_1}}{F'_{\lambda_2}} = \frac{B_{\lambda_1}(T)}{B_{\lambda_2}(T)} = \frac{\lambda_2^5}{\lambda_1^5} \frac{\exp(hc/\lambda_2 kT)-1}{\exp(hc/\lambda_1 kT)-1} . \tag{6.8}$$

The temperature T solved from this equation is a colour temperature.

The observed flux densities correspond to certain magnitudes, m_{λ_1} and m_{λ_2}. The definition of magnitudes gives

$$m_{\lambda_1} - m_{\lambda_2} = -2.5 \lg \frac{F'_{\lambda_1}}{F'_{\lambda_2}} + \text{const} ,$$

where the constant term is a consequence of the different zero points of the magnitude scales. If the temperature is not too high, we can use the Wien approximation in the optical part of the spectrum:

$$m_{\lambda_1} - m_{\lambda_2} = -2.5 \lg \frac{B_{\lambda_1}}{B_{\lambda_2}} + \text{const} = -2.5 \lg \left(\frac{\lambda_2}{\lambda_1}\right)^5 + 2.5 \frac{hc}{kT}\left(\frac{1}{\lambda_1} - \frac{1}{\lambda_2}\right) \lg e + \text{const} .$$

This can be written as

$$m_{\lambda_1} - m_{\lambda_2} = a + b/T_c , \tag{6.9}$$

where a and b are constants. This shows that there is a simple relationship between the difference of two magnitudes and the colour temperature.

Strictly speaking, the magnitudes in (6.9) are monochromatic, but the same relation can be also used with broadband magnitudes, like B and V. In that case, the two wavelengths are essentially the effective wavelengths of the B and V bands. The constant is chosen so that $B - V = 0$ for stars of the spectral type A0 (see Chap. 9). Thus the colour index $B - V$ also gives a colour temperature.

The *kinetic temperature*, T_k, is related to the average speed of gas molecules. The kinetic energy of an ideal gas molecule as a function of temperature follows from the kinetic gas theory:

Kinetic energy $= \frac{1}{2}mv^2 = \frac{3}{2}kT_k$.

Solving for T_k we get

$$T_k = \frac{mv^2}{3k} , \tag{6.10}$$

where m is the mass of the molecule, v its average velocity (or rather its r.m.s. velocity), and k, the Boltzmann constant. For ideal gases, the pressure is directly proportional to the kinetic temperature (cf. *Gas Pressure and Radiation Pressure, p. 267):

$$P = nkT_k , \tag{6.11}$$

where n is the number density of the molecules (molecules per unit volume).

We previously defined the *excitation temperature* T_{exc} as a temperature which, if substituted into the Boltzmann distribution (5.9), gives the observed population numbers. If the distribution of atoms in different levels is a result of mutual collisions of the atoms only, the excitation temperature equals the kinetic temperature, $T_{exc} = T_k$. The *ionization temperature* T_i is found by comparing the number of atoms in different states of ionization. Since stars are not exactly blackbodies, the values of excitation and ionization temperatures usually vary, depending on the element whose spectral lines were used for temperature determination.

In thermodynamic equilibrium, all these various temperatures are equal.

6.1 Examples

6.1.1 *Determination of Effective Temperature*

The observed flux density of Arcturus is

$$F' = 4.5 \times 10^{-8} \mathrm{W\,m}^{-2} .$$

Interferometric measurements give an angular diameter of $\alpha = 0.020''$. Thus, $\alpha/2 = 4.85 \times 10^{-8}$ radians. From (6.2) we get

$$T_e = \left(\frac{4.5 \times 10^{-8}}{(4.85 \times 10^{-8})^2 \times 5.669 \times 10^{-8}} \right)^{1/4} \mathrm{K} = 4300\,\mathrm{K} .$$

6.1.2 The absolute bolometric magnitude of ε Eri is 5.4 and its effective temperature is 4600 K. Find its radius.

By solving the ratio of the radii from (5.18) we get

$$\frac{R}{R_\odot} = \left(\frac{T_\odot}{T} \right)^2 10^{-0.2(M_{bol} - M_{bol,\odot})} .$$

Thus, the radius of ε Eri is

$$R = R_\odot \left(\frac{5800}{4600}\right)^2 10^{-0.2(5.4-4.7)} \approx 1.2 R_\odot \quad .$$

6.1.3 Flux densities at wavelengths 440 nm and 550 nm are 1.30 and 1.00 W m^{-2} m^{-1}, respectively. Find the colour temperature.

If the flux densities at wavelengths λ_1 and λ_2 are F_1 and F_2, respectively, the colour temperature can be solved from the equation

$$\frac{F_1}{F_2} = \frac{B_{\lambda_1}(T)}{B_{\lambda_2}(T)}$$

$$= \frac{(2hc^2/\lambda_1^5)/(\exp(hc/\lambda_1 kT)-1)}{(2hc^2/\lambda_2^5)/(\exp(hc/\lambda_2 kT)-1)}$$

$$= \left(\frac{\lambda_2}{\lambda_1}\right)^5 \frac{\exp(hc/\lambda_1 kT)-1}{\exp(hc/\lambda_2 kT)-1} \quad .$$

If we denote

$$A = \frac{F_1}{F_2}\left(\frac{\lambda_1}{\lambda_2}\right)^5 \quad , \quad B_1 = \frac{hc}{\lambda_1 k} \quad , \quad B_2 = \frac{hc}{\lambda_2 k} \quad ,$$

we get the equation

$$A = \frac{e^{B_2/T}-1}{e^{B_1/T}-1}$$

for the colour temperature T. This equation must be solved numerically.
    In our example the constants have the following values:

$$A = \frac{1.00}{1.30}\left(\frac{550}{440}\right)^5 = 2.348 \quad , \quad B_1 = 32\,700 \text{ K} \quad , \quad B_2 = 26\,160 \text{ K} \quad .$$

By substituting different values for T, we find that $T = 7545$ K satisfies our equation.

6.2 Exercises

6.2.1 The effective temperature of Sirius is 10 000 K, apparent visual magnitude -1.5, distance 2.67 pc and bolometric correction 0.5. What is the radius of Sirius?

6.2.2 The observed flux density of the Sun at $\lambda = 300$ nm is 0.59 W m^{-2} m^{-1}. Find the brightness temperature of the Sun at this wavelength.

6.2.3   The colour temperature can be determined from two magnitudes corresponding to two different wavelengths. Show that

$$T = \frac{7000\,\text{K}}{(B-V)+0.47}\,.$$

The wavelengths of the B and V bands are 440 nm and 548 nm, respectively, and we assume that $B = V$ for stars of the spectral class A0, the colour temperature of which is about 15 000 K.

6.2.4   The kinetic temperature of the plasma in the solar corona can reach 10^6 K. Find the average speed of the electrons in such plasma.

Chapter 7 **Celestial Mechanics**

Celestial mechanics, the study of motions of celestial bodies, together with spherical astronomy, was the main branch of astronomy until the end of the 19th century, when astrophysics began to evolve rapidly. The primary task of classical celestial mechanics was to explain and predict the motions of planets and their satellites. Several empirical models, like epicycles and Kepler's law, had been employed to describe these motions. But none of these models explained why the planets moved the way they did. It was only in the 1680's that a simple explanation was found for all these motions – Newton's law of universal gravitation. In this chapter, we will derive some properties of orbital motion. The physics we need for this is simple indeed, just Newton's laws. (For a review, see *Newton's Laws, p. 153.)

This chapter is mathematically slightly more involved than the rest of the book. We shall use some vector calculus to derive our results, which, however, can be easily understood with very elementary mathematics. A summary of the basic facts of vector calculus is given in Appendix A.3.

7.1 Equations of Motion

We shall concentrate on the systems of only two bodies. In fact, this is the most complicated case that allows a neat analytical solution. For simplicity, let us call the bodies the Sun and a planet, although they could quite as well be a planet and its moon, or the two components of a binary star.

Let the masses of the two bodies be m_1 and m_2 and the radius vectors in some fixed inertial coordinate frame, r_1 and r_2 (Fig. 7.1). The position of the planet relative to the Sun, is denoted by $r = r_2 - r_1$. According to Newton's law of gravitation, the planet feels a gravitational pull proportional to the masses m_1 and m_2 and inversely proportional to the square of the distance r. Since the force is directed towards the Sun, it can be expressed as

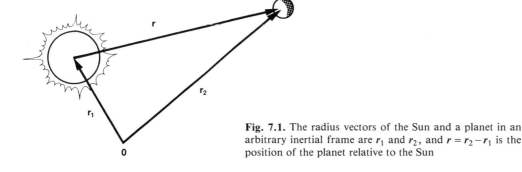

Fig. 7.1. The radius vectors of the Sun and a planet in an arbitrary inertial frame are r_1 and r_2, and $r = r_2 - r_1$ is the position of the planet relative to the Sun

$$F = -Gm_1m_2\frac{\boldsymbol{r}}{r^3},\tag{7.1}$$

where G is the *gravitational constant*. (More about this in Sect. 7.5.)

Newton's second law tells us that the acceleration $\ddot{\boldsymbol{r}}_2$ of the planet is proportional to the applied force:

$$F = m_2\ddot{\boldsymbol{r}}_2.\tag{7.2}$$

Combining (7.1) and (7.2), we get the *equation of motion* of the planet

$$m_2\ddot{\boldsymbol{r}}_2 = -Gm_1m_2\frac{\boldsymbol{r}}{r^3}.\tag{7.3}$$

Since the Sun feels the same gravitational pull, but in the opposite direction, we can immediately write the equation of motion of the Sun:

$$m_1\ddot{\boldsymbol{r}}_1 = Gm_1m_2\frac{\boldsymbol{r}}{r^3}.\tag{7.4}$$

We are mainly interested in the relative motion of the planet with respect to the Sun. To find the equation of the relative orbit, we cancel the masses appearing on both sides of (7.3) and (7.4), and subtract (7.4) from (7.3) to get

$$\ddot{\boldsymbol{r}} = -\mu\frac{\boldsymbol{r}}{r^3},\tag{7.5}$$

where we have denoted

$$\mu = G(m_1+m_2).\tag{7.6}$$

The solution of (7.5) now gives the relative orbit of the planet. The equation involves the radius vector and its second time derivative. In principle, the solution should yield the radius vector as a function of time, $\boldsymbol{r} = \boldsymbol{r}(t)$. Unfortunately, things are not this simple in practice; in fact, there is no way to express the radius vector as a function of time in closed form (i.e. as a finite expression involving familiar elementary functions). Although there are several ways to solve the equation of motion, we must resort to mathematical manipulation in one form or another to figure out the essential properties of the orbit. Next, we shall study one possible method.

7.2 Solution of the Equation of Motion

The equation of motion (7.5) is a second-order (i.e. contains second derivatives) vector valued differential equation. Therefore, we need six integration constants or *integrals*, for the complete solution. The solution is an infinite family of orbits with different sizes, shapes and orientations. A particular solution (e.g. the orbit of Jupiter) is selected by fixing the values of the six integrals. The fate of a planet is unambiguously determined by its position and velocity at any given moment; thus we could take the position and velocity vectors at some moment as our integrals. Although they do not tell us anything about the geometry of the orbit, they can be used as initial values when integrating the orbit numerically with a computer or an abacus. Another set of integrals, the *orbital elements*, contain geometric quantities describing the orbit in a very clear and concrete way. We shall return to these later. A third possible set involves certain physical quantities, which we shall derive next.

We begin by showing that the angular momentum remains constant. The angular momentum of the planet in the heliocentric frame is

$$\boldsymbol{L} = m_2 \boldsymbol{r} \times \dot{\boldsymbol{r}} .$$

Celestial mechanicians usually prefer to use the angular momentum divided by the planet's mass,

$$\boldsymbol{k} = \boldsymbol{r} \times \dot{\boldsymbol{r}} . \tag{7.7}$$

Let us find the time derivative of this:

$$\dot{\boldsymbol{k}} = \boldsymbol{r} \times \ddot{\boldsymbol{r}} + \dot{\boldsymbol{r}} \times \dot{\boldsymbol{r}} .$$

The latter term vanishes as a vector product of two parallel vectors. The former term contains $\ddot{\boldsymbol{r}}$, which is given by the equation of motion:

$$\dot{\boldsymbol{k}} = \boldsymbol{r} \times (-\mu \boldsymbol{r}/r^3) = -(\mu/r^3)\boldsymbol{r} \times \boldsymbol{r} = 0 .$$

Thus $\boldsymbol{k}$ is a constant vector independent of time (as is $\boldsymbol{L}$, of course).

Since the angular momentum vector is always perpendicular to the motion [this follows from (7.7)], the motion is at all times restricted to the invariable plane perpendicular to $\boldsymbol{k}$.

To find another constant vector, we compute the vector product $\boldsymbol{k} \times \ddot{\boldsymbol{r}}$:

$$\boldsymbol{k} \times \ddot{\boldsymbol{r}} = (\boldsymbol{r} \times \dot{\boldsymbol{r}}) \times (-\mu \boldsymbol{r}/r^3) = -\frac{\mu}{r^3}[(\boldsymbol{r} \cdot \boldsymbol{r})\dot{\boldsymbol{r}} - (\boldsymbol{r} \cdot \dot{\boldsymbol{r}})\boldsymbol{r}] .$$

The time derivative of the distance r is equal to the projection of $\dot{\boldsymbol{r}}$ in the direction of $\boldsymbol{r}$ (Fig. 7.2); thus, using the properties of the scalar product, we get $\dot{r} = \boldsymbol{r} \cdot \dot{\boldsymbol{r}}/r$, which gives

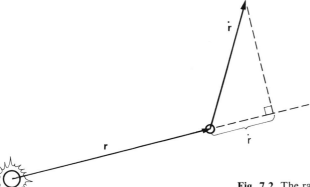

Fig. 7.2. The radial velocity $\dot{r}$ is the projection of the velocity vector $\dot{\boldsymbol{r}}$ in the direction of the radius vector $\boldsymbol{r}$

$$\boldsymbol{r} \cdot \dot{\boldsymbol{r}} = r\dot{r} . \tag{7.8}$$

Hence,

$$\boldsymbol{k} \times \ddot{\boldsymbol{r}} = -\mu \left[\frac{\dot{\boldsymbol{r}}}{r} - \frac{r\dot{r}}{r^2} \right] = \frac{d}{dt}\left(-\mu \frac{\boldsymbol{r}}{r} \right) .$$

The vector product can also be expressed as

$$\boldsymbol{k} \times \ddot{\boldsymbol{r}} = \frac{d}{dt}(\boldsymbol{k} \times \dot{\boldsymbol{r}}) ,$$

since $\boldsymbol{k}$ is a constant vector. Combining this with the previous equation, we have

$$\frac{d}{dt}\left(\boldsymbol{k} \times \dot{\boldsymbol{r}} + \frac{\mu \boldsymbol{r}}{r} \right) = 0 \quad \text{and}$$

$$\boldsymbol{k} \times \dot{\boldsymbol{r}} + \frac{\mu \boldsymbol{r}}{r} = \text{const} = -\mu \boldsymbol{e} . \tag{7.9}$$

Since $\boldsymbol{k}$ is perpendicular to the orbital plane, $\boldsymbol{k} \times \dot{\boldsymbol{r}}$ must lie in that plane. Thus, $\boldsymbol{e}$ is a linear combination of two vectors in the orbital plane; so $\boldsymbol{e}$ itself must be in the orbital plane (Fig. 7.3). Later, we shall see that it points to the direction where the planet is closest to the Sun in its orbit. This point is called the *perihelion*.

One more constant is found by computing $\dot{\boldsymbol{r}} \cdot \ddot{\boldsymbol{r}}$:

$$\dot{\boldsymbol{r}} \cdot \ddot{\boldsymbol{r}} = -\mu \dot{\boldsymbol{r}} \cdot \boldsymbol{r}/r^3 = -\mu r\dot{r}/r^3 = -\mu \dot{r}/r^2 = \frac{d}{dt}\left(\frac{\mu}{r} \right) .$$

Since we also have

$$\dot{\boldsymbol{r}} \cdot \ddot{\boldsymbol{r}} = \frac{d}{dt}\left(\frac{1}{2} \dot{\boldsymbol{r}} \cdot \dot{\boldsymbol{r}} \right) ,$$

we get

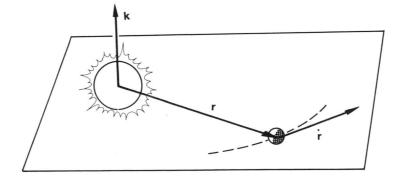

Fig. 7.3. The angular momentum vector $\boldsymbol{k}$ is perpendicular to the radius and velocity vectors of the planet. Since $\boldsymbol{k}$ is a constant vector, the motion of the planet is restricted to the plane perpendicular to $\boldsymbol{k}$

$$\frac{d}{dt}\left(\frac{1}{2}\dot{r}\cdot\dot{r}-\frac{\mu}{r}\right)=0 \quad \text{or}$$

$$\frac{1}{2}v^2-\frac{\mu}{r}=\text{const}=h.\tag{7.10}$$

Here, v is the speed of the planet relative to the Sun. The constant h is called the *energy integral*; the total energy of the planet is $m_2 h$. We must not forget that energy and angular momentum depend on the coordinate frame used. Here we have used a heliocentric frame, which in fact is in accelerated motion.

So far, we have found two constant vectors and one constant scalar. It looks as though we already have seven integrals, i.e. one too many. But not all of these constants are independent; specifically, the following two relations hold:

$$k\cdot e=0,\tag{7.11}$$

$$\mu^2(e^2-1)=2hk^2,\tag{7.12}$$

where e and k are the lengths of e and k. The first equation is obvious from the definition of e and k. To prove (7.12), we square both sides of (7.9) to get

$$\mu^2 e^2=(k\times\dot{r})\cdot(k\times\dot{r})+\mu^2\frac{r\cdot r}{r^2}+2(k\times\dot{r})\cdot\frac{\mu r}{r}.$$

Since k is perpendicular to $\dot{r}$, the length of $k\times\dot{r}$ is $|k|\,\|\dot{r}|$, and $(k\times r)\cdot(k\times r)=k^2 v^2$. Thus, we have

$$\mu^2 e^2=k^2 v^2+\mu^2+\frac{2\mu}{r}[k\times\dot{r}\cdot r].$$

Inside the brackets we have a scalar triple product, where we can exchange the dot and cross to get $k\cdot\dot{r}\times r$. Next, we reverse the order of the two last factors. Because the vector product is anticommutative, we have to change the sign of the product:

$$\mu^2(e^2-1)=k^2 v^2-\frac{2\mu}{r}[k\cdot r\times\dot{r}]=k^2 v^2-\frac{2\mu}{r}k^2=2k^2\left(\frac{1}{2}v^2-\frac{\mu}{r}\right)$$

$$=2k^2 h.$$

This completes the proof of (7.12).

The relations (7.11) and (7.12) reduce the number of independent integrals by two, so we still need one more. The constants we have describe the size, shape and orientation of the orbit completely, but we do not yet know where the planet is! To fix its position in the orbit, we have to determine where the planet is at some time $t=t_0$, or alternatively, at what time it is in some given direction. We use the latter method by specifying the time of perihelion passage, the *time of perihelion*, τ.

7.3 Equation of the Orbit and Kepler's First Law (Fig. 7.4)

In order to find the geometric shape of the orbit, we now derive the equation of the orbit. Since e is a constant vector lying in the orbital plane, we choose it as a reference direction. We denote the angle between the radius vector r and e by f. The angle f is called the *true anomaly*[1]. Using the properties of the scalar product we get

$$r \cdot e = re \cos f.$$

But the product $r \cdot e$ can also be evaluated using the definition of e:

$$r \cdot e = -\frac{1}{\mu}(r \cdot k \times \dot{r} + \mu r \cdot r/r) = -\frac{1}{\mu}(k \cdot \dot{r} \times r + \mu r) = -\frac{1}{\mu}(-k^2 + \mu r)$$

$$= k^2/\mu - r.$$

Thus,

$$r = \frac{k^2/\mu}{1 + e \cos f}. \tag{7.13}$$

This is the general equation of *conic sections* in polar coordinates (see Appendix A.4 for a brief summary of conic sections). The magnitude of e gives the eccentricity of the conic:

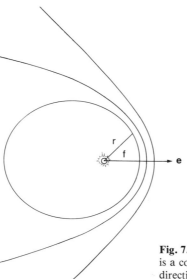

Fig. 7.4. The orbit of an object in the gravitational field of another object is a conic section: ellipse, parabola or hyperbola. Vector e points to the direction of the pericentre, where the orbiting object is closest to central body. If the central body is the Sun, this direction is called the perihelion; if some other star, periastron; if the Earth, perigee, etc. The true anomaly f is measured from the pericentre

[1] There is nothing false or anomalous in this and other anomalies we shall meet later. Angles measured from the perihelion point are called anomalies to distinguish them from longitudes measured from some other reference point (usually the vernal equinox).

$e = 0$	sphere ,	$e = 1$	parabola ,
$0 < e < 1$	ellipse ,	$e > 1$	hyperbola .

Inspecting (7.13), we find that r attains its minimum when $f = 0$, i.e. in the direction of the vector e. Thus, e indeed points to the direction of the perihelion.

Starting with Newton's laws, we have now managed to prove Kepler's first law:

The orbit of a planet is an ellipse, one focus of which is in the Sun.

Without any extra effort, we have also shown that other conic sections, the parabola and hyperbola, are possible orbits.

7.4 Orbital Elements (Fig. 7.5)

We have derived a set of integrals convenient for studying the dynamics of orbital motion. We now turn to another collection of constants more appropriate for describing the geometry of the orbit. The following six quantities are called the *orbital elements*:

- semi-major axis a ;
- eccentricity e ;
- inclination ι ;
- longitude of the ascending node Ω ;
- argument of the perihelion ω ;
- time of the perihelion τ .

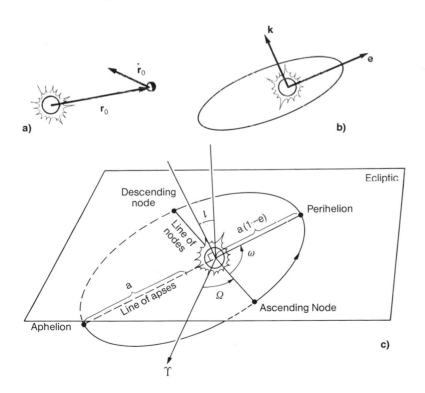

Fig. 7.5a – c. Six integration constants are needed to describe a planet's orbit. These constants can be chosen in various ways. **(a)** If the orbit is to be computed numerically, the simplest choice is the initial value of the radius and velocity vectors. **(b)** Another possibility is to use angular momentum k, direction of the perihelion e (the length of which gives the eccentricity), and the perihelion time τ. **(c)** The third method best describes the geometry of the orbit. The constants are the longitude of the ascending node Ω, the argument of perihelion ω, the inclination ι, the semi-major axis a, the eccentricity e and the time of perihelion τ

The eccentricity is obtained readily as the length of the vector e. From the equation of the orbit (7.13), we see that the *parameter* of the orbit is $p = k^2/\mu$. But the parameter of a conic section is always $a\,|\,1 - e^2\,|$, which gives the semi-major axis, if e and k are known:

$$a = \frac{k^2/\mu}{|\,1 - e^2\,|}\,.$$
(7.14)

By applying (7.12), we get an important relation between the size of the orbit and the energy integral h:

$$a = \begin{cases} -\mu/2h\,, & \text{if the orbit is an ellipse} \\ \ \ \mu/2h\,, & \text{if the orbit is a hyperbola}\,. \end{cases}$$
(7.15)

For a bound system (elliptical orbit), the total energy and the energy integral are negative. For a hyperbolic orbit, h is positive; the kinetic energy is so high that the particle can escape the system (or more correctly, recede without any limit). The parabola, with $h = 0$, is a limiting case between elliptical and hyperbolic orbits. In reality, parabolic orbits do not exist, since hardly any object can have an energy integral exactly zero. However, if the eccentricity is very close to one (as with many comets), the orbit is usually considered parabolic to simplify calculations.

The orientation of the orbit is determined by the directions of the two vectors k (perpendicular to the orbital plane) and e (pointing towards the perihelion). The three angles ι, Ω and ω contain the same information.

The inclination ι gives the obliquity of the orbital plane relative to some fixed reference plane. For bodies in the solar system, the reference plane is usually the ecliptic. For objects moving in the usual fashion, i.e. counterclockwise, the inclination is in the interval $[0°, 90°]$; for retrograde orbits (clockwise motion), the inclination is in the range $[90°, 180°]$. For example, the inclination of Halley's comet is $162°$, which means that the motion is retrograde and the angle between its orbital plane and the ecliptic is $180° - 162° = 18°$.

The longitude of the ascending node, Ω, indicates where the object crosses the ecliptic from south to north. It is measured counterclockwise from the vernal equinox. The orbital elements ι and Ω together determine the orientation of the orbital plane, and they correspond to the direction of k, i.e. the ratios of its components.

The argument of the perihelion, ω, gives the direction of the perihelion, measured from the ascending node in the direction of motion. The same information is contained in the direction of e. Very often, another angle, the *longitude of the perihelion $\tilde{\omega}$* (called pi), is used instead of ω. It is defined as

$$\tilde{\omega} = \Omega + \omega\,.$$
(7.16)

This is a rather peculiar angle, as it is measured partly along the ecliptic, partly along the orbital plane.

We have assumed up to this point that each planet forms a separate two-body system with the Sun. In reality, planets interfere with each other by disturbing each other's orbits. Still, their motions do not deviate very far from the shape of conic sections, and we can use orbital elements to describe the orbits. But the elements are no longer constant; they vary slowly with time. Moreover, their geometric interpretation is no longer quite as obvious as before. But we need not worry about this; the published elements are evaluated in such a way that we can use them to find the positions of the planets exactly as if the elements were constant. The only difference is that we have to use different elements for each moment of time.

Table E.9 (at the end of the book) gives the orbital elements for the nine planets for August 1993. If no higher accuracy is required, these elements can be used for other times, too. Table E.10 gives the elements as polynomials, in which the variable T is the number of Julian centuries elapsed since 1900. Because one Julian century is exactly 36 525 days, this variable for the Julian date JD is simply

$$T = \frac{JD - 2415020}{36525} . \tag{7.17}$$

Even these elements do not include periodic variations, which would make the expressions rather complicated. Instead of the time of perihelion, Table E.10 lists the mean longitude $L = M + \omega + \Omega$, which gives directly the mean anomaly M (which will be defined in Sect. 7.7).

7.5 Kepler's Second and Third Law

The radius vector of a planet in polar coordinates is simply

$$r = r\hat{e}_r , \tag{7.18}$$

where $\hat{e}_r$ is a unit vector parallel with r (Fig. 7.6). If the planet moves with angular velocity $\dot{f}$, the direction of this unit vector also changes at the same rate:

$$\dot{\hat{e}}_r = \dot{f}\hat{e}_f , \tag{7.19}$$

where $\hat{e}_f$ is a unit vector perpendicular to $\hat{e}_r$. The velocity of the planet is found by taking the time derivative of (7.18):

$$\dot{r} = \dot{r}\hat{e}_r + r\dot{\hat{e}}_r = \dot{r}\hat{e}_r + r\dot{f}\hat{e}_f . \tag{7.20}$$

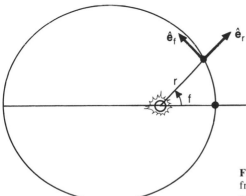

Fig. 7.6. Unit vectors e_r and e_f of the polar coordinate frame. The directions of these change while the planet moves along its orbit

The angular momentum k can now be evaluated using (7.18) and (7.20):

$$k = r \times \dot{r} = r^2 \dot{f} \hat{e}_z \,,$$

where $\hat{e}_z$ is a unit vector perpendicular to the orbital plane. The magnitude of k is

$$k = r^2 \dot{f} \,. \tag{7.21}$$

The surface velocity of a planet means the area swept by the radius vector per unit of time. This is obviously the time derivative of some area, so let us call it $\dot{A}$. In terms of the distance r and true anomaly f, the surface velocity is

$$\dot{A} = \tfrac{1}{2} r^2 \dot{f} \,.$$

By comparing this with the length of k (7.21), we find that

$$\dot{A} = \tfrac{1}{2} k \,. \tag{7.22}$$

Since k is constant, so is the surface velocity. Hence we have Kepler's second law:

The radius vector of a planet sweeps equal areas in equal amounts of time. (See Fig. 7.7.)

Since the Sun – planet distance varies, the orbital velocity must also vary. From Kepler's second law it follows that a planet must move fastest when it is closest to the Sun (near perihelion). Motion is slowest when the planet is farthest from the Sun, at *aphelion*.

We can write (7.22) in the form

$$dA = \tfrac{1}{2} k \, dt \,,$$

and integrate over one complete period:

$$\int\limits_{\text{orbital ellipse}} dA = \frac{1}{2} k \int\limits_0^P dt \,,$$

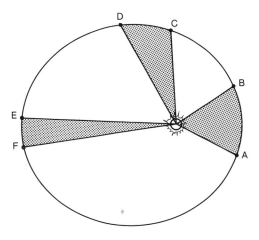

Fig. 7.7. The areas of the shaded sectors of the ellipse are equal. According to Kepler's second law, it takes equal times to travel distances *AB, CD* and *EF*

where P is the orbital period. Since the area of the ellipse is

$$\pi a b = \pi a^2 \sqrt{1 - e^2} \,,$$

where a and b are the semi-major and semi-minor axes and e the eccentricity, we get

$$\pi a^2 \sqrt{1 - e^2} = \tfrac{1}{2} k P \,. \tag{7.23}$$

To find the length of k, we substitute the energy integral h as a function of semi-major axis (7.15) into (7.12) to get

$$k = \sqrt{G(m_1 + m_2) a (1 - e^2)} \,. \tag{7.24}$$

When this is substituted into (7.23) we have

$$P^2 = \frac{4 \pi^2}{G(m_1 + m_2)} a^3 \,. \tag{7.25}$$

This is the exact form of Kepler's third law as derived from Newton's laws. The original version was

The ratio of the cubes of the semi-major axes of the orbits of two planets is equal to the ratio of the squares of their orbital periods.

In this form, the law is not exactly valid, even for planets of the solar system, since their own masses influence their periods. The errors due to ignoring this effect are very small, however.

Kepler's third law becomes remarkably simple if we express distances in astronomical units (AU), times in sidereal years (the abbreviation is unfortunately a, not to be confused with semi-major axis, denoted by a very similar symbol) and masses in solar masses ($m_\odot$). Then $G = 4 \pi^2$ and

$$a^3 = (m_1 + m_2) P^2 \,. \tag{7.26}$$

The masses of objects orbiting around the Sun can safely be ignored (except for the largest planets), and we have the original law $P^2 = a^3$. This is very useful for determining distances of various objects whose periods have been observed. For absolute distances, we have to measure at least one distance in metres to find the length of one AU. Earlier, triangulation was used to measure the parallax of the Sun or a minor planet, such as Eros, that comes very close to Earth. Nowadays, radiotelescopes are used as radar to very accurately measure, for example, the distance to Venus. Since changes in the value of one AU also change all other distances, the International Astronomical Union decided in 1968 to adopt the value

$$1 \, \mathrm{AU} = 1.496{,}000 \times 10^{11} \mathrm{m} \,. \tag{7.27a}$$

The semi-major axis of Earth's orbit is then slightly over one AU. But constants tend to change. And so, after 1984, the astronomical unit has a new value,

$$1 \, \mathrm{AU} = 1.495{,}978{,}70 \times 10^{11} \mathrm{m} \,. \tag{7.27b}$$

Another important application of Kepler's third law is the determination of masses. By observing the period of a natural or artificial satellite, the mass of the central body can be obtained immediately. The same method is used to determine masses of binary stars (more about this subject in Chap. 10).

Although the values of AU and year are accurately known in SI-units, the gravitational constant is known only approximately. Astronomical observations give the product $G(m_1 + m_2)$, but there is no way to distinguish between the contributions of the gravitational constant and those of the masses. The gravitational constant must be measured in the laboratory; this is very difficult because of the weakness of gravitation. Therefore, if a precision higher than $2-3$ significant digits is required, the SI-units cannot be used. Instead we have to use the solar mass as a unit of mass (or, for example, the Earth's mass after $Gm_{\oplus}$ has been determined from observations of satellite orbits).

7.6 Orbit Determination

Celestial mechanics has two very practical tasks: to determine orbital elements from observations and to predict positions of celestial bodies with known elements. Planetary orbits are already known very accurately, but new comets and minor planets are found frequently, requiring orbit determination.

The first practical methods for orbit determination were developed by Johann Karl Friedrich Gauss (1777–1855) at the beginning of the 19th century. By that time, the first minor planets had been discovered, and thanks to Gauss's orbit determinations, they could be found and observed at any time.

At least three observations are needed for computing the orbital elements. The directions are usually measured from plates taken a few nights apart. Using these directions, it is possible to find the corresponding absolute positions (the rectangular components of the radius vector). To be able to do this, we need some additional constraints on the orbit; we must assume that the object moves along a conic section lying in a plane that passes through the Sun. When the three radius vectors are known, we can find the ellipse (or some other conic section) going through these three points. In practice, more observations are used. The elements determined are more accurate if we have more observations and if they cover the orbit more completely.

Although the calculations for orbit determination are not too involved mathematically, they are relatively long and laborious; thus we will omit the details here. Our next task is to find the direction of an object, given its orbital elements.

7.7 Position in the Orbit

Although we already know everything about the geometry of the orbit, we still cannot find the planet at a given time, since we do not know the radius vector r as a function of time. The variable in the equation of the orbit is an angle, the true anomaly f, measured from the perihelion. From Kepler's second law, it follows that f cannot increase at a constant rate with time. Therefore we need some preparations before we can find the radius vector at given instant.

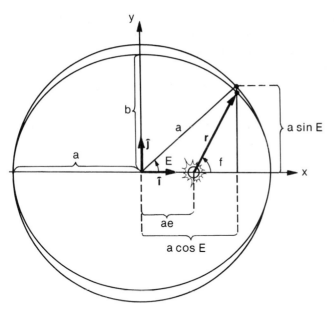

Fig. 7.8. Definition of the eccentric anomaly E. r is the radius vector of the planet

Fig. 7.9. The area of the shaded sector equals b/a times the area $SQ'X$. S = the Sun, Q = the planet, X = the perihelion

The radius vector can be expressed as

$$r = a(\cos E - e)\,\hat{i} + b \sin E \hat{j}\,, \tag{7.28}$$

where $\hat{i}$ and $\hat{j}$ are unit vectors parallel with the major and minor axes, respectively. The angle E is the *eccentric anomaly*; its slightly eccentric definition is shown in Fig. 7.8. Many formulae of elliptical motion become very simple if either time or true anomaly is replaced by the eccentric anomaly. As an example, we take the square of (7.28) to find the distance from the Sun:

$$r = a(1 - e \cos E)\,. \tag{7.29}$$

Our next problem is to find how to calculate E for a given moment of time. According to Kepler's second law, the surface velocity is constant. Thus the area of the shaded sector in Fig. 7.9 is

$$A = \pi a b \frac{t - \tau}{P}\,,$$

where $t - \tau$ is the time elapsed since perihelion, and P is the orbital period. But the area of a part of an ellipse is obtained by reducing the area of the corresponding part of the circumscribed circle by the axial ratio b/a. (As the mathematicians say, an ellipse is an affine transformation of a circle.) Hence the area of SQX is

$$A = \frac{b}{a}\,(\text{area of } SQ'X)$$

$$A = \frac{b}{a} \; (\text{area of the sector } OQ'X - \text{area of the triangle } OQ'S)$$

$$A = \frac{b}{a} \left(\frac{1}{2} a \cdot aE - \frac{1}{2} ae \cdot a \sin E \right) = \frac{1}{2} ab(E - e \sin E) \,.$$

By equating these two expressions for the area A, we get the famous *Kepler's equation*,

$$E - e \sin E = M \,, \qquad \text{where} \tag{7.30}$$

$$M = \frac{2\pi}{P}(t - \tau) \tag{7.31}$$

is the *mean anomaly* of the planet at time t. The mean anomaly increases at a constant rate with time. It indicates where the planet would be if it moved in a circular orbit of radius a. For circular orbits, all three anomalies, f, E, and M are always equal.

If we know the period and the time elapsed after perihelion, we can use (7.31) to find the mean anomaly. Next we must solve for the eccentric anomaly from Kepler's equation (7.30). Finally, the radius vector is given by (7.28). Since the components of r expressed in terms of the true anomaly are $r \cos f$ and $r \sin f$, we find

$$\cos f = \frac{a(\cos E - e)}{r} = \frac{\cos E - e}{1 - e \cos E} \,, \qquad \sin f = \frac{b \sin E}{r} = \sqrt{1 - e^2} \, \frac{\sin E}{1 - e \cos E} \,.$$
$$\tag{7.32}$$

These determine the true anomaly, should it be of interest.

Now we know the position in the orbital plane. This must usually be transformed to some other previously selected reference frame. For example, we may want to know the ecliptic longitude and latitude, which can later be used to find the right ascension and declination. These transformations belong to the realm of spherical astronomy and are briefly discussed in Exercises 7.11.5 – 7.11.7.

7.8 Escape Velocity (Fig. 7.10)

If an object moves fast enough, it can escape from the gravitational field of the central body (to be precise: the field extends to infinity, so the object never really escapes, but is able to recede without any limit). If the escaping object has the minimum velocity allowing escape, it will have lost all its velocity at infinity. There, its kinetic energy is zero, since $v = 0$, and the potential energy is also zero, since the distance r is infinite. At infinite distance, the total energy as well as the energy integral h are zero. The law of conservation of energy gives, then:

$$\frac{1}{2}v^2 - \frac{\mu}{R} = 0 \,,$$

where R is the initial distance, at which the object is moving with velocity v. From this we can solve the *escape velocity*:

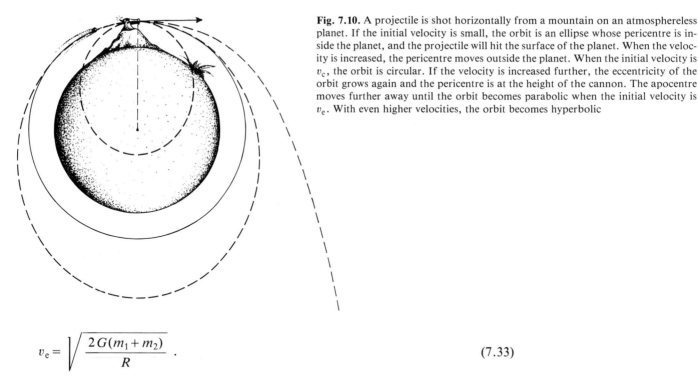

Fig. 7.10. A projectile is shot horizontally from a mountain on an atmosphereless planet. If the initial velocity is small, the orbit is an ellipse whose pericentre is inside the planet, and the projectile will hit the surface of the planet. When the velocity is increased, the pericentre moves outside the planet. When the initial velocity is v_c, the orbit is circular. If the velocity is increased further, the eccentricity of the orbit grows again and the pericentre is at the height of the cannon. The apocentre moves further away until the orbit becomes parabolic when the initial velocity is v_e. With even higher velocities, the orbit becomes hyperbolic

$$v_e = \sqrt{\frac{2G(m_1 + m_2)}{R}} \; . \tag{7.33}$$

For example on the surface of the Earth, v_e is about 11 km/s (if $m_2 \ll m_\oplus$).

The escape velocity can also be expressed using the orbital velocity of a circular orbit. The orbital period P as a function of the radius R of the orbit and the orbital velocity v_c is

$$P = \frac{2\pi R}{v_c} \; .$$

Substitution into Kepler's third law yields

$$\frac{4\pi^2 R^2}{v_c^2} = \frac{4\pi^2 R^3}{G(m_1 + m_2)} \; .$$

From this we can solve the velocity v_c in a circular orbit of radius R:

$$v_c = \sqrt{\frac{G(m_1 + m_2)}{R}} \; . \tag{7.34}$$

Comparing this with the expression of the escape velocity, we see that

$$v_e = \sqrt{2}\, v_c \; . \tag{7.35}$$

Earth's orbit is very nearly circular and the orbital velocity is $v_c \approx 30$ km/s. Thus, to escape from the gravitational pull of the Sun, a spaceship departing from Earth's orbit needs a velocity $v_e \approx 42$ km/s. If the spaceship is initially orbiting the Earth, we also have to add the escape velocity from the Earth.

7.9 Virial Theorem

If a system consists of more than two objects, the equations of motion cannot in general be solved analytically (Fig. 7.11). Given some initial values, the orbits can, of course, be found by numerical integration, but this does not tell us anything about the general properties of all possible orbits. The only integration constants available for an arbitrary system are the total momentum, angular momentum and energy. In addition to these, it is possible to derive certain statistical results, like the virial theorem. It concerns time averages only, but does not say anything about the actual state of the system at some specified moment.

Suppose we have a system of n point masses m_i with radius vectors r_i and velocities $\dot{r}_i$. We define a quantity A (the "virial" of the system) as follows:

$$A = \sum_{i=1}^{n} m_i \dot{r}_i \cdot r_i .$$

The time derivative of this is

$$\dot{A} = \sum_{i=1}^{n} (m_i \dot{r} \cdot \dot{r}_i + m_i \ddot{r}_i \cdot r_i) .$$

The first term equals twice the kinetic energy of the ith particle, and the second term contains a factor $m_i \ddot{r}_i$ which, according to Newton's laws, equals the force applied to the ith particle. Thus we have

$$\dot{A} = 2T + \sum_{i=1}^{n} F_i \cdot r_i ,$$

where T is the total kinetic energy of the system. If $\langle x \rangle$ denotes the time average of x in the time interval $[0, \tau]$, we have

$$\langle \dot{A} \rangle = \frac{1}{\tau} \int_0^{\tau} \dot{A} \, dt = \langle 2T \rangle + \left\langle \sum_{i=1}^{n} F_i \cdot r_i \right\rangle .$$

If the system remains bounded, i.e. none of the particles escapes, all r_i's as well as all velocities will remain bounded. In such a case, A does not grow without limit, and the integral of the previous equation remains finite. When the time interval becomes longer ($\tau \to \infty$), $\langle \dot{A} \rangle$ approaches zero, and we get

$$\langle 2T \rangle + \left\langle \sum_{i=1}^{n} F_i \cdot r_i \right\rangle = 0 . \tag{7.36}$$

This is the general form of the virial theorem. If the forces are due to mutual gravitation only, they have the expressions

$$F_i = -G m_i \sum_{j=1, j \neq i}^{n} m_j \frac{r_i - r_j}{r_{ij}^3} , \quad \text{where} \quad r_{ij} = |r_i - r_j| .$$

Fig. 7.11. When a system consists of more than two bodies, the equations of motion cannot be solved analytically. In the solar system, the mutual disturbances of the planets are usually small and can be taken into account as small perturbations in the orbital elements. K. F. Sundman (1873–1949) designed a machine to carry out the tedious integration of the perturbation equations. This machine, called the perturbograph, is one of the earliest analogue computers; unfortunately, it was never built. Shown is a design for one component that evaluates a certain integral occurring in the equations. Sundman found a semi-analytical solution in the case of three bodies. The solution is a very slowly converging series and cannot be used for any practical calculations. From a purely mathematical point of view, however, the achievement is remarkable. (The picture appeared in K. F. Sundman's paper in *Festskrift tillegnad Anders Donner* in 1915.)

The latter term in the virial theorem is now

$$\left\langle \sum_{i=1}^{n} \boldsymbol{F}_i \cdot \boldsymbol{r}_i \right\rangle = -G \sum_{i=1}^{n} \sum_{j=1, j \neq i}^{n} m_i m_j \frac{\boldsymbol{r}_i - \boldsymbol{r}_j}{r_{ij}^3} \cdot \boldsymbol{r}_i$$

$$= -G \sum_{i=1}^{n} \sum_{j=i+1}^{n} m_i m_j \frac{\boldsymbol{r}_i - \boldsymbol{r}_j}{r_{ij}^3} \cdot (\boldsymbol{r}_i - \boldsymbol{r}_j) ,$$

where the latter form is obtained by rearranging the double sum, combining the terms

$$m_i m_j \frac{r_i - r_j}{r_{ij}^3} \cdot r_i \quad \text{and} \quad m_j m_i \frac{r_j - r_i}{r_{ji}^3} \cdot r_j = m_i m_j \frac{r_i - r_j}{r_{ij}^3} \cdot (-r_j) \, .$$

Since $(r_i - r_j) \cdot (r_i - r_j) = r_{ij}^2$ the sum reduces to

$$-G \sum_{i=1}^{n} \sum_{j=i+1}^{n} \frac{m_i m_j}{r_{ij}} = U \, ,$$

where U is the potential energy of the system. Thus, the virial theorem becomes simply

$$\langle T \rangle = -\tfrac{1}{2} \langle U \rangle \, . \tag{7.37}$$

7.10 The Jeans Limit

We shall later study the birth of stars and galaxies. The initial stage is, roughly speaking, a gas cloud that begins to collapse due to its own gravitation. If the mass of the cloud is high enough, its potential energy exceeds the kinetic energy and the cloud collapses. From the virial theorem we can deduce that the potential energy must be at least twice the kinetic energy. This provides a criterion for the critical mass necessary for the cloud of collapse. This criterion was first suggested by Sir James Jeans in 1902.

The critical mass will obviously depend on pressure P and density ϱ. Since gravitation is the compressing force, the gravitational constant G will probably also enter our expression. Thus the critical mass is of the form

$$M = C P^a G^b \varrho^c \, ,$$

where C is a dimensionless constant, and the constants a, b and c are determined so that the right-hand side has the dimension of mass. The dimension of pressure is $\mathrm{kg\,m^{-1}s^{-2}}$, of gravitational constant $\mathrm{kg^{-1}m^3 s^{-2}}$ and of density $\mathrm{kg\,m^{-3}}$. Thus the dimension of the right-hand side is

$$\mathrm{kg}^{(a-b+c)} \, \mathrm{m}^{(-a+3b-3c)} \, \mathrm{s}^{(-2a-2b)} \, .$$

Since this must be kilograms ultimately, we get the following set of equations:

$$a - b + c = 1, \quad -a + 3b - 3c = 0, \quad -2a - 2b = 0 \, .$$

The solution of this is $a = 3/2$, $b = -3/2$ and $c = -2$. Hence the critical mass is

$$M_{\mathrm{J}} = C \frac{P^{3/2}}{G^{3/2} \varrho^2} \, . \tag{7.38}$$

This is called the *Jeans mass*. In order to determine the constant C, we naturally must calculate both kinetic and potential energy. Another method based on the propagation of waves determines the diameter of the cloud, the *Jeans length* λ_{J}, by requiring that a disturbance of size λ_{J} grow unbounded. The value of the constant C depends on the exact form of the perturbation, but its typical values are in the range $[1/\pi, 2\pi]$. We can take $C = 1$ as well, in which case (7.38) gives a correct order of magnitude for the critical mass. If the mass of a cloud is much higher than M_J, it will collapse by its own gravitation.

In (7.38) the pressure can be replaced by the kinetic temperature T_k of the gas (see Chap. 6 for a definition). According to the kinetic gas theory, the pressure is

$$P = nkT_k ,$$

where n is the number density (particles per unit volume) and k is the Boltzmann constant. The number density is obtained by dividing the density of the gas ϱ by the average molecular weight μ:

$$n = \varrho/\mu , \quad \text{whence}$$

$$P = \varrho k T_k/\mu .$$

By substituting this into (7.38) we get

$$M_J = C\left(\frac{kT_k}{\mu G}\right)^{3/2} \frac{1}{\sqrt{\varrho}} . \tag{7.39}$$

*Newton's Laws

1) In the absence of external forces, a particle will remain at rest or move along a straight line with constant speed.

2) The rate of change of the momentum of a particle is equal to the applied force F:

$$\dot{p} = \frac{d}{dt}(mv) = F \text{ (see footnote } ^2) .$$

3) If particle A exerts a force F on another particle B, B will exert an equal but opposite force $-F$ on A.

If several forces $F_1, F_2,\ldots$ are applied on a particle, the effect is equal to that caused by one force F, which is the vector sum of the individual forces ($F = F_1 + F_2 + \ldots$).

Law of gravitation: If the masses of particles A and B are m_A and m_B and their mutual distance r, the force exerted on A by B is directed towards B and has the magnitude $Gm_A m_B/r^2$, where G is a constant depending on the units chosen.

---

[2] Newton denoted the derivative of a function f by $\dot{f}$ and the integral function by f'. The corresponding notation used by Leibniz was df/dt and $\int f\,dx$. Of Newton's notations, only the dot ($\dot{\ }$) is still used, always signifying the time derivative: $\dot{f} \equiv df/dt$. For example, the velocity $\dot{r}$ is the time derivative of r, the acceleration $\ddot{r}$, its second derivative, etc.

7.11 Examples

7.11.1   Find the orbital elements of Jupiter on September 1, 1981.

The Julian date is 2,444,849, hence from (7.17), $T = 0.8167$. By substituting this into the expressions of Table E.10, we get

$$L = 238°\,2'\,57.32'' + 10{,}930{,}687.148''\,T + 1.20486''\,T^2 - 0.0059367''\,T^3$$

$$= 197.7166°$$

$$\tilde{\omega} = 14.0340° , \qquad \Omega = 100.2689° , \qquad e = 0.0485 ,$$

$$\iota = 1.3041° , \qquad a = 5.202561\,\text{AU} .$$

From these we can compute the argument of perihelion and mean anomaly:

$$\omega = \tilde{\omega} - \Omega = 273.7675° , \qquad M = L - \tilde{\omega} = 183.6802° .$$

7.11.2   *Orbital Velocity*

Comet Austin (1982 g) moves in a parabolic orbit. Find its velocity on October 8, 1982, when the distance from the Sun was 1.10 AU.

The energy integral for a parabola is $h = 0$. Thus (7.10) gives the velocity v:

$$v = \sqrt{\frac{2\mu}{r}} = \sqrt{\frac{2Gm_\odot}{r}}$$

$$\Rightarrow v = \sqrt{\frac{2 \times 4\pi^2 \times 1}{1.10}} = 8.47722\,\text{AU/a}$$

$$= \frac{8.47722 \times 1.496 \times 10^{11}\,\text{m}}{365.2564 \times 24 \times 3600\,\text{s}} = 40{,}162\,\text{m\,s}^{-1} \approx 40\,\text{km\,s}^{-1} .$$

The semi-major axis of the minor planet 1982 RA is 1.568 AU and the distance from the Sun on October 8, 1982, was 1.17 AU. Find its velocity.

The energy integral (7.15) is now

$$h = -\mu/2a .$$

Hence

$$h = \frac{1}{2}v^2 - \frac{\mu}{r} = -\frac{\mu}{2a}$$

$$\Rightarrow v = \sqrt{\mu\left(\frac{2}{r} - \frac{1}{a}\right)}$$

$$= \sqrt{4\pi^2\left(\frac{2}{1.17} - \frac{1}{1.568}\right)} = 6.5044\,\text{AU/a} = 30{,}834\,\text{m\,s}^{-1} \approx 31\,\text{km\,s}^{-1} .$$

7.11.3   In an otherwise empty universe, two rocks of 5 kg each orbit each other at a distance of 1 m. What is the orbital period?

The period is obtained from Kepler's third law:

$$P^2 = \frac{4\pi^2 a^3}{G(m_1 + m_2)} = \frac{4\pi^2 \times 1}{6.67 \times 10^{-11}(5+5)} s^2 = 5.9 \times 10^{10} s^2$$

$$\Rightarrow P \approx 243{,}000\,s \approx 2.8\,d \;.$$

7.11.4   The period of the Martian moon Phobos is 0.3189 d and the radius of the orbit, 9370 km. What is the mass of Mars?

First we change to more appropriate units:

$$P = 0.3189\,d = 0.0008731 \text{ sidereal years}$$

$$a = 9370\,km = 6.2634 \times 10^{-5}\,AU \;.$$

Equation (7.26) gives (it is safe to assume that $m_{\text{Phobos}} \ll m_{\text{Mars}}$):

$$m_{\text{Mars}} = a^3/P^2 - 0.000000322\,m_\odot\ (\approx 0.107\,m_\oplus) \;.$$

7.11.5   Derive formulae for a planet's heliocentric longitude and latitude, given its orbital elements and true anomaly.

We apply the sine formula to the spherical triangle of the figure:

$$\frac{\sin\beta}{\sin i} = \frac{\sin(\omega+f)}{\sin(\pi/2)} \quad \text{or} \quad \sin\beta = \sin i \sin(\omega+f) \;.$$

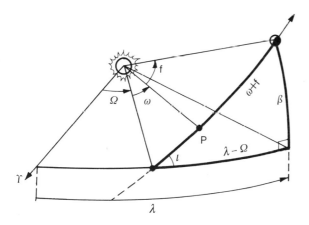

The cosine formula gives

$$\cos(\pi/2) = -\cos i \sin(\omega+f)\cos(\lambda-\Omega) + \cos(\omega+f)\sin(\lambda-\Omega) \;,$$

whence

$$\tan(\lambda-\Omega) = \cos i \tan(\omega+f) \;.$$

7.11.6   Find the radius vector and heliocentric longitude and latitude of Jupiter on September 1, 1981.

The orbital elements were computed in Exercise 7.11.1:

$$a = 5.202561 , \qquad \Omega = 100.2689° ,$$

$$e = 0.0485 , \qquad \omega = 273.7675° ,$$

$$\iota = 1.3041° , \qquad M = 183.6802° .$$

Since the mean anomaly was obtained directly, we need not compute the time elapsed since perihelion.

Now we have to solve Kepler's equation. It cannot be solved analytically, and we are obliged to take the brute force approach (also called numerical analysis) in the form of iteration. For iteration, we write the equation as

$$E_{n+1} = M + e \sin E_n,$$

where E_n is the value found in the nth iteration. The mean anomaly is a reasonable initial guess, E_0. (N.B.: Here, all angles must be in radians; otherwise, nonsense results!) The iteration proceeds as follows:

$$E_0 = M = 3.205824 , \qquad E_3 = M + e \sin E_2 = 3.202854 ,$$

$$E_1 = M + e \sin E_0 = 3.202711 , \qquad E_4 = M + e \sin E_3 = 3.202855 ,$$

$$E_2 = M + e \sin E_1 = 3.202862 , \qquad E_5 = M + e \sin E_4 = 3.202855 ,$$

after which successive approximations no longer change, which means that the solution, accurate to six decimal places, is

$$E = 3.202855 = 183.5101° .$$

The radius vector is

$$r = a(\cos E - e) \hat{\imath} + a\sqrt{1 - e^2} \sin E \hat{\jmath} = -5.4451 \hat{\imath} - 0.3181 \hat{\jmath}$$

and the distance from the Sun,

$$r = a(1 - e \cos E) = 5.4544 \, \text{AU} .$$

Because both components of r are negative, the planet is in the third quadrant. The true anomaly is

$$f = \arctan \frac{-0.3181}{-5.4451} = 183.3438° .$$

Applying the results of the previous example, we find the latitude and longitude:

$$\sin \beta = \sin \iota \sin(\omega + f) = \sin 1.3041° \sin(273.7675° + 183.3438°) = 0.02258$$

$$\Rightarrow \beta = 1.2941° \approx 1.3° .$$

$$\tan(\lambda - \Omega) = \cos i \tan(\omega + f) = \cos 1.3041° \tan(273.7675° + 183.3438°)$$
$$= -8.0135$$
$$\Rightarrow \lambda = \Omega + 97.1131° = 100.2689° + 97.1131° = 197.3820° \approx 197.4° \;.$$

(We must be careful here; the equation for $\tan(\lambda - \Omega)$ allows two solutions. If necessary, a figure can be drawn to decide which is the correct one.)

7.11.7 Find Jupiter's right ascension and declination on September 1, 1981.

In Exercise 7.11.6, we found the longitude and latitude, $\lambda = 197.4°$, $\beta = 1.3°$. The corresponding rectangular (heliocentric) coordinates are:

$$x = r \cos \lambda \cos \beta = -5.2040 \,\text{AU} \;,$$
$$y = r \sin \lambda \cos \beta = -1.6290 \,\text{AU} \;,$$
$$z = r \sin \beta = 0.1232 \,\text{AU} \;.$$

To find the direction relative to Earth, we have to find where the Earth is. In principle, we could repeat the previous procedure with Earth's orbital elements. Or, if we are lazy, we could pick up the nearest Astronomical Almanac, which lists the equatorial coordinates of the Earth:

$$X_{\oplus} = 0.9442 \,\text{AU} \;,$$
$$Y_{\oplus} = -0.3422 \,\text{AU} \;,$$
$$Z_{\oplus} = -0.1487 \,\text{AU} \;.$$

Jupiter's ecliptic coordinates must be transformed to equatorial ones by rotating them around the x-axis by an angle ε [the obliquity of the ecliptic, (8.2)]:

$$X_J = x = -5.2040 \,\text{AU} \;,$$
$$Y_J = y \cos \varepsilon - z \sin \varepsilon = -1.5436 \,\text{AU} \;,$$
$$Z_J = y \sin \varepsilon + z \cos \varepsilon = -0.5350 \,\text{AU} \;.$$

Then the position relative to the Earth is

$$X_0 = X_J - X_{\oplus} = -6.1482 \,\text{AU} \;,$$
$$Y_0 = Y_J - Y_{\oplus} = -1.2014 \,\text{AU} \;,$$
$$Z_0 = Z_J - Z_{\oplus} = -0.3863 \,\text{AU} \;.$$

And finally, the right ascension and declination are

$$\alpha = \arctan(Y_0/X_0) = 191.0567° = 12 \,\text{h} \, 44 \,\text{min} \, 14 \,\text{s} \;,$$
$$\delta = \arctan \frac{Z_0}{\sqrt{X_0^2 + Y_0^2}} = -3.5286° \approx -3° \, 32' \;.$$

Comparison with the Astronomical Almanac shows that our declination is correct within 2 minutes of arc and right ascension within 7 minutes of arc. This is good enough, considering that we have neglected all short-period perturbations in Jupiter's orbital elements.

7.11.8 An interstellar hydrogen cloud contains 10 atoms per cm^3. How big must the cloud be to collapse due to its own gravitation? The temperature of the cloud is 100K.

The mass of one hydrogen atom is 1.67×10^{-27}kg, which gives a density

$$\varrho = n\mu = 10^7 \text{m}^{-3} \times 1.67 \times 10^{-27} \text{kg} = 1.67 \times 10^{-20} \text{kg/m}^3 .$$

The critical mass is

$$M_{\rm J} = \left(\frac{1.38 \times 10^{-23} \text{JK}^{-1} \times 100\text{K}}{1.67 \times 10^{-27} \text{kg} \times 6.67 \times 10^{-11} \text{N m}^2 \text{kg}^{-2}} \right)^{3/2} \frac{1}{\sqrt{1.67 \times 10^{-20} \text{kg/m}^3}}$$

$$\approx 10^{34} \text{kg} \approx 5000 \, m_{\odot} .$$

The radius of the cloud is

$$R = \sqrt[3]{\frac{3}{4\pi} \frac{M}{\varrho}} \approx 5 \times 10^{17} \text{m} \approx 20 \, \text{pc} .$$

7.12 Exercises

7.12.1 Find the ratio of the orbital velocities at aphelion and perihelion, v_a/v_p. What is this ratio for the Earth?

7.12.2 The perihelion and aphelion of the orbit of Eros are 1.1084 and 1.8078 astronomical units from the Sun. What is the velocity of Eros when its distance from the Sun is equal to the mean distance of Mars?

7.12.3 Find the radius of the orbit of a geostationary satellite; such a satellite remains always in the same direction as seen from the Earth. Are there areas on the surface of the Earth that cannot be seen from any geostationary satellite? If so, what fraction of the total surface area?

7.12.4 From the angular diameter of the Sun and the length of the year, derive the mean density of the Sun.

7.12.5 Find the mean, eccentric and true anomalies of the Earth one quarter of a year after the perihelion.

7.12.6 The velocity of a comet is $5\,\mathrm{m\,s^{-1}}$, when it is very far from the Sun. If it moved along a straight line, it would pass the Sun at a distance of 1 AU. Find the eccentricity, semimajor axis and perihelion distance of the orbit. What will happen to the comet?

Chapter 8   The Solar System

The solar system consists of a central star, called *the Sun*, nine *planets*, dozens of *moons*, thousands of *minor planets*, and myriads of *comets* and *meteoroids*. The planets in order from the Sun are: *Mercury, Venus, Earth, Mars, Jupiter, Saturn, Uranus, Neptune,* and *Pluto*.

8.1 An Overview

Gravity controls motions of the solar system bodies. The planetary orbits around the Sun (Fig. 8.1) are almost coplanar ellipses which deviate only slightly from circles. The innermost and the outermost planets, Mercury and Pluto, have the largest deviations. The *asteroids*, or minor planets, circle the Sun mainly between the orbits of Mars and Jupiter, and their orbital planes are often more tilted than those of the planetary orbits. All minor planets revolve in the same direction as the major planets do; comets, however, may move in the opposite direction. Cometary orbits are very elongated, even hyperbolic. Most of the moons circle their parent planets in the same direction as the planet itself around the Sun.

    The planets can be divided into two physically different groups. Mercury, Venus, the Earth, and Mars care called *terrestrial* (Earth-like) planets; they have a solid

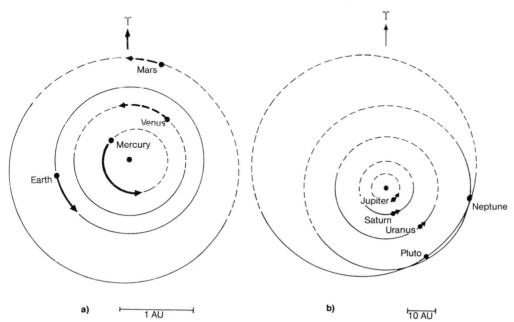

Fig. 8.1. (a) Planetary orbits from Mercury to Mars. The dashed line represents the part of the orbit below the ecliptic; the arrows show the distances travelled by the planets during one month. **(b)** Planets from Jupiter to Pluto. The arrows indicate the distances travelled by the planets during one year

surface, are of almost equal size (diameters from 5000 to 12000 km), and have quite a high mean density (4000 – 5000 kg m^{-3}; the density of water is 1000 kg m^{-3}). The planets from Jupiter to Neptune are called *Jovian* (Jupiter-like) or *giant planets*. The densities of the giant planets are about 1000 – 2000 kg m^{-3}, and most of their volume is liquid. Pluto is a unique object falling outside this classification.

Since the 1960's, the planets have been studied with space probes. The beginning of this era meant a revolution in planetary science, which had almost disappeared during the decades of extensive growth of other branches of astronomy. Most of the data we have today were collected by probes. Pluto is the only planet which is not studied by probes, and there are no firm plans for Pluto mission.

8.2 Planetary Configurations

The apparent motions of the planets are quite complicated, partly because they reflect the Earth's motion around the Sun (Figs. 8.2, 8.3). Normally, the planets move eastward (*direct* motion, counterclockwise as seen from the Northern hemisphere) when compared with the stars. Sometimes, the motion reverses to the opposite direction, or *retrograde*. After a few weeks of retrograde motion, the direction is changed again, and the planet continues in the original direction. It is quite understandable that the ancient astronomers had great difficulties in explaining and modelling such complicated turns and loops.

Figure 8.4 explains some basic planetary configurations. A *superior planet* (planet outside the orbit of the Earth) is said to be in *opposition* when it is exactly opposite to the Sun, i.e. when the Earth is between the planet and the Sun. When the planet is behind the Sun, it is in *conjunction*. In practice, the planet may not be exactly opposite or behind the Sun because the orbits of the planet and the Earth are not in the same plane. In astronomical almanacs, oppositions and conjunctions are defined in terms of longitudes. The longitudes of a body and the Sun differ by 180° at the moment of

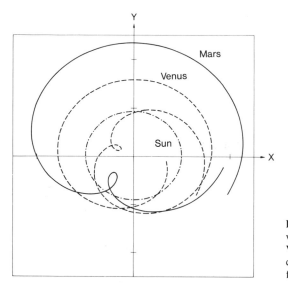

Fig. 8.2. Planetary motions are complex when viewed from the Earth. The orbits of the Sun, Venus and Mars are shown here in the coordinate system fixed to the Earth and viewed from the north celestial pole

opposition; in conjunction, the longitudes are equal. However, the right ascension is used if the other body is not the Sun. Those points at which the apparent motion of a planet turns towards the opposite direction are called *stationary points*. Opposition occurs in the middle of the retrograde loop.

Inferior planets (Mercury and Venus) are never in opposition. The configuration occurring when either of these planets is between the Earth and the Sun is called *inferior conjunction*. The conjunction corresponding to that of a superior planet is called *superior conjunction*. The maximum (eastern or western) *elongation*, i.e. the angular distance of the planet from the Sun is 28° for Mercury and 47° for Venus.

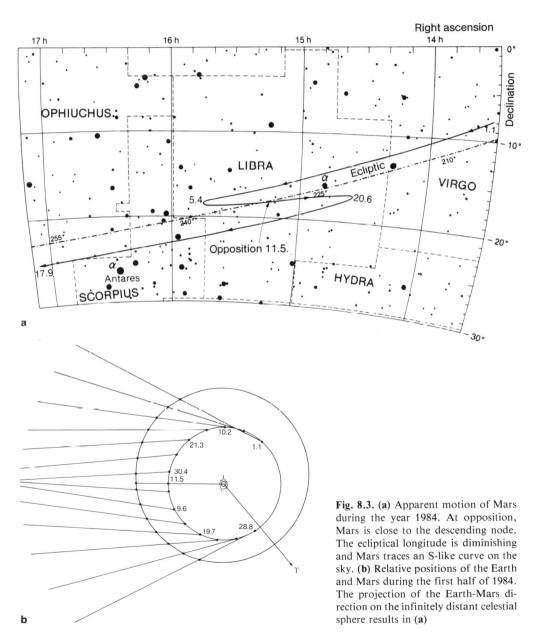

Fig. 8.3. (a) Apparent motion of Mars during the year 1984. At opposition, Mars is close to the descending node. The ecliptical longitude is diminishing and Mars traces an S-like curve on the sky. **(b)** Relative positions of the Earth and Mars during the first half of 1984. The projection of the Earth-Mars direction on the infinitely distant celestial sphere results in **(a)**

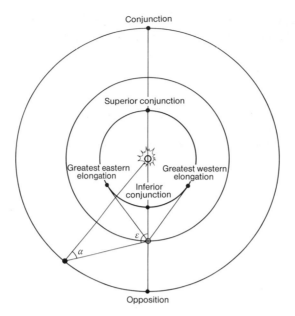

Fig. 8.4. Planetary configurations. The angle α (Sun-object-Earth) is the phase angle and ε (Sun-Earth-object) is the elongation

Elongations are called eastern or western, depending on which side of the Sun the planet is seen on. The planet is an "evening star" and sets after the Sun when it is in eastern elongation; in western elongation, the planet is seen in the morning sky as a "morning star".

The *synodic period* is the time interval between two successive events (e.g. oppositions). The period which we used in the previous chapters is the *sidereal period*, the true time of revolution around the Sun, unique for each object. The synodic period depends on the difference of the sidereal periods of two bodies.

Let the sidereal periods of two planets be P_1 and P_2 (assume that $P_1 < P_2$). Their mean angular velocities (mean motions) are $2\pi/P_1$ and $2\pi/P_2$. After one synodic period $P_{1,2}$, the inner planet has made one full revolution more than the outer planet:

$$P_{1,2}\,\frac{2\pi}{P_1} = 2\pi + P_{1,2}\,\frac{2\pi}{P_2}, \quad \text{or}$$

$$1/P_{1,2} = 1/P_1 - 1/P_2. \tag{8.1}$$

The angle Sun-planet-Earth is called the *phase angle*, often denoted by the Greek letter α. The phase angle is between $0°$ and $180°$ in the case of Mercury and Venus. This means that we can see "full Venus", "half Venus", and so on, exactly as in the phases of the Moon. The phase angle range for the superior planets is more limited. For Mars, the maximum phase is $41°$, for Jupiter $11°$, and for Pluto only $2°$.

8.3 Orbit of the Earth

The *sidereal year* is the real orbital period of the Earth around the Sun. After 1 sidereal year, the Sun is seen at the same position relative to the stars. The length of the sidereal year is 365.2564 mean solar days (in fact, ephemeris days, but the difference is very small, and we may neglect it here). We noted earlier that, owing to precession, the direction of the vernal equinox moves along the ecliptic at about 50″ per year. This means that the Sun returns to the vernal equinox before one complete sidereal year has elapsed. This time interval, called the *tropical year*, is 365.2422 days. A third definition of the year is based on the perihelion passages of the Earth. Planetary perturbations cause a gradual change in the direction of the Earth's perihelion. The time interval between two perihelion passages is called the *anomalistic year*, the length of which is 365.2596 days, a little longer than the sidereal year.

The equator of the Earth is tilted about 23.5° with respect to the ecliptic. Owing to perturbations, this angle changes with time. If periodic terms are neglected, the *obliquity of the ecliptic* can be expressed as:

$$\varepsilon = 23.452294° - 0.00013° \, t \, , \tag{8.2}$$

where t is the time (in years) elapsed since the beginning of the year 1900. The expression is valid for a few centuries before and after the year 1900.

The declination of the Sun varies between $-\varepsilon$ and $+\varepsilon$ during the year. At any given time, the Sun is seen at zenith from one point on the surface of the Earth. The latitude of this point is the same as the declination of the Sun. At the latitudes $-\varepsilon$ (the *Tropic of Capricorn*) and $+\varepsilon$ (the *Tropic of Cancer*), the Sun is seen at zenith once every year, and between these latitudes, twice a year. In the Northern hemisphere, the Sun will not set if the latitude is greater than $90-\delta$, where δ is the declination of the Sun.

The southernmost latitude where the midnight Sun can be seen is thus $90-\varepsilon$ = 66.55°. This is called the *Arctic Circle*. (The same holds true in the Southern hemisphere.) The Arctic Circle is the southernmost place where the Sun is (in theory) below the horizon during the whole day at the winter solstice. The sunless time lasts longer and longer when one goes north (south in the Southern hemisphere). At the poles, day and night last half a year each. In practice, refraction and location of the observing site will have a large influence on the visibility of the midnight Sun and the number of sunless days. Because refraction raises objects seen at the horizon, the midnight Sun can be seen a little further south than at the Arctic Circle. For the same reason, the Sun can be seen simultaneously at both poles around the time of vernal and autumnal equinox.

The eccentricity of the Earth's orbit is about 0.0167, so the distance from the Sun varies between 147 and 152 million km. The flux density of solar radiation varies somewhat at different parts of the Earth's orbit, but this has practically no effect on the seasons. In fact, the Earth is at perihelion in the beginning of January, in the middle of the northern hemsiphere's winter. The seasons are due to the obliquity of the ecliptic.

The flux density of solar radiation depends on three factors. First, the flux per unit area is proportional to sin a, where a is the elevation of the Sun. During the summer, the elevation can have greater values than in the winter, giving more energy per unit area. Second, during the summer, the Sun stays longer above the horizon. A third effect is due to the atmosphere: When the Sun is near the horizon, the radiation must penetrate thick atmospheric layers. This means large extinction and less radiation at the surface. These effects are discussed in detail in Exercise 8.23.2.

8.4 Orbit of the Moon

The Earth's satellite, *the Moon*, circles the Earth counterclockwise. One revolution, the *sidereal month*, takes about 27.322 days. In practice, a more important period is the *synodic month*, the duration of the Lunar phases (e.g., from full moon to full moon). In the course of one sidereal month, the Earth has travelled almost 1/12 of its orbit around the Sun. The Moon still has about 1/12 of its orbit to go before the Earth-Moon-Sun configuration is again the same. This takes about 2 days, so the phases of the Moon are repeated every 29 days. More exactly, the length of the synodic month is 29.531 days.

The *new moon* is that instant when the Moon is in conjunction with the Sun. Almanacs define the phases of the Moon in terms of ecliptical longitudes; the longitudes of the new moon and the Sun are equal. Usually the new moon is slightly north or south of the Sun because the lunar orbit is tilted 5° with respect to the ecliptic.

About 2 days after the new moon, the waxing crescent moon can be seen in the western evening sky. About 1 week after the new moon, the *first quarter* follows, when the longitudes of the Moon and the Sun differ by 90°. The right half of the Moon is seen lit (note: left half when seen from the Southern hemisphere!). The *full moon* appears a fortnight after the new moon, and 1 week after this, the *last quarter*. Finally, the waning crescent moon disappears in the glory of the morning sky.

The orbit of the Moon is approximately elliptic. The length of the semi-major axis is 384 400 km and the eccentricity, 0.055. Owing to perturbations caused mainly by the Sun, the orbital elements vary with time. The minimum distance of the Moon from the centre of the Earth is 356 400 km, and the maximum distance, 406 700 km. This range is larger than the one calculated from the semi-major axis and the eccentricity. The apparent angular diameter is in the range 29.4′ − 33.5′.

The rotation time of the Moon is equal to the sidereal month, so the same side of the Moon always faces the Earth. Such *synchronous rotation* is common among the satellites of the solar system: almost all large moons rotate synchronously.

The orbital speed of the Moon varies according to Kepler's second law; The rotation period, however, remains constant. This means that, at different phases of the lunar orbit, we can see slightly different parts of the surface. When the Moon is close to its perigee, its speed is greater than average (and thus greater than the mean rotation rate), and we can see more of the right-hand edge of the Moon's limb (as seen from the Northern hemisphere). Correspondingly, at the apogee we see "behind" the left-hand edge. Owing to these *librations*, a total of 59% of the surface area can be seen from the Earth (Fig. 8.5). The libration is quite easy to see if one follows some detail at the edge of the lunar limb.

The orbital plane of the Moon is tilted only about 5° to the ecliptic. However, the orbital plane changes gradually with time, owing mainly to the perturbations caused by the Earth and the Sun. These perturbations cause the nodal line (the intersection of the plane of the ecliptic and the orbital plane of the Moon) to make one full revolution in 18.6 years. We have already encountered the same period in the nutation. When the ascending node of the lunar orbit is close to the vernal equinox, the Moon can be 23.5° + 5° = 28.5° north or south of the equator. When the descending node is close to the vernal equinox, the zone where the Moon can be found is only 23.5° − 5° = 18.5° wide.

The *nodical* or *draconitic month* is the time in which the Moon moves from one ascending node back to the next one. Because the line of nodes is rotating, the nodical

Fig. 8.5. Librations of the Moon can be seen in this pair of photographs taken when the Moon was close to the perigee and the apogee, respectively. (Helsinki University Observatory)

month is 3 hours shorter than the sidereal month, i.e. 27.212 days. The orbital ellipse itself also precesses slowly. The orbital period from perigee to perigee, the *anomalistic month*, is 5.5 h longer than the sidereal month, or about 27.555 days.

Gravitational differences caused by the Moon on different parts of the Earth's surface give rise to the *tides* (Fig. 8.6). Gravity is greatest at the sublunar point and smallest at the opposite side of the Earth. At these points, the surface of the seas is highest (high tide, *flood*). About 6 h after flood, the surface is lowest (low tide, *ebb*). The Sun generates a similar effect, but its magnitude is less than half that of the lunar tide. When the Sun and the Moon are in the same direction with respect to the Earth (new moon) or opposite each other (full moon), the tidal effect reaches its maximum; this is called spring tide.

The sea level typically varies 1 m, but in some narrow straits, the difference can be as great as 15 m. The solid surface of the Earth also suffers tidal effects, but the amplitude is much smaller, about 30 cm.

Fig. 8.6. Tidal forces acting on the Earth generate tides (*see text*)

Tides generate friction, which dissipates the rotational and orbital kinetic energy of the Earth-Moon system. This energy loss induces some changes in the system. First, the rotation of the Earth slows down until the Earth also rotates synchronously, i.e. the same side of the Earth will always face the Moon. Secondly, the orbital speed of the Moon decreases, and the Moon drifts about 1 cm away from the Earth with every revolution.

8.5 Eclipses and Occultations

An *eclipse* is an event in which a body goes through the shadow of another body. The most frequently observed eclipses are the lunar eclipses and the eclipses of the large satellites of Jupiter. An *occultation* takes place when an occulting body goes in front of another object; typical examples are stellar occultations caused by the Moon. Generally, occultations can be seen only in a narrow strip; an eclipse is visible wherever the body is above the horizon.

Solar and lunar eclipses are the most spectacular events in the sky. A solar eclipse occurs when the Moon is between the Earth and the Sun (Fig. 8.7). (According to the

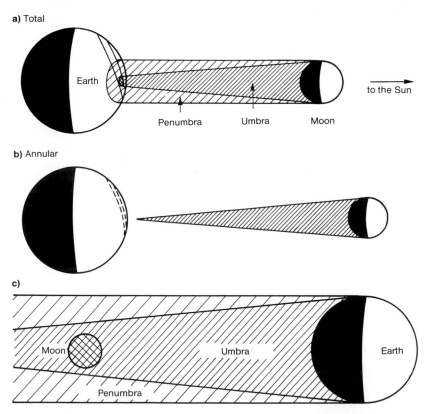

Fig. 8.7. (a) A total solar eclipse can be seen only inside a narrow strip; outside the zone of totality the eclipse is partial. **(b)** An eclipse is annular if the Moon is at apogee from where the shadow of the Moon does not reach the Earth. **(c)** A lunar eclipse is visible everywhere where the Moon is above the horizon

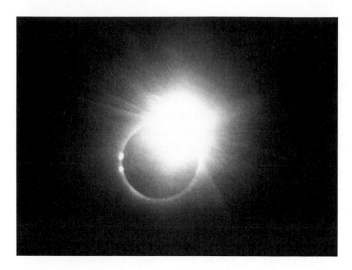

Fig. 8.8. The last total eclipse of the Sun in Europe occurred in 1990 over Finland. The next one will be over Southern Europe in 1999. (Photo by Matti Martikainen)

definition, a solar eclipse is not an eclipse but an occultation!) If the whole disc of the Sun is behind the Moon, the eclipse is *total* (Fig. 8.8); otherwise, it is *partial*. If the Moon is close to its apogee, the apparent diameter of the Moon is smaller than that of the Sun, and the eclipse is *annular*.

A lunar eclipse is total if the Moon is entirely inside the umbral shadow of the Earth; otherwise, the eclipse is partial. A partial eclipse is difficult to see with the unaided eye because the lunar magnitude remains almost unchanged. During the total phase, the Moon is coloured deep red because some red light is refracted through the Earth's atmosphere.

If the orbital plane of the Moon coincided with the plane of the ecliptic, one solar and one lunar eclipse would occur every synodic month. However, the plane is tilted about 5°; therefore, at full moon, the Moon must be close to the nodes in order for an eclipse to occur. The angular distance of the Moon from the node must be smaller than 4.6° for a total lunar eclipse, and 10.3° for a total solar eclipse.

Two to seven eclipses occur annually. Usually eclipses take place in a set of $1-3$ eclipses, separated by an interval of 173 days. In one set there can be just one solar eclipse or a succession of solar, lunar and another solar eclipse. In one year, eclipses belonging to 2 or 3 such sets can take place.

The Sun and the (ascending or descending) node of the lunar orbit are in the same direction once every 346.62 days. Nineteen such periods ($=6585.78$ days $=18$ y 11 d) are very close to the length of 223 synodic months. This means that the Sun-Moon configuration and the eclipses are repeated in the same order after this period. The cycle was already known by the ancient Babylonians and, for historical reasons, it is known as the *Saros*.

During a solar eclipse, the shadow of the Moon on the Earth's surface is always less than 270 km wide. The shadow moves at least 34 km/min; thus the maximum duration of an eclipse is $7\frac{1}{2}$ minutes. The maximum duration of a lunar eclipse is 3.8 h, and the duration of the total phase is always shorter than 1.7 h.

Observations of the stellar occultations caused by the Moon formerly served as an accurate method for determining the lunar orbit. Radio astronomers used to use occultations for determining the exact positions of some radio sources. Nowadays, occultations are used for studying the Earth's rotation, measuring stellar diameters and searching for binary stars. Because the Moon has no atmosphere, the star disappears abruptly in less than 1/50 s. If a fast photometer is used for recording the event, the typical diffraction pattern can be seen. The shape of the diffraction is different for a binary object.

Because the Moon moves eastwards, stars are occulted by the dark edge of the Moon during the first quarter. Therefore occultation is easier to observe, and photometric measurements are possible; at the same time, it is much more difficult to observe the appearance of an object. There are some bright stars inside the zone where the Moon moves, but the occultation of a bright, naked-eye object is quite rare.

Occultations are also caused by planets and asteroids. Accurate predictions are complicated because such an event is visible only in a very narrow path. Great advances have been made since the 1970's, and now occultations are frequently observed. The Uranian rings were found during an occultation in 1977, and the shape of some asteroids has been studied during some favourable events, timed exactly by several observers located along the predicted path.

A *transit* is an event in which Mercury or Venus moves across the Solar disc. A transit can occur only when the planet is close to its orbital node at the time of inferior conjunction. Transit of Mercury occurs about 13 times per century; transit of Venus, only twice. The next transits of Mercury are: Nov 15, 1999; May 7, 2003; Nov 8, 2006; May 9, 2016 and Nov 11, 2019. The next three transits of Venus are: Jun 8, 2004; Jun 6, 2012 and Dec 11, 2117. In the 18th century, the two transits of Venus were used for determining the value of the astronomical unit.

8.6 Albedos

The planets and all other bodies of the solar system only reflect the radiation of the Sun (we may neglect here the thermal and radio wave radiation and concentrate mainly on the visual wavelengths). The brightness of a body depends on its distance from the Sun and the Earth, and on the albedo of its surface. The term *albedo* defines the ability of a body to reflect light.

Let the radiation flux density on the surface of the Sun be $F_\odot$ [W m^{-2}]. The flux density at the distance r is

$$F_p = F_\odot \, \frac{R_\odot^2}{r^2} \, , \tag{8.3}$$

where $R_\odot / r$ is the angular diameter of the Sun at a distance r. If the radius of the planet is R, the area of the cross section is πR^2, and the total flux on the surface of the planet is

$$L = F_p \pi R^2 = F_\odot \, \frac{R_\odot^2}{r^2} \, \pi R^2 \, . \tag{8.4}$$

The *Bond albedo A* (or spherical albedo) is defined as the ratio of the emergent flux to the incident flux. The flux reflected by the planet is thus

$$L' = A \cdot L = A F_\odot \, \frac{R_\odot^2}{r^2} \, \pi R^2 \, . \tag{8.5}$$

The planet is observed at the distance Δ. If radiation is reflected isotropically, the observed flux density should be

$$F = L'/4\pi\Delta^2 . \tag{8.6}$$

However, radiation is reflected anisotropically, and the flux density must be corrected by a factor $C\Phi(\alpha)$, depending on the phase angle. $\Phi(\alpha)$ is called the *phase function* and is normalized so that $\Phi(\alpha = 0°) = 1$. The normalization constant C is such that

$$\frac{C\int_S \Phi(\alpha)\,dS}{4\pi\Delta^2} = 1 , \tag{8.7}$$

where the integration is extended over the surface of the sphere of radius Δ. This simply means that all the radiation reflected from the planet is found somewhere on the surface of the sphere. The observed flux density at the distance Δ is

$$F = \frac{C\Phi(\alpha)L'}{4\pi\Delta^2} = \frac{C\Phi(\alpha)}{4\pi\Delta^2} A F_\odot \frac{R_\odot^2}{r^2} \pi R^2 . \tag{8.8}$$

The normalization constant C is calculated as follows: The surface element of a sphere of radius Δ is $dS = \Delta\,d\alpha\Delta\sin\alpha\,d\phi$, and the integral in (8.7) can be expressed as

$$\int_S \Phi(\alpha)\,dS = \Delta^2 \int_{\alpha=0}^{\pi}\int_{\phi=0}^{2\pi} \Phi(\alpha)\sin\alpha\,d\alpha\,d\phi = \Delta^2 2\pi \int_0^\pi \Phi(\alpha)\sin\alpha\,d\alpha$$

and

$$C = \frac{4\pi\Delta^2}{\int_S \Phi(\alpha)\,dS} = \frac{2}{\int_0^\pi \Phi(\alpha)\sin\alpha\,d\alpha} . \tag{8.9}$$

If p is defined as

$$p = CA/4 \tag{8.10}$$

then the Bond albedo A can be expressed as

$$A = p \cdot 2\int_0^\pi \Phi(\alpha)\sin\alpha\,d\alpha \equiv p \cdot q . \tag{8.11}$$

p is called the *geometric albedo* and

$$q = 2\int_0^\pi \Phi(\alpha)\sin\alpha\,d\alpha \tag{8.12}$$

is the *phase integral*.

The physical interpretation of the geometric albedo may not be obvious. The reason for introducing p can be seen as follows:

A *Lambertian surface* is defined as an absolutely white, diffuse surface which reflects all radiation, i.e. its $A = 1$. Moreover, the phase function is $\Phi(\alpha) = \cos\alpha$, when $\alpha \in [0, \pi/2]$ and $\Phi(\alpha) = 0$, when $\alpha \in (\pi/2, \pi]$. (In reality, no such surface exists but there are some materials which behave almost like a Lambertian surface.) With these definitions, the constant C can be obtained by considering such a surface perpendicular to the incident light.

$$C = \frac{2}{\int\limits_0^\pi \Phi(\alpha) \sin\alpha \, d\alpha} = \frac{2}{\int\limits_0^{\pi/2} \cos\alpha \sin\alpha \, d\alpha} = 4 \, .$$

Inserting this into (8.10), we obtain $p = A = 1$ for a Lambertian surface.

Next we shall compare the ratio of the flux density reflected by a planet to the flux reflected by a Lambertian surface which has the same surface area as the cross section of the planet. Moreover, the Lambertian surface is assumed to be perpendicular to the incident light and the observer is observing at the phase angle $\alpha = 0°$. The observed flux density can be found from (8.8):

$$F(\alpha=0) = \frac{C}{4\pi\Delta^2} A F_\odot \frac{R_\odot^2}{r^2} \pi R^2 \, .$$

Similarly, the flux density for the Lambertian surface is

$$F_\mathrm{L}(\alpha=0) = \frac{4}{4\pi\Delta^2} \cdot 1 \cdot F_\odot \frac{R_\odot^2}{r^2} \pi R^2 \, .$$

The ratio of these two is

$$\frac{F(\alpha=0)}{F_\mathrm{L}(\alpha=0)} = \frac{CA}{4} = p \, . \tag{8.13}$$

Now we have found the physical interpretation for p: the geometric albedo is the ratio of the flux densities at phase angle $\alpha = 0°$, reflected by a planet and a Lambertian surface. It turns out that p can be derived from the observations, but the Bond albedo can be determined only if the phase integral q is also known.

8.7 Planetary Photometry, Polarimetry and Spectroscopy

By using (8.8), Fig. 8.9 and the definition of p [(8.13)], the observed flux density can be expressed as

$$F = \frac{p}{\pi} \frac{\Phi(\alpha)}{\Delta^2} F_\odot \frac{R_\odot^2}{r^2} \pi R^2 \, . \tag{8.14}$$

The flux density of the Sun at the Earth is

$$F_\oplus = F_\odot \frac{R_\odot^2}{r_\oplus^2} \, . \tag{8.15}$$

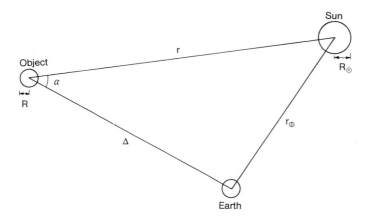

Fig. 8.9. Symbols used in the photometric formulae

Using these expressions and the definition of magnitudes, we find the magnitude difference of the planet and the Sun to be

$$m - m_\odot = -2.5 \lg \frac{F}{F_\oplus}$$

$$= -2.5 \lg \frac{p\,\Phi(\alpha)\,R^2 r_\oplus^2}{\Delta^2 r^2} \tag{8.16}$$

$$= -2.5 \lg p \frac{R^2 r_\oplus^2}{r_\oplus^4} - 2.5 \lg \frac{r_\oplus^4}{\Delta^2 r^2} - 2.5 \lg \Phi(\alpha)$$

$$= -2.5 \lg p \frac{R^2}{r_\oplus^2} + 5 \lg \frac{r\Delta}{r_\oplus^2} - 2.5 \lg \Phi(\alpha) \,. \tag{8.17}$$

If

$$V(1,0) = m_\odot - 2.5 \lg p \frac{R^2}{r_\oplus^2} \,, \tag{8.18}$$

then the magnitude of a planet can be expressed as

$$m = V(1,0) + 5 \lg \frac{r\Delta}{r_\oplus^2} - 2.5 \lg \Phi(\alpha) \,. \tag{8.19}$$

In planetary science, $V(1,0)$ is called the *absolute magnitude* (not to be confused with the absolute magnitude in stellar astronomy!). By looking at (8.19), we can see that $V(1,0)$ is the magnitude of a body if it is at the distance of 1 AU from the Earth and the Sun at a phase angle $\alpha = 0°$. As will be immediately noticed, this is physically impossible because the observer would be in the very centre of the Sun! Thus $V(1,0)$ can never be observed.

The phase function $\Phi(\alpha)$ is also unknown. This means that from the observations, one can calculate only

$$V(1,\alpha) = m - 5 \lg \frac{r\Delta}{r_\oplus^2} = V(1,0) - 2.5 \lg \Phi(\alpha) \,, \tag{8.20}$$

which is the absolute magnitude at phase angle α. $V(1,\alpha)$, plotted as a function of the

Fig. 8.10. The phase curves of the Moon and Venus. The absolute magnitude $V(1, \alpha)$ is normalized so that $V(1, 0) = 0$

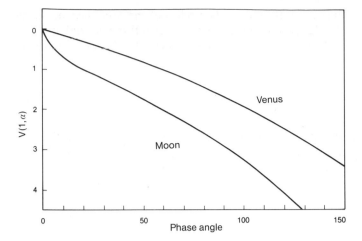

phase angle, is called the *phase curve* (Fig. 8.10). The phase curve extrapolated to $\alpha = 0°$ gives $V(1, 0)$.

By using (8.16) at $\alpha = 0°$, the geometric albedo can be solved in terms of observed values:

$$p = \left(\frac{r \Delta}{r_\oplus R} \right)^2 10^{-0.4(m_0 - m_\odot)}, \tag{8.21}$$

where $m_0 = m(\alpha = 0°)$. As can easily be seen, p can be greater than unity but in the real world, it is normally well below that. A typical value for p is $0.1 - 0.5$.

The Bond albedo can be determined only if the phase function Φ is known. Superior planets (and other bodies orbiting outside the Earth's orbit) can be observed only in a limited phase angle range, and therefore Φ is poorly known. The situation is somewhat better for the inferior planets. In spite of this, and especially in popular texts, the Bond albedo is listed instead of p (naturally without mentioning the exact names!). A good excuse for this is the obvious physical meaning of the former, and also the fact that the Bond albedo is normalized to $[0, 1]$.

Opposition Effect. The brightness of an atmosphereless body increases rapidly when the phase angle approaches zero. When the phase is larger than about 10°, the changes are smaller. This rapid brightening close to the opposition is called the *opposition effect*. A qualitative explanation is that close to the opposition, no shadows are visible. When the phase angle increases, the shadows become visible and the brightness drops. An atmosphere destroys the opposition effect. This can be seen in the phase curve of Venus (Fig. 8.10).

The shape of the phase curve depends on the geometric albedo. It is possible to estimate the geometric albedo if the phase curve is known. This requires at least a few observations at different phase angles. Most critical is the range $0° - 10°$. A known phase curve can be used for determining the diameter of a body, e.g. for determining sizes of asteroids. Apparent diameters of asteroids are so small that they can not be directly measured; instead some indirect method has to be used. Other available methods are polarimetric and radiometric (thermal radiation) observations. The accuracy achieved is about 10%.

Polarimetric Observations. The light reflected by the bodies of the solar system is usually polarized. The amount of polarization depends on the reflecting material and also on the geometry: polarization is a function of the phase angle. The *degree of polarization P* is defined as

$$P = \frac{F_\perp - F_\|}{F_\perp + F_\|} , \qquad (8.22)$$

where $F_\perp$ is the flux density of radiation, perpendicular to a fixed plane, and $F_\|$ is the flux density parallel to the plane. In solar system studies, polarization is usually referred to the plane defined by the Earth, the Sun, and the object. According to (8.22), P can be positive or negative; thus the terms "positive" and "negative" polarization are used.

The degree of polarization as a function of the phase angle depends on the surface structure and the atmosphere. The degree of polarization of the light reflected by the surface of an atmosphereless body is positive when the phase angle is greater than about 20°. Closer to opposition, polarization is negative. When light is reflected from an atmosphere, the degree of polarization as a function of the phase angle is more complicated. By combining observations with a theory of radiative transfer, one can compute atmosphere models. For example, the composition of Venus' atmosphere could be studied before any probes were sent to the planet.

Planetary Spectroscopy. The photometric and polarimetric observations discussed above were monochromatic. However, the studies of the atmosphere of Venus also used spectral information. Broadband UBV photometry or polarimetry is the simplest example of spectrophotometry (spectropolarimetry). The term spectrophotometry usually means observations made with several narrowband filters. Naturally, solar system objects are also observed by means of "classical" spectroscopy.

Spectrophotometry and polarimetry give information at discrete wavelengths only. In practice, the number of points of the spectrum (or the number of filters available) is often limited to 20 – 30. This means that no details can be seen in the spectra. On the other hand, in ordinary spectroscopy, the limiting magnitude is smaller, although the situation is rapidly improving with the new generation detectors, such as the CCD camera.

The spectrum observed is the spectrum of the Sun. Generally, the planetary contribution is relatively small, and these differences can be seen when the solar spectrum is subtracted. The Uranian spectrum is a typical example (Fig. 8.11). There are strong absorption bands in the near-infrared. Laboratory measurements have shown that these are due to methane. A portion of the red light is also absorbed, causing the greenish colour of the planet. The general techniques of spectral observations are discussed in the context of stellar spectroscopy in Chap. 9.

Wavelength [nm]

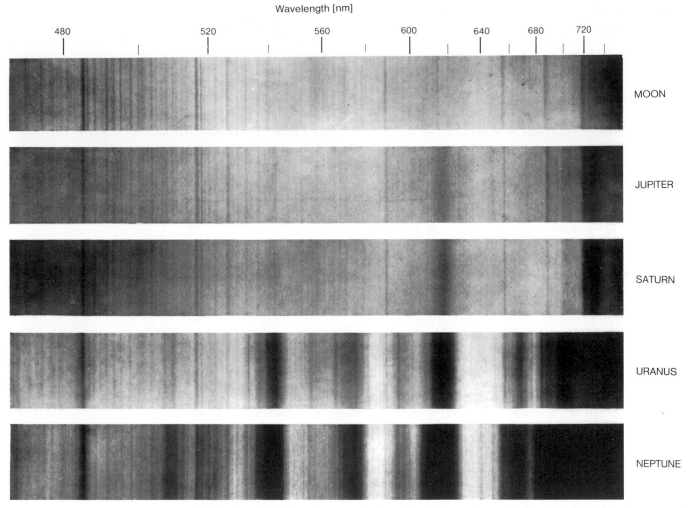

Fig. 8.11. Spectra of the Moon and the giant planets. Strong absorption bands can be seen in the spectra of Uranus and Neptune. [Lowell Observatory Bulletin **42** (1909)]

8.8 Thermal Radiation of the Planets

Thermal radiation of the solar system bodies depends on the albedo and the distance from the Sun, i.e. on the amount of absorbed radiation. Internal heat is important in Jupiter and Saturn, but we may neglect it at this point.

By using the Stefan-Boltzmann law, the flux on the surface of the Sun can be expressed as

$$L = 4\pi R_\odot^2 \, \sigma T_\odot^4 \ .$$

If the Bond albedo of the body is A, the fraction of the radiation absorbed is $(1-A)$. This is later emitted as heat. If the body is at a distance r from the Sun, the absorbed flux is

$$L_{\mathrm{abs}} = \frac{R_\odot^2 \sigma T_\odot^4 \pi R^2}{r^2} (1-A) \ . \tag{8.23}$$

There are good reasons to assume that the body is in *thermal equilibrium*, i.e. the emitted and the absorbed fluxes are equal. If not, the body will warm up or cool down until equilibrium is reached.

Let us first assume that the body is rotating slowly. The dark side has had time to cool down, and the thermal radiation is emitted mainly from one hemisphere. The flux emitted is

$$L_{\mathrm{em}} = 2\pi R^2 \sigma T^4 \ , \tag{8.24}$$

where T is the temperature of the body and $2\pi R^2$ is the area of one hemisphere. In thermal equilibrium, (8.23) and (8.24) are equal:

$$\frac{R_\odot^2 T_\odot^4}{r^2} (1-A) = 2 T^4 \tag{8.25}$$

$$\Rightarrow T = T_\odot \left(\frac{1-A}{2} \right)^{1/4} \left(\frac{R_\odot}{r} \right)^{1/2} \ .$$

A body rotating quickly emits an approximately equal flux from all parts of its surface. The emitted flux is

$$L_{\mathrm{em}} = 4\pi R^2 \sigma T^4$$

and the temperature,

$$T = T_\odot \left(\frac{1-A}{4} \right)^{1/4} \left(\frac{R_\odot}{r} \right)^{1/2} \ . \tag{8.26}$$

The theoretical temperatures obtained above are not valid for most of the major planets. The main "culprits" responsible here are the atmosphere and the internal heat. Measured and theoretical temperatures of some major planets are compared in Table 8.1. Venus is an extreme example of the disagreement between theoretical and actual

Table 8.1. Theoretical and observed temperatures of some planets

Planet	Albedo	Distance from the Sun [AU]	Theoretical temperature [K]		Observed maximum temperature [K]
			(8.25)	(8.26)	
Mercury	0.06	0.39	525	440	615
Venus	0.76	0.72	270	230	750
Earth	0.36	1.00	290	250	310
Mars	0.16	1.52	260	215	290
Jupiter	0.73	5.20	110	90	130

figures. The reason is the so-called "greenhouse effect": radiation is allowed to enter, but not to exit. The same effect is at work in the Earth's atmosphere. Without the greenhouse effect, the mean temperature could be well below the freezing point and the whole Earth would be ice-covered.

8.9 The Structure of Planets

Terrestrial Planets (Fig. 8.12). The structure of the terrestrial planets can be studied with seismic waves. The waves formed in an earthquake are reflected and refracted inside a planet like any other wave at the boundary of two different layers. The waves are longitudinal or transversal (*P* and *S* waves, respectively). Both can propagate in solid materials such as rock. However, only the longitudinal wave can penetrate liquids. One can determine whether a part of the interior material is in the liquid state and where the boundaries of the layers are by studying the recordings of seismometers placed on the surface of a planet. Naturally, the Earth is the best-known body, but quakes of the Moon, Venus, and Mars have also been observed.

The terrestrial planets have an iron-nickel core. Mercury has the relatively largest core; Mars, the smallest. Around this Fe – Ni core is a mantle, composed of silicates (compounds of silicon). The density of the outermost layers is about $3000\,\mathrm{kg\,m^{-3}}$. The mean density of the terrestrial planets is $3500 - 5500\,\mathrm{kg\,m^{-3}}$.

Giant Planets (Fig. 8.13). The internal structure of the giant planets cannot be observed with seismic waves since the planets do not have a solid surface. An alternative is to

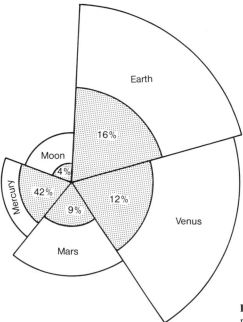

Fig. 8.12. Internal structure and relative sizes of the terrestrial planets. The size of the core is expressed in %

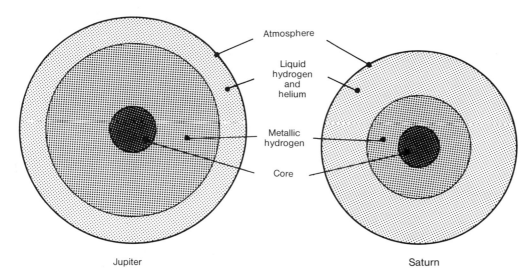

Atmosphere

Liquid
hydrogen
and
helium

Metallic
hydrogen

Core

Jupiter   Saturn

Fig. 8.13. Internal structure of Jupiter and Saturn. Differences in size and distance from the Sun generate differences in the chemical composition and internal structure

study the shape of the gravity field by observing the orbit of a probe when it passes (or circles) the planet. This will give some information on the internal structure, but the details depend on the mathematical and physical models used for interpretation.

The mean densities of the giant planets are quite low; the density of Saturn, for example, is only $700 \, \text{kg m}^{-3}$. (If Saturn were put in a bathtub, it would float on the water!) Most of the volume of a giant planet is a mixture of hydrogen and helium. In the centre, there is possibly a silicate core, the mass of which is perhaps a few Earth masses. The core is surrounded by a layer of metallic hydrogen. Due to the extreme pressure, hydrogen is not in its normal molecular form, H_2, but dissociated into atoms. In this state, hydrogen is electrically conducting. The magnetic fields of the giant planets may originate in the layer of metallic hydrogen.

Closer to the surface, the pressure is lower and hydrogen is in molecular form. The relative thickness of the layers of metallic and molecular hydrogen vary from planet to planet. On top is a gaseous atmosphere, only a few hundred kilometres thick. The clouds at the top of the atmosphere form the visible "surface" of the planet.

Flattening. A rotating planet is always flattened. The amount of flattening depends on the rotation rate and the strength of the material; a liquid drop is more easily deformed than a rock. The shape of the surface can be derived from the equation of motion. If the rotation rate is moderate, the equilibrium shape of a liquid body is an ellipsoid of revolution (a Maclaurin sphcroid). The shortest axis is the axis of rotation. If R_e and R_p are the equatorial and polar radii (Fig. 8.14), respectively, the shape of the planet can be expressed as

$$\frac{x^2}{R_e^2} + \frac{y^2}{R_e^2} + \frac{z^2}{R_p^2} = 1 \ .$$

The dynamical flattening, denoted by ε, is defined as

$$\varepsilon = \frac{R_e - R_p}{R_e} \ .$$

Fig. 8.14. A spheroid is the equilibrium shape of a slowly rotating body. The equatorial radius is denoted by R_e and the polar radius by R_p

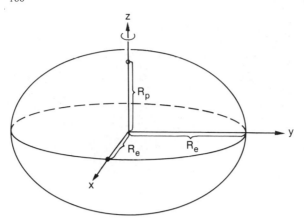

Because $R_e > R_p$, the flattening ε is always positive.

Asteroids are so small that they are not flattened by rotation. Bodies larger than about 1000 km in diameter are deformed into a symmetric shape due to gravity; the internal strength of the material keeps smaller bodies irregular in shape.

8.10 Planetary Surfaces

Planetary surfaces are modified by several geological processes (Fig. 8.15). These include continental drift, volcanism, meteorite impacts and climate. The Earth is an example of a body whose surface has been renewed many times during past aeons. Continental drift gives rise, for example, to mountain formation. Similar formations have been observed on Mars and Venus. However, on Mercury and the Moon, no continental drift has been detected.

Fig. 8.15. Ages of the surfaces of Mercury, the Earth, the Moon and Mars. The curve represents the fraction of the surface which existed at a certain time. Most of the surface of the Moon, Mercury and Mars are more than 3500 million years old, whereas the surface of the Earth is mostly younger than 200 million years

Fig. 8.16. The number of meteorite impact craters is a good indicator of the age of the surface. The shape of the craters contains information on the strength of the material. Objects from upper left are: Mercury, Moon, Ganymede and Enceladus. (NASA)

Volcanism is a minor factor on the Earth (at least now), but the surface of the Jovian moon, Io, is changing rapidly, due to violent volcanic eruptions. Volcanos have been observed on Mars and Venus, but not on the Moon.

Lunar craters are meteorite impact craters, common on almost every body with a solid surface. Meteorites are bombarding the planets continuously, but the rate has been reduced since the beginnings of the solar system. The number of impact craters reflects the age of the surface (Fig. 8.16). The Jovian moon, Callisto, is a typical example of a body with an ancient surface. An opposite example is the Earth, whose atmosphere protects the surface and destroys the traces of impacts. All smaller meteorites are burned to ashes in the atmosphere (one need only note the number of shooting stars), and some larger bodies are bounced back to outer space. The traces on the surface are destroyed very quickly by erosion in less than a few million years.

Climate has the greatest influence on the Earth and Venus. Both planets have a thick atmosphere. On Mars, powerful dust storms deform the landscape, too, often covering the planet with yellowish dust clouds.

8.11 Atmospheres and Magnetospheres

Atmospheres. Excluding Mercury (and possibly Pluto), all major planets have an atmosphere. The composition, thickness, density and structure of the atmosphere vary from planet to planet, but some common features can be found (see, e.g., Figs. 8.17, 8.18).

Let us first study the dependence of the temperature T, pressure P, and density ρ on the height, h. Let us consider a cylinder with a length dh. The change in the pressure dP, from the height h to $(h+dh)$ is proportional to the mass of the gas in the cylinder:

$$dP = -g\rho\,dh \ , \tag{8.27}$$

where g is the acceleration of gravity. Equation (8.27) is the *equation of hydrostatic equilibrium*. (It is discussed in detail in Chap. 11 "Stellar Structure".)

As a first approximation, we may assume that g does not depend on height. In the case of the Earth, the error is only about 3% if g is considered constant from the surface to a height of 100 km.

Fig. 8.17. (a) Temperature as a function of height in the atmospheres of Venus, the Earth, and Mars. **(b)** Temperature profiles of the atmospheres of Jupiter and Saturn. The zero height is chosen to be the point where the pressure is 100 mbar. Numbers along the curves are pressures in millibars

a

b

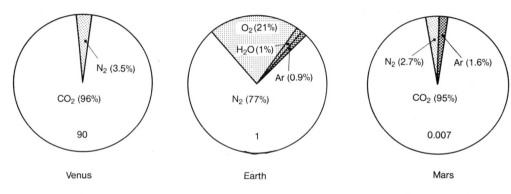

Fig. 8.18. Relative abundances of the most abundant gases in the atmospheres of Venus, the Earth, and Mars. The number at the bottom of each circle denotes the surface pressure in atms

The equation of state of the ideal gas

$$PV = NkT \tag{8.28}$$

gives the expression for the pressure P

$$P = \frac{\rho k T}{\mu} \, , \tag{8.29}$$

where N is the number of atoms or molecules, k is the Boltzmann constant, μ is the mass of one atom or molecule and

$$\rho = \frac{\mu N}{V} \, . \tag{8.30}$$

By using the equation of hydrostatic equilibrium (8.27), and the equation of state (8.29), we obtain

$$\frac{dP}{P} = -g \frac{\mu}{kT} dh \, .$$

Integration yields P as a function of height:

$$P = P_0 \exp\left(-\int_0^h \frac{\mu g}{kT} \, dh \right) = P_0 \exp\left(-\int_0^h \frac{dh}{H} \right) \, . \tag{8.31}$$

The variable H, which has the dimension of length, is called the *scale height*:

$$H = \frac{kT}{\mu g} \, . \tag{8.32}$$

The scale height defines the height at which the pressure has decreased by a factor e. H is a function of height, but here we may assume that it is constant. With this approximation, we obtain

$$-\frac{h}{H} = \ln\frac{P}{P_0}$$

Table 8.2. Scale heights of some gases in the atmospheres of Venus, the Earth, and Mars

Gas	Molecular weight [amu]	The Earth H [km]	Venus H [km]	Mars H [km]
H	2	120	360	290
O	32	7	23	81
HO	18	13	40	32
CO	44	5	16	13
N	28	8	26	20
Temperature [K]		275	750	260
Acceleration of gravity [m s]		9.81	8.61	3.77

or, using (8.29),

$$\frac{\rho\, T(h)}{\rho_0 T_0} = e^{-h/H} \; . \tag{8.33}$$

The scale height is an important parameter in many formulae describing the structure of the atmosphere (Table 8.2). For example, if the change of the pressure or the density is known as a function of height, the mean molecular weight of the atmosphere can be computed. The scale height of the Jovian atmosphere was determined in 1952 when Jupiter occulted a star. With these observations, the scale height was calculated to be 8 km, and the mean molecular weight, $3-5$ amu (atomic mass unit, 1/12 of the mass of carbon-12). Thus the main components are hydrogen and helium, a result later confirmed by probes.

If the scale height of the Earth's atmosphere is known, one can estimate the height at which observations can be made without excessive atmospheric perturbations. For example, infrared observations are limited by water vapour and carbon dioxide. The scale height of CO_2 is 5 km, which means that the partial pressure is halved already at a height of 3.5 km. Thus infrared observations can be made from some moderately high mountains (like Manua Kea in Hawaii). The scale height of water vapour is 13 km, but the relative humidity and hence the actual water content is very site- and time-dependent.

The scale height and the temperature of the atmosphere define the permanence of the atmosphere. If the speed of a molecule is greater than the escape velocity, the molecule will escape into space. The whole atmosphere could disappear in a relatively short time.

According to the kinetic gas theory, the mean velocity $\bar{v}$ of a molecule depends both on the kinetic temperature T_k of the gas, and on the mass m of the molecule [see (6.10)]:

$$\bar{v} = \sqrt{\frac{3 k T_k}{m}} \; .$$

If the mass of a planet is M and its radius R, the escape velocity is given by (7.33)

$$v_e = \sqrt{\frac{2 G M}{R}} \; .$$

Even if the mean velocity is smaller than the escape velocity, the atmosphere can evaporate into space if there is enough time, since some molecules will always have velocities exceeding v_e. Assuming a velocity distribution, one can calculate the probability for $v > v_e$. Hence it is possible to estimate what fraction of the atmosphere will disappear in, say, 10^9y. As a rule of thumb, it can be said that at least half of the atmosphere will remain over 1000 million years if the mean velocity $\bar{v} < 0.2 v_e$.

The probability that a molecule close to the surface will escape is insignificantly small. The free mean path of a molecule is very small when the gas density is high (Fig. 8.19). Thus the escaping molecule is most probably leaving from the uppermost layers. The *critical layer* is defined as a height at which a molecule, moving upwards, has a probability $1/e$ of hitting another molecule. The part of the atmosphere above the critical layer is called the *exosphere*. The Earth's exosphere begins at a height of 500 km, where the kinetic temperature of the gas is $1500 - 2000$K and the pressure is lower than in the best terrestrial vacuums.

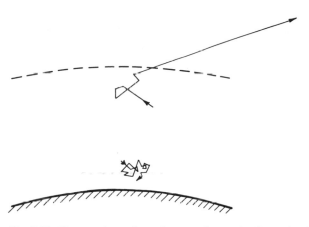

Fig. 8.19. Close to the surface, the mean free path of a molecule is smaller than higher in the atmosphere where the gas density is smaller. The escaping molecules originate close to the critical layer

Magnetospheres. The magnetosphere is the "outer boundary" of a planet. Size and shape depend on the strength of the magnetic field of the planet and on the solar wind.

On the solar side is a *bow shock* (Fig. 8.20), typically at a distance of a few tens of planetary radii. At the bow shock, particles of the solar wind first hit the magnetosphere. The magnetosphere is limited by the *magnetopause*, flattened on the solar side and extended to a long tail on the opposite side. Charged particles inside the magnetopause are captured by the magnetic field and some particles are accelerated to great velocities. If the velocities are interpreted according to the kinetic gas theory, these velocities correspond to even millions of kelvins. However, the density, and thus the total energy, is very small. The "hottest" places found are around Jupiter and Saturn.

The region of space containing trapped charged particles, the so-called radiation belts around the Earth, are named van Allen's belts. These radiation zones were discovered by the first US satellite, Explorer 1, in 1958. The number of charged particles increases after strong solar bursts. Some of these particles "leak" to the atmosphere, resulting in strong aurorae. Similar effects have also been detected in Jupiter, Saturn and Uranus.

Fig. 8.20. Structure of the magnetosphere of the Earth. (A. Nurmi/Tiede 2000)

8.12 Mercury

Mercury is the innermost planet of the solar system. Its diameter is 4800 km and its mean distance from the Sun, 0.39 AU. The eccentricity of the orbit is 0.21, which means that the distance varies between 0.31 and 0.47 AU. Because of the high eccentricity, the surface temperature of the subsolar point varies substantially: at the perihelion, the temperature is about 700 K; at the aphelion, it is 100 K lower.

Mercury's relatively small size and proximity to the Sun, resulting in low gravity and high temperature, are the reasons for its lack of atmosphere. Due to the absence of an atmosphere, the temperature on Mercury drops very rapidly after sunset. The length of a day is 176 Earth days, longer than that of any other planet. The rotational axis is almost perpendicular to the orbital plane; therefore it is possible that, close to the poles, there are areas where the temperature is moderate.

As late as the beginning of the 1960's, it was believed that Mercury always turns the same side towards the Sun. However, measurements of the thermal radio emission showed that the temperature of the night side is too high, about 100 K, instead of almost absolute zero. Finally, the rotation period was established by radar. One revolution around the Sun takes 88 days. The rotation period is two-thirds of this, or 59 days. This means that every second time the planet is in, say, perihelion, the same hemisphere faces the Sun. This kind of spin-orbit coupling can result from tidal forces exerted by a central body on an object moving in a fairly eccentric orbit.

Re-examination of old observations revealed why Mercury had been presumed to rotate synchronously. Owing to its geometry, Mercury is easiest to observe in spring and autumn. In six months, Mercury orbits twice around the Sun, rotating exactly three times around its own axis. Consequently, during observations, the same side was always facing the Sun! The details visible on the surface are very obscure and the few exceptional observations made were interpreted as observational errors.

Mercury is always found in the vicinity of the Sun; its maximum elongation is only 28°. Observations are difficult because Mercury is always seen in a bright sky and close to the horizon. Moreover, when closest to the Earth, the dark side of the planet is towards us.

The first maps of Mercury were drawn at the end of the 19th century but the reality of the details was not confirmed. The best (and thus far unique) data were received in 1974 and 1975, when the US space probe Mariner 10 passed Mercury three times. The orbital period of Mariner 10 around the Sun was exactly twice the period of Mercury.

Fig. 8.21. Surface details on Mercury. The diameter of the large crater is about 120 km. (NASA)

Fig. 8.22. A mosaic picture of Mercury. (NASA)

Fig. 8.21   **Fig. 8.22**

And again the two-thirds-factor meant that the same side of the planet was illuminated during every fly-by! The other side is still unknown.

The Mariner 10 data revealed a moon-like landscape (Figs. 8.21, 22). Mercury's surface is marked by craters and larger circular areas, caused by impacts of minor planets. Most of the craters are 3000 – 4000 million years old, indicating that the surface is old and undisturbed by continental drift or volcanic eruptions. The largest lava-filled circular area is the 1300-km-wide Caloris Basin.

The interior of Mercury is obscure. The only relevant data were obtained during the Mariner 10 fly-bys when the gravity field was measured. Since Mercury has no satellite, the mass (and mass distribution) and density could not be determined before the force exerted by the gravitational field on a spacecraft was measured. According to theoretical models, the internal structure is similar to that of the Earth. It has been said that Mercury looks like the Moon from the outside but is terrestrial from the inside.

Due to the vicinity of the Sun, the temperature of the primeval nebula at the distance of Mercury was quite high during planetary formation. Thus the relative abundances of the volatile elements are smaller than on any other terrestrial planet. This is discussed in more detail in the chapter on planetary formation (see Sect. 8.21).

Mercury has a weak magnetic field, about 1% as strong as that of the Earth. The presence of the magnetic field is unexpected because Mercury rotates slowly. According to theory, a field is generated by flows in a liquid, electrically conducting core. These flows should cause continental drift, which has not been observed on Mercury. On the other hand, the magnetic field cannot be a remnant from ancient times, since the internal temperature of the planet must have exceeded the critical Curie point (which is the temperature at which material becomes demagnetized).

8.13 Venus

Venus is the brightest object in the sky, after the Sun and the Moon. Like Mercury, Venus can be seen only in the morning or in the evening sky. (It is sometimes possible to see Venus even if the Sun is above the horizon, if its exact position is known.) In antiquity, Venus was thought to be two different planets, Hesperos and Phosphorus, evening star and morning star.

The maximum elongation of Venus is about 47°. Venus is a remarkable object when shining in the dark sky at its brightest, 35 days before or after the inferior conjunction, when one-third of the surface is seen lit (Fig. 8.23). At the inferior conjunction, the Earth-Venus distance is only 42 million km. The diameter of Venus is about 12000 km, which means that the apparent diameter can be as large as one arc minute. Under favourable conditions it is even possible to see the shape of the crescent Venus with binoculars. At the superior conjunction, the apparent diameter is only 10 arc seconds.

Venus is covered by clouds. Its surface is nowhere visible; only featureless yellowish cloud tops can be seen. The rotation period was long unknown, and the measured 4-day period was the rotation time of the clouds. Finally, in 1962, radar measurements revealed that the rotation period is 243 days in a retrograde direction, i.e. opposite to the rotation of other planets. The axis of rotation is almost perpendicular to the orbital plane; the inclination is 177°.

The temperature at the cloud tops is about 250 K. Because the Bond albedo is as high as 75%, the surface temperature was believed to be moderate, even suitable for

Fig. 8.23. The phases of Venus were discovered by Galileo Galilei in 1610. This drawing illustrates how the apparent size of Venus changes with phase. The planet is far behind the Sun when the illuminated side faces the Earth

life. Opinions changed dramatically when thermal radio emission was measured at the end of the 1950's. This emission originates on the surface of the planet and can penetrate the clouds. The surface temperature turned out to be 750 K, well above the melting point of lead. As mentioned in Sect. 8.8, the reason for this is the greenhouse effect. The infrared radiation is blocked by atmospheric carbon dioxide, the main component of the atmosphere.

The chemical composition of the Venusian atmosphere was known prior to the space age. Spectroscopic observations revealed CO_2, and some clues to the cloud composition were obtained from polarimetric observations. The famous French planetary astronomer, Bernard Lyot, made polarimetric observations already in the 1920's, but not until decades later was it realized that his observations could be explained by assuming that light was scattered by spherical particles whose index of refraction is 1.44. This is significantly higher than the index of refraction of water, 1.33. Moreover, if the particles were spherical, they should be liquid, which is impossible for water at that temperature. A good candidate was sulphuric acid H_2SO_4. Later, probes confirmed this interpretation.

Mariner 2 (1962) was the first space probe to encounter the planet. Five years later, Soviet Venera 4 sent the first data from below the clouds, and the first pictures of the surface were sent by Venera 9 and 10 in 1975. The first radar map was completed in 1980, after 18 months of mapping by US Pioneer Venus 1. The first radar map was completed in 1980, after 18 months of mapping by US Pioneer Venus 1. The best and most the complete maps (about 99% of the planet's surface) are made from the radar observations of Magellan probe in the beginning of 90's. The resolution of the maps is as high as 100 metres.

Earth and Venus are almost equal in size, and their interiors are assumed to be similar. Probably due to its slow rotation, however, Venus has no magnetic field. The analyses made by the Veneras have shown that the surface material of Venus is similar to terrestrial granite and basalt (Fig. 8.24).

Fig. 8.24. The surface of Venus photographed by Venera 14

Fig. 8.25. One peculiar volcanic feature on Venus are round coronae. They are probably formed as a result of heat flow from the interior of the planet. This corona has a diameter of 375 km. (Photo NASA)

Radar mapping revealed canyons, mountains, craters, volcanos and other volcanic formations (Fig. 8.25). Contrary to the Earth, volcanic features are quite evenly distributed all over the surface of Venus. There are no evidences on massive crustal movements although local deformations can exist. Most of impact craters are undeformed. The reason for this is unclear because erosion, volcanism and tectonic forces should affect the craters, too. Resurfacing processes may frequently cover the old craters.

There are only two major highland areas, the area of which is only about 15% of the Venusian surface. The largest continent, Aphrodite Terra, is close to Venus' equator, it is the size of South America. Another large continent at the latitude 70° north is called Ishtar Terra, where the highest mountain on Venus, the 12-km-high Maxwell Montes is situated. (IAU decided that the Venusian nomenclature has to be feminine. Maxwell Montes, after the famous physicist James Clerk Maxwell, is an exception.)

Venus' atmosphere is mainly composed of carbon dioxide. It is very dry: the amount of water vapour present is only $1/1\,000\,000$ of that in the Earth's atmosphere. One possible explanation is that, due to solar UV-radiation, the water has dissociated to hydrogen and oxygen in the upper layers of the atmosphere, the latter escaping into interplanetary space.

About 1% of the incident light reaches the surface of Venus; this light is deep red after travelling through the clouds and the thick atmosphere. The pressure of the atmosphere at the surface is 90 atm. The visibility is several kilometres, and even in the clouds, a few hundred metres. The densest clouds are at a height of 50 km (Fig. 8.26a), but their thickness is only $2-3$ km. Above this, there are haze-like layers which form the visible "surface" of the planet (Fig. 8.26b). The uppermost clouds move rapidly;

they rotate around the planet in about 4 days, pushed by strong winds powered by the Sun.

8.14 The Earth and the Moon

The Earth. The third planet from the Sun, the Earth, and its satellite, the Moon, form almost a double planet. The relative size of the Moon is larger than that of any other satellite, excluding the moon of Pluto. Usually satellites are much smaller than their parent planets.

The Earth is a unique body, since a considerable amount of free water is found on its surface. This is possible only because the temperature is above the freezing point and below the boiling point of water. The Earth is also the only planet where some kind of life is known to exist. (Whether or not it is intelligent is yet to be resolved . . . .) The moderate temperature and the water are essential for terrestrial life forms.

The diameter of the Earth is 12000 km. At the centre, there is an iron-nickel core where the temperature is 5000 K, the pressure $3 \times 10^{11} \mathrm{Nm}^{-2}$ and the density $12000 \mathrm{kg m}^{-3}$ (Fig. 8.27). The nonexistence of the seismic S-waves below a depth of 3000 km indicates that the core is molten. However, the innermost core is possibly solid; the speed of the P-wave changes rapidly at a depth of 5000 km.

A silicate mantle extends from 3000 km below the surface upwards. Under pressure, the material behaves like a viscous liquid or an amorphous medium, resulting in slow vertical flows. A thin crust floats on the mantle. The thickness of the crust is only 12 – 65 km; it is thickest below high mountain ranges such as the Himalayas, and thinnest below the mid-ocean basins.

Continental drift is powered by the motion of the material in the mantle. About 200 million years ago, all the continents were joined; then the flow pattern in the mantle changed and the "protocontinent", *Pangaea*, broke up. The Atlantic Ocean is still growing and new material is flowing up at the mid-Atlantic ridge. North America is drifting away from Europe at the rate of a few centimetres per year.

New material is flowing up at the mid-sea ridges, pushing the tectonic plates apart. Mountains are formed where two plates collide. At the border of continents, one plate can be pushed below another, resulting in earthquakes as deep as 600 km. At the mid-sea ridges, the quakes are at a depth of only tens of kilometres.

The solid surface is also deformed by climate. Mountains and craters are eroded in tens of millions of years, due to the combined effects of temperature changes, rain and wind. Human activity is also becoming an increasingly important factor, changing the landscape in some areas.

Most of the Earth's surface is covered by water which condensed from the water vapour released in volcanic eruptions. The primordial atmosphere of the Earth was very different from the modern one; there was, for example, no oxygen. When organic chemical processes started in the oceans, the amount of oxygen rapidly increased (and was poison to the first forms of life!). The original carbon dioxide is now mainly concentrated in carbonate rocks, such as limestone, and the methane was dissociated by solar UV-radiation.

The Earth's main atmospheric constituents are nitrogen (77% by volume), and oxygen (21%). Other gases, such as argon, carbon dioxide, and water vapour occur in minor amounts. The chemical composition is unchanged in the lower part of the atmosphere, called the *troposphere*. Most of the climatic phenomena occur in the tropo-

Fig. 8.26a. Most of the Venusian clouds are at a height of 50 – 70 km. No clouds exist close to the surface

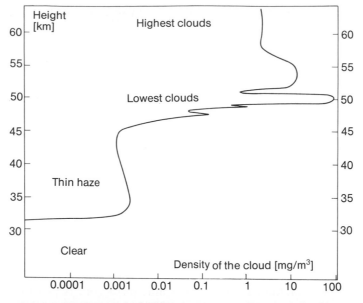

Fig. 8.26b. Ultraviolet images reveal cloud patterns in the upper atmosphere. (NASA)

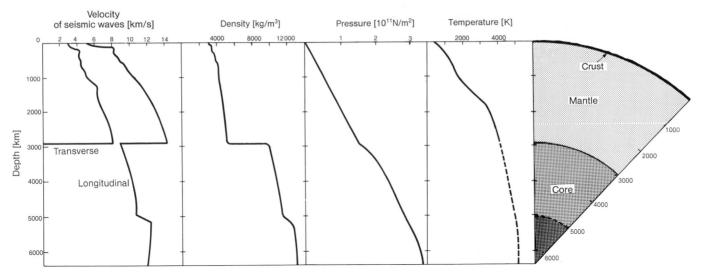

Fig. 8.27. Internal structure of the Earth. The speed of the seismic waves, density, pressure, and temperature are shown as a function of depth

sphere, which reaches up to 8 – 10 km. The limit is variable, being lowest at the poles, and highest at the equator, where it can extend up to 18 km.

The layer above the troposphere is the *stratosphere*, extending up to 60 km. The boundary between the troposphere and the stratosphere is called the *tropopause*. In the troposphere, the temperature decreases 5 – 7 K/km, but in the stratosphere it begins to rise, due to the absorption of solar radiation by carbon dioxide, water vapour and ozone. The ozone layer, which shields the Earth from the solar UV-radiation, is at a height of 20 – 25 km.

Above the stratosphere is the *ionosphere*, the playground of meteors and aurorae. It also plays an important role in radio communications, since radio waves are reflected by the ionosphere. The ionosphere goes over into the *exosphere* at about a height of 500 km. There, the air pressure is much lower than in the best laboratory vacuums.

The magnetic field of the Earth is generated by flows in its core. The field is almost a dipole but there are considerable local and temporal variations. The mean field strength close to the equator is 3.1×10^{-5} Tesla (0.31 Gauss). The dipole is tilted 11° with respect to the Earth's axis, but the direction gradually changes with time. Moreover, the magnetic north and south poles have exchanged places several times during the past million years.

The Moon. Our nearest neighbour in space is the Moon. Dark and light areas are visible even with the naked eye. For historical reasons, the former are called seas or *maria* (from Latin, mare, sea, *pl.* maria). The lighter areas are uplands but the maria have nothing to do with terrestrial seas, since there is no water on the Moon. Numerous craters, all meteorite impacts, can be seen, even with binoculars or a small telescope (Fig. 8.29). The lack of atmosphere, volcanism, and tectonic activity help to preserve these formations.

The Moon is the best-known body after the Earth. The first man landed on the Moon in 1969. A total of over 2000 samples, weighing 382 kg, were collected during the six Apollo flights. Moreover, three unmanned Soviet Luna spacecraft collected and re-

Fig. 8.28. A cyclone on the Earth. (This should be compared with the Great Red Spot of Jupiter in Fig. 8.43.) (NASA)

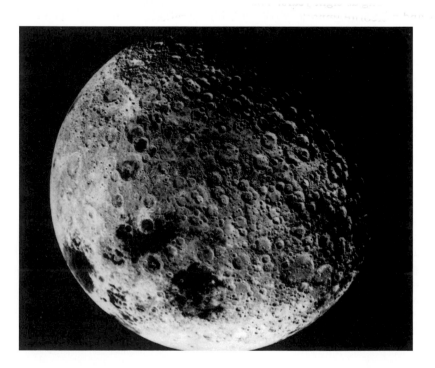

Fig. 8.29. Far side of the Moon. (NASA)

Fig. 8.30. Apollo 17 astronaut Harrison Schmitt on the Moon in 1972. (NASA)

turned about 310 grams of lunar soil. Instruments placed on the Moon by the Apollo astronauts operated as long as eight years. These included seismometers, which detected moonquakes and meteorite impacts, and passive laser reflectors which made exact Earth-Moon distance measurements possible.

The origin of the Moon is still somewhat obscure; it has, however, not been torn off from the Earth at the Pacific Ocean, as sometimes believed. Firstly, the Pacific is less than 200 million years old, and was formed as a result of continental drift. Secondly, the chemical composition of lunar soil is different from that of terrestrial material. On the Moon, there is no water in the rocks, contrary to on the Earth, where water is common (crystal water). Also, the relative abundance of volatile elements, or elements condensing at very low temperatures, is lower on the Moon than on the Earth. The Moon formed somewhere close to the Earth, but whether it has always been a satellite or was captured afterwards is unknown.

Seismometric and gravimetric measurements have supplied basic information on the internal structure of the moon. Moonquakes take place at a depth of 800 – 1000 km, considerably deeper than terrestrial quakes, and they are also much weaker than on the Earth. Most of the quakes occur at the boundary of the solid mantle, or *lithosphere*, and the core, or *asthenosphere* (Fig. 8.31). The *S*-waves cannot penetrate the asthenosphere, indicating that it is at least partially molten. Tidal forces may generate at least some of the moonquakes.

Lunar orbiters have observed local mass concentrations, *mascons*, beneath the maria. These are large basaltic blocks, formed after the huge impacts which produced the maria. The craters were filled by lava flows during the next billion years or so in several phases. This can be seen, e.g. in the area of Mare Imbrium. Large maria were

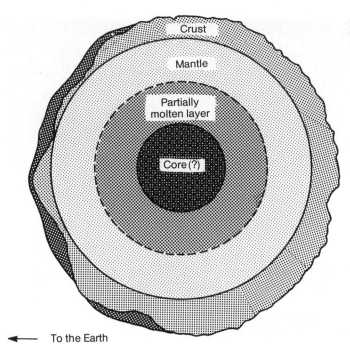

Crust

Mantle

Partially
molten layer

Core (?)

◄── To the Earth

formed about 4000 million years ago when meteorite bombardment was much heavier than today. The last 3000 million years have been quite peaceful, without any major events.

*Atmospheric Phenomena

The best-known atmospheric phenomenon is the *rainbow*, which is due to the refraction of light from water droplets. The radius of the arc of the rainbow is about 41° and the width, 1.7°. The centre of the arc is opposite the Sun (or any other source of light). When the light is refracted inside a water droplet, it is divided into a spectrum, where the red colour is at the outer edge and blue, at the inner edge. Light can be reflected twice inside the droplet, resulting in a secondary rainbow outside the primary one. The colours of the secondary rainbow are in reversed order and its radius is 52°. A rainbow caused by the Moon is usually very weak and colourless, since the human eye is incapable of resolving colours from a dim object.

A *halo* results when the solar or lunar light is reflected from atmospheric ice crystals. The most common halo is a 22° arc or circle around the Sun or the Moon. Usually the halo is white, but occasionally even bright colours can be seen. Another common form is the side lobes which are at the same height as the Sun but at a distance of 22° from it. All other forms of halo are less common. The best "weather" for halos is when there are cirrostratus or cirrus clouds or an icy fog in the sky.

Noctilucent clouds are thin formations of cloud, at a height of approximately 80 km. The clouds contain particles which are less than one micron in diameter, and become visible only when the Sun (which is below the horizon) illuminates the clouds. Most favourable conditions are at the northern latitudes during the summer nights when the Sun is only a few degrees below the horizon.

The night sky is never absolutely dark. One reason (in addition to light pollution) is the *airglow*, or light emitted by excited atmospheric molecules. Most of the radiation is in the infrared domain, but e.g. the forbidden line of oxygen, at 558 nm, has also been detected.

The same greenish oxygen line is clearly seen in *aurorae* which are formed at a height of 80 – 300 km. Aurorae can be seen mainly from relatively northern or southern latitudes because the Earth's magnetic field forces charged particles, coming from the Sun, close towards the magnetic poles. Alaska and northern Scandinavia are the best places to observe aurorae. Occasionally, aurorae can be seen as far south as 40°. They are usually greenish or yellow-green, but red aurorae have been observed, too. They most commonly appear as

(a) A typical halo   (b) Aurorae. (Photos by P. Parviainen)

so-called arcs, which are often dim and motionless, or as belts, which are more active and may contain rapidly varying vertical rays.

Meteors or shooting stars are small grains of sand, a few micrograms or grams in weight, which hit the Earth's atmosphere. Due to friction, the body heats up and starts to glow at a height of 100 km. Some 20 – 40 km lower, the whole grain has burnt to ashes. The duration of a typical meteor is less than a second. The brightest meteors are called *bolides* (magnitude smaller than about − 2). Even larger particles may survive down to the Earth. Meteors are further discussed in Sect. 8.20.

8.15 Mars

Mars is the outermost of the terrestrial planets. Its diameter is only half that of the Earth. Seen through a telescope, Mars seems to be a reddish disk with dark spots and white polar caps. The polar caps wax and wane with the Martian seasons, indicating that they are composed of ice. Darker areas were suspected to be vegetation. At the end of the 19th century, an Italian astronomer, Giovanni Schiaparelli claimed that there are canals on Mars. In the United States, the famous planetary astronomer Percival Lowell studied the canals and even published books on the topic. Martians were also very popular in the science fiction literature. Now the canals are known to be nonexistent, an optical illusion when the obscure details at the limit of visibility seem to form straight lines, canals. Finally, the first clear pictures by Mariner 4 in 1965 buried even the most optimistic hopes concerning life on Mars.

Mars is a superior planet, which means that it is easily observable when close to the Earth, i.e. during opposition, when the planet is above the horizon all night long. Thus much was known about Mars prior to the space age. Mars has seasons like the Earth.

Fig. 8.32. Details of the Martian southern hemisphere at Argyre Planitia. Thin haze is visible at the horizon. (NASA)

Its rotation axis is tilted some 25° to the ecliptic, about the same amount as the Earth's axis. A Martian day is only half an hour longer than a terrestrial day. Huge dust storms are occasionally seen on Mars.

Probes revealed details of the planet. Craters were found in the first pictures sent by the probes. The southern hemisphere is especially marked by craters (Fig. 8.32), indicating that the original surface is still visible there. On the other hand, the northern hemisphere has an abundance of large lava basins and volcanos. The largest volcano, *Olympus Mons* (Fig. 8.33), protrudes more than 20 km above the surrounding terrain. The diameter at the bottom is about 600 km. There are also several canyons, the largest of which is *Valles Marineris*. Its length is 5000 km, width, 200 km, and depth, about 6 km. Compared with Valles Marineris, the Grand Canyon is merely a scratch on the surface.

Ancient riverbeds (Fig. 8.34), too small to be seen from the Earth, were also discovered by spacecraft. (These have nothing to do with the famous canals.) At present, the temperature and air pressure on Mars are too low for free water to exist. The mean temperature is $-53\,°C$ and, on a warm summer day, the temperature can rise close to zero near the equator. The pressure is only $5-8$ mbar. Rivers were probably formed soon after the formation of Mars itself, when there was a great deal of water and the atmospheric pressure and temperature were higher. Nowadays most of the water is contained in kilometres-deep permafrost below the surface and in the polar caps.

US Viking 1 and 2 landed on Mars in 1976 (Fig. 8.35). The Viking orbiters mapped the planet and relayed the data collected by the landing modules to the Earth. The temperature measurements of the orbiters showed that the polar caps are composed both of water and carbon dioxide ice. The northern cap is almost season-independent, ex-

Fig. 8.33   Fig. 8.34

Fig. 8.33. Olympus Mons, one of the largest known volcanos in the solar system. (Photo Lunar and Planetary Institute)

Fig. 8.34. Riverbeds and "islands" around some craters indicate that there has been water on the Martian surface in the past. (NASA)

tending down to latitude 70°. On the other hand, the southern cap, which reaches to the latitude −60° in the southern winter, disappears almost totally during the summer. The southern cap consists mostly of CO_2 ice. The permanent parts are of ordinary water ice, since the temperature, −73 °C, is too high for CO_2 ice. The water ice layers are hundreds of metres thick.

The dark areas are not vegetation, but loose dust, moved around by strong winds. These winds raise the dust high into the atmosphere, colouring the Martian sky red. The cameras of the Vikings revealed a reddish regolithic surface, scattered with boulders. The red colour is caused mainly by iron oxide; already in the 1950's, the existence of limonite ($2FeO_3 3H_2O$) was deduced from polarization measurements. The on-site analysis showed that the soil consists of 13% iron and 21% silicon. The abundance of sulphur was found to be ten times that found on the Earth.

Three biological experiments of the Viking landers searched for signs of life. No organic compounds were found − however, the biological tests did give some unexpected results. A closer look at the results indicated no life, but some uncommon chemical reactions.

The atmosphere of Mars is mainly composed of carbon dioxide (95%), not nitrogen, as was formerly believed; it contains only 2% nitrogen and 0.1−0.4% oxygen. The atmosphere is very dry: if all the moisture were condensed on the surface, the water layer would be thinner than 0.1 mm. The atmospheric pressure on the surface is only

1/200 atm. A part of the atmosphere has escaped but it is probable that Mars never had a thick atmosphere.

Mars has two moons, *Phobos* and *Deimos* (Fig. 8.36). The size of Phobos is roughly 27 km × 21 km × 19 km, and the orbital period around Mars is only 7 h 39 min. In the Martian sky, Phobos rises in the west and sets in the east. Deimos is smaller. Its diameter is 15 km × 12 km × 11 km. There are craters on both moons. Polarimetric and photometric results show that they are composed of material resembling carbonaceous chondrite meteorites.

8.16 Asteroids

Asteroids, or minor planets, orbit the Sun mainly between Mars and Jupiter. Most of the asteroids are in the asteroid belt with distances of 2.2 – 3.3 AU from the Sun (Fig. 8.37a). The first asteroid was discovered in 1801, and there are about 5000 numbered asteroids in the 1993 minor planet catalog. Nowadays, the number of catalogued asteroids increases by dozens every year. It has been estimated that almost half a million asteroids larger than 500 metres exist in the asteroid belt. Yet the total mass of the asteroids is less than 1/1000 of the Earth's mass. The first and the biggest is 1 Ceres; its diameter is 1000 km.

Fig. 8.36. Phobos (*left*) and Deimos, the two moons of Mars. (NASA)

The centre of the asteroid belt is at a distance of approximately 2.8 AU, as predicted by the Titius-Bode law (Sect. 8.21). According to a formerly popular theory, asteroids are debris from the explosion of a planet. This theory, like catastrophe theories in general, has been abandoned. The currently accepted theory assumes that asteroids were formed simultaneously with the major planets. The primeval asteroids were large chunks, orbiting between the orbits of Mars and Jupiter. Due to collisions and fragmentation, the present asteroids are debris of those primordial bodies which were never able to form a large planet. Some of the biggest asteroids may be those original chunks.

The distribution of asteroids inside the asteroid belt is uneven (Fig. 8.37b); they seem to avoid some areas known as the *Kirkwood gaps*. The most prominent void areas are at distances where the orbital period of an asteroid around the Sun (given by Kepler's third law) is in the ratio 1:3, 2:5, 3:7, or 1:2 to the orbital period of Jupiter. The motion of an asteroid orbiting in such a gap would be in resonance with Jupiter, and even small perturbations would tend to grow with time. The body would eventually be moved to another orbit. However, the resonance effects are not so simple: sometimes an orbit is "locked" to a resonance, e.g. the *Trojans* move along the same orbit as Jupiter (1:1 resonance), and the *Hilda* group is in 2:3 resonance.

Some minor families of asteroids move outside the main belt. These include the above-mentioned Trojans, which orbit 60° behind and ahead of Jupiter. The Trojans,

Fig. 8.37. (a) Most of the asteroids orbit the Sun in the asteroid belt between Mars and Jupiter. The figure shows the positions of 6200 catalogued asteroids on 1 January, 1995, and the orbits and positions of the Earth, Mars, Jupiter, and Saturn. **(b)** The total number of asteroids as a function of the distance from the Sun. Each bin corresponds to 0.01 AU. The empty areas, the Kirkwood gaps, are at those points, where the orbital period of an asteroid is in a simple ratio to the orbital period of Jupiter

which are close to the special points L4 and L5 of the solution of the so-called restricted three-body problem. At these Lagrangian points, a massless body can remain stationary with respect to the massive primaries (in this case, Jupiter and the Sun). In fact, the asteroids are oscillating around the stationary points, but such orbits can be shown to be stable against perturbations.

Another large family is the *Apollo-Amor* asteroids. The perihelia of Apollo and Amor are inside the Earth's orbit and between the orbits of the Earth and Mars, respectively. These asteroids are all small, less than 30 km in diameter. The most famous is 433 Eros, which was used for determining the length of the astronomical unit. When closest to the Earth, Eros is at a distance of only 20 million km. There is a marginal probability that some Earth-crossing asteroids will collide with the Earth. It has been estimated that on the average, two such collisions take place in one million years.

The most distant asteroid, 2060 Chiron, was discovered in 1977. The aphelion is close to the orbit of Uranus and the perihelion is slightly inside the orbit of Saturn. It is possible that a family of asteroids exists beyond the orbit of Jupiter; they are very faint and thus difficult to find.

An asteroid observer needs a telescope, since even the brightest asteroids are too faint to be seen with the naked eye. Asteroids are points of light like a star, even if seen through a large telescope; only their slow motion against the stellar background reveals that they are members of the solar system. The rotation of an asteroid gives rise to a regular light variation (Fig. 8.38). However, the amplitude is generally so small that a photometer is needed for accurate measurement. The rotation period of an asteroid can be calculated if the light curve is well observed, preferably within the span of many nights. The exact periods of more than 300 asteroids are known. Typical periods range from 4 to 15 hours but shorter and longer periods have been discovered.

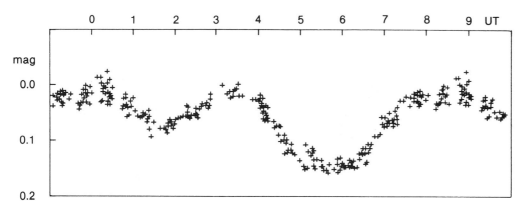

Fig. 8.38. Composite light curve of asteroid 64, Angelina. The curve has been constructed from observations made during several nights. The rotation period is slightly less than 9 hours

The exact diameters of asteroids were long unknown. Edward E. Barnard of the Lick Observatory determined visually the diameters of 1 Ceres, 2 Vesta, 3 Juno, and 4 Pallas in the 1890's. Practically no other reliable results existed prior to the 1960's, when the methods described in Sects. 8.7 and 8.8 were successfully applied to asteroids. Moreover, an ever increasing number of asteroid occultations are being observed.

The first images of asteroids were obtained in the early 1990's. In 1991 Galileo probe passed asteroid 951 Gaspra, and in 1993 asteroid 243 Ida, on its long way to Jupiter. The images of Gaspra (Fig. 8.39) showed an irregular, 17 km long body with pulverized rock and dust on its surface. The shape gave hint that once the asteroid may have been two separate objects that merged into one. At the end of 1992 asteroid 4179 Toutatis passed the Earth only by 4 million km. Radar images revealed a two-body system, where the components were touching each other. Double asteroids may be quite common, and there exist light curves of some asteroids which have been interpreted as results of twin bodies.

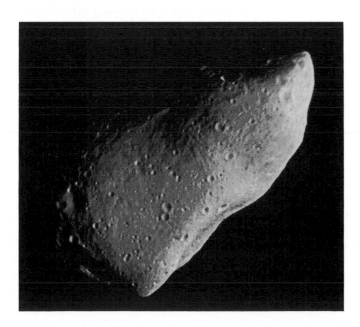

Fig. 8.39. Asteroid Gaspra was photographed by Galileo spacecraft in October 1991. The illuminated part of the asteroid is about 16×12 km. The smallest craters in this view are about 300 m across. (Photo NASA)

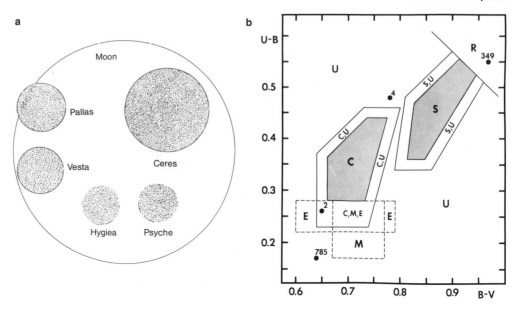

Fig. 8.40. (a) Sizes of some asteroids compared with the Moon. **(b)** The classification of asteroids can be based, for example, on UBV photometry. The colour indices B-V and U-B are different for different types of asteroids. (E. Bowell)

The composition of asteroids is similar to that of iron-, stone-, and iron-stone meteorites. The taxonomy of asteroids was developed in the mid-1970's, when enough observational data had become available (Fig. 8.40). Most asteroids can be divided into two groups, according to their photometric and polarimetric properties. 95% of the classified asteroids are either C- or S-types. C-asteroids are dark (geometric albedo $p \sim 0.06$), and they contain a considerable amount of carbon (mnemonic C from carbon). They resemble stony meteorites. Traces of silicates have been found in their spectra, such as olivine Mg_2SiO_4 or Fe_2SiO_4. The reflectivity of S-asteroids is higher and their spectra are close to those of stone-iron meteorites. All other classes are quite rare, containing possibly only a few members. Some asteroids, having features of many groups, are classified as "unclassified".

8.17 Jupiter

The realm of terrestrial planets ends at the asteroid belt. Outside this, the relative abundance of volatile elements is higher and the original composition of the solar nebula is still preserved in the giant planets. The first and largest is Jupiter. Its mass is 2.5 times the total mass of all other planets, almost 1/1000 of the Solar mass. The bulk of Jupiter is mainly hydrogen and helium. The relative abundance of these elements are approximately the same as on the Sun, and the density is of the same order of magnitude, namely $1330 \, \text{kg m}^{-3}$.

During oppositions, the angular size of Jupiter is as large as 50″. The dark *belts* and lighter *zones* are visible even with a small telescope. These are cloud formations, parallel to the equator. The most famous detail is the *Great Red Spot*, a huge cyclone, rotating counterclockwise once every six days. The spot was discovered by Giovanni Cassini in 1655; it has survived for centuries, but its true age is unknown.

The rotation of Jupiter is quite rapid; one revolution takes 9 h 55 min 29.7 s. This is the period of the magnetic field, also reflecting the rotation of the innermost core, since the field is frozen in the core. As might be expected, Jupiter does not behave like a rigid body. The rotation period of the surface layers (clouds) is considerably longer at the poles than at the equator. Due to its rapid rotation, Jupiter is nonspherical; flattening is as large as 1/15.

There is possibly an iron-nickel core in the centre of Jupiter. The mass of the core is probably equal to a few tens of Earth masses. The core is surrounded by a layer of metallic liquid hydrogen, where the temperature is over 10000 K and the pressure, three million atm. Owing to this huge pressure, the hydrogen is dissociated into single atoms, a state unknown in ordinary laboratory environments. In this exotic state, hydrogen has many features typical of metals. This layer is electrically conductive, giving rise to a strong magnetic field. Closer to the surface where the pressure is lower, the hydrogen is present as normal molecular hydrogen, H_2. At the top is a 1000-km-thick atmosphere.

Jupiter radiates twice the amount of heat that it receives from the Sun. This heat is a remnant of the energy released in the gravitational contraction during the formation of the planet. Thus Jupiter is still gradually cooling. The internal heat is transferred outwards by convection; this gives rise to flows in the metallic hydrogen, causing the magnetic field.

Belts and zones are stable cloud formations (Fig. 8.41). Their width and colour may vary with time, but the semi-regular pattern can be seen up to the latitude 50°. The colour of the polar areas is close to that of the belts. The belts are reddish or brownish, and the motion of the gas inside a belt is downwards. The gas flows upwards in the white zones. The clouds in the zones are slightly higher and have a lower temperature than those in the belts. Strong winds or jet streams blow along the zones and belts. The speed of the wind reaches almost 150 m/s at some places in the atmosphere, according to the measurements of the Galileo probe at the end of the year 1995.

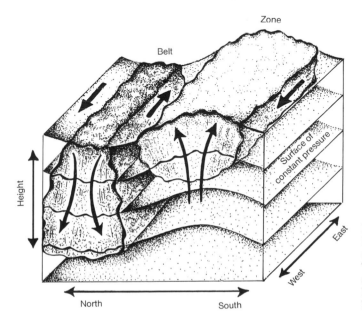

Fig. 8.41. Light bands and dark belts are cloud formations with opposite vertical gas motions. The gas is rising in the bands and sinking in the belts. Also the horizontal winds in bands and belts are in opposite directions

Fig. 8.42. Jupiter and its moon,
Io. (NASA)

Fig. 8.43. The Great Red Spot of Jupiter. (NASA)

The colour of the Great Red Spot (GRS) resembles the colour of the belts (Fig. 8.43). Sometimes, it is almost colourless, but shows no signs of decrepitude. The GRS is 14000 km wide and 30000 – 40000 km long. Some smaller red and white spots can also be observed on Jupiter, but their lifetime is generally much less than a few years.

The atmospheric composition of Jupiter was accurately measured by the US Pioneer 10 and 11, Voyager 1 and 2, and Galileo probes. The latest results by Galileo show that the helium abundance in the higher atmosphere is only about half of that in the Sun. The situation is somewhat similar in Saturn. The helium depletion may be due to differentiation, i.e. the heavier helium is gradually sinking to the interior of the planet thus causing over-abundance of hydrogen. Other compounds found in the atmosphere include methane, ethane and ammonia. The temperature in the cloud tops is about 130 K.

Jupiter is an intense radio source. Its radio emission can be divided into three components, namely thermal millimetre and centimetre radiation, nonthermal decimetric radiation and burstal-decametric radiation. The nonthermal emission is most interesting; it is partly synchrotron radiation, generated by relativistic electrons in the Jovian magnetosphere. Its intensity varies in phase with Jupiter's rotation; thus the radio emission can be used for determining the exact rotation rate. The decametric bursts are related to the position of the innermost large moon, Io, and are possibly generated by the strong electric current observed between Io and Jupiter.

The ring of Jupiter (Fig. 8.44) was discovered in 1979. It is sparse and narrow, about 6500 km wide and less than a kilometre thick. The ring particles are small, a few microns only, and they scatter light forwards much more effectively than backwards. A ring consisting of such small particles cannot be stable, and new material must enter the ring continuously. The most probable source is Io.

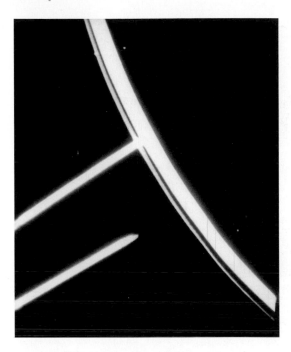

Fig. 8.44. The ring of Jupiter. The Sun is behind the limb of the planet (the light band from *top to bottom*) and the ring (the loop at *left*) seems to end abruptly when it enters the shadow of Jupiter. (NASA)

Jupiter has 16 moons. The four largest, *Io, Europa, Ganymede* and *Callisto* are called the *Galilean satellites* (Fig. 8.45), in honour of Galileo Galilei, who discovered them in 1610. The Galilean satellites can be seen with ordinary binoculars. They are the size of the Moon or even Mercury. The other moons are small, less than about 200 km in diameter.

Owing to tidal forces, the orbit of Io, Europa and Ganymede have been locked into a resonance, so that their longitudes λ strictly satisfy the equation

$$\lambda_{Io} - 3\lambda_{Europa} + 2\lambda_{Ganymede} = 180° \qquad (8.34)$$

Hence the moons can never be in the same direction when seen from Jupiter.

Io is the innermost Galilean satellite. It is a little larger than the Moon. Its surface is spotted by numerous *calderas*, volcanoes without a mountain. The molten material is ejected up to a height of 250 km. The volcanic activity on Io is much stronger than on the Earth. The eruptions are possibly powered by tidal forces caused by Jupiter, Europa and Ganymede. The tides generate friction, which is transformed to heat. This heat keeps the sulphur compounds molten beneath the colourful surface of Io. No traces of impact craters are visible. The whole surface is new, being renewed continuously by eruptions. There is no water on Io.

Europa is the smallest of the Galilean satellites, a little smaller than the Moon. The surface is ice-covered and the geometric albedo is as high as 0.6. The surface is extremely smooth and without any traces of impact craters. The surface is renewed continuously by fresh water, trickling from the internal ocean. At the centre, there is a silicate core.

Ganymede is the largest moon in the solar system. Its diameter is 5300 km; it is larger than the planet Mercury. About 50% of the mass is water or ice, the other half being silicates (rocks). The density of craters on the surface varies, indicating that there are areas of different ages.

Fig. 8.45 a – d. The Galilean satellites: (**a**) Io, (**b**) Ganymede, (**c**) Europa, (**d**) Callisto. (NASA)

Callisto is the outermost of the large moons. It is dark; its geometric albedo is less than 0.2. The ancient surface is peppered by meteorite craters; no signs of geological activity are visible. The surface is a mixture of stone and ice; the internal structure is much the same as Ganymede's.

The outer moons are divided into two groups. The orbits of the inner group are inclined about 35° to the equator of Jupiter. The four outermost moons are in eccentric retrograde orbits. It is possible that these are asteroids captured by Jupiter.

8.18 Saturn

Saturn is the second largest planet. Its diameter is about 120000 km, ten times the diameter of the Earth, and the mass, 95 times the Earth's mass. The density is only 700 kg m^{-3}, less than the density of water. The rotation axis is tilted about 27° with respect to the orbital plane, so every 15 years, the northern or the southern pole is well observable. The rotation period is 10 h 30 m.

The internal structure of Saturn resembles that of Jupiter. Due to its smaller size, the metallic hydrogen layer is not so thick as on Jupiter. Saturn's appearance is also somewhat different. Viewed from the Earth, Saturn is a yellowish disk without any conspicuous details. The clouds have fewer features than those on Jupiter, partly because a thin haze floats above the cloudtops. Furthermore, Saturn is further from the Sun than Jupiter and thus has a different energy budget.

The winds, or jet streams, are similar to those of Jupiter. The temperature at the cloudtops is about 94 K, and Saturn radiates twice as much energy as it receives from the Sun.

Saturn's most remarkable feature is a thin ring system (Figs. 8.46, 8.47), lying in the planet's equatorial plane. The Saturnian rings can be seen even with a small telescope. The rings were discovered by Galileo Galilei in 1610; only 45 years later, did Christian

Fig. 8.46. Saturn and its rings. (NASA)

Fig. 8.47. A schematic drawing of the structure of the Saturnian rings

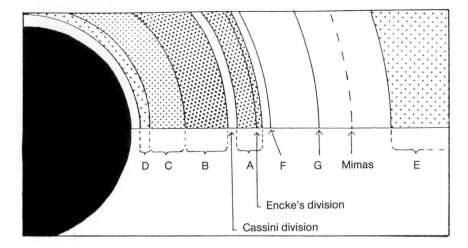

Huygens establish that the formation observed was actually a ring, and not two oddly behaving bulbs, as they appeared to Galileo.

The rings are made of normal water ice. The size of the ring particles ranges from microns to truck-size chunks. Most of the particles are in range of centimetres to metres. The width of the whole ring system is 60 000 km (about the radius of Saturn) and the thickness, at most, 100 m, and possibly only a few metres.

According to Earth-based observations, the rings are divided into three parts, called simply A, B, and C. The innermost C ring is 17 000 km wide and consists of very thin material. There is some material even inside this (sometimes referred to as the D ring), and a haze of particles may extend down to the clouds of Saturn.

The B ring is the brightest ring. Its total width is 26 000 km, but the ring is divided into thousands of narrow ringlets, seen only by the probes (Fig. 8.48). From the Earth,

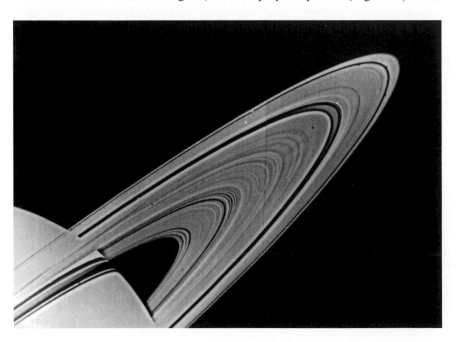

Fig. 8.48. At a close distance, the rings can be seen to be divided into thousands of narrow ringlets. (NASA)

the ring seems more or less uniform. Between B and A, there is a 3000-km-wide gap, the *Cassini division*. It is not totally void, as was previously believed; some material and even narrow ringlets have been found in the division by the probes.

The A ring is not divided into narrow ringlets as clearly as the B ring. There is one narrow but obvious gap, *Encke's division*, close to the outer edge of the ring. The outer edge is very sharp, due to the "shepherd" moon, some 800 km outside the ring. The moon prevents the ring particles from spreading out to larger orbits. It is possible that the appearance of B is due to yet-undiscovered moonlets inside the ring.

The F ring, discovered in 1979, is about 3000 km outside A. The ring is only a few hundred kilometres wide. On both sides there is a small moon; these shepherds prevent the ring from spreading. An interior moon passing a ring particle causes the particle to move to a larger orbit. Similarly, at the outer edge of the ring, a second moon forces the particles inwards. The net result is that the ring is kept narrow.

Outside the F ring, there are some zones of very sparse material, sometimes referred to as the G and E rings. These are merely collections of small particles, not real rings.

The Saturnian rings were possibly formed together with Saturn and are not debris from some cosmic catastrophe, like remnants of a broken moon. The total mass of the rings is 1/10 000 000 of the mass of Saturn. If all ring particles were collected together, they would form an ice ball, 600 km in diameter.

Most of the Saturnian moons (Fig. 8.49) were observed by Pioneer 11 and Voyager 1 and 2. Three new moons were discovered by Voyager 1, and 5 more by Earth-based observations in 1980. Some moons are dynamically interesting; some have an exotic geological past. Most of the moons are made up of almost pure ice, since their mean densities are close to that of water. The temperature of the primeval nebula at the distance of Saturn was so low that bodies of pure ice could form and survive.

Outside the F ring, there are two moonlets almost in the same orbit; the difference of the semi-major axes is less than the radii of the moons. The inner moon is gaining on the outer one. The moons will not collide, since the speed of the trailing moon increases and the moon moves outwards. The speed of the leading moon decreases and it drops inwards. The moons exchange their roles and "the show goes on".

The innermost of the "old" moons is *Mimas*. There is a huge crater on Mimas' surface with a diameter of 100 km and a depth of 9 km. Bigger craters exist but, relative to the size of the parent body, this is almost the biggest crater there could be room for (otherwise the crater would be bigger than Mimas itself). On the opposite side, some grooves can be seen, possibly signifying that impact has almost torn apart the moon.

The surface of the next moon, *Enceladus*, consists of almost pure ice, and one side is nearly craterless. Craters and grooves can be found on the other hemisphere. Tidal forces result in volcanic activity where water (not lava or other "hot" material) is discharged to the surface.

Titan is the largest of the Saturnian moons. Its diameter is 5100 km, so it is only slightly smaller than Jupiter's moon, Ganymede. Titan is the only moon with an atmosphere. The atmosphere is mainly nitrogen (99%), and the pressure at the surface is 1.5 − 2 bar; the temperature is about 90 K. Reddish clouds form the visible surface some 200 km above the solid body.

Iapetus is one of the strangest objects in the solar system. One side of Iapetus is extremely dark (albedo only 3%). The other side is more than five times brighter. The ultimate reason for the dark colour is unknown.

Fig. 8.49 a – c. Saturnian moons: (a) Mimas, (b) Dione, (c) Hyperion. The irregular shape of Hyperion can be seen from the three separate pictures. (NASA)

8.19 Uranus, Neptune and Pluto

The planets from Mercury to Saturn were known already in antiquity. Uranus, Neptune and Pluto can only be observed with a telescope. Uranus and Neptune are giants, similar to Jupiter and Saturn. Pluto is a small icy world, totally different from any other planet.

Uranus. The famous German-English amateur astronomer William Herschel discovered Uranus in 1781. Herschel himself first thought that the new object was a comet. However, the extremely slow motion revealed that the body was far beyond the orbit of Saturn. Based on the first observations, the Finnish astronomer Anders Lexell calculated a circular orbit. He was one of the first to propose that the newly discovered object was a planet. Johann Bode of the Berlin Observatory suggested the name Uranus but more than five decades passed before the name was unanimously accepted.

The mean distance of Uranus is 19 AU, and the orbital period is 84 years. The inclination of the rotation axis is 98°, which is totally different from the other planets. Due to this uncommon geometry, the poles are either lit or in darkness for decades. The rotation period, confirmed by the Voyager 2 magnetometric measurements in 1986, is 17.3 hours; the exact period had been uncertain prior to the fly-by.

Uranus is greenish, as viewed through a telescope. Its colour is due to the strong methane absorption bands in the near-infrared. A part of the red light is also absorbed, leaving the green and blue part of the spectrum untouched. Uranus is almost featureless (Fig. 8.50) because its clouds are below a thick haze or smog. The temperature at the cloud tops is 84 K.

The internal structure of Uranus is thought to be slightly different from that of other giant planets. Above the innermost rocky core, there is a layer of water, which, in turn, is surrounded by a mantle of hydrogen and helium. The mixture of water and ammonia and methane therein are dissociated to ions under the heavy pressure. This mixture behaves more like a molten salt than like water. The convection flows in this electrically conductive "sea" give rise to the Uranian magnetic field. The strength of the magnetic field at the cloud tops is comparable to the terrestrial field. However, Uranus is much larger than the Earth, so the true strength is 50 times greater than the Earth's field. The Uranian magnetic field is tilted 60° with respect to the rotation axis. No other planet has such a highly inclined magnetic field.

The Uranian rings (Fig. 8.51) were discovered in 1977, during a stellar occultation. Some secondary occultations were observed before and after the main event. A total of 10 rings are known. Nine were discovered by Earth-based observations and one during the Voyager 2 fly-by. All the rings are dark and very narrow, only a few hundred metres or a few kilometres wide. The Voyager 2 results showed that the rings contain very little dust, unlike those of Jupiter and Saturn. The mean size of the ring particles is

Fig. 8.50. Uranus is a featureless body, even if viewed from a close distance. (NASA)

Fig. 8.51. The rings of Uranus are very narrow and composed of a dark material. The tenth ring was discovered by Voyager 2 in 1986. (NASA)

more than 1 metre. The ring particles are darker than practically any known material in the solar system; the cause of this dark colour is unknown.

There are 15 moons orbiting around Uranus, ten of which were discovered by Voyager 2. The geological history of some moons is puzzling, and many features reminiscent of an active past can be found. The composition is more like that of Jovian; the mean density is well above the density of water.

The innermost of the large moons, *Miranda*, is one of the most peculiar objects discovered (Fig. 8.52). It has several geological formations also found elsewhere (but here they are all mixed together), in addition to the quite unique V-shaped formations. It is possible that Miranda's present appearance is the result of a vast collision that broke the moon apart; some pieces may have later settled down, inside out. Another peculiar object is *Umbriel*. It belongs to the ever increasing family of unusual dark bodies (such as the Uranian rings, one side of Iapetus and the Halley comet). The dark surface of Umbriel is covered by craters without any traces of geological activity.

Neptune. The orbit of Uranus was well known already in the beginning of the 19th century. However, some unknown perturbations displaced Uranus from its predicted orbit. Based on these perturbations, John Couch Adams, of Cambridge, and Urbain Jean-Joseph Le Verrier, of Paris, independently predicted the position of the unknown perturbing planet.

The new planet was discovered in 1846 by Johann Gottfried Galle at the Berlin Observatory; Le Verrier's prediction was found to be only 1° off. The discovery gave rise to a heated controversy as to who should be given the honour of the discovery, since Adams' calculations were not published outside the Cambridge Observatory. When the quarrel was settled years later, both men were equally honoured. The discovery of Neptune was also a great triumph of the Newtonian theory of gravity.

The semi-major axis of the orbit of Neptune is 30 AU and the orbital period around the Sun, 165 years. The internal rotation period, confirmed by Voyager 2 in 1989, is 16 hours 7 minutes and the rotation period of the outer layers of the clouds is about

Fig. 8.52a – c. Uranian moons: **(a)** Miranda, **(b)** Ariel, **(c)** Titania. (NASA)

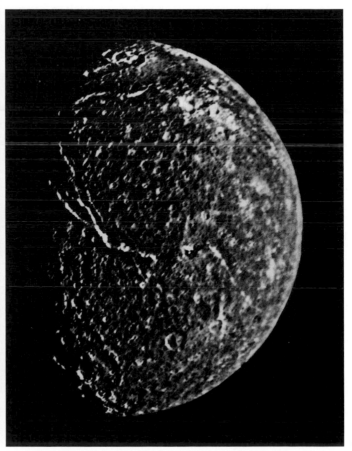

17 hours. The obliquity of the rotation axis is 29° but the magnetic field is tilted some 50° with respect to the rotation axis. The magnetic field is tilted like in Uranus, but the field strength is much smaller.

The density of Neptune is 1660 kg m⁻³, and the diameter 48 600 km. Thus the density of Neptune is higher than that of other giant planets. The internal structure is quite simple: The core, composed of silicates (rocks) is about 16 000 km in diameter. This is surrounded by a layer of water and liquid methane and the outermost gaseous layer, atmosphere, is mainly composed of hydrogen and helium, methane and ethane being a minor components.

Cloud structures are more complicated than on Uranus, and some dark spots, like in Jupiter, were visible during the Voyager fly-by (Fig. 8.53). The speed of the winds are quite high, up to 400 m/s.

Like other giants, also Neptune has rings (Fig. 8.54). The rings were discovered by Voyager 2, although their existence was expected already prior the fly-by. Two, relatively bright but very narrow rings are at the distance of 53 000 and 62 000 km from the centre of the planet. Moreover, there are some faint areas of fine dust.

There are 8 known moons, six of which were discovered by Voyager 2. The largest of the moons, Triton, is 2700 km in diameter, and it has a thin atmosphere, mainly composed of nitrogen. The albedo is high: Triton reflects 60 – 80% of the incident light. The surface is relatively young, without any considerable impact craters (Fig. 8.55). There are some active "geysirs" of liquid nitrogen, which partly explains the high albedo and the lack of the craters. On Triton the low temperature, 37 K, means that the nitrogen covers the surface like snow.

Fig. 8.53. Clouds of Neptune and the Great Dark Spot as seen from Voyager 2 in August 1989. (NASA)

Fig. 8.54. The Neptunian ring system seen from behind the planet. (NASA)

Fig. 8.55. The southern hemisphere of Triton, Neptune's largest satellite. The dark spots may indicate eruptions of "icy vulcanoes". (NASA)

Pluto. Pluto is the outermost planet of the solar system. It was discovered in 1930 at the Lowell Observatory, Arizona, after an extensive photographic search (Fig. 8.56). This search had already been initiated in the beginning of the century by Percival Lowell, on the basis of the perturbations observed in the orbits of Uranus and Neptune. Finally, Clyde Tombaugh discovered Pluto less than 6° off the predicted position. However, Pluto turned out to be far too small to cause any perturbations on Uranus or Neptune. Thus the discovery was purely accidental, and the perturbations observed were not real, but caused by minor errors of observation.

Pluto has no visible disk; instead, it resembles a point, like a star. This fact gave an upper limit for the diameter of Pluto, which turned out to be about 3000 km. The exact mass was unknown until the discovery of the Plutonian moon, *Charon,* in 1978. The mass of Pluto is only 0.2% of the mass of the Earth.

Mutual occultations of Pluto and Charon in 1985 – 87 gave accurate diameters of each body: The diameter of Pluto is 2300 km and the diameter of Charon is 1500 km. The density of Pluto turned out to be about $2100 \, \mathrm{kg \, m^{-3}}$, somewhat more than previously presumed. Thus Pluto is not a huge methane snowball but a substantial amount of its mass is composed of rocks.

The orbit of Pluto is quite peculiar. The eccentricity is 0.25 and the inclination, 17°. During its 250-year orbit, Pluto is closer to the Sun than Neptune for 20 years; one such period lasts from 1979 to 1999. There is no danger of Pluto and Neptune colliding, since Pluto is high above the plane of ecliptic, when at the distance of Neptune.

A "Trans-Plutonian planet" has been sought, more or less seriously. According to Tombaugh, no object brighter than magnitude 15 can exist. The discovery of such a faint object is very unlikely.

Fig. 8.56. Pluto is so distant that even large telescopes see it only as a point of light. Pluto and its moon Charon can be barely separated by ground-based telescopes. The best views have been obtained from the Hubble Space Telescope. (Photo NASA)

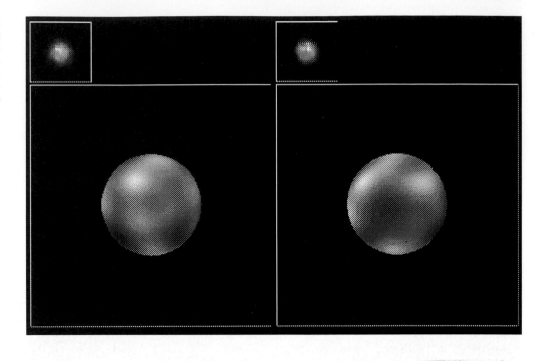

8.20 Minor Bodies of the Solar System

So far we have considered only planets, planetary satellites and asteroids. There are also other bodies in the solar system, like comets, meteoroids and interplanetary dust.

Comets (Fig. 8.57). Small agglomerates of ice, snow, and dust can form comets; a typical diameter is of the order of 10 km or less. A comet is invisible when far from the Sun; when it gets closer than about 2 AU, the heat of the Sun starts to melt the ice and snow. The outflowing gas and dust form an envelope around the core. Radiation pressure and the solar wind push gas and dust away from the Sun, resulting in the typical long-tailed shape of a comet.

The tail is always pointing away from the Sun, a fact which was noticed already in the 16th century. Usually, there are two tails, a gas tail and a dust tail. The ions in the gas tail are driven by the solar wind. Some of the light is reflected solar light, but the brightness of the gas tail is mostly due to emission by the excited atoms. The dust tail is caused by the radiation pressure. Because the velocities of the particles of the dust tail are lower than the velocities in the gas tail, the dust tail is often more curved than the gas tail.

Cometary material is very loose. The whole comet may break apart; in 1975 Comet West disintegrated into 4–5 pieces after perihelion and Comet Shoemaker-Levy 9 which collided into Jupiter in 1994 was torn apart two years earlier when it passed Jupiter at a distance of 21 000 km. Comets are rather ephemeral things, surviving only a few thousand revolutions around the Sun or less. The so-called short-period comets are all newcomers and can survive only a short time here, in the central part of the solar system.

Since comets in the central solar system are rapidly destroyed, there has to be some source of new short-period comets. According to the most commonly accepted hypothesis, a huge cloud of comets, the *Oort cloud*, surrounds the solar system at a distance of about ten thousand AU. Occasionally perturbations from passing stars send some of these comets into orbits, which bring them into the central parts of the solar system, where they are seen as *long-period comets*. Some of these are put into short-period orbits by the perturbations of Jupiter and Saturn, whereas others are ejected from the solar system.

The most famous (and also best known) periodic comet is *Halley's comet*. Its orbital period is about 76 years; it was last in perihelion in 1986. During this apparition, the comet was also observed by probes, revealing the solid cometary body itself for the first time. Halley is a 13×7 km, peanut-shaped chunk whose surface is covered by an extremely black layer of a possibly tarlike organic or other similar material. Violent out-

Fig. 8.57. Comet Mrkos 27.8.1957. (Palomar Observatory)

Fig. 8.58. Meteors or shooting stars are easy to capture on film: one just leaves the field of view the camera loaded with a sensitive film on a tripod with the shutter open for an hour or so. Stars make curved trails on the film. (L. Häkkinen)

Fig. 8.57   **Fig. 8.58**

bursts of gas and dust make an exact prediction of its brightness impossible, as often noticed when cometary magnitudes have been predicted.

The observations revealed that the "classical" dirty snowball model is not quite accurate; at least the surface is more dirt than snow. Several chemical compounds were observed, including water. Spectroscopically, the most abundant constituents are hydrogen, carbon, nitrogen, oxygen and their compounds.

Meteoroids. Solid bodies smaller than asteroids are called meteoroids. The boundary between asteroids and meteoroids, however, is diffuse; it is a matter of taste whether a 10-m body is called an asteroid or a meteoroid. When a meteoroid hits the Earth's atmosphere, an optical phenomenon, called a *meteor* ("shooting star") is seen (Fig. 8.58). The smallest bodies causing meteors are about 1 gram in mass; the (micro)meteoroids smaller than this do not result in optical phenomena. However, even these can be observed with radar which is able to detect the column of ionized air. Micrometeoroids can also be studied with particle detectors installed in artificial satellites and space probes. The number of meteoroids increases rapidly as their size diminishes.

Due to perspective, all meteors coming from the same direction seem to radiate from the same point. Such meteor streams are, e.g. the Perseids in August and the Geminides in December; the names are given according to the constellation in which the radiation point seems to be. Many meteor streams are in the same orbit as a known comet, so at least some meteoroids are of cometary origin.

Most of the meteoroids are small and burn to ashes at a height of 100 km. However, larger bodies may come through and fall to the Earth. These are called *meteorites*.

The relative speed of a typical meteoroid varies in the range 10 – 70 km/s. The speed of the largest bodies does not diminish in the atmosphere; thus, they hit the Earth at their cosmical speeds, resulting in large impact craters. Smaller bodies slow down and drop like stones.

Thousands of meteorites have been found. One of the best places to find meteorites is the Antarctic, where the pieces are carried by ice to the edge of the continent. (In suitable places, meteorites can be picked like berries in a forest!)

Iron-meteorites or irons, composed of almost pure nickel-iron, comprise about one-quarter of all meteorites. Actually the irons are in a minority among meteoroids, but they survive their violent voyage through the atmosphere more easily than other weaker bodies. Three-quarters are stony meteorites, or stones. Meteoroids themselves can be divided into three groups of roughly equal size. One-third is ordinary stones, chondrites. The second class contains weaker carbonaceous chondrites and the third class includes cometary material, loose bodies of ice and snow which are unable to survive down to the Earth.

Interplanetary Dust. Two weak-light phenomena, namely *zodiacal light* (Fig. 8.59) and *gegenschein* (counterglow) make it possible to observe interplanetary dust, small dust particles reflecting the light of the Sun. This weak glow can be seen above the rising or setting Sun (zodiacal light) or exactly opposite the Sun (gegenschein).

Elementary particles hitting the Earth originate both in the Sun and outside the solar system. Charged particles, mainly protons, electrons and alpha particles (helium nuclei) flow continuously out of the Sun. At the distance of the Earth, the speed of this *solar wind* is 300 – 500 km/s. The particles interact with the solar magnetic field. The strength of the solar magnetic field at the Earth's distance is about 1/1000 of that of the Earth. Particles coming from outside the solar system are called *cosmic rays*. In addi-

Fig. 8.59. Zodiacal light

tion to the light particles, cosmic rays contain also heavy nuclei. Some of these particles may have been born in supernova explosions, others can be corpuscular stellar wind.

8.21 Cosmogony

Cosmogony is a branch of astronomy which studies the origin of the solar system. The first steps of the planetary formation processes are closely connected to star formation.

The solar system has some very distinct features which have to be explained by any serious cosmogonical theory. These include:

– planetary orbits are almost coplanar and also parallel to the Solar equator;
– orbits are almost circular;
– planets orbit the Sun counterclockwise, which is also the direction of solar rotation;
– planets also rotate around their axes counterclockwise (excluding Venus, Uranus and Pluto);

– planetary distances approximately obey the empirical Titius-Bode law, i.e.

$$a = 0.4 + 0.3 \times 2^n \,, \qquad n = -\infty, 0, 1, 2, \ldots \tag{8.35}$$

where the semi-major axis a is expressed in AU;
– planets have 98% of the angular momentum of the solar system but only 0.15% of the total mass;
– terrestrial and giant planets exhibit physical and chemical differences;
– the structure of planetary satellite systems resembles miniature solar systems.

The first modern cosmogonical theories were introduced in the 18th century. One of the first cosmogonists was Immanuel Kant, who in 1755, presented his *nebular hypothesis*. According to this theory, the solar system condensed from a large rotating nebula. Kant's nebular hypothesis is surprisingly close to the basic ideas of modern cosmogonical models. In a similar vein, Pierre Simon de Laplace suggested in 1796 that the planets have formed from gas rings ejected from the equator of the collapsing Sun.

The main difficulty of the nebular hypothesis was its inability to explain the distribution of angular momentum in the solar system. Although the planets represent less than 1% of the total mass, they possess 98% of the angular momentum. There appeared to be no way of achieving such an unequal distribution. A second objection to the nebular hypothesis was that it provided no mechanism to form planets from the postulated gas rings.

To overcome the difficulties of the nebular hypothesis, various catastrophe theories were introduced. Such theories were popular in the 19th century and in the first decades of the 20th century. Already in 1745, Georges Louis Leclerc de Buffon had proposed that the planets were formed from a vast outflow of solar material, ejected upon the impact of a large comet.

Later, the cometary impact idea was replaced with the theory of a close encounter with another star. Such theories were developed, e.g. by Forest R. Moulton (1905) and James Jeans (1917). Strong tidal forces during the closest approach would tear some gas out of the Sun; this material would later accrete into planets. Such a close encounter would be an extremely rare event. Assuming a typical star density of 0.15 stars per cubic parsec and an average relative velocity of 20 km/s, we find that only a few encounters would have taken place in the whole Galaxy during the last 5000 million years. The solar system could be a unique specimen.

The main objection to the collision theory is that according to calculations, most of the material torn off the Sun would be captured by the passing star, rather than remaining in orbit around the Sun. There also was no obvious reason why the material would form a planetary system.

In the face of the dynamical and statistical difficulties of the collision theory, the nebular hypothesis was revised and modified in the 1940's. In particular, it became clear that magnetic forces could efficiently transfer angular momentum from the Sun to the planetary nebula. The main principles of planetary formation are now thought to be reasonably well understood.

The oldest rocks found on the Earth are about 3700 million years old; some lunar and meteorite samples are somewhat older. When all the facts are put together, it can be estimated that the Earth and other planets were formed about 4600 million years ago. On the other hand, the age of the Galaxy is at least twice this, so the overall conditions have not changed significantly during the lifetime of the solar system. Thus, present star formation may shed some light on cosmogonical problems.

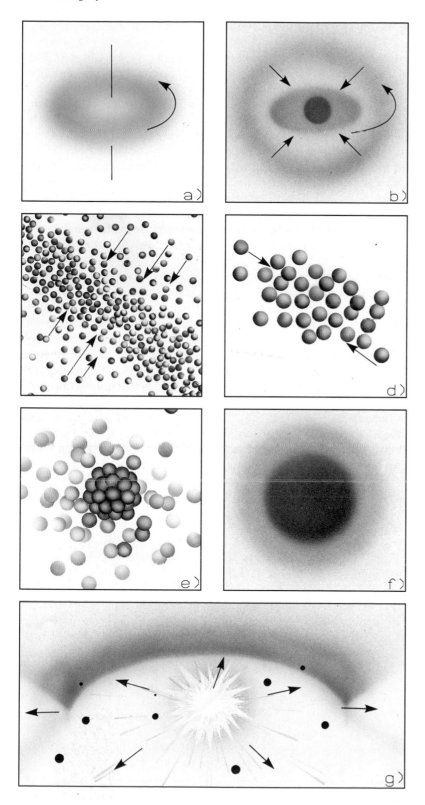

Fig. 8.60a – g. Formation of the Solar system. (**a**) A large rotating cloud, the mass of which was 3 – 4 solar masses, began to condense. (**b**) The innermost part condensed most rapidly and a disk of gas and dust formed around the proto-sun. (**c**) Dust particles in the disk collided with each other forming larger particles and sinking rapidly to a single plane. (**d**) Particles clumped together into planetesimals which were of the size of present asteroids. (**e**) These clumps drifted together, forming planet-size bodies which began (**f**) to collect gas and dust from the surrounding cloud. (**g**) The strong solar wind "blew" away extra gas and dust; the planet formation was finished

The Sun and practically the whole solar system simultaneously condensed from a rotating collapsing cloud of dust and gas (Fig. 8.60). The elements heavier than helium were formed in the interiors of stars of preceding generations, as explained in Sect. 12.8. The collapse of the cloud was initiated by a shock wave emanating from an earlier supernova explosion. (This protostar formation will be discussed in detail in Sect. 16.4)

When the originally spherical cloud collapsed, particles inside the cloud collided with each other. Rotation of the cloud allowed the particles to sink towards the same plane, perpendicular to the rotation axis of the cloud, but prevented them from moving towards the axis. This explains why the planetary orbits are in the same plane.

The mass of the proto-Sun was larger than the mass of the modern Sun. The flat disk in the plane of the ecliptic contained perhaps 1/10 of the total mass. Moreover, far outside, the remnants of the outer edges of the original cloud were still moving towards the centre. The Sun was losing its angular momentum to the surrounding gas by means of the magnetic field. When nuclear reactions were ignited, a strong solar wind carried away more angular momentum from the Sun. The final result was the modern, slowly rotating Sun.

The small particles in the disk were accreting to larger clumps by means of continuous mutual collisions, resulting finally in asteroid-size bodies, *planetesimals*. The gravity of the clumps pulled them together, forming ever growing seeds of planets. When these protoplanets were large enough, they started to accrete gas and dust from the surrounding cloud. Some minor clumps were orbiting planets; these became moons. Mutual perturbations may have prevented planetesimals in the current asteroid belt from ever being able to become "grown-up" planets. Morever, resonances could explain the Titius-Bode law: the planets were able to accrete in very limited zones only (Table 8.3).

The temperature distribution of the primordial cloud explains the differences of the chemical composition of the planets (Fig. 8.61). The volatile elements (such as hydrogen and helium, and ices) are almost totally absent in the innermost planets. The planets from Mercury to Mars are composed of "rocks", relatively heavy material which condenses above 500 K or so. The relative abundance of this material in the primeval

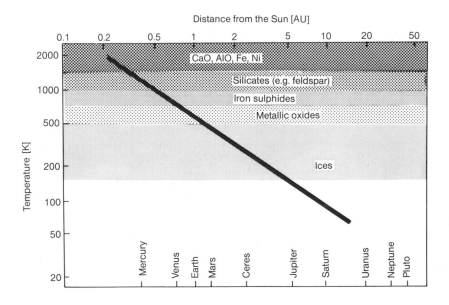

Fig. 8.61. Temperature distribution in the solar system during planet formation. The present chemical composition of the planets reflects this temperature distribution. The approximate condensing temperatures of some compounds have been indicated

Table 8.3. True distances of the planets from the Sun and distances according to the Titius-Bode law

Planet	n	Calculated distance [AU]	True distance [AU]
Mercury	–	0.4	0.4
Venus	0	0.7	0.7
Earth	1	1.0	1.0
Mars	2	1.6	1.5
Ceres	3	2.8	2.8
Jupiter	4	5.2	5.2
Saturn	5	10.0	9.2
Uranus	6	19.6	19.2
Neptune	7	38.8	30.1
Pluto	8	77.2	39.5

nebula was only 0.4%. Thus the masses of terrestrial planets are relatively small. More than 99% of the material was left over.

At the distance of Mercury, the temperature was about 1400 K. At this temperature, iron and nickel compounds begin to condense. The relative abundance of these compounds is greatest on Mercury and smallest on Mars, where the temperature was only 450 K. Thus the amount of iron(II)oxide, FeO, is relatively high on Mars, whereas there is practically no FeO on Mercury.

At the distance of Saturn, the temperature was so low that bodies of ice could form; e.g. some moons of Saturn are almost pure water ice. Because 98.2% of the primordial material was hydrogen and helium, the abundance of hydrogen and helium in Jupiter and Saturn is close to those values. However, the relative importance of ices became more prominent at the distance of Uranus and Neptune. A considerable amount of the mass of these planets can be water.

Meteorite bombardment, contraction and radioactive decay produced a great deal of "extra" heat after the planetary formation. This gave rise to the partial melting of some terrestrial planets, resulting in differentiation of material: the heavy elements sank to the centre and the light dross was left to float on the surface.

The left over material wandered among the planets. Planetary perturbations caused bodies in unstable orbits to collide with planets or to be slung out of the solar system. Asteroids remained in their stable orbits. On the outskirts of the solar system, bodies of ice and dust, such as comets, could also survive.

The beginning of the solar nuclear reactions meant the end of planetary formation. The Sun was in the T Tauri phase (see Sect. 14.3) losing mass to a strong solar wind. The mass loss was as high as $10^{-7} M_\odot$/y. However, this phase was relatively short and the total mass loss did not exceed $0.1 M_\odot$. The solar wind "blew" away the interplanetary gas and dust, and the gas accretion to the planets was over.

The solar wind or radiation pressure has no effect on millimetre- and centimetre-sized particles. However, they will drift into the Sun because of the *Poynting-Robertson effect*, first introduced by John P. Poynting in 1903. Later, H. P. Robertson derived the effect by using the theory of relativity. When a small body absorbs and emits radiation, it loses its orbital angular momentum and the body spirals to the Sun. At the distance of the asteroid belt, this process takes only a million years or so.

8.22 Other Solar Systems

According to modern cosmogonical theories, planetary systems are common around stars. However, such systems are difficult to observe, since the brightness of a planet is much less than the brightness of the nearby star.

Theoretical models predict that planetesimals are accreting to planets much like in our solar system. Most calculations have dealt with single stars around which stable planetary orbits are easier to find. However, stable orbits can also be found around binary stars, either close to one of the components, or far from both.

There are over 10^{11} stars in the Galaxy, and possibly $10^9 - 10^{10}$ stars with a planetary system. Inside a sphere of 20 pc around the Sun, there are less than 20 stars where spectral class and total brightness are close to those of the Sun. Most of these are members of binary or multiple systems, but 5 are single stars (ε Eridani, ε Indi, τ Ceti, σ Dragonis, and δ Pavonis). Planets have been looked for around these stars but without success.

There is also another way to detect possible planets, namely by studying the proper motion of a star. A massive planet orbiting a star also disturbs the motion of the star, and small periodic perturbations can be seen in its proper motion. However, such deviations are extremely small, even when the planet orbiting the star is heavier than Jupiter. There are indications of one or more such planets around several stars, but all the results are still controversial.

Some primordial form of planetary systems have been discovered. There are sparse material disks around some young stars but it is uncertain if any planets can form from these disks. So far, we have only one example of planetary systems, our own.

8.23 Examples

8.23.1 *Sidereal and Synodic Period*
The time interval between two successive oppositions of Mars is 779.9 d. Calculate the semi-major axis of Mars' orbit.

The synodic period is 779.9 d = 2.14 years. We obtain from (8.1)

$$\frac{1}{P_2} = \frac{1}{1} - \frac{1}{2.14} = 0.53 \Rightarrow P_2 = 1.88 \,\text{y} \ .$$

By using Kepler's third law (7.26), ($m_{\mars} \ll M_\odot$), the semi-major axis, is found to be

$$a = P^{2/3} = 1.88^{2/3} = 1.52 \,\text{AU} \ .$$

8.23.2 *Solar Energy Flux on the Earth*
Calculate the diurnal solar energy flux per unit area at the distance of the Earth.

The solar flux density outside the Earth's atmosphere (the solar constant) is $S_0 = 1370 \,\text{W/m}^2$. Consider a situation at latitude ϕ, when the solar declination is δ. If the atmospheric extinction is neglected, the flux density on the surface is

$$S = S_0 \sin a \ ,$$

where a is the elevation of the Sun. Equation (2.15) yields $\sin a$ as a function of latitude, declination, and hour angle h:

$$\sin a = \sin \delta \sin \phi + \cos \delta \cos \phi \cos h \ .$$

On a cloudless day, the energy is received between sunrise and sunset. The corresponding hour angles can be obtained from the equation above, when $a = 0$:

$$\cos h_0 = -\tan \delta \tan \phi \ .$$

In the course of one day, the energy received on a unit area is

$$W = \int_{-h_0}^{h_0} S \, dt \ .$$

The hour angle h is expressed in radians, so the time t is

$$t = \frac{h}{2\pi} P \ ,$$

where $P = 1\,\mathrm{d} = 24\,\mathrm{h}$. The total energy is thus

$$W = \int_{-h_0}^{h_0} S_0 (\sin \delta \sin \phi + \cos \delta \cos \phi \cos h) \frac{P}{2\pi} \, dh$$

$$= \frac{S_0 P}{\pi} (h_0 \sin \delta \sin \phi + \cos \delta \cos \phi \sin h_0) \ , \qquad \text{where}$$

$$h_0 = \arccos(-\tan \delta \tan \phi) \ .$$

For example near the equator ($\phi = 0°$) $\cos h_0 = 0$ and

$$W(\phi = 0°) = \frac{S_0 P}{\pi} \cos \delta \ .$$

At those latitudes where the Sun will not set, $h_0 = \pi$ and

$$W_{\mathrm{circ}} = S_0 P \sin \delta \sin \phi \ .$$

Near the poles, the Sun is always circumpolar when above the horizon, and so

$$W(\phi = 90°) = S_0 P \sin \delta \ .$$

Interestingly enough, during the summer when the declination of the Sun is large, the polar areas receive more energy than the areas close to the equator. This is true when

$$W(\phi = 90°) > W(\phi = 0°)$$

$$\Leftrightarrow S_0 P \sin \delta > S_0 P \cos \delta / \pi$$

$$\Leftrightarrow \tan \delta > 1/\pi$$

$$\Leftrightarrow \delta > 17.7° \ .$$

The declination of the Sun is greater than this about two months every summer.

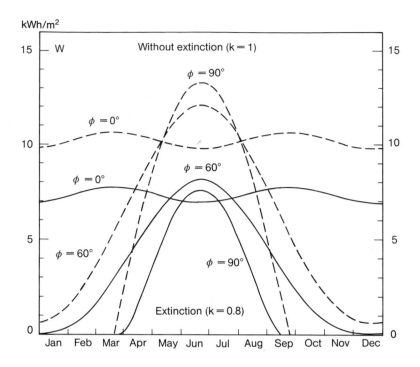

However, atmospheric extinction diminishes these values, and the loss is at its greatest at the poles, where the elevation of the Sun is always relatively small. Radiation must penetrate thick layers of the atmosphere and the path length is comparable to $1/\sin a$. If it is assumed that the fraction k of the flux density reaches the surface when the Sun is at zenith, the flux density when the Sun is at the elevation a is

$$S' = S k^{1/\sin a} \; ,$$

where $S = S_0 \sin a$. The total energy received during one day is thus

$$W = \int_{-h_0}^{h_0} S' dt = \int_{-h_0}^{h_0} S_0 \sin a \, k^{1/\sin a} \, dt \; .$$

This cannot be solved in a closed form and numerical methods must be used.

The figure shows the received energy $W[\mathrm{kWh/m^2}]$ during a year at latitudes $\phi = 0°$, $60°$, and $90°$ without extinction, and when $k = 0.8$, which is close to the real value.

8.23.3 Magnitude of a Planet

The apparent magnitude of Mars during the 1975 opposition was $m_1 = -1.6$ and the distance to the Sun, $r_1 = 1.55\,\mathrm{AU}$. During the 1982 opposition, the distance was $r_2 = 1.64\,\mathrm{AU}$. Calculate the apparent magnitude in the 1982 opposition.

At opposition, the distance of Mars from the Earth is $\Delta = r - 1$. The observed flux density depends on the distances to the Earth and the Sun,

$$F \propto \frac{1}{r^2 \Delta^2} \ .$$

According to (4.9) we obtain

$$m_1 - m_2 = -2.5 \lg \frac{r_2^2 (r_2 - 1)^2}{r_1^2 (r_1 - 1)^2}$$

$$\Rightarrow m_2 = m_1 + 5 \lg \frac{r_2 (r_2 - 1)}{r_1 (r_1 - 1)} = -1.6 + 5 \lg \frac{1.64 \cdot 0.64}{1.55 \cdot 0.55} \approx -1.1 \ .$$

The same result is obtained if (8.19) is separately written for both oppositions.

8.23.4 *The Brightness of Venus*
Find the instant when Venus is brightest if the brightness is a function of the size of the illuminated surface and the distance from the Earth. The orbits are assumed to be circular.

The size of the illuminated surface is the area of the semicircle $ACE \pm$ half the area of the ellipse $ABCD$. The semiaxes of the ellipse are R and $R \cos \alpha$. If the radius of the planet is R, the illuminated area is

$$\pi \frac{R^2}{2} + \frac{1}{2} \pi R \cdot R \cos \alpha = \frac{\pi}{2} R^2 (1 + \cos \alpha) \ ,$$

where α is the phase angle. On the other hand, the flux density is comparable to the square of the distance Δ. Thus

$$F \propto \frac{1 + \cos \alpha}{\Delta^2} \ . \tag{1}$$

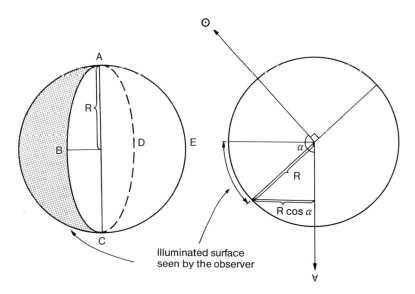

Illuminated surface
seen by the observer

The cosine formula yields

$$r_{\oplus}^2 = r^2 + \Delta^2 - 2\Delta r \cos \alpha \ . \tag{2}$$

When $\cos \alpha$ is solved and inserted in (1) we obtain

$$F \propto \frac{2\Delta r + r^2 + \Delta^2 - r_{\oplus}^2}{2r\Delta^3} \ . \tag{3}$$

The minimum of (3) yields the distance where Venus is brightest:

$$\frac{\partial F}{\partial \Delta} = -\frac{4r\Delta + 3r^2 - 3r_{\oplus}^2 + \Delta^2}{2r\Delta^4} = 0$$

$$\Rightarrow \Delta = -2r \pm \sqrt{r^2 + 3r_{\oplus}^2} \ .$$

If $r = 0.723$ AU and $r_{\oplus} = 1$ AU, the distance is $\Delta = 0.43$ AU and the corresponding phase angle from (2), $\alpha = 118°$. Thus Venus is brightest shortly after the largest eastern elongation and before the largest western elongation. From the sine formula we obtain

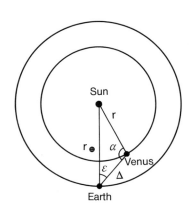

$$\frac{\sin \varepsilon}{r} = \frac{\sin \alpha}{r_{\oplus}} \ .$$

The corresponding elongation is $\varepsilon = 40°$, and

$$\frac{1 + \cos \alpha}{2} \cdot 100\% = 27\%$$

of the surface is seen lit.

8.23.5 Size of an Asteroid

The apparent visual magnitude V of a minor planet at the moment of opposition was observed to be $V_0 = 10.0$. Give an estimate of the asteroid's size when the geometric albedo is assumed to be $p = 0.15$. Give an error estimate if the geometric albedo is wrong by a factor of two. The visual magnitude of the Sun is $V_{\odot} = -26.8$, and the distance of the asteroid from the Earth was $\Delta = 3$ AU.

At opposition, the distance of a body from the Sun is $r = \Delta + 1$. Assuming that $r_{\oplus} = 1$ AU, the radius R can be solved from (8.21)

$$R = \frac{r\Delta}{r_{\oplus}} \sqrt{\frac{10^{-0.4(V_0 - V_{\odot})}}{p}} = \frac{3 \cdot 4}{1} \sqrt{\frac{10^{-0.4(10 + 26.8)}}{0.15}} \cdot 1.5 \cdot 10^8 \text{km}$$

$$= 200 \text{ km} \ .$$

If $p = 0.075$, the size of the asteroid is $R = 290$ km and if $p = 0.3$, the radius will be $R = 140$ km.

8.24 Exercises

8.24.1 The interval between two oppositions of a planet was 398.9 d. The angular diameter of the planet at the opposition was 47.2″. Find the sidereal period, semimajor axis, and the true diameter (in kilometers) of the planet. Which planet was it?

8.24.2 a) Assume that three bodies move along circular orbits with angular velocities (mean motions) n_1, n_2 and n_3. Show that these bodies have a common synodic period if and only if there are nonzero integers k_1, k_2 and k_3 such that

$$k_1 n_1 + k_2 n_2 + k_3 n_3 = 0 \ ,$$
$$k_1 + k_2 + k_3 = 0 \ .$$

b) The resonance of the Galilean satellites can be expressed in terms of their mean motions as

$$n_{\text{Io}} - 3 n_{\text{Europa}} + 2 n_{\text{Ganymede}} = 0 \ .$$

Find the synodic period of these three moons.

8.24.3 a) Find the daily retrograde apparent motion of an exterior planet at its opposition. Assume that the planet and the Earth have circular orbits. b) Pluto was found in 1930 from two plates, exposed 6 days apart during the opposition of the planet. On those plates one degree corresponded to 3 cm. How much (in cm) had Pluto moved between the exposures? How much does a typical main belt asteroid move in the same time?

8.24.4 The angular diameter of the Moon is 0.5°. The full moon has an apparent magnitude of -12.5 and the Sun -26.7. Find the geometric and Bond albedos of the Moon, assuming that the reflected light is isotropic (into solid angle 2π sterad).

8.24.5 An asteroid with a diameter of 100 m approaches the Earth at a velocity of 30 km s^{-1}. Find the apparent magnitude of the asteroid a) one week, b) one day before the collision. Assume that the phase angle is $\alpha = 0°$ and the geometric albedo of the asteroid is $p = 0.1$. What do you think about the chances of finding the asteroid well in advance before the crash?

8.24.6 Find the temperature of a comet when its distance from the Sun is a) 10 AU, b) 1 AU, c) 0.5 AU. Assume the Bond albedo of the comet is 0.05.

Chapter 9 Stellar Spectra

All our information about the physical properties of stars comes more or less directly from studies of their spectra. In particular, by studying the strength of various absorption lines, stellar masses, temperatures and compositions can be deduced. The line shapes contain detailed information about atmospheric processes.

As we have seen in Chap. 3, the light of a star can be dispersed into a spectrum by means of a prism or a diffraction grating. The distribution of the energy flux density over frequency can then be derived. The spectra of stars consist of a *continuous spectrum* or *continuum* with narrow *spectral lines* superimposed (Fig. 9.1). The lines in stellar spectra are mostly dark *absorption lines*, but in some objects, bright *emission lines* also occur.

In a very simplified way the continuous spectrum can be thought of as coming from the hot surface of the star. Atoms in the atmosphere above the surface absorb certain characteristic wavelengths of this radiation, leaving dark "gaps" at the corresponding points in the spectrum. In reality there is no such sharp separation between surface and atmosphere. All layers emit and absorb radiation, but the net result of these processes is that less energy is radiated at the wavelengths of the absorption lines.

The spectra of stars are classified on the basis of the strengths of the spectral lines. *Isaac Newton* observed the solar spectrum in 1666, but, properly speaking, spectroscopy began in 1814 when *Joseph Fraunhofer* observed the dark lines in the spectrum of the Sun. The absorption lines are also known as Fraunhofer lines. In 1860, *Gustav Robert Kirchhoff* and *Robert Bunsen* identified the lines as the characteristic lines produced by various elements in an incandescent gas.

9.1 Measuring Spectra

The most important methods of forming a spectrum are by means of an *objective prism* or a *slit spectrograph*. In the former case one obtains a photograph, where each stellar image has been spread into a spectrum. Up to several hundred spectra can be photographed on a single plate and used for spectral classification. The amount of detail that can be seen in a spectrum depends on its *dispersion*, the range of wavelengths per millimetre on the plate. The dispersion of an objective prism is a few tens of nanometres per millimetre. More detailed observations require a slit spectrograph, which can reach a dispersion $1 - 0.01$ nm/mm. The detailed shape of individual spectral lines can then be studied.

Fig. 9.1. Typical stellar spectrum. The continuous spectrum is brightest at about 550 nm and gets fainter towards smaller and larger wavelengths. Dark absorption lines are superimposed on the continuum. The star shown, η Pegasi, has a spectrum very similar to that of the Sun. (Mt. Wilson Observatory)

654.2   656.0

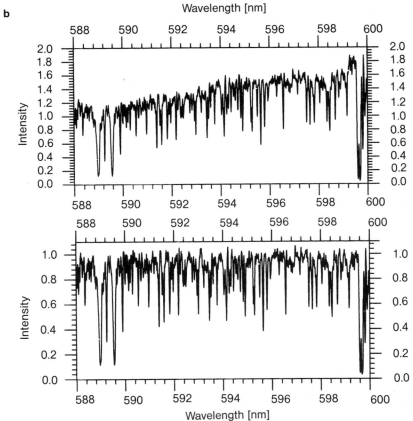

Fig. 9.2. (a) This pair of pictures shows a section of a photograph of a stellar spectrum and the corresponding rectified microdensitometer intensity tracing. The original spectrum was taken at the Crimean Observatory. (b) This pair of pictures shows a more extensive part of the spectrum. The intensity curve of the first picture has been rectified in the second one by normalizing the value of the continuum intensity to one. (Pictures by J. Kyröläinen and H. Virtanen, Helsinki Observatory)

The photograph of the spectrum is converted to an intensity tracing showing the flux density as a function of wavelength. This is done by means of a microdensitometer, measuring the amount of light transmitted by the recorded spectrum. Since the blackening of a photographic plate is not linearly related to the amount of energy it has received, the measured blackening has to be calibrated by comparison with known exposures. In modern CCD spectrographs, the intensity curve is determined directly without the intervening step of a photographic plate. For measurements of line strengths, the spectrum is usually rectified by dividing by the continuum intensity.

Figure 9.2 shows a photograph of the spectrum of a star and the intensity curve obtained from a calibrated and rectified microdensitometer tracing. The second pair of pictures show the intensity curve before and after the normalization. The absorption lines appear as troughs of various sizes in the curve. In addition to the clear and deep lines, there are large numbers of weaker lines that can barely be discerned. The graininess of the photographic emulsion is a source of noise which appears as irregular fluctuations of the intensity curve. Some lines are so close together that they appear blended at this dispersion.

The detailed shape of a spectral line is called the *line profile*. The true shape of the line reflects the properties of the stellar atmosphere, but the observed profile is also spread out by the measuring instrument. However, the total absorption in the line, usually expressed in terms of the *equivalent width*, is less sensitive to observational effects. The equivalent width is the width of a totally dark rectangular line with the same total absorption as the observed line (see Fig. 9.3).

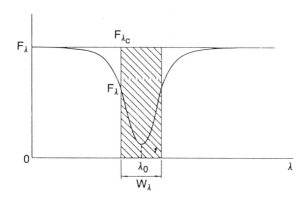

Fig. 9.3. The equivalent width W of a spectral line is defined so that the line and the rectangle have the same area in the picture. For example, the typical equivalent width of a metal line (Γc) in the solar spectrum is about 10 pm

The equivalent width of a spectral line depends on how many atoms in the atmosphere are in a state in which they can absorb the wavelength in question. The more atoms there are, the stronger and broader the spectral line is.

The chemical composition of the atmosphere can be determined from the strengths of the spectral lines. With the introduction of large computers, it has become feasible to construct quite detailed models of the structure of stellar atmospheres, and to compute the emergent spectrum for a given model. The computed synthetic spectrum can be compared with the observations and the theoretical model modified until a good fit is obtained. The theoretical models then give the number of absorbing atoms, and hence the element abundances, in the atmosphere. The construction of model atmospheres will be discussed in Sect. 9.6.

9.2 The Harvard Spectral Classification

The spectral classification scheme in present use was developed at Harvard Observatory in the United States in the early 20th century. The work was begun by *Henry Draper*, who in 1872 took the first photograph of the spectrum of Vega. Later Draper's widow donated the observing equipment and a sum of money to Harvard Observatory to continue the work of classification.

The main part of the classification was done by *Annie Jump Cannon*, using objective prism spectra. The *Henry Draper Catalogue* (HD) was published in 1918 – 1924. It contains 225,000 stars extending down to ninth magnitude. Altogether, more than 390,000 stars have been classified at Harvard.

The Harvard classification is based on lines that are mainly sensitive to the stellar temperature, rather than to gravity or luminosity. Important lines are the hydrogen Balmer lines, the lines of neutral helium, the iron lines, the H and K doublet of ionized calcium at 396.8 and 393.3 nm, the G band due to the CH molecule and some metals around 431 nm, the neutral calcium line at 422.7 nm and the lines of titanium oxide (TiO).

The main types in the Harvard classification are denoted by capital letters. They were initially ordered in alphabetical sequence, but subsequently it was noticed that they could be ordered according to temperature. With the temperature decreasing towards the right the sequence is

$$O - B - A - F - G - K - M \begin{smallmatrix} C \\ \\ S \end{smallmatrix} .$$

Additional notations are Q for novae, P for planetary nebulae and W for Wolf-Rayet stars. The class C consists of the earlier types R and N. The spectral classes C and S represent parallel branches to types G – M, differing in their surface chemical composition. There is a well-known mnemonic for the spectral classes: "Oh be a fine girl kiss me (right now, sweetheart)!"

The spectral classes are divided into subclasses denoted by the numbers 0...9; sometimes decimals are used, e.g. B 0.5 (Figs. 9.4, 9.5). The main characteristics of the different classes are:

Type O. Blue stars, surface temperature 20,000 – 35,000 K. Spectrum with lines from multiply ionized atoms, e.g. He II, C III, N III, O III, Si V. He I visible, H I lines weak.

Type B. Blue-white stars, surface temperature about 15,000 K. He II lines have disappeared, He I (403 nm) lines are strongest at B2, then get weaker and have disappeared at type B9. The K line of Ca II becomes visible at type B3. H I lines getting stronger. O II, Si II and Mg II lines visible.

Type A. White stars, surface temperature about 9000 K. The H I lines are very strong at A 0 and dominate the whole spectrum, then get weaker. H and K lines of Ca II getting stronger. He I no longer visible. Neutral metal lines begin to appear.

Type F. Yellow-white stars, surface temperature about 7000 K. H I lines getting weaker, H and K of Ca II getting stronger. Many other metal lines, e.g. Fe I, Fe II, Cr II, Ti II, clear and getting stronger.

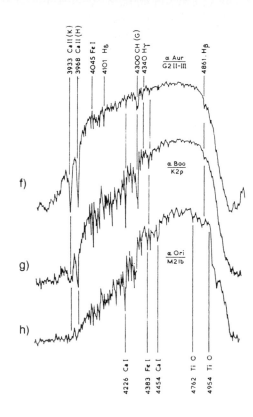

Fig. 9.4. Spectra of early and late spectral type stars between 375 and 390 nm. (**a**) The upper star is Vega, of spectral type A0, and (**b**) the lower one is Aldebaran, of spectral type K5. The hydrogen Balmer lines are strong in the spectrum of Vega; in that of Aldebaran, there are many metal lines. (Lick Observatory)

Fig. 9.5a – h. Intensity curves for various spectral classes showing characteristic spectral features. The name of the star and its spectral and luminosity class are given next to each curve, and the most important spectral features are identified. (Drawing by J. Dufay)

Type G. Yellow stars like the Sun, surface temperature about 5500 K. The H I lines still getting weaker, H and K lines very strong, strongest at G0. Metal lines getting stronger. G band clearly visible. CN lines seen in giant stars.

Type K. Orange-yellow stars, surface temperature about 4000 K. Spectrum dominated by metal lines. H I lines insignificant. Ca I 422.7 nm clearly visible. Strong H and K lines and G band. TiO bands become visible at K5.

Type M. Red stars, surface temperature about 3000 K. TiO bands getting stronger. Ca I 422.7 nm very strong. Many neutral metal lines.

Type C. Carbon stars, previously R and N. Very red stars, surface temperature about 3000 K. Strong molecular bands, e.g. C_2, CN and CH. No TiO bands. Line spectrum like the K and M types.

Type S. Red low-temperature stars (about 3000 K). Very clear ZrO bands. Also other molecular bands, e.g. YO, LaO and TiO.

The main characteristics of the classification scheme can be seen in Fig. 9.6 showing the variations of some typical absorption lines in the different spectral classes. (The stellar colours correspond to the actual visual sensation only at the extremes of the classification, most clearly for the red stars, which really do appear red.)

The *early,* i.e. hot, spectral classes are characterized by the lines of ionized atoms, whereas the cool, or *late,* spectral types have lines from neutral atoms. The absorption bands of molecules also appear in the late spectral types. These differences are essentially due to temperature differences. Apart from the parallel branches C and S, differences in pressure and chemical composition are not taken into account in the spectral classification.

To see how the strengths of the spectral lines are determined by the temperature, we consider, for example, the neutral helium lines at 402.6 nm and 447.2 nm. These are only seen in the spectra of hot stars. The reason for this is that the lines are due to absorption by excited atoms, and that a high temperature is required to produce any appreciable excitation. As the stellar temperature increases, more atoms are in the required excited state, and the strength of the helium lines increases. When the temperature becomes even higher, helium begins to be ionized, and the strength of the neutral helium lines begins to decrease. In a similar way one can understand the variation with temperature of other important lines, such as the calcium H and K lines and the hydrogen Balmer lines.

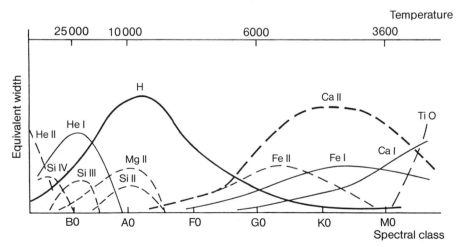

Fig. 9.6. Equivalent width of some important spectral lines in the various spectral classes. [Struve, O. (1959): *Elementary Astronomy* (Oxford University Press, New York) p. 259]

9.3 The Yerkes Spectral Classification

The Harvard classification only takes into account the effect of the temperature on the spectrum. For a more precise classification, one also has to take into account the luminosity of the star, since two stars with the same effective temperature may have widely different luminosities. A two-dimensional system of spectral classification was introduced by *William W. Morgan, Philip C. Keenan* and *Edith Kellman* of Yerkes Observatory. This system is known as the MKK or *Yerkes classification*. (The MK classification is a modified, later version.) The MKK classification is based on the visual scrutiny of slit spectra with a dispersion of 11.5 nm/mm. It is carefully defined on the basis of standard stars and the specification of luminosity criteria. Six different *luminosity classes* are distinguished:

Ia   most luminous supergiants;
Ib   less luminous supergiants;
II   luminous giants;
III   normal giants;
IV   subgiants;
V   main sequence stars (dwarfs).

The luminosity class is determined from spectral lines that depend strongly on the stellar surface gravity, which is closely related to the luminosity. The masses of giants and dwarfs are roughly similar, but the radii of giants are much larger than those of dwarfs. Therefore the gravitational acceleration $g = GM/R^2$ at the surface of a giant is much smaller than for a dwarf. In consequence, the gas density and pressure in the atmosphere of a giant is much smaller. This gives rise to *luminosity effects* in the stellar spectrum, which can be used to distinguish between stars of different luminosities.

1) For spectral types B   F, the lines of neutral hydrogen are deeper and narrower for stars of higher luminosities. The reason for this is that the metal ions give rise to a fluctuating electric field near the hydrogen atoms. This field leads to shifts in the hydrogen energy levels (the Stark effect), appearing as a broadening of the lines. The effect becomes stronger as the density increases. Thus the hydrogen lines are narrow in absolutely bright stars, and become broader in main sequence stars and even more so in white dwarfs (Fig. 9.7).

2) The lines from ionized elements are relatively stronger in high-luminosity stars. This is because a higher density makes it easier for electrons and ions to recombine to neutral atoms. On the other hand, the rate of ionization is essentially determined by the radiation field, and is not appreciably affected by the gas density. Thus a given radiation field can maintain a higher degree of ionization in stars with more extended atmospheres. For example, in the spectral classes F − G, the relative strengths of the ionized strontium (Sr II) and neutral iron (Fe I) lines can be used as a luminosity indicator. Both lines depend on the temperature in roughly the same way, but the Sr II lines become relatively much stronger as the luminosity increases.

3) Giant stars are redder than dwarfs of the same spectral type. The spectral type is determined from the strengths of spectral lines, including ion lines. Since these are stronger in giants, a giant will be cooler, and thus also redder, than a dwarf of the same spectral type.

4) There is a strong cyanogen (CN) absorption band in the spectra of giant stars, which is almost totally absent in dwarfs. This is partly a temperature effect, since the cooler atmospheres of giants are more suitable for the formation of cyanogen.

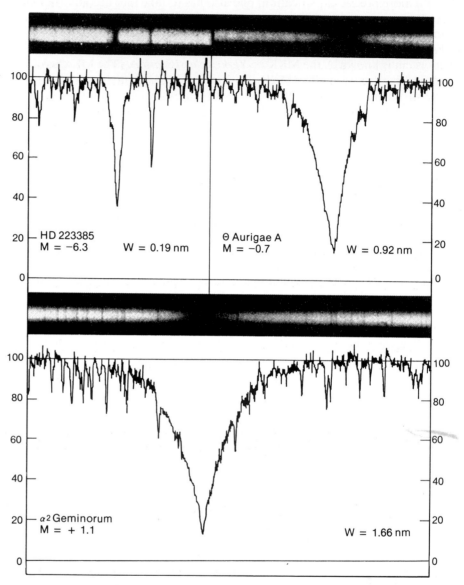

Fig. 9.7. Luminosity effects in the hydrogen Hγ line in A stars. The vertical axis gives the normalized intensity. HD 223385 (*upper left*) is an A2 supergiant, where the line is very weak, θ Aurigae A is a giant star and α² Geminorum is a main sequence star, where the line is very broad. [Aller, L. H. (1953): *Astrophysics. The Atmospheres of the Sun and Stars* (The Ronald Press Company, New York) p. 318]

9.4 Peculiar Spectra

The spectra of some stars differ from what one would expect on the basis of their temperature and luminosity (see, e.g., Fig. 9.8). Such stars are called *peculiar*. The most common peculiar spectral types will now be considered.

The *Wolf-Rayet stars* are very hot stars; the first examples were discovered by *Charles Wolf* and *Georges Rayet* in 1867. The spectra of Wolf-Rayet stars have broad emission lines of hydrogen and ionized helium, carbon, nitrogen and oxygen. There are hardly any absorption lines. The Wolf-Rayet stars are thought to be members of binary systems, where they have lost their outer layers to a companion star. This has exposed the stellar interior, which gives rise to a different spectrum than the normal outer layers.

In some O and B stars, the hydrogen absorption lines have weak emission components either at the line centre or in its wings. These stars are called *Be* and *shell stars* (the letter e after the spectral type indicates that there are emission lines in the spectrum). The emission lines are formed in a rotationally flattened gas shell around the star. The shell and Be stars show irregular variations, apparently related to structural changes in the shell. About 15% of all O and B stars have emission lines in their spectra.

The strongest emission lines are those of the *P Cygni* stars, which have one or more sharp absorption lines on the short wavelength side of the emission line. It is thought that the lines are formed in a thick expanding envelope. The P Cygni stars are often variable. For example, P Cygni itself has varied between third and sixth magnitude during the past centuries. At present its magnitude is about 5.

The peculiar A stars or *Ap stars* (p = peculiar) are usually strongly magnetized stars, where the lines are split into several components by the Zeeman effect. The lines of certain elements, such as magnesium, silicon, europium, chromium and strontium, are exceptionally strong in the Ap stars. Lines of rarer elements such as mercury, galli-

Fig. 9.8 a, b. Peculiar spectra. **(a)** R Geminorum (*above*) is an emission line star, with bright emission lines, indicated by arrows, in its spectrum; **(b)** the spectrum of a normal star is compared with one in which the zirconium lines are unusually strong. (Mt. Wilson Observatory and Helsinki Observatory)

um or krypton may also be present. Otherwise, the Ap stars are like normal main sequence stars.

The *Am stars* (m = metallic) also have anomalous element abundances, but not to the same extent as the Ap stars. The lines of e.g. the rare earths and the heaviest elements are strong in their spectra; those of calcium and scandium are weak.

We have already mentioned the S and C stars, which are special classes of K and M giants with anomalous element abundances. In the S stars, the normal lines of titanium, scandium and vanadium oxide are replaced with oxides of heavier elements, zirconium, yttrium and barium. A large fraction of the S stars are irregular variables. The name of the C stars refers to carbon. The metal oxide lines are almost completely absent in their spectra; instead, various carbon compounds (CN, C_2, CH) are strong. The abundance of carbon relative to oxygen is $4-5$ times greater in the C stars than in normal stars. The C stars are divided into two groups, hotter R stars and cooler N stars.

Another type of giant stars with abundance anomalies are the *barium stars*. The lines of barium, strontium, rare earths and some carbon compounds are strong in their spectra. Apparently nuclear reaction products have been mixed up to the surface in these stars.

9.5 The Hertzsprung-Russell Diagram

Around 1910, *Ejnar Hertzsprung* and *Henry Norris Russell* studied the relation between the absolute magnitudes and the spectral types of stars. The diagram showing these two variables is now known as the *Hertzsprung-Russell diagram* (HR diagram) (Fig. 9.9). It has turned out to be an important aid in studies of stellar evolution.

In view of the fact that stellar radii, luminosities and surface temperatures vary widely, one might have expected the stars to be uniformly distributed in the HR diagram. In fact it is found that most stars are located along a roughly diagonal curve called the *main sequence*. The Sun is situated about the middle of the main sequence.

The HR diagram also shows that the yellow and red stars (spectral types G-K-M) are clustered into two clearly separate groups: the *main sequence* of *dwarf stars* and the *giants*. The giant stars fall into several distinct groups. The *horizontal branch* is an almost horizontal sequence, about absolute visual magnitude zero. The *red giant branch* rises almost vertically from the main sequence at spectral types K and M in the HR diagram. Finally, the *asymptotic branch* rises from the horizontal branch and approaches the bright end of the red giant branch. These various branches represent different phases of stellar evolution (cf. Sects. 12.3, 12.4).

A typical horizontal branch giant is about a hundred times brighter than the Sun. Since giants and dwarfs of the same spectral class have nearly the same surface temperature, the difference in luminosity must be due to a difference in radius according to (5.18). For example Arcturus, which is one of the brightest stars in the sky, has a radius about thirty times that of the Sun.

The brightest red giants are the *supergiants*, with magnitudes up to $M_v = -7$. One example is Betelgeuze in Orion, with a radius of 400 solar radii and 20,000 times more luminous than the Sun.

About 10 magnitudes below the main sequence are the *white dwarfs*. They are quite numerous in space, but faint and difficult to find. The best-known example is Sirius B, the companion of Sirius.

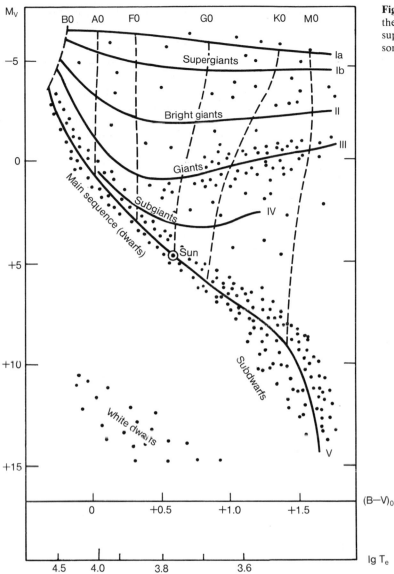

Fig. 9.9. Hertzsprung-Russell diagram. The main sequence and the horizontal giant branch can be clearly distinguished. The supergiants are scattered above the giants. To the lower left are some white dwarfs

There are some stars in the HR diagram which are located below the giant branch, but still clearly above the main sequence. These are known as *subgiants*. Similarly, there are stars below the main sequence, but brighter than the white dwarfs, known as *subdwarfs*.

When interpreting the HR diagram, one has to take into account *selection effects:* absolutely bright stars are more likely to be included in the sample, since they can be discovered at greater distances. If only stars within a certain distance from the Sun are included, the distribution of stars in the HR diagram looks quite different. This can be seen in Fig. 9.10, showing the HR diagram for stars within 10 pc from the Sun (marked by dots): there are no giant or bright main sequence stars among these.

The HR diagrams of star clusters are particularly important for the theory of stellar evolution. They will be discussed in Sect. 17.2.

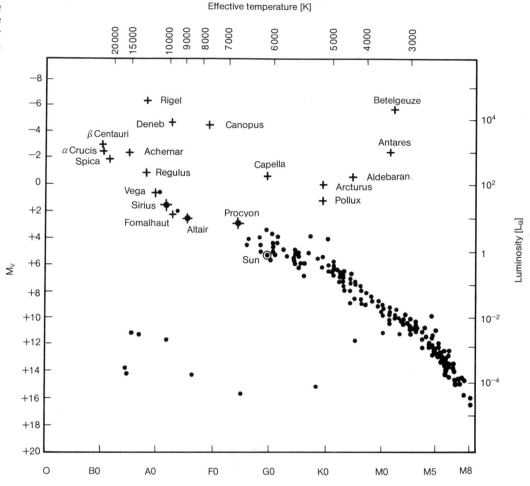

9.6 Model Atmospheres

The stellar atmosphere consists of those layers of the star where the radiation that is transmitted directly to the observer originates. Thus in order to interpret stellar spectra, one needs to be able to compute the structure of the atmosphere and the emerging radiation.

In actual stars, there are many factors, such as rotation and magnetic fields, that complicate the problem of computing the structure of the atmosphere. We shall only consider the classical problem of finding the structure, i.e. the distribution of pressure and temperature with depth, in a static, unmagnetized atmosphere. In that case, a model atmosphere is completely specified by giving the chemical composition, the gravitational acceleration at the surface, g, and the energy flux from the stellar interior, or equivalently, the effective temperature T_{eff}.

The basic principles involved in computing a model stellar atmosphere are the same as for stellar interiors and will be discussed in Sects. 11.1, 2. Essentially, there are two differential equations to be solved: the equation of hydrostatic equilibrium, which fixes the distribution of pressure, and an equation of energy transport, which will have a different form, depending on whether the atmosphere is radiative or convective, and which determines the temperature distribution.

The values of the various physical quantities in an atmosphere are usually given as functions of some suitably defined continuum optical depth τ. Thus pressure, temperature, density, ionization and the occupation numbers of various energy levels can all be obtained as functions of τ. When these are known, the intensity of radiation emerging from the atmosphere can be computed. In *The Intensity Emerging from a Stellar Atmosphere (p. 248), it is shown that approximately the emergent spectrum originates at unit optical depth, measured along each light ray. On this basis, one can predict whether a given spectral line will be present in the spectrum.

Consider a spectral line formed when an atom (or ion) in a given energy state absorbs a photon. From the model atmosphere, the occupation number of the absorbing level is known as a function of the (continuum) optical depth τ. If now there is a layer above the depth $\tau = 1$ where the absorbing level has a high occupancy, the optical depth in the line will become unity before $\tau = 1$, i.e. the radiation in the line will originate higher in the atmosphere. Because the temperature increases inwards, the intensity in the line will correspond to a lower temperature, and the line will appear dark. On the other hand, if the absorbing level is unoccupied, the optical depth at the line frequency will be the same as the continuum optical depth. The radiation at the line frequency will then come from the same depth as the adjacent continuum, and no absorption line will be formed.

The expression for the intensity derived in *The Intensity Emerging from a Stellar Atmosphere (p. 248) also explains the phenomenon of *limb darkening* seen in the Sun (Scct. 13.2). The radiation that reaches us from near the edge of the solar disc emerges at a very oblique angle (θ near 90°), i.e. $\cos \theta$ is small. Thus this radiation originated at small values of τ, and hence at low temperatures. In consequence, the intensity coming from near the edge will be lower, and the solar disc will appear darker towards the limb. The amount of limb darkening also gives an empirical way of determining the temperature distribution in the solar atmosphere.

Our presentation of stellar atmospheres has been highly simplified. In practice, the spectrum is computed numerically for a range of parameter values. The values of T_{eff}, g and element abundances for various stars can then be found by comparing the observed line strengths and other spectral features with the theoretical ones. We shall not go into details on the procedures used.

9.7 What Do the Observations Tell Us?

To conclude this chapter, we shall give a summary of the properties of stars revealed by the observations. At the end of the book, there are tables of the brightest and of the nearest stars. The HR diagram of both groups has already been presented (Fig. 9.10).

Of the brightest stars, four have a negative magnitude. Some of the apparently bright stars are absolutely bright supergiants, others are simply nearby.

In the list of nearby stars, the dominance of faint dwarf stars, already apparent in the HR diagram, is worth noting. Most of these belong to the spectral types K and M. There are indications that some nearby stars may also have very faint companions, with masses about that of Jupiter, i.e. planets. The existence of such companions is still controversial and they have not been included in the table.

Stellar spectroscopy offers an important way of determining fundamental stellar parameters, in particular mass and radius. However, the spectral information needs to be calibrated by means of direct measurements of these quantities. These will be considered next.

The *masses* of stars can be determined in the case of double stars orbiting each other. (The details of the method will be discussed in Chap. 10, "Binary Stars and Stellar Masses".) These observations have shown that the mass of main sequence stars becomes larger, the higher on the main sequence a star is located. One thus obtains an empirical *mass – luminosity relation*, which can be used to estimate stellar masses on the basis of the spectral type.

The observed relation between mass and luminosity is shown in Fig. 9.11. For large masses (above three solar masses), the luminosity is roughly proportional to the cube of the mass, $L \propto M^3$. For small masses (less than half a solar mass), the dependence is $L \propto M^{2.5}$. These relations are only approximate. According to them, a ten solar mass star is a thousand times brighter than the Sun, corresponding to a magnitude difference of 7.5 mag.

The smallest observed stellar masses are about 1/20 of a solar mass, corresponding to stars in the lower right-hand part of the HR diagram. The masses of white dwarfs are less than one solar mass. The masses of the most massive main sequence and supergiant stars are between 10 and $50 M_\odot$.

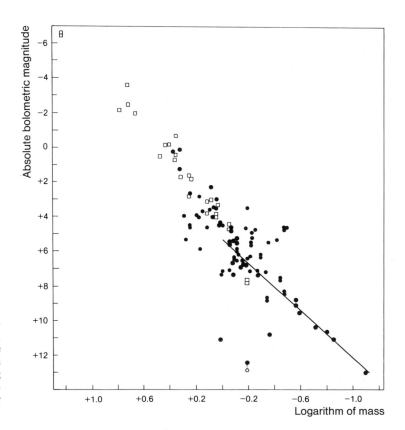

Fig. 9.11. Mass-luminosity relation. The picture is based on binaries with known masses. The filled circles are visual binaries and the open squares eclipsing variables. The straight line corresponds to the mass-luminosity relation for low-mass stars

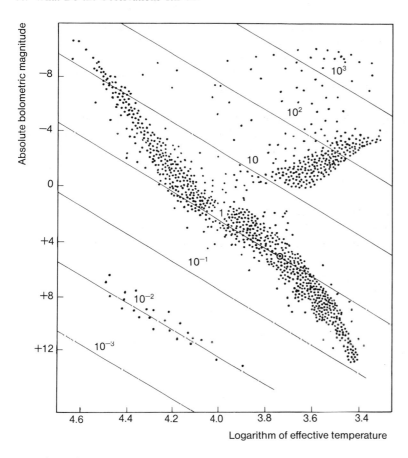

Fig. 9.12. Straight lines in the HR diagram corresponding to constant stellar radii. The radii are given in terms of the solar radius. The Sun's position in the diagram is indicated

Direct interferometric measurements of stellar angular diameters have been made for only a few dozen stars. When the distances are known, these immediately yield the value of the radius. In eclipsing binaries, the radius can also be directly measured (see Sect. 10.4). Altogether, close to a hundred stellar radii are known from direct measurements. In other cases, the radius must be estimated from the absolute luminosity and effective temperature.

In discussing stellar radii, it is convenient to use a version of the HR diagram with $\lg T_e$ on the horizontal and M_{bol} or $\lg (L/L_\odot)$ on the vertical axis. If the value of the radius R is fixed, then (5.18) yields a linear relation between the bolometric magnitude and $\lg T_e$. Thus lines of constant radius in the HR diagram are straight. Lines corresponding to various values of the radius are shown in Fig. 9.12. The smallest stars are the white dwarfs with radii of about one per cent of the solar radius, whereas the largest supergiants have radii several thousand times larger than the Sun. Not included in the figure are the compact stars (neutron stars and black holes) with typical radii of a few tens of kilometres.

Since the stellar radii vary so widely, so do the densities of stars. The density of giant stars may be only 10^{-4}kg/m^3, whereas the density of white dwarfs is about 10^9kg/m^3.

The range of values for stellar effective temperatures and luminosities can be immediately seen in the HR diagram. The range of effective temperature is $2000 - 40{,}000\,\text{K}$, and that of luminosity, $10^{-4} - 10^6 L_\odot$. The *rotation* of stars appears as a broadening of the spectral lines. One edge of the stellar disc approaches us, the other edge is receding, and the radiation from the edges is Doppler shifted accordingly. The rotational velocity observed in this way is only the component along the line of sight. The true velocity is obtained by dividing with $\sin i$, where i is the angle between the line of sight and the rotational axis. A star seen from the direction of the pole will show no rotation. Assuming the axes of rotation to be randomly oriented, the distribution of rotational velocities can be statistically estimated. The hottest stars appear to rotate faster than the cooler ones. The rotational velocity at the equator varies from $200 - 250$ km/s for O and B stars to about 2 km/s for spectral type G. In shell stars, the rotational velocity may reach 500 km/s.

The *chemical composition* of the outer layers is deduced from the strengths of the spectral lines (Sect. 9.6). About three-fourths of the stellar mass is hydrogen. Helium comprises about one-fourth, and the abundance of other elements is very small. The abundance of heavy elements in young stars (about 2%) is much larger than in old ones, where it is less than 0.02%.

*The Intensity Emerging from a Stellar Atmosphere

The intensity of radiation emerging from an atmosphere is given by the expression (2) in *The Equation of Radiative Transfer in an Atmosphere (p. 125), i.e.

$$I_\nu(0, \theta) = \int_0^\infty S_\nu(\tau_\nu) \exp(-\tau_\nu \sec \theta) \, d\tau_\nu \sec \theta. \tag{1}$$

If a model atmosphere has been computed, the source function S_ν is known.

An approximate formula for the intensity can be derived as follows. Let us expand the source function in a Taylor series about some arbitrary point τ^*, thus

$$S_\nu = S_\nu(\tau^*) + (\tau_\nu - \tau^*) S_\nu'(\tau^*) + \dots,$$

where the dash denotes a derivative. With this expression, the integral in (1) can be performed, yielding

$$I_\nu(0, \theta) = S_\nu(\tau^*) + (\cos \theta - \tau^*) S_\nu'(\tau^*) + \dots.$$

If we now choose $\tau^* = \cos \theta$, the second term will vanish. In local thermodynamic equilibrium, the source function will be the Planck function $B_\nu(T)$. We thus obtain the *Eddington-Barbier approximation*

$$I_\nu(0, \theta) = B_\nu(T[\tau_\nu = \cos \theta]).$$

According to this expression, the radiation emerging in a given direction originated at unit optical depth along that direction.

9.8 Exercises

9.8.1 Arrange the following spectra in the order of decreasing temperature:

Chapter 10 Binary Stars and Stellar Masses

Quite often, two stars may appear to be close together in the sky, although they are really at very different distances. Such chance pairs are called *optical binary stars*. However, many close pairs of stars really are at the same distance and form a physical system in which two stars are orbiting around each other. Less than half of all stars are single stars like the Sun. More than 50% belong to systems containing two or more members. In general, the multiple systems have a hierarchical structure: a star and a binary orbiting around each other in triple systems, two binaries orbiting around each other in quadruple systems. Thus most multiple systems can be described as binaries with several levels.

Binaries are classified on the basis of the method of their discovery. *Visual binaries* can be seen as two separate components, i.e. the separation between the stars is larger than about 0.1 arc sec. The relative position of the components changes over the years as they move in their orbits (Fig. 10.1). In *astrometric binary stars* only one component is seen, but its variable proper motion shows that a second invisible component must be present. The *spectroscopic binary stars* are discovered on the basis of their spectra. Either two sets of spectral lines are seen or else the Doppler shift of the observed lines varies periodically, indicating an invisible companion. The fourth class of binaries are the *photometric binary stars* or the *eclipsing variables*. In these systems, the components of the pair regularly pass in front of each other, causing a change in the total apparent magnitude.

Binary stars can also be classified on the basis of their mutual separation. In distant binaries, the separation between the components is tens or hundreds of astronomical units and their orbital periods are from tens to thousands of years. In close binaries, the

Fig. 10.1. When a visual binary is followed for a long time, the components can be seen to move with respect to each other. Picture of Krüger 60. (Yerkes Observatory)

separation is from about one AU down to the radius of the stars. The orbital period ranges from a few hours to a few years.

The stars in a binary system move in an elliptical orbit around the centre of mass of the system. In Chap. 7 it was shown that the relative orbit, too, is an ellipse, and thus the observations are often described as if one component remained stationary and the other orbited around it.

10.1 Visual Binaries

We consider a visual binary, assuming initially that the brighter primary component is stationary and the fainter secondary component is orbiting around it. The angular separation of the stars and the angular direction to the secondary can be directly observed. Making use of observations extending over many years or decades, the relative orbit of the secondary can be determined. The first binary orbit to be determined was that of ξ UMa, in 1830 (Fig. 10.2).

The observations of visual binaries only give the projection of the orbital ellipse on the plane of the sky. The shape and position of the true orbit are not known. However, they can be calculated if one makes use of the fact that the primary should be located at a focal point of the relative orbit. The deviation of the projected position of the primary from the focus of the projected orbit allows one to determine the orientation of the true orbit.

The absolute size of the orbit can only be found if the distance of the binary is known. Knowing this, the total mass of the system can be calculated from Kepler's third law.

The masses of the individual components can be determined by observing the motions of both components relative to the centre of mass (Fig. 10.3). Let the major semiaxes of the orbital ellipses of the primary and the secondary be a_1 and a_2. Then, according to the definition of the centre of mass,

$$\frac{a_1}{a_2} = \frac{m_2}{m_1} , \qquad (10.1)$$

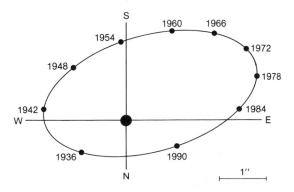

Fig. 10.2. In 1830 the orbit of ξ Ursae Majoris was the first binary orbit determined observationally

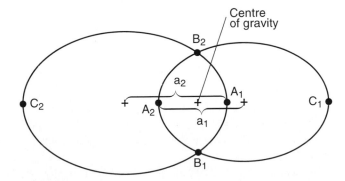

Fig. 10.3. The components of a binary system move around their common centre of mass. A_1, A_2 denote the positions of the stars at a given time A, and similarly for B and C

where m_1 and m_2 are the component masses. The major semiaxis of the relative orbit is

$$a = a_1 + a_2 . \tag{10.2}$$

For example, the masses of the components of ξ UMa have been found to be 1.3 and 1.0 solar masses.

10.2 Astrometric Binary Stars

In astrometric binaries, only the orbit of the brighter component about the centre of mass can be observed. If the mass of the visible component is estimated, e.g. from its luminosity, the mass of the invisible companion can also be estimated.

The first astrometric binary was Sirius, which in the 1830's was observed to have an undulating proper motion. It was concluded that it had a small companion, which was visually discovered a few decades later (Fig. 10.4 and Fig. 15.1). The companion, Sirius B, was a completely new type of object, a white dwarf.

The proper motions of nearby stars have been carefully studied in the search for planetary systems. Although e.g. Barnard's star may have unseen companions, the existence of planetary systems around other stars cannot yet be considered established.

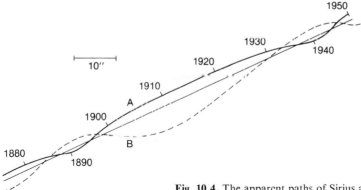

Fig. 10.4. The apparent paths of Sirius and its companion in the sky

10.3 Spectroscopic Binaries (Fig. 10.5)

The spectroscopic binaries appear as single stars in even the most powerful telescopes, but their spectra show a regular variation. The first spectroscopic binary was discovered in the 1880's, when it was found that the spectral lines of ζ UMa or Mizar split into two at regular intervals.

The Doppler shift of a spectral line is directly proportional to the radial velocity. Thus the separation of the spectral lines is largest when one component is directly approaching and the other is receding from the observer. The period of the variation is the orbital period of the stars. Unfortunately, there is no general way of determining

Fig. 10.5. Spectrum of the spectroscopic binary κ Arietis. In the upper spectrum the spectral lines are doubled, in the lower one single. (Lick Observatory)

the position of the orbit in space. The observed velocity v is related to the true velocity v_0 according to

$$v = v_0 \sin i ,\tag{10.3}$$

where the inclination i is the angle between the line of sight and the normal of the orbital plane.

Consider a binary where the components move in circular orbit about the centre of mass. Let the radii of the orbits be a_1 and a_2. From the definition of the centre of mass $m_1 a_1 = m_2 a_2$ and writing $a = a_1 + a_2$, one obtains

$$a_1 = \frac{a m_2}{m_1 + m_2} .\tag{10.4}$$

The true orbital velocity is

$$v_{0,1} = \frac{2 \pi a_1}{P} ,$$

where P is the orbital period. The observed orbital velocity according to (10.3) is thus

$$v_1 = \frac{2 \pi a_1 \sin i}{P} .\tag{10.5}$$

Substituting (10.4), one obtains

$$v_1 = \frac{2 \pi a}{P} \frac{m_2 \sin i}{m_1 + m_2} .$$

Solving for a and substituting it in Kepler's third law, the so-called *mass function equation* is obtained:

$$\frac{m_2^3 \sin^3 i}{(m_1 + m_2)^2} = \frac{v_1^3 P}{2 \pi G} .\tag{10.6}$$

If one component in a spectroscopic binary is so faint that its spectral lines cannot be observed, only P and v_1 are observed. Equation (10.6) then gives the value of the mass function. Neither the masses of the components nor the total mass can be determined.

If the spectral lines of both components can be observed, v_2 is also known. Then (10.5) gives

$$\frac{v_1}{v_2} = \frac{a_1}{a_2}$$

and furthermore the definition of the centre of mass gives

$$m_1 = \frac{m_2 v_2}{v_1} \ .$$

When this is substituted in (10.6), the value of $m_2 \sin^3 i$, and correspondingly, $m_1 \sin^3 i$, can be determined. However, the actual masses cannot be found without knowing the inclination.

The size of the binary orbit (the major semiaxis a) is obtained from (10.5) apart from a factor $\sin i$. In general the orbits of binary stars are not circular and the preceding expressions cannot be applied as they stand. For an eccentric orbit, the shape of the velocity variation departs more and more from a simple sine curve as the eccentricity increases. From the shape of the velocity variation, both the eccentricity and the longitude of the periastron can be determined. Knowing these, the mass function or the individual masses can again be determined to within a factor $\sin^3 i$.

10.4 Photometric Binary Stars

In the photometric binaries, a periodic variation in the total brightness is caused by the motions of the components in a double system. Usually the photometric binaries are eclipsing variables, where the brightness variations are due to the components passing in front of each other. A class of photometric binaries where there are no actual eclipses are the *ellipsoidal variables*. In these systems, at least one of the components has been distorted into an ellipsoidal shape by the tidal pull of the other one. At different phases of the orbit, the projected surface area of the distorted component varies. The surface temperature will also be lower at the ends of the tidal bulges. Together these factors cause a small variation in brightness.

The inclination of the orbit of an eclipsing binary must be very close to 90°. These are the only spectroscopic binaries where the inclination is known and thus where the masses can be uniquely determined.

The variation of the magnitude of eclipsing variables as a function of time is called the *light curve*. According to the shape of the light curve, they are grouped into three main types: *Algol-, β Lyrae-* and *W Ursae Majoris* type (Fig. 10.6).

Algol Stars. The Algol-type eclipsing variables have been named after β Persei or Algol. During most of the period, the light curve is fairly constant. This corresponds to phases during which the stars are seen separate from each other and the total magnitude remains constant. There are two different minima in the light curve, one of which, the primary minimum, is usually much deeper than the other one. This is due to the brightness difference of the stars. When the larger star, which is usually a cool giant, eclipses the smaller and hotter component, there is a deep minimum in the light curve.

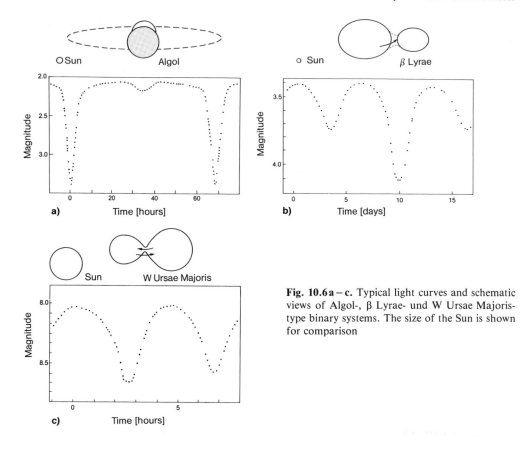

Fig. 10.6a – c. Typical light curves and schematic views of Algol-, β Lyrae- und W Ursae Majoris-type binary systems. The size of the Sun is shown for comparison

When the small, bright star passes across the disc of the giant, the total magnitude of the system does not change by much.

The shape of the minima depends on whether the eclipses are partial or total. In a partial eclipse, the light curve is smooth, since the brightness changes smoothly as the depth of the eclipse varies. In a total eclipse, there is an interval during which one component is completely invisible. The total brightness is then constant and the light curve has a flat-bottomed minimum. The shape of the minima in Algol variables thus gives information on the inclination of the orbit.

The duration of the minima depends on the ratio of the stellar radii to the size of the orbit. If the star is also a spectroscopic binary, the true dimensions of the orbit can be obtained. In that case, the masses and the size of the orbit, and thus also the radii can be determined, without having to know the distance of the system.

β Lyrae Stars. In the β Lyrae-type binaries, the total magnitude varies continuously. The stars are so close to each other that one of them has been pulled into ellipsoidal shape. Thus the brightness varies also outside the eclipses. The β Lyrae variables can be described as eclipsing ellipsoidal variables. In the β Lyrae system itself, one star has overfilled its Roche lobe (see Sect. 12.6) and is steadily losing mass to its companion. The mass transfer causes additional features in the light curve.

W UMa Stars. In W UMa stars, the light curve minima are almost identical, very round and broad. These are close binary systems where both components overfill their Roche lobes, forming a so-called *contact binary* system.

The observed light curves of photometric binaries may contain many additional features that confuse the preceding classification.

- The shape of the star may be distorted by the tidal force of the companion. The star may be ellipsoidal or fill its Roche surface, in which case it becomes drop-like in shape.
- The limb darkening (Sect. 9.6 and 13.2) of the star may be considerable. If the radiation from the edges of the stellar disc is fainter than that from the centre, it will tend to round off the light curve.
- In elongated stars there is gravitational darkening: the parts most distant from the centre are cooler and radiate less energy.
- There are also reflection phenomena in stars. If the stars are close together, they will heat the sides facing each other. The heated part of the surface will then be brighter.
- In systems with mass transfer, the material falling onto one of the components will change the surface temperature.

All these additional effects cause difficulties in interpreting the light curve. Usually, one computes a theoretical model and the corresponding light curve, which is then compared with the observations. The model is varied until a satisfactory fit is obtained.

So far we have been concerned solely with the properties of binary systems in the optical domain. Recently, many double systems that radiate strongly at other wavelengths have been discovered. Particularly interesting are the binary pulsars, where the velocity variation can be determined from radio observations. Many different types of binaries have also been discovered at x-ray wavelengths. These systems will be discussed in Chap. 15.

The binary stars are the only stars with accurately known masses. The masses for other stars are estimated from the mass-luminosity relation (Sect. 9.7), but this has to be calibrated by means of binary observations.

10.5 Examples

10.5.1 *The Mass of a Binary Star*

The distance of a binary star is 10 pc and the largest angular separation of the components is 7$''$ and the smallest is 1$''$. The orbital period is 100 years. The mass of the binary is to be determined, assuming that the orbital plane is normal to the line of sight.

From the angular separation and the distance, the major semiaxis is

$$a = 4'' \times 10\,\mathrm{pc} = 40\,\mathrm{AU} \; .$$

According to Kepler's third law

$$m_1 + m_2 = \frac{a^3}{P^2} = \frac{40^3}{100^2} M_\odot = 6.4 M_\odot \; .$$

Let the major semiaxis for one component be $a_1 = 3''$ and for the other $a_2 = 1''$. Now the masses of the components can be determined separately:

$$m_1 a_1 = m_2 a_2 \Rightarrow m_1 = \frac{a_2}{a_1} m_2 = \frac{m_2}{3}$$

$$m_1 + m_2 = 6.4 \Rightarrow m_1 = 1.6, \quad m_2 = 4.8 .$$

10.5.2   The Light Curve of a Binary

Let us suppose that the line of sight lies in the orbital plane of an Algol type binary, where both components have the same radius. The light curve is essentially as shown in the figure. The primary minimum occurs when the brighter component is eclipsed. The depth of the minima in magnitudes will be calculated.

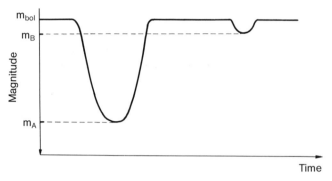

If the effective temperatures of the stars are T_A and T_B and their radius is R, their luminosities are given by

$$L_A = 4\pi R^2 \sigma T_A^4 , \quad L_B = 4\pi R^2 \sigma T_B^4 .$$

The flat part of the curve corresponds to the total luminosity

$$L_{\text{tot}} = L_A + L_B .$$

The luminosities may be expressed as absolute bolometric magnitudes by means of (4.14). Since the distance moduli of the components are the same, the apparent bolometric magnitude at the primary minimum will be

$$m_A - m_{\text{tot}} = M_A - M_{\text{tot}} = -2.5 \lg \frac{L_A}{L_{\text{tot}}} = +2.5 \lg \frac{L_{\text{tot}}}{L_A}$$

$$= 2.5 \lg \frac{4\pi R^2 \sigma T_A^4 + 4\pi R^2 \sigma T_B^4}{4\pi R^2 \sigma T_A^4} = 2.5 \lg \left[1 + \left(\frac{T_B}{T_A} \right)^4 \right] .$$

Similarly, the depth of the secondary minimum is

$$m_B - m_{\text{tot}} = 2.5 \lg \left[1 + \left(\frac{T_A}{T_B} \right)^4 \right] .$$

Let the effective temperatures of the stars be $T_A = 5000\,\text{K}$ and $T_B = 12000\,\text{K}$. The depth of the primary minimum is then

$$m_A - m_{\text{tot}} = 2.5\lg\left[1 + \left(\frac{12\,000}{5000}\right)^4\right] \approx 3.8\,\text{mag} \ .$$

The secondary minimum is

$$m_B - m_{\text{tot}} = 2.5\lg\left[1 + \left(\frac{5000}{12\,000}\right)^4\right] \approx 0.03\,\text{mag} \ .$$

10.6 Exercises

10.6.1   The components of a binary move along circular orbits. The mutual distance is 1 AU, and the mass of each component is $1\,M_\odot$. An observer in the plane of the orbit will see periodic splitting of the spectral lines. What is the maximum separation of the components of the $H\gamma$ line?

10.6.2   A planet (mass m) is orbiting a star (mass M) at a distance α. The distance of the star from the centre of gravity of the system is α'. Show that

$$MP^2 = \alpha^2(\alpha - \alpha') \ ,$$

where P is period in years, distances are in AU's and masses in solar masses.

10.6.3   The distance of Barnard's star is 1.83 pc and mass $0.135\,M_\odot$. It has been suggested that it oscillates with an amplitude of $0.026''$ in 25 year periods. Assuming this oscillation is caused by a planet, find the mass and radius of the orbit of this planet.

Chapter 11 **Stellar Structure**

The stars are huge gas spheres, hundred thousands or millions of times more massive than the Earth. A star such as the Sun can go on shining steadily for thousands of millions of years. This is shown by studies of the prehistory of the Earth, which indicate that the energy radiated by the Sun has not changed by much during the last four thousand million years. The equilibrium of a star must remain stable for such periods.

11.1 Internal Equilibrium Conditions

Mathematically, the conditions for the internal equilibrium of a star can be expressed as four differential equations governing the distribution of mass, gas pressure and energy production and transport in the star. These equations will now be derived.

Hydrostatic Equilibrium. The force of gravity pulls the stellar material towards the centre. It is resisted by the pressure force due to the thermal motions of the gas molecules. The first equilibrium condition is that these forces be in equilibrium.

Consider a cylindrical volume element at the distance r from the centre of the star (Fig. 11.1). Its volume is $dV = dA\,dr$, where dA is its base area and dr, its height; its mass $dm = \rho\,dA\,dr$, where $\rho(r)$ is the gas density at the radius r. If the mass inside radius r is M_r, the gravitational force on the volume element will be

$$dF_g = -\frac{GM_r\,dm}{r^2} = -\frac{GM_r\rho}{r^2}\,dA\,dr\,,$$

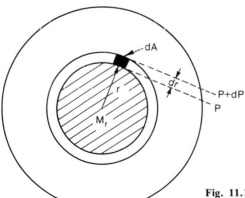

Fig. 11.1. In hydrostatic equilibrium the sum of the gravitational and pressure force acting on a volume element is zero

where G is the gravitational constant. The minus sign in this expression means that the force is directed towards the centre of the star. If the pressure at the lower surface of the volume element is P and at its upper surface $P + dP$, the net force of pressure acting on the element is

$$dF_P = P\, dA - (P + dP)\, dA = -dP\, dA \; .$$

Since the pressure decreases outwards, dP will be negative and the force dF_P, positive. The equilibrium condition is that the total force acting on the volume element should be zero, i.e.

$$0 = dF_g + dF_P = -\frac{GM_r \rho}{r^2}\, dA\, dr - dP\, dA \; , \qquad \text{or}$$

$$\frac{dP}{dr} = -\frac{GM_r \rho}{r^2} \; . \tag{11.1}$$

This is the *equation of hydrostatic equilibrium*.

Mass Distribution. The second equation gives the mass contained within a given radius. Consider a spherical shell of thickness dr at the distance r from the centre (Fig. 11.2). Its mass is

$$dM_r = 4\pi r^2 \rho\, dr \; ,$$

giving the *mass continuity equation*

$$\frac{dM_r}{dr} = 4\pi r^2 \rho \; . \tag{11.2}$$

Energy Production. The third equilibrium condition expresses the conservation of energy, in the requirement that any energy produced in the star has to be carried to the surface and radiated away. We again consider a spherical shell of thickness dr and mass dM_r at the radius r (Fig. 11.3). Let L_r be the energy flux, i.e. the amount of energy

Fig. 11.2. The mass of a thin spherical shell is the product of its volume and its density

Fig. 11.3. The energy flowing out of a spherical shell is the sum of the energy flowing into it and the energy generated within the shell

Fig. 11.2

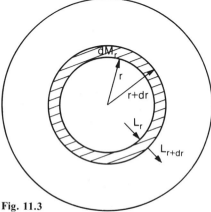

Fig. 11.3

passing through the surface r per unit time. If ε is the energy production coefficient, i.e. the amount of energy released in the star per unit time and mass, then

$$dL_r = L_{r+dr} - L_r = \varepsilon\, dM_r = 4\pi r^2 \rho \varepsilon\, dr .$$

Thus the energy conservation equation is

$$\frac{dL_r}{dr} = 4\pi r^2 \rho \varepsilon . \qquad (11.3)$$

The rate at which energy is produced depends on the distance to the centre. Essentially all of the energy radiated by the star is produced in the hot and dense core. In the outer layers, the energy production is negligible and L_r is almost constant.

The Temperature Gradient. The fourth equilibrium equation gives the temperature as a function of radius or the temperature gradient dT/dr. The form of the equation depends on how the energy is transported: by *conduction, convection* or *radiation*.

In the interiors of normal stars, conduction is very inefficient, since the electrons carrying the energy can only travel a short distance without colliding with other particles. Conduction only becomes important in compact stars, white dwarfs and neutron stars, where the free path of the electrons becomes very large. In normal stars, conductive energy transport can be neglected.

In radiative energy transport, photons emitted in hotter parts of the star are absorbed in cooler regions, which they heat. The star is said to be in radiative equilibrium, when the energy released in the stellar interior is carried outwards entirely by radiation.

The radiative temperature gradient is related to the energy flux L_r according to

$$\frac{dT}{dr} = \left(-\frac{3}{4ac}\right)\left(\frac{\kappa\rho}{T^3}\right)\left(\frac{L_r}{4\pi r^2}\right), \qquad (11.4)$$

where a is the radiation constant (see p. 267), c, the speed of light, and ρ, the density. The mass absorption coefficient κ gives the amount of absorption per unit mass. Its value depends on the temperature, density and chemical composition.

In order to derive (11.4), we consider the equation of radiative transfer, (1), on p. 126. In terms of the variables used in the present chapter, it may be written

$$\cos\theta \frac{dI_v}{dr} = -\kappa_v \rho I_v + j_v .$$

In this equation, κ_v is replaced with a suitable mean value κ. The equation is then multiplied with $\cos\theta$ and integrated over all directions and frequencies. On the left hand side, I_v may be approximated with the Planck function B_v. The frequency integral may then be evaluated by means of (5.14). On the right-hand side, the first term can be expressed in terms of the flux density according to (4.2) and the integral over directions of the second gives zero, since j_v does not depend on θ. One thus obtains

$$\frac{4\pi}{3}\frac{d}{dr}\left(\frac{ac}{4\pi}T^4\right) = -\kappa\rho F_r .$$

Finally, using the relation

$$F_r = \frac{L_r}{4\pi r^2}$$

between the flux density F_r and the energy flux L_r, one obtains (11.4).

The derivative dT/dr is negative, since the temperature increases inwards. Clearly, there has to be a temperature gradient, if energy is to be transported by radiation: otherwise the radiation field would be the same in all directions and the net flux F_r would vanish.

If the radiative transfer of energy becomes inefficient, the absolute value of the radiative temperature gradient becomes very large. In that case, motions are set up in the gas, which carry the energy outwards more efficiently than the radiation. In these convective motions, hot gas rises upwards into cooler layers, where it loses its energy and sinks again. The rising and sinking gas elements also mix the stellar material, and the composition of the convective parts of a star becomes homogeneous.

The expression for the convective temperature gradient is

$$\frac{dT}{dr} = \left(1 - \frac{1}{\gamma}\right) \frac{T}{P} \frac{dP}{dr}, \tag{11.5}$$

where P is the gas pressure and the *adiabatic exponent*

$$\gamma = C_P/C_V$$

is the ratio of the specific heats at constant pressure and volume, respectively. Equation (11.5) is called the *adiabatic temperature gradient*. It says that the moving gas masses obey the adiabatic equation of state,

$$T \propto P^{1-1/\gamma}.$$

The ratio of specific heats γ depends on the state of ionization of the gas and can be calculated if the temperature, density and composition are known.

The convective motions set in when the radiative temperature gradient becomes larger in absolute value than the adiabatic gradient, i.e. if either the radiative gradient becomes steep, or if the convective gradient becomes shallow. From (11.4) it can be seen that a steep radiative gradient is expected, if either the energy flux density or the mass absorption coefficient becomes large. The convective gradient may become small, if the adiabatic exponent approaches 1.

In the practical computation of stellar structure, one uses either (11.4) or (11.5), depending on which equation gives a less steep temperature gradient. In the outermost layers of stars, (11.5) is no longer a good approximation. The convective temperature gradient is then calculated using the so-called mixing-length theory. The theory of convection is a difficult and still imperfectly understood problem, which is beyond the scope of this presentation.

Boundary Conditions. In order to obtain a well-posed problem, four boundary conditions have to be prescribed for the preceding differential equations. First, there are no sources of energy or mass at the centre; thus $M = L = 0$ when $r = 0$. Furthermore, the total mass within the radius R of the star is fixed, $M_R = M$, which defines the value of the radius for a given mass. Finally, the temperature and pressure at the stellar surface have some determinate values, T_R and P_R. These will be very small compared to those in the interior, and thus it is usually sufficient to take $T_R = P_R = 0$.

In addition to these boundary conditions, one needs an expression for the pressure, which is given by the equation of state, as well as expressions for the mass absorption coefficient and the energy generation rate, which will be considered later.

The solution of the basic differential equations give the mass, temperature, density and energy flux as functions of the radius. The stellar radius and luminosity can then be calculated and compared with the observations.

The properties of a stellar equilibrium model are essentially determined once the mass and the chemical composition have been given. This result is known as the *Vogt-Russell theorem*.

11.2 Physical State of the Gas

Due to the high temperature, the gas in the stars is almost completely ionized. The interactions between individual particles are small, so that, to a good approximation, the gas obeys the perfect gas equation of state,

$$P = \frac{k}{\mu m_H} \rho T , \tag{11.6}$$

where k is Boltzmann's constant, μ, the mean molecular weight in units of m_H, and m_H, the mass of the hydrogen atom.

The mean molecular weight can be approximately calculated assuming complete ionization. An atom with nuclear charge Z then produces $Z+1$ free particles (the nucleus and Z electrons). Hydrogen gives rise to two particles per atomic mass unit; helium gives rise to three particles per four atomic mass units. For all elements heavier than hydrogen and helium, it is usually sufficient to take $Z+1$ to be half the atomic weight. (Exact values could easily be calculated, but the abundance of heavy elements is so small that this is usually not necessary.) In astrophysics, the relative mass fraction of hydrogen is conventionally denoted by X, that of helium by Y and that of all heavier elements by Z, so that

$$X + Y + Z = 1 . \tag{11.7}$$

(The Z occurring in this equation should not be confused with the nuclear charge, which is unfortunately denoted by the same letter.) Thus the mean molecular weight will be

$$\mu = 1/(2X + \tfrac{3}{4}Y + \tfrac{1}{2}Z) . \tag{11.8}$$

At high temperatures, the radiation pressure has to be added to the gas pressure described by the perfect gas equation. The pressure exerted by radiation is (see p. 267)

$$P_{\text{rad}} = \tfrac{1}{3} a T^4 , \tag{11.9}$$

where a is the radiation constant. Thus the total pressure is

$$P = \frac{k}{\mu m_{\mathrm{H}}} \rho T + \frac{1}{3} a T^4 . \tag{11.10}$$

The perfect gas law does not apply at very high densities. When the density inside a star exceeds $10^7 \, \mathrm{kg/m^3}$, the electrons become degenerate and their pressure is then that of a degenerate gas. In normal stars, the gas is usually nondegenerate, but in white dwarfs and in neutron stars, degeneracy is of central importance. The pressure of a degenerate electron gas is (see p. 268)

$$P \approx (h^2/m_{\mathrm{e}})(N/V)^{5/3} , \tag{11.11}$$

where m_{e} is the electron mass and N/V, the number of electrons per unit volume. This equation may be written in terms of the density

$$\rho = N \mu_{\mathrm{e}} m_{\mathrm{H}} / V ,$$

where μ_{e} is the mean molecular weight per free electron. An expression for μ_{e} may be derived in analogy with (11.8):

$$\mu_{\mathrm{e}} = 1/(X + \tfrac{2}{4} Y + \tfrac{1}{2} Z) = 2/(X + 1) .$$

For the solar hydrogen abundance, this yields

$$\mu_{\mathrm{e}} = 2/(0.71 + 1) = 1.17 .$$

The final expression for the pressure is

$$P \approx (h^2/m_{\mathrm{e}})(\mu_{\mathrm{e}} m_{\mathrm{H}})^{-5/3} \rho^{5/3} . \tag{11.12}$$

This is the equation of state of a degenerate electron gas. In contrast to the perfect gas law, the pressure no longer depends on the temperature, only on the density and on the particle masses.

In normal stars, the degenerate gas pressure is negligible, but in the central parts of giant stars and in white dwarfs, where the density is of the order of $10^8 \, \mathrm{kg/m^3}$, the degenerate gas pressure is dominant, in spite of the high temperature.

At even higher densities, the electron momenta become so large that their velocities approach the speed of light. In this case, the formulae of the special theory of relativity have to be used. The pressure of a relativistic degenerate gas is

$$P \approx h c (N/V)^{4/3} = h c (\mu_{\mathrm{e}} m_{\mathrm{H}})^{-4/3} \rho^{4/3} . \tag{11.13}$$

In the relativistic case, the pressure is proportional to the density to the power 4/3, rather than 5/3, as for the nonrelativistic case. The transition to the relativistic situation takes place roughly at the density $10^9 \, \mathrm{kg/m^3}$.

In general, the pressure inside a star depends on the temperature (except for a completely degenerate gas), density and chemical composition. In actual stars, the gas will never be totally ionized or completely degenerate. The pressure will then be given by more complicated expressions. Still, it can be calculated for each case of interest. One may then write

$$P = P(T, \rho, X, Y, Z) , \tag{11.14}$$

giving the pressure as a known function of the temperature, density and chemical composition.

The *opacity* of the gas describes how difficult it is for radiation to propagate through it (see Sect. 5.8). The opacity depends on the temperature, density and chemical composition of the gas. It is usually written as $\alpha = \kappa \rho$, where ρ is the density of the gas and κ, the mass absorption coefficient.

The inverse of the opacity represents the mean free path of radiation in the medium, i.e. the distance it can propagate without being scattered or absorbed. The different types of absorption processes (bound-bound, bound-free, free-free) have been described in Sect. 5.1. The opacity of the stellar material due to each process can be calculated for relevant values of temperature and density.

*Gas Pressure and Radiation Pressure

Let us consider noninteracting particles in a rectangular box. The particles may also be photons. Let the sides of the box be Δx, Δy and Δz, and the number of particles, N. The pressure is caused by the collisions of the particles with the sides of the box. When a particle hits a wall perpendicular to the x-axis, its momentum in the x-direction, p_x, changes by $\Delta p = 2p_x$. The particle will return to the same wall after the time $\Delta t = 2\Delta x / v_x$. Thus the pressure exerted by the particles on the wall (surface area $A = \Delta y \Delta z$) is

$$P = \frac{F}{A} = \frac{\sum \Delta p / \Delta t}{A} = \frac{\sum p_x v_x}{\Delta x \Delta y \Delta z} = \frac{N \langle p_x v_x \rangle}{V} ,$$

where $V = \Delta x \Delta y \Delta z$ is the volume of the box and the angular brackets represent the average value. The momentum is $p_x = m v_x$ (where $m = h\nu / c^2$ for photons), and hence

$$P = \frac{Nm \langle v_x^2 \rangle}{V} .$$

Suppose the velocities of the particles are isotropically distributed. Then

$$\langle v_x^2 \rangle = \langle v_y^2 \rangle = \langle v_z^2 \rangle , \quad \text{i.e.}$$

$$\langle v^2 \rangle = \langle v_x^2 \rangle + \langle v_y^2 \rangle + \langle v_z^2 \rangle = 3 \langle v_x^2 \rangle \quad \text{and}$$

$$P = \frac{Nm \langle v^2 \rangle}{3V} .$$

If the particles are gas molecules, the energy of a molecule is $\varepsilon = m v^2 / 2$. The total energy of the gas is $E = N\langle \varepsilon \rangle = \frac{1}{2} Nm \langle v^2 \rangle$, and the pressure may be written

$$P = \frac{2}{3} \frac{E}{V} \quad \text{(gas)} .$$

If the particles are photons, they move with the speed of light and their energy is $\varepsilon = mc^2$. The total energy of a photon gas is thus $E = N\langle \varepsilon \rangle = Nmc^2$ and the pressure is

$$P = \frac{1}{3} \frac{E}{V} \quad \text{(radiation)} .$$

According to (4.7), (4.4) and (5.15), the energy density of blackbody radiation is

$$\frac{E}{V} = u = \frac{4\pi}{c} I = \frac{4}{c} F = \frac{4}{c} \sigma T^4 \equiv a T^4 ,$$

where $a = 4\sigma/c = 7.564 \times 10^{-16}$ J m^{-3} K^{-4} is the radiation constant. Thus the radiation pressure is

$$P_{\text{rad}} = a T^4/3 \ .$$

*The Pressure of a Degenerate Gas

A gas where all available energy levels up to a limiting momemtum p_0, known as the Fermi momentum, are filled is called degenerate. We shall determine the pressure of a completely degenerate electron gas.

Let the volume of the gas be V. We consider electrons with momenta in the range $(p, p+dp)$. Their available phase space volume is $4\pi p^2 dp \, V$. According to the Heisenberg uncertainty relation, the elementary volume in phase space is h^3 and, according to the Pauli exclusion principle, this volume can contain two electrons with opposite spins. Thus the number of electrons in the momentum interval $(p, p+dp)$ is

$$dN = 2 \, \frac{4\pi p^2 \, dp \, V}{h^3} \ .$$

The total number of electrons with momenta smaller than p_0 is

$$N = \int dN = \frac{8\pi V}{h^3} \int_0^{p_0} p^2 \, dp = \frac{8\pi V}{3h^3} \, p_0^3 \ .$$

Hence the Fermi momentum p_0 is

$$p_0 = \left(\frac{3}{\pi}\right)^{1/3} \frac{h}{2} \left(\frac{N}{V}\right)^{1/3} \ .$$

Nonrelativistic Gas. The kinetic energy of an electron is $\varepsilon = p^2/2m_e$. The total energy of the gas is

$$E = \int \varepsilon \, dN = \frac{4\pi V}{m_e h^3} \int_0^{p_0} p^4 \, dp = \frac{4\pi V}{5 m_e h^3} \, p_0^5 \ .$$

Introducing the expression for the Fermi momentum p_0, one obtains

$$E = \frac{\pi}{40} \left(\frac{3}{\pi}\right)^{5/3} \frac{h^2}{m_e} V \left(\frac{N}{V}\right)^{5/3} \ .$$

The pressure of the gas according to p. 267 is

$$P = \frac{2}{3} \frac{E}{V} = \frac{1}{20} \left(\frac{3}{\pi}\right)^{2/3} \frac{h^2}{m_e} \left(\frac{N}{V}\right)^{5/3} \ .$$

Here, N/V is the number density of electrons.

Relativistic Gas. If the density becomes so large that the electron kinetic energy ε corresponding to the Fermi momentum exceeds the rest energy $m_e c^2$, the relativistic expression for the electron energy has to be used. In the extreme relativistic case, $\varepsilon = cp$ and the total energy

$$E = \int \varepsilon \, dN = \frac{8\pi c V}{h^3} \int_0^{p_0} p^3 \, dp = \frac{2\pi c V}{h^3} \, p_0^4 \ .$$

The expression for the Fermi momentum remains unchanged, and hence

$$E = \frac{\pi}{8} \left(\frac{3}{\pi}\right)^{4/3} h c V \left(\frac{N}{V}\right)^{4/3} \ .$$

The pressure of the relativistic electron gas is obtained from the formula derived for a photon gas on p. 267,

$$P = \frac{1}{3}\frac{E}{V} = \frac{1}{8}\left(\frac{3}{\pi}\right)^{1/3} hc \left(\frac{N}{V}\right)^{4/3}.$$

We have obtained the nonrelativistic and extreme relativistic approximations to the pressure. In intermediate cases, the exact expression for the electron energy,

$$\varepsilon = (m_e^2 c^4 + p^2 c^2)^{1/2}$$

has to be used.

The preceding derivations are rigorously valid only at zero temperature. However, the densities in compact stars are so high that the effects of a nonzero temperature are negligible, and the gas may be considered completely degenerate.

11.3 Stellar Energy Sources

When the equations of stellar structure were derived, the character of the source of stellar energy was left unspecified. Knowing a typical stellar luminosity, one can calculate how long different energy sources would last. For instance, normal chemical burning could produce energy for only a few thousand years. The energy released by the contraction of a star would last slightly longer, but after a few million years, this energy source would also run out.

Terrestrial biological and geological evidence shows that the solar luminosity has remained fairly constant for at least a few thousand million years. Since the age of the Earth is about 5000 million years, the Sun has presumably existed at least for that time. Since the solar luminosity is 4×10^{26} W, it has radiated about 6×10^{43} J in 5×10^9 years. The Sun's mass is 2×10^{30} kg; thus it must be able to produce at least 3×10^{13} J/kg.

The general conditions in the solar interior are known, regardless of the exact energy source. Thus, in Exercise 11.5.5, it will be estimated that the temperature at half the radius is about 5 million degrees. The central temperature must be about ten million kelvins, which is large enough for *thermonuclear fusion reactions* to take place.

In fusion reactions, light elements are transformed into heavier ones. The final reaction products have a smaller total mass than the initial nuclei. This mass difference is released as energy, according to Einstein's relation $E - mc^2$. Thermonuclear reactions are commonly referred to as burning, although they have no relation to the chemical burning of ordinary fuels.

The atomic nucleus consists of protons and neutrons, together referred to as nucleons. We define

m_p: proton mass,
m_n: neutron mass,
Z: nuclear charge $\triangleq$ atomic number,
N: neutron number,
$A = Z + N$: atomic weight,
$m(Z, N)$: mass of the nucleus.

The mass of the nucleus is smaller than the sum of the masses of all its nucleons. The difference is called the *binding energy*. The binding energy per nucleon is

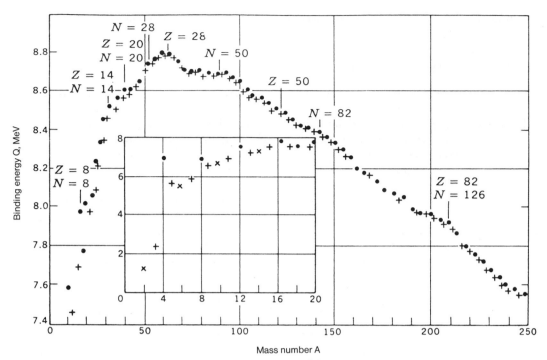

Fig. 11.4. The nuclear binding energy per nucleon as a function of the atomic weight. Among isotopes with the same atomic weight, the one with the largest binding energy is shown. The points correspond to nuclei with even proton and neutron numbers, the crosses to nuclei with odd mass numbers. [Preston, M. A. (1962): *Physics of the Nucleus* (Addison-Wesley Publishing Company, Inc., Reading, Mass.]

$$Q = \frac{1}{A}\,[Zm_{\mathrm{p}} + Nm_{\mathrm{n}} - m(Z, N)]\,c^2 . \tag{11.15}$$

It turns out that Q increases towards heavier elements, up to iron ($Z = 26$). Beyond iron, the binding energy again begins to decrease (Fig. 11.4).

It is known that the stars consist mostly of hydrogen. Let us consider how much energy would be released by the fusion of four hydrogen nuclei into a helium nucleus. The mass of a proton is 1.6725×10^{-27} kg and that of a helium nucleus is 6.644×10^{-27} kg. The mass difference, 4.6×10^{-29} kg, corresponds to an energy difference $E = 4.1 \times 10^{-12}$ J. Thus 0.7% of the mass is turned into energy in the reaction, corresponding to an energy release of 6×10^{14} J per kg hydrogen. This should be compared with our previous estimate that 3×10^{13} J/kg is needed.

Already in the 1930's, it was generally accepted that stellar energy had to be produced by nuclear fusion. In 1938, *Hans Bethe* and independently *Carl Friedrich von Weizsäcker* put forward the first detailed mechanism for energy production in the stars, the *carbon-nitrogen-oxygen (CNO) cycle*. The other important energy generation processes (the *proton-proton chain* and the *triple-alpha reaction*) were not proposed until the 1950's.

The Proton-Proton Chain (Fig. 11.5). In stars with masses of about that of the Sun or smaller, the energy is produced by the proton-proton (pp) chain. It consists of the following steps:

ppI:

(1) $^1H + {}^1H \rightarrow {}^2H + e^+ + \nu_e$
(2) $^2H + {}^1H \rightarrow {}^3He + \gamma$
(3) $^3He + {}^3He \rightarrow {}^4He + 2\,{}^1H$.

For each reaction (3), the reactions (1) and (2) have to take place twice. The first reaction step has a very small probability, which has not been measured in the laboratory. At the central density and temperature of the Sun, the expected time for a proton to collide with another one to form a deuteron is 10^{10} years, on the average. It is only thanks to the slowness of this reaction that the Sun is still shining. If it were faster, the Sun would have burnt out long ago. The neutrino produced in reaction (1) can

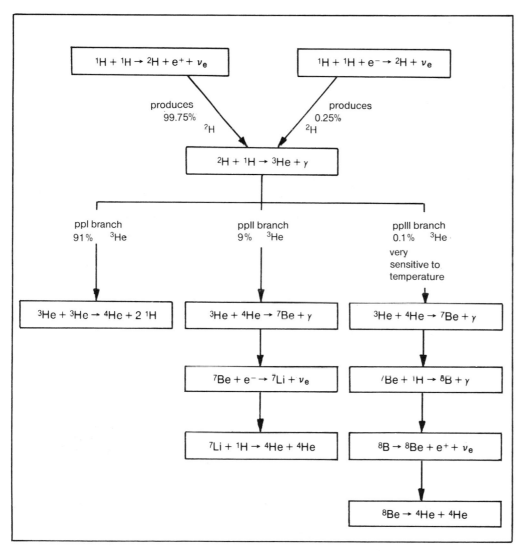

Fig. 11.5. The proton-proton chain. In the ppI branch, four protons are transformed into one helium nucleus, two positrons, two neutrinos and radiation. The relative weights of the reactions are given for conditions in the Sun. The pp chain is the most important energy source in stars with mass below $1.5\,M_\odot$

escape freely from the star and carries away some of the energy released. The positron e^+ is immediately annihilated together with an electron, giving rise to two gamma quanta.

The second reaction, where a deuteron and a proton unite to form the helium isotope ^{3}He, is very fast compared to the preceding one. Thus the abundance of deuterons inside stars is very small.

The last step in the pp chain can take three different forms. The ppI chain shown above is the most probable one. In the Sun, 91% of the energy is produced by the ppI chain.

It is also possible for ^{3}He nuclei to unite into ^{4}He nuclei in two additional branches of the pp chain.

ppII:

(3) ^{3}He + ^{4}He → ^{7}Be + γ
(4) ^{7}Be + e^- → ^{7}Li + ν_e
(5) ^{7}Li + ^{1}H   → ^{4}He + ^{4}He

ppIII:

(3) ^{3}He + ^{4}He → ^{7}Be + γ
(4) ^{7}Be + ^{1}H   → ^{8}B + γ
(5) ^{8}B   → ^{8}Be + e^+ + ν_e
(6) ^{8}Be   → ^{4}He + ^{4}He .

The Carbon Cycle (Fig. 11.6). At temperatures below 20 million degrees, the pp chain is the main energy production mechanism. At higher temperatures, corresponding to

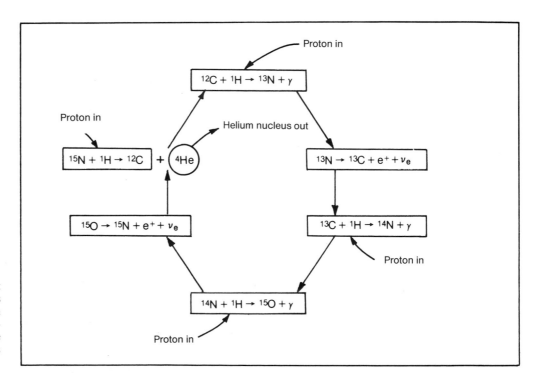

Fig. 11.6. The CNO cycle ist catalysed by ^{12}C. It transforms four protons into a helium nucleus, two positrons, two neutrinos and radiation. It is the dominant energy source for stars more massive than $1.5\,M_\odot$

stars with masses above $1.5\,M_\odot$, the *carbon (CNO) cycle* becomes dominant, because its reaction rate increases more rapidly with temperature. In the CNO cycle, carbon, oxygen and nitrogen act as catalysts.

The reaction cycle is the following:

(1) $^{12}C + {}^1H \rightarrow {}^{13}N + \gamma$
(2) $^{13}N \qquad \rightarrow {}^{13}C + e^+ + \nu_e$
(3) $^{13}C + {}^1H \rightarrow {}^{14}N + \gamma$
(4) $^{14}N + {}^1H \rightarrow {}^{15}O + \gamma$
(5) $^{15}O \qquad \rightarrow {}^{15}N + e^+ + \nu_e$
(6) $^{15}N + {}^1H \rightarrow {}^{12}C + {}^4He$.

Reaction (4) is slowest, and thus determines the rate of the CNO cycle. At a temperature of 20 million degrees, the reaction time for reaction (4) is a million years.

The fraction of energy released as radiation in the CNO cycle is slightly smaller than in the pp chain, because more energy is carried away by neutrinos.

The Triple Alpha Reaction. As a result of the preceding reactions, the abundance of helium in the stellar interior increases. At a temperature above 10^8 degrees, the helium can be transformed into carbon in the triple alpha reaction:

(1) $^4He + {}^4He \leftrightarrows {}^8Be$
(2) $^8Be + {}^4He \rightarrow {}^{12}C + \gamma$.

Here, ^8Be is unstable and decays into two helium nuclei or alpha particles in 2.6×10^{-16} seconds. The production of carbon thus requires the almost simultaneous collision of three α particles. The reaction is often written

$$3\,{}^4He \rightarrow {}^{12}C + \gamma\;.$$

Once helium burning has been completed, at higher temperatures other reactions become possible, in which heavier elements up to iron and nickel are built up. Examples of such reactions are various alpha reactions and oxygen-, carbon- and silicon burning.

Alpha Reactions. During helium burning, some of the carbon nuclei produced react with helium nuclei to form oxygen, which in turn reacts to form neon, etc. These reactions are fairly rare and thus are not important as stellar energy sources. Examples are

$$
\begin{aligned}
^{12}C + {}^4He \ &\rightarrow \ {}^{16}O + \gamma \\
^{16}O + {}^4He \ &\rightarrow \ {}^{20}Ne + \gamma \\
^{20}Ne + {}^4He \ &\rightarrow \ {}^{24}Mg + \gamma \;.
\end{aligned}
$$

Carbon Burning. After the helium is exhausted, carbon burning sets in at the temperature $(5-8) \times 10^8$ K:

$$
\begin{aligned}
^{12}C + {}^{12}C \ &\rightarrow \ {}^{24}Mg + \gamma \\
&\rightarrow \ {}^{23}Na + p \\
&\rightarrow \ {}^{20}Ne + {}^4He \\
&\rightarrow \ {}^{23}Mg + n \\
&\rightarrow \ {}^{16}O + 2\,{}^4He \;.
\end{aligned}
$$

Oxygen Burning. Oxygen is consumed at slightly higher temperatures in the reactions

$$
\begin{aligned}
{}^{16}\text{O} + {}^{16}\text{O} &\rightarrow {}^{32}\text{S} + \gamma \\
&\rightarrow {}^{31}\text{P} + \text{p} \\
&\rightarrow {}^{28}\text{Si} + {}^{4}\text{He} \\
&\rightarrow {}^{31}\text{S} + \text{n} \\
&\rightarrow {}^{24}\text{Mg} + 2\,{}^{4}\text{He} .
\end{aligned}
$$

Silicon Burning. After several intermediate steps, the burning of silicon produces nickel and iron. The total process may be expressed as

$$
\begin{aligned}
{}^{28}\text{Si} + {}^{28}\text{Si} &\rightarrow {}^{56}\text{Ni} + \gamma \\
{}^{56}\text{Ni} &\rightarrow {}^{56}\text{Fe} + 2\,\text{e}^{+} + 2\,\nu_{\text{e}} .
\end{aligned}
$$

When the temperature becomes larger than 10^9 K, the energy of the photons becomes large enough to destroy certain nuclei. Such reactions are called *photonuclear reactions* or *photodissociations*.

The production of elements heavier than iron requires an input of energy, and therefore such elements cannot be produced by thermonuclear reactions. Elements heavier than iron are almost exclusively produced by *neutron capture* during the final violent stages of stellar evolution (Sect. 12.5).

The rates of the reactions presented above can be determined by laboratory experiments or by theoretical calculations. Knowing them, one can calculate the rate at which energy is released per unit mass and time as a function of the density, temperature and chemical composition:

$$
\varepsilon = \varepsilon (T, \rho, X, Y, Z) . \tag{11.16}
$$

In reality, the relative abundance of each of the heavier nuclei needs to be known, not just their total abundance Z.

11.4 Stellar Models

A theoretical stellar model is obtained if one solves the differential equations for stellar structure. As we have already noted, the model is uniquely defined, once the chemical composition and the mass of the star have been given.

Stars just formed out of the interstellar medium are chemically homogeneous. When stellar models for homogeneous stars are plotted in the HR diagram, they fall along the lower edge of the main sequence. The theoretical sequence obtained in this way is called the *zero age main sequence*, ZAMS. The exact position of the ZAMS depends on the initial chemical composition. For stars with an initial abundance of heavy elements like that in the Sun, the computed ZAMS is in good agreement with the observations. If the initial value of Z is smaller, the theoretical ZAMS falls below the main sequence in the subdwarf region of the HR diagram. This is related to the classification of stars into populations I and II, which is discussed in Sect. 18.2.

The theoretical models also provide an explanation of the mass-luminosity relation. The computed properties of zero age main sequence stars of different masses are given in Table 11.1. The chemical composition assumed is $X = 0.71$ (hydrogen mass

Table 11.1. Properties of zero age main sequence stars

$M/M_\odot$	$L/L_\odot$	T_e [10^3 K]	$R/R_\odot$	T_c [10^6 K]	ρ_c [g/cm^3]	M_{ci}/M	M_{ce}/M
30	140 000	44	6.6	36	3.0	0.60	0
15	21 000	32	4.7	34	6.2	0.39	0
9	4 500	26	3.5	31	7.9	0.26	0
5	630	20	2.2	27	26	0.22	0
3	93	14	1.7	24	42	0.18	0
1.5	5.4	8.1	1.2	19	95	0.06	0
1.0	0.74	5.8	0.87	14	89	0	0.01
0.5	0.038	3.9	0.44	9.1	78	0	0.41

(T_c = central temperature; ρ_c = central density; M_{ci}/M = relative mass of convective interior; M_{ce}/M = relative mass of convective envelope)

fraction), $Y = 0.27$ (helium) and $Z = 0.02$ (heavier elements), except for the 30 M star, which has $X = 0.70$ and $Y = 0.28$. The luminosity of a one solar mass star is $0.74\,L_\odot$ and the radius $0.87\,R_\odot$. Thus the Sun has brightened and expanded to some extent during its evolution. However, these changes are small and do not conflict with the evidence for a steady solar energy output. In addition, the biological evidence only goes back about 3000 million years.

The model calculations show that the central temperature in the smallest stars ($M = 0.08\,M_\odot$) is about 4×10^6 K, which is the minimum temperature required for the onset of thermonuclear reactions. In the biggest stars ($M = 50\,M_\odot$), the central temperature reaches 4×10^7 K.

The changes in chemical composition caused by the nuclear reactions can be computed, since the rates of the various reactions at different depths in the star are known. For example, the change ΔX of the hydrogen abundance in the time interval Δt is proportional to the rate of energy generation ε and to Δt:

$$\Delta X \propto -\varepsilon \Delta t . \tag{11.17}$$

The constant of proportionality is clearly the amount of hydrogen consumed per unit energy [kg/J]. The value of this constant of proportionality is different for the pp chain and the CNO cycle. Therefore the contribution from each reaction chain must be calculated separately in (11.17). For elements that are produced by the nuclear reactions, the right-hand side contribution in (11.17) is positive. If the star is convective, the change in composition is obtained by taking the average of (11.17) over the convection zone.

11.5 Examples

11.5.1 *The Gravitational Acceleration at the Solar Surface*

The expression for the gravitational acceleration is

$$g = \frac{GM_\odot}{R^2} .$$

Using the solar mass $M = 1.989 \times 10^{30}$ kg and radius $R = 6.96 \times 10^8$ m, one obtains

$$g = 274 \text{ m s}^{-2} = 28\, g_0 ,$$

where $g_0 = 9.81$ m s^{-2} is the gravitational acceleration at the Earth's surface.

11.5.2 The Average Density of the Sun

The volume of a sphere with radius R is

$$V = \tfrac{4}{3}\pi R^3 ;$$

thus the average density of the Sun is

$$\bar\rho = \frac{M}{V} = \frac{3M}{4\pi R^3} \approx 1410 \text{ kg m}^{-3} .$$

11.5.3 Estimate of the Pressure at Half the Solar Radius

The pressure can be estimated from condition for the hydrostatic equilibrium, (11.1). Suppose the density is constant and equal to the average density $\bar\rho$. Then the mass within the radius r is

$$M_r = \tfrac{4}{3}\pi\bar\rho r^3$$

and the hydrostatic equation can be written

$$\frac{dP}{dr} = -\frac{GM_r\rho}{r^2} = -\frac{4\pi G\bar\rho^2 r}{3} .$$

This can be integrated from half the solar radius, $r = R_\odot/2$, to the surface, where the pressure vanishes:

$$\int_P^0 dP = -\frac{4}{3}\pi G\bar\rho^2 \int_{R_\odot/2}^{R_\odot} r\, dr$$

$$\Rightarrow P = \tfrac{1}{2}\pi G\bar\rho^2 R_\odot^2 \approx \tfrac{1}{2}\pi\, 6.67 \times 10^{-11} \times 1410^2 \times (6.96 \times 10^8)\text{ N m}^{-2}$$

$$\approx 10^{14} \text{ Pa} .$$

This estimate is extremely rough, since the density increases strongly inwards.

11.5.4 The Mean Molecular Weight of the Sun

In the outer layers of the Sun, the initial chemical composition has not been changed by nuclear reactions. In this case, one can use the values $X = 0.71$, $Y = 0.27$ and $Z = 0.02$. The mean molecular weight (11.8) is then

$$\mu = 1/(2 \times 0.71 + 0.75 \times 0.27 + 0.5 \times 0.02) \approx 0.61 .$$

When the hydrogen is completely exhausted, $X = 0$ and $Y = 0.98$, and hence

$$\mu = 1/(0.75 \times 0.98 + 0.5 \times 0.02) \approx 1.34 .$$

11.5.5   *The Temperature of the Sun at $r = R_\odot/2$*

Using the density from Exercise 11.5.2 and the pressure from Exercise 11.5.3, the temperature can be estimated from the perfect gas law (11.6). Assuming the surface value for the mean molecular weight (Exercise 11.5.4), one obtains the temperature

$$T = \frac{\mu m_H P}{k \rho} = \frac{0.61 \times 1.67 \times 10^{-27} \times 1.0 \times 10^{14}}{1.38 \times 10^{-23} \times 1410} \approx 5 \times 10^6 \, \text{K} \,.$$

11.5.6   *The Radiation Pressure in the Sun at $r = R_\odot/2$*

The radiation pressure at the temperature estimated in the previous exercise [see (11.9)] is

$$P_{\text{rad}} = \tfrac{1}{3} a T^4 = \tfrac{1}{3} \times 7.564 \times 10^{-16} \times (5 \times 10^6)^4 \approx 2 \times 10^{11} \, \text{Pa} \,.$$

This is about a thousand times smaller than the gas pressure estimated in Exercise 11.5.3. Thus it confirms that the use of the perfect gas law in Exercise 11.5.5 was correct.

11.5.7   *The Path of a Photon from the Centre of a Star to Its Surface*

Radiative energy transport can be described as a random walk, where a photon is repeatedly absorbed and re-emitted in a random direction. Let the step length of the walk (the mean free path) be d. Consider, for simplicity, the random walk in a plane. After one step, the photon is absorbed at

$$x_1 = d \cos \theta_1 \,, \quad y_1 = d \sin \theta_1 \,,$$

where θ_1 is an angle giving the direction of the step. After N steps, the coordinates are

$$x = \sum_1^N d \cos \theta_i \,, \quad y = \sum_1^N d \sin \theta_i \,,$$

and the distance from the starting point is

$$r^2 = x^2 + y^2 = d^2 \left[\left(\sum_1^N \cos \theta_i \right)^2 + \left(\sum_1^N \sin \theta_i \right)^2 \right] \,.$$

The first term in square brackets can be written

$$\left(\sum_1^N \cos \theta_i \right)^2 = (\cos \theta_1 + \cos \theta_2 + \ldots + \cos \theta_N)^2$$
$$= \sum_1^N \cos^2 \theta_i + \sum_{i \neq j} \cos \theta_i \cos \theta_j \,.$$

Since the directions θ_i are randomly distributed and independent,

$$\sum_{i \neq j} \cos \theta_i \cos \theta_j = 0 \,.$$

The same result applies for the second term in square brackets. Thus

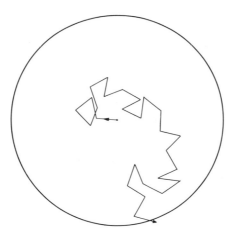

$$r^2 = d^2 \sum_1^N (\cos^2\theta_i + \sin^2\theta_i) = Nd^2 .$$

After N steps, the photon is at the distance $r = dN^{1/2}$ from the starting point. Similarly, a drunkard taking a hundred one-metre steps at random will have wandered ten metres from his/her starting point. The same result applies in three dimensions.

The time taken by a photon to reach the surface from the centre depends on the mean free path $d = 1/\alpha = 1/\kappa\rho$. The value of κ at half the solar radius can be estimated from the values of density and temperature obtained in Exercises 11.5.2 and 11.5.5. The mass absorption coefficient in these conditions is found to be $\kappa = 10 \ \mathrm{m^2/kg}$. (We shall not enter on how it is calculated.) The photon mean free path is then $d = 1/\kappa\rho$ $= 10^{-4}$ m. This should be a reasonable estimate in most of the solar interior. Since the solar radius $r = 10^9$ m, the number of steps needed to reach the surface will be $N = (r/d)^2 = 10^{26}$. The total path travelled by the photon is $s = Nd = 10^{22}$ m, and the time taken is $t = s/c = 10^6$ years; a more careful calculation gives $t = 10^7$ years. Thus it takes 10 million years for the energy generated at the centre to radiate into space. Of course the radiation that leaves the surface does not consist of the same gamma photons that were produced near the centre. The intervening scattering, emission and absorption processes have transformed the radiation into visible light (as can easily be seen).

11.6 Exercises

11.6.1 How many hydrogen atoms are there in the Sun per each helium atom?

11.6.2 a) How many pp reactions take place in the Sun every second? The luminosity of the Sun is 3.9×10^{26} W, the mass of a proton is 1.007280 amu, and that of the α particle 4.001514 amu (1 amu is 1.6604×10^{-27} kg). b) How many neutrinos produced in these pp reactions will hit the Earth in one second?

11.6.3 The mass absorption coefficient of a neutrino is $\kappa = 10^{-21} \ \mathrm{m^2 \ kg^{-1}}$. Find the mean free path at the centre of the Sun.

Chapter 12 **Stellar Evolution**

In the preceding chapter, we have seen how one can compute the evolution of a star by starting from a homogeneous model representing a newly formed system. When the chemical composition of the star changes with time, a new model is computed each time. In this chapter, we shall consider the theoretical evolutionary paths of systems with various masses and see how the computed evolution explains the observational data.

12.1 Evolutionary Time Scales

Changes in a star may take place on quite different time scales at different evolutionary phases. There are three important basic time scales: the *nuclear time scale* t_n, the *thermal time scale* t_t and the *dynamical* or *free fall time scale* t_d.

The Nuclear Time Scale. The time in which a star radiates away all the energy that can be released by nuclear reactions is called the nuclear time scale. An estimate of this time can be obtained if one calculates the time in which all available hydrogen is turned into helium. On the basis of theoretical considerations and evolutionary computations, it is known that only just over 10% of the total mass of hydrogen in the star can be consumed before other, more rapid evolutionary mechanisms set in. Since 0.7% of the rest mass is turned into energy in hydrogen burning, the nuclear time scale will be

$$t_n \approx \frac{0.007 \times 0.1 \, M c^2}{L} . \tag{12.1}$$

For the Sun one obtains the nuclear time scale 10^{10} y, and thus

$$t_n \approx \frac{M/M_\odot}{L/L_\odot} \times 10^{10} \, \text{y} . \tag{12.2}$$

This gives the nuclear time scale as a function of the mass M and luminosity L of a given star. For example, if the mass is $30 \, M_\odot$, one obtains t_n about 2 million years. The reason for the shorter time scale is that the stellar luminosity strongly increases for higher masses (Table 11.1).

The Thermal Time Scale. The time in which a star would radiate away all its thermal energy if the nuclear energy production were suddenly turned off is called the thermal time scale. This is also the time it takes for radiation from the centre to reach the surface. The thermal time scale may be estimated as

$$t_t \approx \frac{0.5 \, G M^2/R}{L} \approx \frac{(M/M_\odot)^2}{(R/R_\odot)(L/L_\odot)} \times 2 \times 10^7 \, \text{y} , \tag{12.3}$$

where G is the constant of gravity and R, the stellar radius. For the Sun, the thermal time scale is about 20 million years or 1/500 of the nuclear time scale.

The Dynamical Time Scale. The third and shortest time scale is the time it would take a star to collapse if the pressure supporting it against gravity were suddenly removed. It can be estimated from the time it would take for a particle to fall freely from the stellar surface to the centre. This is half of the period given by Kepler's third law, where the major semiaxis of the orbit corresponds to half the stellar radius R:

$$t_d = \frac{2\pi}{2} \sqrt{\frac{(R/2)^3}{GM}} \approx \sqrt{\frac{R^3}{GM}} \,. \tag{12.4}$$

The dynamical time scale of the Sun is about half an hour.

The ordering of the time scales is normally like that in the Sun, i.e. $t_d \ll t_t \ll t_n$.

12.2 The Contraction of Stars Towards the Main Sequence

The formation and subsequent gravitational collapse of condensations in the interstellar medium will be considered in a later chapter. Here we shall follow the behaviour of such a *protostar*, when it is already in the process of contraction.

When a cloud contracts, gravitational potential energy is released and transformed into thermal energy of the gas and into radiation. Initially the radiation can propagate freely through the material, because the density is low and the opacity small. Therefore most of the liberated energy is radiated away and the temperature does not increase. The contraction takes place on the dynamical time scale; the gas is falling freely inwards.

The density and the pressure increase most rapidly near the centre of the cloud. As the density increases, so does the opacity. A larger fraction of the liberated energy is then turned into heat, and the temperature begins to rise. This leads to a further increase in the pressure that is resisting the free fall. The contraction of the central part of the cloud slows down. The outer parts, however, are still falling freely.

At this stage, the cloud may already be considered a protostar. It consists mainly of hydrogen in molecular form. When the temperature reaches 1800 K, the hydrogen molecules are dissociated into atoms. The dissociation consumes energy, and the rise in temperature is slowed down. The pressure then also grows more slowly and this in turn means that the rate of contraction increases. The same sequence of events is repeated, first when hydrogen is ionized at 10^4 K, and then when helium is ionized. When the temperature has reached about 10^5 K, the gas is essentially completely ionized.

The contraction of a protostar only stops when a large fraction of the gas is ionized. The star then settles into hydrostatic equilibrium. Its further evolution takes place on the thermal time scale, i.e. much more slowly. The radius of the protostar has shrunk from its original value of about 100 AU to about 1/4 AU. It will usually be located inside a larger gas cloud and will be accreting material from its surroundings. Its mass therefore grows, and the central temperature and density increase.

The temperature of a star that has just reached equilibrium is still low and its opacity correspondingly large. Thus it will be convective in its centre. The convective

Fig. 12.1. The paths in the HR diagram of stars contracting to the main sequence on the thermal time scale. After a rapid dynamical collapse the stars settle on the Hayashi track and evolve towards the main sequence on the thermal time scale. [Models by Iben, I. (1965): Astrophys. J. **141**, 993]

energy transfer is quite efficient and the surface of the protostar will therefore be relatively bright.

We now describe the evolution in the HR diagram. Initially, the protostar will be faint and cool, and it will reside at the lower far right in the HR diagram (outside Fig. 12.1). During the collapse, its surface rapidly heats up and brightens and it moves to the upper right of Fig. 12.1. At the end of the collapse, the star will settle at a point corresponding to its mass on the *Hayashi track*. The Hayashi track (Fig. 12.1) gives the location in the HR diagram of completely convective stars. Stars to its right cannot be in equilibrium and will collapse on the dynamic time scale.

The star will now evolve almost along the Hayashi track on the thermal time scale. In the HR diagram, it moves almost vertically downwards, its radius decreases and its luminosity drops (Fig. 12.1). As the temperature goes on increasing in its centre, the opacity diminishes and energy begins to be transported by radiation. The mass of the radiative region will gradually grow until finally most of the star is radiative. By then, the central temperature will have become so large that nuclear reactions begin. Previously, all the stellar energy had been released potential energy, but now the nuclear reactions make a growing contribution and the luminosity increases. The stellar surface temperature will also increase and the star will move slightly upwards to the left in the HR diagram. In massive stars, this turn to the left occurs much earlier, because their central temperatures are higher and the nuclear reactions are initiated earlier.

For solar mass stars, the rapid collapse of the protostellar cloud only lasts for a few hundred years. The final stage of condensation is much slower, lasting several tens of millions of years. This length of time strongly depends on the stellar mass because of the luminosity dependence of the thermal time scale. A $15 M_\odot$ star condenses to the main sequence in 60 000 years, whereas for a $0.1 M_\odot$ star, the time is hundreds of millions of years.

Some of the hydrogen burning reactions start already at a few million degrees. For example, lithium, beryllium and boron burn to helium in the ppII and ppIII branches of the pp chain long before the complete chain has turned on. Because the star is convective and thus well mixed during the early stages, even its surface material will have been processed in the centre. Although the abundances of the above-mentioned elements are small, they give important information on the central temperature.

The beginning of the main sequence phase is marked by the start of hydrogen burning in the pp chain at a temperature of about 4 million degrees. The new form of energy production completely dominates the energy release due to the contraction. As the contraction is halted, the star makes a few oscillations in the HR diagram, but soon settles in equilibrium and the long, quiet main sequence phase begins.

It is difficult to observe stars during contraction, because the new-born stars are usually hidden among dense clouds of dust and gas. However, some condensations in interstellar clouds have been discovered and near them, very young stars. One example are the T Tauri stars. Their lithium abundance is relatively high, which indicates that they are newly formed stars in which the central temperature has not yet become large enough to destroy lithium. Near the T Tauri stars, small, bright, star-like nebulae, the *Herbig-Haro objects*, have been discovered. These are thought to be produced in the interaction between a stellar wind and the surrounding interstellar medium.

12.3 The Main Sequence Phase

The *main sequence phase* is that evolutionary stage in which the energy released by the burning of hydrogen in the core is the only source of stellar energy. During this stage, the star is in stable equilibrium, and its structure changes only because its chemical composition is gradually altered by the nuclear reactions. Thus the evolution takes place on a nuclear time scale, which means that the main sequence phase is the longest part of the life of a star. For example, for a solar mass star, the main sequence phase lasts for about 10 000 million years. More massive stars evolve more rapidly, because they radiate much more energy. Thus the main sequence phase of a 15 solar mass star is only about 10 million years. On the other hand, less massive stars have a longer main sequence lifetime: a $0.25 M_\odot$ star spends about 70 000 million years on the main sequence.

Since stars are most likely to be found in the stage of steady hydrogen burning, the main sequence in the HR diagram is richly populated, in particular at its low-mass end. The more massive upper main sequence stars are less abundant because of their shorter main sequence lifetimes.

If the mass of a star becomes too large, the force of gravity can no longer resist the radiation pressure. Stars more massive than this upper limit cannot form, because they cannot accrete additional mass during the contraction phase. Theoretical computations give a limiting mass of about $100 M_\odot$; the most massive stars observed are about $70 M_\odot$.

There is also a lower-mass limit of the main sequence. Stars below $0.08 M_\odot$ never become hot enough for hydrogen burning to begin. The smallest protostars therefore contract to planet-like dwarfs. During the contraction phase they radiate because potential energy is released, but eventually they begin to cool. In the HR diagram, such stars first move almost vertically downwards and then further downwards to the right.

The Upper Main Sequence. The stars on the *upper main sequence* are so massive and their central temperature so high that the CNO cycle can operate. On the *lower main sequence*, the energy is produced by the pp chain. The pp chain and the CNO cycle are equally efficient at a temperature of 18 million degrees, corresponding to the central temperature of a $1.5 M_\odot$ star. The boundary between the upper and the lower main sequence corresponds roughly to this mass.

The energy production in the CNO cycle is very strongly concentrated at the core. The outward energy flux will then become very large, and can no longer be maintained by radiative transport. Thus the upper main sequence stars have a *convective core*, i.e. the energy is transported by material motions. These keep the material well mixed, and thus the hydrogen abundance decreases uniformly with time within the entire convective region.

Outside the core, there is *radiative equilibrium*, i.e. the energy is carried by radiation and there are no nuclear reactions. Between the core and the envelope, there is a transition region where the hydrogen abundance decreases inwards.

The mass of the convective core will gradually diminish as the hydrogen is consumed. In the HR diagram, the star will slowly shift to the upper right as its luminosity grows and its surface temperature decreases (Fig. 12.2). When the central hydrogen supply becomes exhausted, the core of the star will begin to shrink rapidly. The surface temperature will increase and the star will quickly move to the upper left. Because of the contraction of the core, the temperature in the hydrogen shell just outside the core will increase. It rapidly becomes high enough for hydrogen burning to set in again.

The Lower Main Sequence (Fig. 12.3). On the lower main sequence, the central temperature is lower than for massive stars, and the energy is generated by the pp chain. Since the rate of the pp chain is not as sensitive to temperature as that of the CNO cycle, the energy production is spread over a larger region than in the more massive stars. In consequence, the core never becomes convectively unstable, but remains radiative.

In the outer layers of lower main sequence stars, the opacity is high because of the low temperature. Radiation can then no longer carry all the energy, and convection will set in. The structure of lower main sequence stars is thus opposite that of the upper main sequence: the centre is radiative and the envelope is convective. Since there is no mixing of material in the core, the hydrogen is most rapidly consumed at the very centre, and the hydrogen abundance increases outwards.

As the amount of hydrogen in the core decreases, the star will slowly move upwards in the HR diagram, almost along the main sequence (Fig. 12.2). It becomes slightly brighter and hotter, but its radius will not change by much. The evolutionary track of the star will then bend to the right, as hydrogen in the core nears its end. Eventually the

Fig. 12.2. Stellar evolutionary paths in the HR diagram at the main sequence phase and later. On the main sequence, bounded by dashed curves, the evolution is on the nuclear time scale. The post-main sequence evolution to the red giant phase is on the thermal time scale. The point marked He corresponds to helium ignition, and in low-mass stars, the helium flash. The straight line shows the location of stars with the same radius. [Iben, I. (1967): Annu. Rev. Astron. Astrophys. **5**, 571; data for $30\,M_\odot$ from Stothers, R. (1966): Astrophys. J. **143**, 91]

Fig. 12.3a–c. Energy transport in the main sequence phase. **(a)** The least massive stars ($M < 0.26\,M_\odot$) are convective throughout. **(b)** For $0.26\,M_\odot < M < 1.5\,M_\odot$, the core is radiative and the envelope convective. **(c)** Massive stars ($M > 1.5\,M_\odot$) have a convective core and a radiative envelope

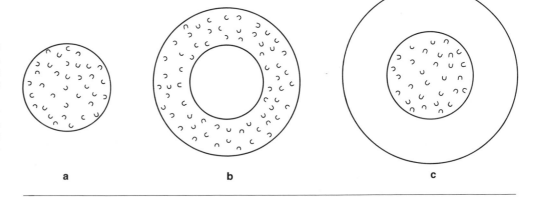

core is almost pure helium. Hydrogen will continue to burn in a thick shell around the core.

Stars with masses between $0.08 M_\odot$ and $0.26 M_\odot$ have a very simple evolution. During their whole main sequence phase, they are fully convective, which means that their entire hydrogen content is available as fuel. These stars evolve very slowly to the upper left in the HR diagram. Finally, when all their hydrogen has burned to helium, they contract to become white dwarfs.

12.4 The Giant Phase

The main-sequence phase of stellar evolution ends when hydrogen is exhausted at the centre. The star then settles in a state in which hydrogen is burning in a shell surrounding a helium core. As we have seen, the transition takes place gradually in low-mass stars, while the high-mass stars make a rapid jump in the HR diagram at this point.

The mass of the helium core is increased by the hydrogen burning in the shell. This leads to an expansion of the envelope of the star, which moves almost horizontally to the right in the HR diagram. As the convective envelope becomes more extensive, the star approaches the Hayashi track. Since it cannot pass further to the right, and since its radius continues to grow, the star has to move upwards along the Hayashi track towards larger luminosities (Fig. 12.2). The star has become a red giant.

In low-mass stars, as the mass of the core grows, its density will eventually become so high that it becomes degenerate. The central temperature will continue to rise. The whole helium core will have a uniform temperature because of the high conductivity of the degenerate gas. If the mass of the star is larger than $0.26 M_\odot$, the central temperature will eventually reach about 100 million degrees, which is enough for helium to burn to carbon in the triple alpha process.

Helium burning will set in simultaneously in the whole central region and will suddenly raise its temperature. Unlike a normal gas, the degenerate core cannot expand, although the temperature increases [cf. (11.12)], and therefore the increase in temperature will only lead to a further acceleration of the rate of the nuclear reactions. When the temperature increases further, the degeneracy of the gas is removed and it will begin to expand violently. Only a few seconds after the ignition of helium, there is an explosion, the *helium flash*.

The energy from the helium flash is absorbed by the outer layers, and thus it does not lead to the complete disruption of the star. In fact the luminosity of the star drops in the flash, because when the centre expands, the outer layers contract. The energy released in the flash is turned into potential energy of the expanded core. Thus after the helium flash, the star will settle into a new state, where helium is steadily burning to carbon in a nondegenerate core.

After the helium flash, the star finds itself on the horizontal giant branch in the HR diagram. However, the route leading there is not straight. The star may make several oscillations back and forth in the HR diagram before finally settling down.

When the helium is exhausted in the core, there are two nuclear burning shells in the star. In the inner one, helium is burning; in the outer one, hydrogen. Such a configuration is unstable and the stellar material may become mixed or matter may be ejected into space in a shell, like that of a planetary nebula.

In high-mass stars, the central temperature is higher and the central density lower, and the core will therefore not be degenerate. Thus helium burning can set in noncatastrophically as the central regions contract. When the central helium supply is exhausted, helium will continue to burn in a shell.

The core will go on contracting and becoming hotter. First, carbon burning and, subsequently, oxygen and silicon burning (see Sect. 11.3) will be ignited. As each nuclear fuel is exhausted in the centre, the burning will continue in a shell. The star will thus contain several nuclear burning shells. At the end, the star will consist of a sequence of layers differing in composition, in massive stars (more massive than $15 M_\odot$) all the way up to iron.

The End of the Giant Phase. The evolution that follows helium burning depends strongly on the stellar mass. The mass determines how high the central temperature can become and the degree of degeneracy, when heavier nuclear fuels are ignited.

Stars less massive than $3 M_\odot$ never become hot enough to ignite carbon burning in the core. At the end of the giant phase, the radiation pressure expels the outer layers, which form a *planetary nebula*. The hot core remains as a white dwarf.

In stars with masses in the range $3-15 M_\odot$, either carbon or oxygen is ignited explosively just like helium in low-mass stars: there is a *carbon* or *oxygen flash*. This is much more powerful than the helium flash, and will make the star explode as a *supernova* (Sect. 14.3). The star will probably be completely disrupted by the explosion.

The central parts of the most massive stars, with masses larger than $15 M_\odot$, burn all the way to iron ^{56}Fe. All nuclear sources of energy will then be completely exhausted. The structure of a 30 solar mass star at this stage is schematically shown in Fig. 12.4. The star is made up of a nested sequence of zones bounded by shells burning silicon ^{28}Si, oxygen ^{16}O and carbon ^{12}C, helium ^{4}He and hydrogen ^{1}H. However, this is not a stable state, since the cessation of nuclear reactions in the core means that the central pressure will fall, and the core will collapse. The energy released in the collapse goes into dissociating the iron nuclei first to helium and then to protons and neutrons. This will further speed up the collapse, just like the dissociation of molecules speeds up the collapse of a protostar. The collapse takes place on a dynamical time scale, which, in

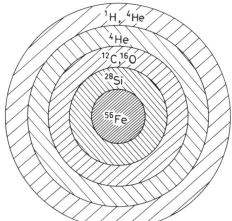

Fig. 12.4. The structure of a massive star (30 $M_\odot$) at a late evolutionary stage. The star consists of layers with different composition separated by nuclear burning shells

the dense stellar core, is only a fraction of a second. The outer parts will also collapse, but more slowly. In consequence, the temperature will increase in layers containing unburnt nuclear fuel. This will burn explosively, releasing immense amounts of energy in a few seconds.

The final result is that the outer layers will explode as a supernova. In the dense central core, the protons and electrons combine to form neutrons. The core will finally consist almost entirely of neutrons, which become degenerate because of the high density. The degeneracy pressure of the neutrons will stop the collapse of a small mass core. However, if the mass of the core is large enough, a black hole will probably be formed.

12.5 The Final Stages of Evolution

The endpoints of stellar evolution can be seen from Fig. 12.5. This shows the relation between mass and central density for a body at zero temperature, i.e. the final equilibrium when a massive body has cooled. There are two maxima on the curve. The mass corresponding to the left-hand maximum is called the *Chandrasekhar mass*, $M_{Ch} = 1.2 - 1.4\,M_\odot$, and that corresponding to the right-hand one, the *Oppenheimer-Volkov mass*, $M_{OV} = 1.5 - 2\,M_\odot$.

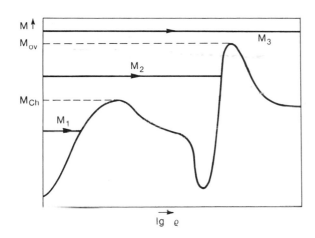

Fig. 12.5. The evolutionary end points of stars with different masses shown as a function of central density. The curve shows the behaviour of the central density of completely degenerate ($T = 0\,\mathrm{K}$) bodies. The Chandrasekhar mass M_{Ch} and the Oppenheimer-Volkov mass M_{OV} correspond to maxima on this curve

Let us first consider a star with mass less than M_{Ch}. Suppose the mass does not change. When the nuclear fuel is exhausted, the star will become a white dwarf, which will gradually cool down and contract. In Fig. 12.5, it moves horizontally to the right. Finally it will reach zero temperature and end up on the left-hand rising part of the equilibrium curve. Its final equilibrium is a completely degenerate black dwarf.

If the mass of the star is larger than M_{Ch} but smaller than M_{OV}, it can continue cooling until it reaches the right-hand rising section of the curve. Again there is a stable final state, this time corresponding to a completely degenerate neutron star.

An even more massive star, with mass larger than M_{OV}, will go on contracting past the density corresponding to a neutron star. There is then no longer any known possible stable equilibrium, and the star must go on contracting to form a black hole.

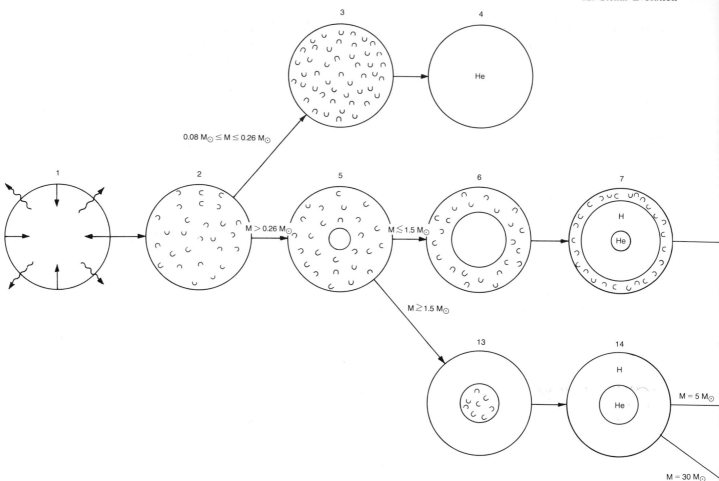

The preceding considerations are purely theoretical. The final evolutionary stages of real stars (Fig. 12.6) involve many imperfectly known factors, which may affect the final equilibrium. Perhaps most important is the question of mass loss, which is very difficult to settle either observationally or theoretically. For example, in a supernova explosion, the whole star may be disrupted, and it is very uncertain whether what remains will be a neutron star, a black hole or nothing at all. (The structure of compact stars will be discussed in Chap. 15.)

Fig. 12.6. Evolution schemes for stars with different masses. The radius is scaled to be the same in all drawings. In reality, there are vast differences in the sizes of different stars and different phases of evolution. In the beginning (*1*) a gas cloud is contracting rapidly in free fall. Because the gas is quite rarefied, radiation escapes easily from the cloud. As the density increases, radiation transport becomes more difficult, and the released energy tends to warm up the gas. The contraction lasts until the gas is completely ionized, and the star, which has become a protostar, is in hydrostatic equilibrium (*2*). The star is convective throughout its interior.

Now evolution continues on a thermal time scale. The contraction is much slower than in the free-fall phase. The phases of further evolution are determined by the mass M of the star. For $M < 0.08\,M_\odot$ the temperature in the centre does not rise high enough for hydrogen burning, and these stars contract to planetlike brown dwarfs. Stars with $M \geq 0.08\,M_\odot$ start hydrogen burning when the temperature has reached about 4×10^6 K. This is the beginning of the main sequence phase. In the main sequence, the lowest-mass stars with $0.08\,M_\odot \leq M \leq 0.26\,M_\odot$ are entirely convective, and thus they remain homogeneous (*3*). Their evolution is very slow, and after all the hydrogen has been burnt to helium, they contract to white dwarfs (*4*).

The increasing temperature makes the stars with $M > 0.26\,M_\odot$ radiative in the centre as the opacity decreases (5). The low-mass stars with $0.26\,M_\odot \le M \le 1.5\,M_\odot$ remain radiative in the centre during the main sequence phase (6) as they burn their hydrogen through the pp-chain. The outer part is convective. At the end of the main sequence phase, hydrogen burning continues in a shell surrounding the helium core (7). The outer part expands, and the giant phase begins. The contracting helium core is degenerate and warms up. At about 10^8 K, the triple-alpha process begins and leads immediately to the helium flash (8). The explosion is damped by the outer parts, and helium burning goes on in the core (9). Hydrogen is still burning in an outer shell.

As the central helium is exhausted, helium burning moves over to a shell (10). At the same time, the outer part expands and the star loses some of its mass. The expanding envelope forms a planetary nebula (11). The star in the centre of the nebula becomes a white dwarf (12).

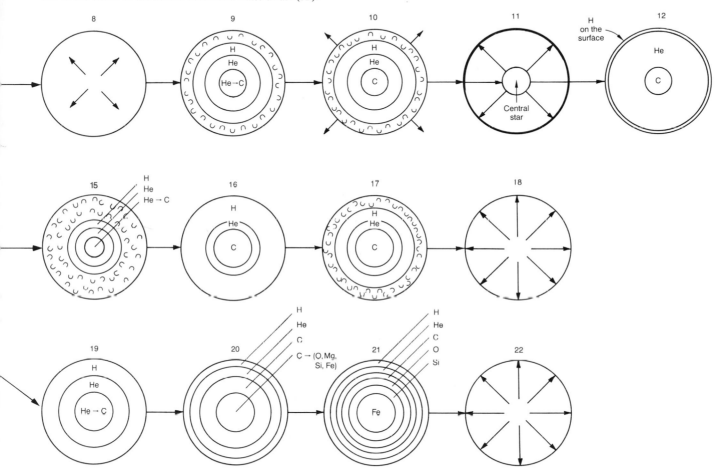

In the upper main sequence with $M \gtrsim 1.5\,M_\odot$, energy is released through the CNO cycle, and the core becomes convective, while the outer part is radiative (13). The main sequence phase ends as the hydrogen in the core is exhausted, and shell burning begins (14). The helium core remains convective and nondegenerate, and helium burning begins without perturbations (15 and 19). Afterwards, helium burning moves over to a shell (16 and 20). For stars with $3\,M_\odot \le M \le 15\,M_\odot$, the carbon in the core is degenerate, and a carbon flash occurs (17). This leads to a supernova explosion (18) and probably to the complete destruction of the star.

For the most massive stars with $M \gtrsim 15\,M_\odot$, the carbon core remains convective, and carbon burns to oxygen and magnesium. Finally, the star consists of an iron core surrounded by shells with silicon, oxygen, carbon, helium and hydrogen (21). The nuclear fuel is now exhausted, and the star collapses on a dynamical time scale. The result is a supernova (22). The outer parts explode, but the remaining core continues to contract to a neutron star or a black hole.

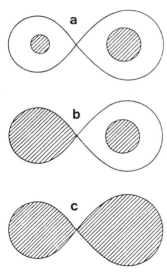

Fig. 12.7a – c. The types of close binary systems: (**a**) detached, (**b**) semidetached and (**c**) contact binaries. [Kopal, Zd. (1959): *Close Binary Systems,* The International Astrophysics Series, Vol. 5 (Chapman & Hall Ltd., London) p. 483]

12.6 The Evolution of Close Binary Stars

If the components of a binary star are well separated, they do not significantly perturb one another. When studying their evolution, one can regard them as two single stars evolving independently, as described above. However, in close binary pairs, this will no longer be the case.

Close binary stars are divided into three classes, as shown in Fig. 12.7: *detached, semidetached* and *contact binaries.* The figure-eight curve drawn in the figure is called the *Roche surface.* If the star becomes larger than this surface, it begins to lose mass to its companion through the waist of the Roche surface.

During the main sequence phase, the stellar radius does not change much, and each component will remain within its own Roche lobe. When the hydrogen is exhausted, the stellar core will rapidly shrink and the outer layers expand, as we have seen. At this stage, a star may exceed its Roche lobe and mass transfer may set in.

Close binary stars are usually seen as eclipsing binaries. One example is Algol, in the constellation Perseus. The components in this binary system are a normal main sequence star and a *subgiant,* which is much less massive than the main sequence star. The subgiant has a high luminosity and thus has apparently already left the main sequence. This is unexpected, since the components were presumably formed at the same time, and the more massive star should evolve more rapidly. The situation is known as the *Algol paradox*: for some reason, the less massive star has evolved more rapidly.

In the 1950's, a solution to the paradox proposed that the subgiant was originally more massive, but that it had lost mass to its companion during its evolution. Since the 1960's, mass transfer in close binary systems has been much studied, and has turned out be a very significant factor in the evolution of close binaries.

As an example, let us consider a close binary, where the initial masses of the components are 1 and 2 solar masses and the initial orbital period, 1.4 days (Fig. 12.8). After evolving away from the main sequence, the more massive component will exceed the Roche limit and begin to lose mass to its companion. Initially, the mass will be transferred on the thermal time scale, and after a few million years, the roles of the components will be changed: the initially more massive component has become less massive than its companion.

The binary is now semidetached and can be observed as an Algol-type eclipsing binary. The two components are a more massive main sequence star and a less massive subgiant filling its Roche surface. The mass transfer will continue, but on the much slower nuclear time scale. Finally, mass transfer will cease and the less massive component will contract to a $0.6\,M_\odot$ white dwarf.

The more massive $2.4\,M_\odot$ star now evolves and begins to lose mass, which will accumulate on the surface of the white dwarf. The accumulated mass may give rise to *nova outbursts,* where material is ejected into space by large explosions. Despite this, the mass of the white dwarf will gradually grow and may eventually exceed the Chandrasekhar mass. The white dwarf will then collapse and explode as a type I supernova.

As a second example, we can take a massive binary, with the initial masses 20 and $8\,M_\odot$ and the initial period, 4.7 days (Fig. 12.9). The more massive component evolves rapidly, and at the end of the main sequence phase, it will transfer more than $15\,M_\odot$ of its material to the secondary. The mass transfer will occur on the thermal time scale,

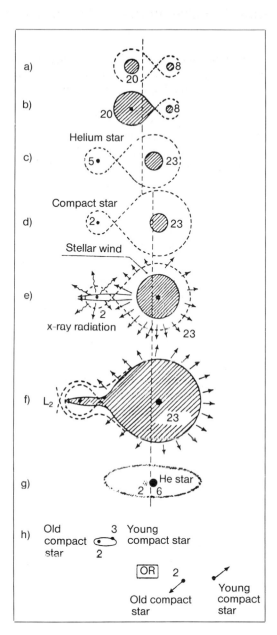

Fig. 12.8a – f. Evolution of a low-mass binary: **(a)** both components on the main sequence; **(b)** mass transfer from the more massive component; **(c)** light subgiant and massive main sequence star; **(d)** white dwarf and main sequence star; **(e)** mass transferred to the white dwarf from the more massive component leads to nova outbursts; **(f)** the white dwarf mass exceeds the Chandrasekhar mass and explodes as a type I supernova

Fig. 12.9a – h. Evolution of a massive binary. It has been assumed that the supernova explosion of a $5\,M_\odot$ helium star leaves a $2\,M_\odot$ compact remnant (neutron star or black hole). **(a)** Main sequence phase; **(b)** beginning of the first mass transfer phase; **(c)** end of the first mass transfer phase; the first Wolf-Rayet phase begins; **(d)** the helium star (Wolf-Rayet star) has exploded as a supernova; **(e)** the $23\,M_\odot$ component becomes a supergiant; the compact component is a strong x-ray source; **(f)** beginning of the second mass transfer phase; the x-ray source is throttled and large-scale mass loss begins; **(g)** second Wolf-Rayet phase; **(h)** the $6\,M_\odot$ helium star has exploded as a supernova; the binary may or may not be disrupted, depending on the remaining mass which, in this case, is only a few ten thousand years. The end result is a *helium star*, having as companion an unevolved main sequence star. The properties of the helium star are like those of a *Wolf-Rayet star*.

Helium continues to burn to carbon in the core of the helium star, and the mass of the carbon core will grow. Eventually the carbon will be explosively ignited, and the star will explode as a supernova. The consequences of this explosion are not known, but let us suppose that a $2\,M_\odot$ compact remnant is left. As the more massive star expands, its stellar wind will become stronger, giving rise to strong x-ray emission as it hits the compact star. This x-ray emission will only cease when the more massive star exceeds its Roche surface.

The system will now rapidly lose mass and angular momentum. A steady state is finally reached when the system contains a $6\,M_\odot$ helium star in addition to the $2\,M_\odot$ compact star. The helium star is seen as a Wolf-Rayet star, which, after about a million years, explodes as a supernova. This will probably lead to the breakup of the binary system. However, for certain values of the mass, the binary may remain bound. Thus a binary neutron star may be formed.

12.7 Comparison with Observations

The most important direct support for the theoretical evolutionary models is obtained from the properties of observed HR diagrams. If the theoretical models are correct, the observed number of stars should reflect the duration of the various evolutionary phases. These are given for stars of different masses in Table 12.1. The stars are most numerous along the main sequence. Giants are also common and, in addition to these, there are white dwarfs, subgiants, etc. The sparsely populated region to the right of the main sequence, the *Hertzsprung gap*, is explained by the rapid transition from the main sequence to the giant phase.

The cepheids provide an important test for the evolutionary models. The pulsations and the relation between period and luminosity for the cepheids can be understood on the basis of theoretical stellar models.

The evolutionary models can also explain the HR diagrams of star clusters. Let us assume that all the stars in a cluster were formed at the same time. In the youngest systems, the associations, the stars will mainly be found on the upper main sequence,

Table 12.1. Stellar lifetimes (unit 10^6 years)

Mass $[M/M_\odot]$	Spectral type on the main sequence	Contraction to main sequence	Main sequence	Main sequence to red giant	Red giant
30	O5	0.02	4.9	0.55	0.3
15	B0	0.06	10	1.7	2
9	B2	0.2	22	0.2	5
5	B5	0.6	68	2	20
3	A0	3	240	9	80
1.5	F2	20	2000	280	
1.0	G2	50	10000	680	
0.5	M0	200	30000		
0.1	M7	500	10^7		

since the most massive stars evolve most rapidly. To the right of the main sequence, there will be less massive T Tauri stars, which are still contracting. In intermediate age open clusters, the main sequence will be well developed and its upper end should bend to the right, since the most massive stars will already have begun to evolve off the main sequence. In the old globular clusters, the giant branch should increase in importance in the older clusters. These predictions are confirmed by the observations, which will be further discussed in Chap. 17 on star clusters.

Of course, the most detailed observations can be made of the Sun, which is therefore a crucial point of comparison for the theoretical models. If a star of one solar mass with an initial composition of 71% hydrogen, 27% helium and 2% heavier elements is allowed to evolve for 5000 million years, it will be very similar to our present Sun. In particular, it will have the same radius, surface temperature and luminosity. According to calculations, about half of the Sun's supply of hydrogen fuel has been consumed. The Sun will go on shining like a normal main sequence star for another 5000 million years, before there will be any dramatic change.

Some problems remain in regard to the observations. One is the so-called solar neutrino problem. The neutrinos produced by solar nuclear reactions have been observed since the beginning of the 1970's by the techniques described in Sect. 3.7. Only the neutrinos formed in the relatively rare ppIII reaction are energetic enough to be observed in this way. Their observed number is too small: whereas the models predict about 5 units, the observations have consistently only registered $1-2$.

The discrepancy may be due to a fault in the observational technique or to some unknown properties of the neutrinos. However, if the solar models are really in error, the central temperature of the Sun would have to be about 20% lower than thought, which would be in serious conflict with the observed solar luminosity. One possibility is that some of the electron neutrinos change to other, unobservable particles during their passage to Earth.

A second problem is the observed abundance of lithium and beryllium. The solar surface contains a normal abundance of beryllium, but very little lithium. This should mean that during its contraction, the Sun was still fully convective when the central temperature was high enough to destroy lithium (3×10^6 K), but not beryllium (4×10^6 K). However, according to the standard solar evolution models, convection ceased in the centre already at a temperature of 2×10^6 K. One suggested explanation is that the convection has later carried down lithium to layers were the temperature is high enough to destroy it.

12.8 The Origin of the Elements

There are just under a hundred naturally occurring elements, and about 300 isotopes in the solar system (Fig. 12.10). In Sect. 12.4, we have seen how the elements up to iron are produced when hydrogen burns to helium and helium further to carbon, oxygen and heavier elements.

Almost all nuclei heavier than helium were produced in nuclear reactions in stellar interiors. In the oldest stars, the mass fraction of heavy elements is only about 0.02%, whereas in the youngest stars, it is a few per cent. Nevertheless, most of the stellar material is hydrogen and helium. According to the standard cosmological models, these were formed in the early stages of the Universe, when the temperature and density

Fig. 12.10. Element abundances in the solar system as a function of the nuclear mass number. The abundance of Si has been normalized as 10^6

were suitable for nuclear reactions. (This will be discussed in Chap. 20.) Although helium is produced during the main sequence stellar evolution, very little of it is actually returned into space to be incorporated into later stellar generations. Most of it is either transformed into heavier elements by further reactions, or else remains locked up inside a white dwarf remnant. Therefore the helium abundance does not increase by much due to stellar processes.

The most important nuclear reactions leading to the build-up of the heavy nuclei up to iron were presented in Sect. 11.3. The probabilities of the various reactions are determined either by experiments or by theoretical calculations. When they are known, the relative abundances of the various nuclei produced can be calculated.

The formation of elements heavier than iron requires an input of energy, and thus they cannot be explained in the same manner. Still, heavy nuclei are continually produced. In 1952, technetium was discovered in the atmosphere of a red giant. The half-life of the most long-lived isotope ^{98}Tc is about 1.5×10^6 years, so that the observed technetium must have been produced in the star.

Most of the nuclei more massive than iron are formed by *neutron capture*. Since the neutron does not have an electric charge, it can easily penetrate into the nucleus. The probability for neutron capture depends both on the kinetic energy of the incoming neutron and on the mass number of the nucleus. For example in the solar system, the abundances of isotopes show maxima at the mass numbers $A = 70-90$, 130, 138, 195 and 208. These mass numbers correspond to nuclei with closed neutron shells at the neutron numbers $N = 50, 82,$ and 126. The neutron capture probability for these nuclei is very small. The closed shell nuclei thus react more slowly and are accumulated in greater abundances.

In neutron capture, a nucleus with mass number A is transformed into a more massive nucleus:

$$(Z, A) + n \rightarrow (Z, A+1) + \gamma .$$

The newly formed nucleus may be unstable to β decay, where one neutron is transformed into a proton:

$$(Z, A+1) \rightarrow (Z+1, A+1) + e^- + \bar{\nu}_e .$$

Two kinds of neutron capture processes are encountered, depending on the value of the neutron flux. In the slow s-process, the neutron flux is so small that any β decays have had time to occur before the next neutron capture reaction takes place. The most stable nuclei up to mass number 210 are formed by the s-process. These nuclei are said to correspond to the β stability valley. The s-process explains the abundance peaks at the mass numbers 88, 138 and 208.

When the neutron flux is large, β decays do not have time to happen before the next neutron capture. One then speaks of the rapid r-process, which gives rise to more neutron-rich isotopes. The abundance maxima produced by the r-process lie at mass numbers about ten units smaller than those of the s-process.

A neutron flux sufficient for the s-process is obtained in the course of normal stellar evolution. For example, some of the carbon and oxygen burning reactions produce free neutrons. If there is convection between the hydrogen and helium burning shells, free protons may be carried into the carbon-rich layers. Then the following neutron-producing reaction chain becomes important:

$$\begin{aligned}
{}^{12}C + p &\rightarrow {}^{13}N + \gamma \\
{}^{13}N &\rightarrow {}^{13}C + e^+ + \nu_e \\
{}^{13}C + {}^4He &\rightarrow {}^{16}O + n .
\end{aligned}$$

The convection can also carry the reaction products nearer to the surface.

The neutron flux required for the r-process is about 10^{22} cm^{-3}, which is too large to be produced during normal stellar evolution. The only presently known site where a large enough neutron flux is expected is near a neutron star forming in a supernova explosion. In this case, the rapid neutron capture leads to nuclei that cannot capture more neutrons without becoming strongly unstable. After one or more rapid β decays, the process continues.

The r-process stops when the neutron flux decreases. The nuclei produced then gradually decay by the β-process towards more stable isotopes. Since the path of the r-process goes about ten mass units below the stability valley, the abundance peaks produced will fall about ten units below those of the s-process. This is shown in Fig. 12.11. The most massive naturally occurring elements, such as uranium, thorium and plutonium, are formed by the r-process.

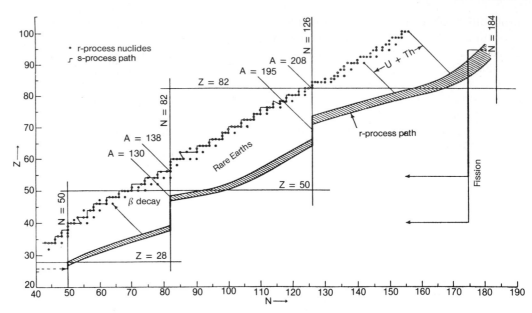

Fig. 12.11. Neutron capture paths for the s-process and r-process (from left to right). The s-process follows a path along the line of beta stability. The stable r-process nuclei (small circles) result from beta decay of their proton-rich progenitors on the shaded path, shown lower. Beta decay occurs along straight lines, $A = $ const. The closed neutron shells in nuclei at $N = 50$, 82 and 126 correspond to abundance peaks in s-process nuclei at $A = 88$, 138 and 208, and in r-process nuclei at $A = 80$, 130 and 195. [Seeger, P. A., Fowler, W. A., Clayton, D. D. (1965): Astrophys. J. Suppl. **11**, 121]

There are about 40 isotopes on the proton-rich side of the β stability valley that cannot be produced by neutron capture processes. Their abundances are very small, relative to the neighbouring isotopes. They are formed in supernova explosions at temperatures higher than 10^9 K by reactions known as the *p-process*. At this temperature, pair formation can take place:

$$\gamma \rightarrow e^+ + e^- \, .$$

The positron may either be annihilated immediately or be consumed in the reaction

$$e^+ + (Z, A) \rightarrow (Z+1, A) + \bar{\nu}_e \, .$$

Another reaction in the p-process is

$$(Z, A) + p \rightarrow (Z+1, A+1) + \gamma \, .$$

Finally, the *fission* of some heavier isotopes may give rise to p-process nuclei. Examples of this are the isotopes ^{184}W, ^{190}Pt and ^{196}Hg, formed by the fission of lead.

All the preceding reaction products are ejected into the interstellar medium in the supernova explosion. Collisions between cosmic rays and heavy nuclei then finally give rise to the light elements lithium, beryllium and boron. Thus the abundances of essentially all naturally occurring isotopes can be explained.

During succeeding generations of stars, the relative abundance of heavy elements increases in the interstellar medium. They can then be incorporated into new stars, planets − and living beings.

12.9 Examples

12.9.1 An interstellar cloud has a mass of one solar mass and density of 10^{10} hydrogen atoms per cm^3. Its rotation period is 1000 years. What is the rotation period after the cloud has condensed into a star of solar size?

The angular momentum is $L = I\omega$, where ω is the angular velocity and I is the moment of inertia. For a homogeneous sphere

$$I = \tfrac{2}{5}MR^2 \ ,$$

where M is the mass and R the radius. From the conservation of the angular momentum we get

$$L = I_1\omega_1 = I_2\omega_2$$
$$\Rightarrow \frac{I_1 2\pi}{P_1} = \frac{I_2 2\pi}{P_2}$$
$$\Rightarrow P_2 = P_1\frac{I_2}{I_1} = P_1\frac{(2/5)MR_2^2}{(2/5)MR_1^2} = P_1\left(\frac{R_2}{R_1}\right)^2 \ ,$$

where P_1 and P_2 are the rotation periods before and after the collapse. The mass of the cloud is

$$M = \tfrac{4}{3}\pi R^3\rho = \tfrac{4}{3}\pi R^3 \times 10^{16} \times 1.6734 \times 10^{-27} \text{ kg} = 1\,M_\odot = 1.989 \times 10^{30}\text{ kg} \ .$$

Solving for the radius we get $R = 3 \times 10^{13}$ m. The rotation period after the collapse is

$$P_2 = 1000\text{y}\left(\frac{6.96 \times 10^8 \text{ m}}{3 \times 10^{13}\text{ m}}\right)^2 = 5.4 \times 10^{-7}\text{y} = 17\text{ s} \ .$$

This is several orders of magnitude shorter than the actual period. Somehow the star has to get rid of most of its angular momentum during the process.

12.10 Exercises

12.10.1 Find the free fall time scale for a hydrogen cloud, if the density of H_2 molecules is 3000 cm^{-3}. Assume that stars condense from such clouds, there are 100 clouds in the Galaxy, the mass of each cloud is $5 \times 10^4 M_\odot$, and 10% of the mass is converted into stars. Also, assume that the average mass of a star is $1 M_\odot$. How many stars are born in one year?

12.10.2 The mass of Vega (spectral class A0V) is $2M_\odot$, radius $3R_\odot$, and luminosity $60 L_\odot$. Find its thermal and nuclear time scales.

12.10.3   Assume that a star remains 10^{10} years in the main sequence and burns 10% of its hydrogen. Then the star will expand into a red giant, and its luminosity will increase by a factor of 100. How long is the red giant stage, if we assume that the energy is produced only by burning the remaining hydrogen?

Chapter 13   The Sun

The Sun is our nearest star. It is important for astronomy because many phenomena which can only be studied indirectly in other stars can be directly observed in the Sun (e.g. stellar rotation, starspots, the structure of the stellar surface). Our present picture of the Sun is based both on observations and on theoretical calculations. Some observations of the Sun disagree with the theoretical solar models. The details of the models will have to be changed, but the general picture should remain valid.

13.1 Internal Structure

The Sun is a typical main sequence star. Its principal properties are:

mass	m	$= 1.989 \times 10^{30} \, \text{kg}$
radius	R	$= 6.960 \times 10^{8} \, \text{m}$
mean density		$= 1409 \, \text{kg m}^{-3}$
central density	ρ_c	$= 1.6 \times 10^{5} \, \text{kg m}^{-3}$
luminosity	L	$= 3.9 \times 10^{26} \, \text{W}$
effective temperature	T_e	$= 5785 \, \text{K}$
central temperature	T_c	$= 1.5 \times 10^{7} \, \text{K}$
absolute bolometric magnitude	M_{bol}	$= 4.72$
absolute visual magnitude	M_v	$= 4.79$
spectral class		G2 V
colour indices	B-V	$= 0.62$
	U-B	$= 0.10$
surface chemical composition	X	$= 0.71$
	Y	$= 0.27$
	Z	$= 0.02$
rotational period		
at the equator		25 d
at 60° latitude		29 d

On the basis of these data, the solar model shown in Fig. 13.1 has been calculated. The energy is produced by the pp chain in a small central region. 99% of the solar energy is produced within a quarter of the solar radius.

The Sun produces energy at the rate of $4 \times 10^{26} \, \text{W}$, which is equivalent to changing about four million tonnes of mass into energy every second. The mass of the Sun is so large, about 330 000 times that of the Earth, that during the whole main sequence lifetime of the Sun, less than 0.1% of its mass is turned into energy.

When the Sun formed about 5000 million years ago, its composition was the same everywhere as its present surface composition. Since energy production is concentrated

Fig. 13.1. The distribution of temperature, pressure, energy production and mass as functions of radius in the Sun

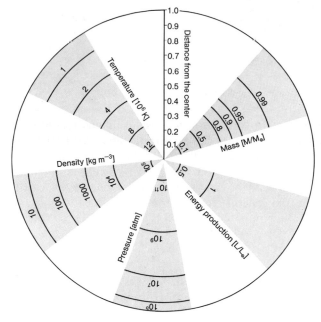

Fig. 13.2. The interior and surface of the Sun. The various kinds of solar phenomena are schematically indicated. [Based on Van Zandt, R. P. (1977): *Astronomy for the Amateur,* Planetary Astronomy, Vol. I, 3rd ed. (published by the author, Peoria, Ill.)]

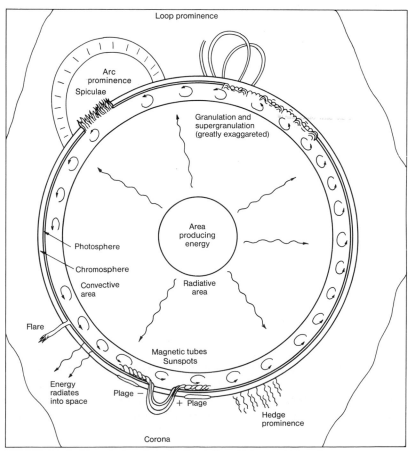

at the very centre, hydrogen is consumed most rapidly there. At about a quarter of the radius, the hydrogen abundance is still the same as in the surface layers, but going inwards from that point, it rapidly decreases. In the central core, only 40% of the material is hydrogen. About 5% of the hydrogen in the Sun has been turned into helium.

The radiative central part of the Sun extends to about 70% of the radius. At that radius, the temperature has dropped so much that the gas is no longer completely ionized. The opacity of the solar material then strongly increases, preventing the propagation of radiation. In consequence, convection becomes a more efficient means of energy transport. Thus the Sun has a convective envelope (Fig. 13.2).

The Solar Neutrino Problem. The central nuclear reactions produce neutrinos at several of the steps in the pp chain (see Fig. 11.5). These neutrinos can propagate freely through the outer layers, and thus give direct information about conditions near the centre of the Sun. When neutrinos from the Sun were first observed in the 1970's, their number was found to be only about a third of what was predicted. This disagreement is called the *solar neutrino problem*.

In the first experiments only neutrinos from the ppII and ppIII branches were observed. Since only a small fraction of the solar luminosity is produced in these reactions, it was not clear what were the consequences of these results for solar models. In the 1990's neutrinos produced in the ppI branch, the main branch of the pp chain, have been observed. Although the disagreement with the standard models is slightly smaller in these observations (about 60% of the predicted flux has been observed), the neutrino problem still remains.

Perhaps the most popular explanation for the solar neutrino problem is based on neutrino oscillations. According to this explanation, if neutrinos have a small mass (about 10^{-2} eV), an electron neutrino could change into a μ or a τ neutrino as it passes through the outer parts of the Sun. Since only electron neutrinos have been observed in the experiments, the small observed flux of neutrinos could thus be understood. Since there are no direct measurements of neutrino masses, it is also possible that there is some flaw in the standard solar models.

The Solar Rotation. As soon as telescopes were introduced, it was observed from the motions of sunspots that the Sun is rotating with a rotational period of about 27 days. As early as 1630 *Christoph Scheiner* showed that there was *differential rotation*: the rotational period near the poles was more than 30 days, while it was only 25 days at the equator. The rotational axis of the Sun is inclined at 7° with respect to the plane of the ecliptic, so that the North Pole of the Sun is best visible from the Earth in September.

The motions of sunspots still give the best information on the rotation near the surface of the Sun. Other surface features also have been used for this purpose. The rotational velocity has also been measured directly from the Doppler effect. The angular velocity is usually written $\Omega = A - B \sin^2 \psi$, where ψ is the latitude with respect to the equator. The measured values of the coefficients are $A = 14.5$ and $B = 2.9$ degrees/day.

The rotational velocity deeper down in the Sun cannot be directly observed. In the 1980's a method to estimate the rotation in the interior became available, when it became possible to measure the frequencies of solar oscillations from the variations in spectral lines. These oscillations are essentially sound waves produced by turbulent gas motions in the convection zone. These sound waves have calculable oscillation periods (about 3 – 12 minutes), which depend on the conditions in the solar interior. By comparing the observed and theoretical values one can get information about the conditions deep inside the Sun. The idea of the method is the same as that used when studying the interior of the Earth by means of waves from earthquakes, and it is therefore called *helioseismology*.

Using helioseismology, models for the solar rotation throughout the convection zone have been deduced. It appears that the angular velocity in the whole convection zone is almost the same as at the surface, although it decreases slightly with radius near the equator, and increases near the poles. The angular velocity of the radiative core is still uncertain, but there are indications that the core is rotating as a solid body with approximately the average surface angular velocity.

The solar differential rotation is maintained by gas motions in the convection zone. Explaining the observed behaviour is a difficult problem that is not yet completely understood.

13.2 The Atmosphere

The solar atmosphere is divided into the *photosphere* and the *chromosphere*. Outside the actual atmosphere, the *corona* extends much further outwards.

The Photosphere. The innermost layer of the atmosphere is the photosphere, which is only about 300 – 500 km thick. The photosphere is the visible surface of the Sun, where the density rapidly increases inwards, hiding the interior from sight. The temperature at the inner boundary of the photosphere is 8000 K and at the outer boundary, 4500 K. Near the edge of the solar disc, the line of sight enters the photosphere at a very small angle and never penetrates to large depths. Near the edges one therefore only sees light from the cooler, higher layers. For this reason, the edges appear darker; this phenomenon is known as *limb darkening*. Both the continuous spectrum and the absorption lines are formed in the photosphere, but the light in the absorption lines comes from higher layers and therefore the lines appear dark.

The solar convection is visible on the surface as the *granulation* (Fig. 13.3), an uneven, constantly changing granular pattern. At the bright centre of each granule, gas is rising upward, and at the darker granule boundaries, it is sinking down again. The size of a granule seen from the Earth is typically 1″, corresponding to about 1000 km on the solar surface. There is also a larger scale convection called *supergranulation* in the photosphere. The cells of the supergranulation may be about 1′ in diameter. The observed velocities in the supergranulation are mainly directed along the solar surface.

The Chromosphere (Fig. 13.4). Outside the photosphere there is a layer, perhaps about 500 km thick, where the temperature increases from 4500 K to about 6000 K, the chromosphere. Outside this layer, there is a transition region of a few thousand kilometres, where the chromosphere gradually goes over into the corona. In the outer parts of the transition region, the kinetic temperature is already about 10^6 K.

Normally the chromosphere is not visible, because its radiation is so much weaker than that of the photosphere. However, during total solar eclipses, the chromosphere shines into view for a few seconds at both ends of the total phase, when the Moon hides the photosphere completely. The chromosphere then appears as a thin reddish sickle or ring.

Fig. 13.3. The granulation of the solar surface. The granules are produced by streaming gas. Their typical diameter is 1000 km. (Photograph Mt. Wilson Observatory)

During eclipses, the chromospheric spectrum, called the flash spectrum, can be observed. It is an emission line spectrum with more than 3000 identified lines. Brightest among these are the lines of hydrogen, helium and certain metals.

One of the strongest chromospheric emission lines is the hydrogen Balmer α line (Fig. 13.5) at a wavelength of 656.3 nm. Since the Hα line in the normal solar spectrum is a very dark absorption line, a photograph taken at this wavelength will show the solar chromosphere. For this purpose, one uses narrow-band filters letting through only the light in the Hα line. The resulting pictures show the solar surface as a mottled, wavy disc. The bright regions are usually the size of a supergranule, and are bounded by *spicules* (Fig. 13.6). These are flamelike structures, rising up to 10000 km above the

Fig. 13.4. Flash spectrum of the solar chromosphere, showing bright emission lines

Fig. 13.5. The solar surface in the hydrogen Hα line. Active regions near the equator appear bright; the dark filaments are prominences

Fig. 13.6. Spicules, flamelike uprisings near the edge of the solar disc. (Photograph Big Bear Solar Observatory)

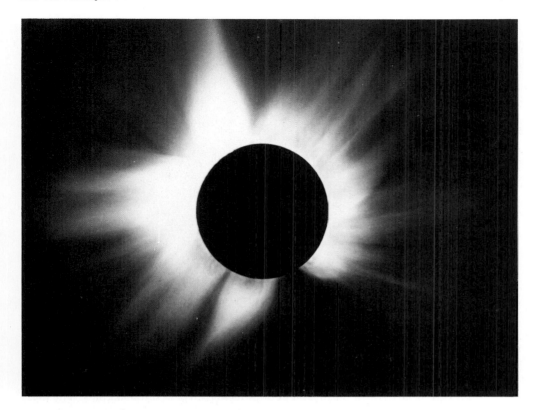

Fig. 13.7. Previously, the corona could be studied only during total solar eclipses. The picture is from the eclipse 7.3.1970. Nowadays the corona can be studied continuously using the so-called coronograph

chromosphere, and lasting for a few minutes. Against the bright surface of the Sun, they look like dark streaks; at the edges, they look like bright flames.

The Corona. The chromosphere gradually goes over into the corona. The corona is also best seen during total solar eclipses (Fig. 13.7). It then appears as a halo of light extending out to a few solar radii. The surface brightness of the corona is about that of the full moon, and it is therefore difficult to see next to the bright photosphere.

The inner part of the corona, the K corona, has a continuous spectrum formed by the scattering of the photospheric light by electrons. Further out, a few solar radii from the surface, is the F corona, which has a spectrum showing Fraunhofer absorption lines. The light of the F corona is sunlight scattered by dust.

In the latter part of the 19th century, strong emission lines, which did not correspond to those of any known element, were discovered in the corona (Fig. 13.8). It was thought that a new element, called coronium, had been found − a little earlier, helium had been discovered in the Sun before it was known on Earth. About 1940, it was established that the coronal lines were due to highly ionized atoms, e.g. thirteen times ionized iron. Much energy is needed to remove so many electrons from the atoms. The entire corona has to have a temperature of about a million degrees.

A continuous supply of energy is needed in order to maintain the high temperature of the corona. According to earlier theories, the energy came in the form of acoustic or magnetohydrodynamic shock waves generated at the solar surface by the convection. Most recently, heating by electric currents induced by changing magnetic fields has been suggested. Heat would then be generated in the corona almost like in an ordinary light bulb.

Fig. 13.8. The presence of lines from highly ionized atoms in the coronal spectrum shows that the temperature of the corona has to be very high

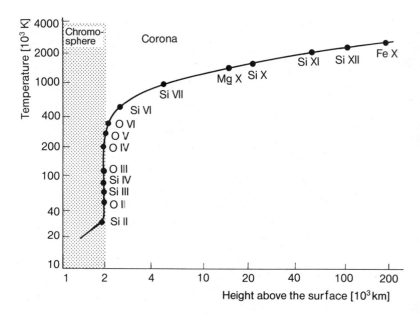

In spite of its high temperature, the coronal gas is so diffuse that the total energy stored in it is small. It is constantly streaming outwards, gradually becoming a *solar wind*, which carries a flux of particles away from the Sun. The gas lost in this way is replaced with new material from the chromosphere. Near the Earth, the density of the solar wind is typically $5-10$ particles/cm^3 and its velocity, about 500 km/s. The mass loss of the Sun due to the solar wind is about $10^{-13} M_\odot$ per year.

13.3 Solar Activity

Sunspots. The clearest visible sign of solar activity are the *sunspots*. The existence of sunspots has been known for long (Fig. 13.9), since the largest ones can be seen with the naked eye by looking at the Sun through a suitably dense layer of fog. More precise observations became available beginning in the 17th century, when Galilei started to use the telescope for astronomical observations.

A sunspot looks like a ragged hole in the solar surface. In the interior of the spot is a dark *umbra* and around it, a less dark *penumbra*. By looking at spots near the edge of the solar disc, it can be seen that the spots are slightly depressed with respect to the rest of the surface. The surface temperature in a sunspot is about 1500 K below that of its surroundings, which explains the dark colour of the spots.

The diameter of a typical sunspot is about 10 000 km and its lifetime is from a few days to several months, depending on its size. The larger spots are more likely to be long-lived. Sunspots often occur in pairs or in larger groups. By following the motions of the spots, the period of rotation of the Sun can be determined.

The variations in the number of sunspots have been followed for almost 250 years. The frequency of spots is described by the Zürich sunspot number Z:

$$Z = C(S + 10G) , \qquad (13.1)$$

where S is number of spots and G, the number of spot groups visible at a particular time. C is a constant depending on the observer and the conditions of observation.

Fig. 13.9. The sunspots are the form of solar activity that has been known for the longest time. (Photograph Mt. Wilson Observatory)

In Fig. 13.10, the variations in the Zürich sunspot number between the 18th century and the present are shown. Evidently the number of spots varies with an average period of about 11 years. The actual period may be between 7 and 17 years. In the past decades, it has been about 10.5 years. Usually the activity rises to its maximum in about 3 – 4 years, and then falls off slightly more slowly. The period was first noted by *Samuel Heinrich Schwabe* in 1843.

The variations in the number of sunspots have been fairly regular since the beginning of the 18th century. However, in the 17th century, there were long intervals when there were essentially no spots at all. This quiescent period is called the *Maunder minimum*. The similar *Spörer minimum* occurred in the 15th century, and other quiet intervals have been inferred at earlier epochs. The mechanism behind these irregular variations in solar activity is not yet understood.

The magnetic fields in sunspots are measured on the basis of the Zeeman effect, and may be as large as 0.45 tesla. (The Earth's magnetic field is 0.06 mT.) The strong magnetic field inhibits convective energy transport, which explains the lower temperature of the spots.

Fig. 13.10. The Zürich sunspot number from 1700 to 1992. The number of sunspots and spot groups varies with a period of about 11 years

Sunspots often occur in pairs where the components have opposite polarity. The structure of such *bipolar groups* can be understood if the field rises into a loop above the solar surface, connecting the components in a pair of spots. If gas is streaming along such a loop, it becomes visible as a *loop prominence* (Fig. 13.11).

The periodic variation in the number of sunspots reflects a variation in the general solar magnetic field. At the beginning of new activity cycle spots first begin to appear at about 40° latitude. As the cycle advances, the spots move closer to the equator. The characteristic pattern in which spots appear, shown in Fig. 13.12, is known as the *butterfly diagram*. Spots of the next cycle begin to appear while those of the old one are still present near the equator. Spots belonging to the new cycle have a polarity opposite that of the old ones. (Spots in opposite hemispheres also have opposite polarity.) Since the field is thus reversed between consecutive 11 year cycles the complete period of solar magnetic activity is 22 years.

The following general qualitative description of the mechanism of the solar cycle was proposed by *Horace W. Babcock*. Starting at a solar minimum, the field will be

Fig. 13.11. In pairs of sunspots, the magnetic field lines form a loop outside the solar surface. Material streaming along the field lines may form loop prominences. (Photograph Mt. Wilson Observatory)

Fig. 13.12. At the beginning of an activity cycle, sunspots appear at high latitudes. As the cycle advances the spots move towards the equator. (Diagram by H. Virtanen, based on Greenwich Observatory observations)

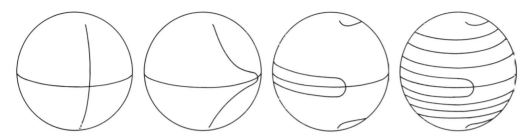

Fig. 13.13. Because the Sun rotates faster at the equator than at the poles, the field lines of the solar magnetic field are drawn out into a tight spiral

of a generally dipolar character. Because a conducting medium, such as the outer layers of the Sun, cannot move across the field lines, these will be frozen into the plasma and carried along by it. The rotation period of the Sun is 25 days near the equator and more than 30 days near the pole. This *differential rotation* will draw the field into a tight spiral (Fig. 13.13). In the process, the field becomes stronger, and this amplification will be a function of latitude.

When the subsurface field becomes strong enough, it gives rise to a "magnetic buoyancy" that lifts ropes of magnetic flux above the surface. This happens first at a latitude about 40°, and later at lower latitudes. These protruding flux ropes expand into loops forming bipolar groups of spots. As the loops continue expanding they make contact with the general dipolar field, which still remains in the polar regions. This leads to a rapid reconnection of the field lines neutralizing the general field. The final result when activity subsides is a dipolar field with a polarity opposite the initial one.

Thus the Babcock model accounts for the butterfly diagram, the formation of bipolar magnetic regions and the general field reversal between activity maxima. Nevertheless, it remains an essentially phenomenological model, and alternative scenarios have been proposed. In *dynamo theory* quantitative models for the origin of magnetic fields in the Sun and other celestial bodies are studied. In these models the field is produced by convection and differential rotation of the gas. A completely satisfactory dynamo model for the solar magnetic cycle has not yet been found. For example, it is not yet known whether the field is produced everywhere in the convection zone, or just in the

Fig. 13.14. The Sun seen in visible light (*right*) and in the hydrogen Hα line (*left*). The bright regions around the sunspots and sunspot groups are called plages. (Photographs Yerkes Observatory)

boundary layer between the convective and radiative regions, as some indications suggest.

Other Activity. The Sun shows several other types of surface activity: *faculae* and *plages; prominences; flares*.

The faculae and plages are local bright regions in the photosphere and chromosphere, respectively. Observations of the plages are made in the hydrogen Hα or the

Fig. 13.15. (a) Quiescent "hedgerow" prominence (Photograph Sacramento Peak Observatory). **(b)** Larger eruptive prominence (Photograph Big Bear Solar Observatory)

Fig. 13.16. A violent flare near some small sunspots. (Photograph Sacramento Peak Observatory)

Fig. 13.17. At radio wavelengths, a flare appears as a bright region, rapidly moving away ▶ from the Sun. (Picture CSIRO)

calcium K lines (Fig. 13.14). The plages usually occur where new sunspots are forming, and disappear when the spots disappear. Apparently they are caused by the enhanced heating of the chromosphere in strong magnetic fields.

The prominences are among the most spectacular solar phenomena. They are glowing gas masses in the corona, easily observed near the edge of the Sun. There are several types of prominences (Fig. 13.15): the quiescent prominences, where the gas is slowly sinking along the magnetic field lines; loop prominences, connected with magnetic field loops in sunspots; and the rarer eruptive prominences, where gas is violently thrown outwards.

The temperature of prominences is about $10\,000 - 20\,000$ K. In Hα photographs of the chromosphere, the prominences appear as dark *filaments* against the solar surface.

The flare outbursts are among the most violent forms of solar activity (Fig. 13.16). They appear as bright flashes, lasting from one second to just under an hour. In the flares, a large amount of energy stored in the magnetic field is suddenly released. The detailed mechanism is not yet known.

Flares can be observed at all wavelengths. The hard x-ray emission of the Sun may increase a hundred times during a flare. Several different types of flares are observed at radio wavelengths (Fig. 13.17). The emission of solar cosmic ray particles also rises.

The flares give rise to disturbances on the Earth. The x-rays cause changes in the ionosphere, which affect short-wave radio communications. The flare particles give rise to strong aurorae when they enter the Earth's magnetic field a few days after the outburst.

Solar Radio Emission. The Sun is the strongest radio source in the sky and has been observed since the 1940's. In contrast to optical emission, the radio picture of the Sun shows a strong *limb brightening*. This is because the radio radiation comes from the upper layers of the atmosphere. Since the propagation of radio waves is obstructed by free electrons, the high electron density near the surface prevents radio radiation from getting out. Shorter wavelengths can propagate more easily, and thus millimetre-wavelength observations give a picture of deeper layers in the atmosphere, whereas the long wavelengths show the upper layers. (The 10-cm emission originates in the upper layers of the chromosphere and the 1-m emission, in the corona.)

The radio emission of the Sun is constantly changing according to solar activity. During large storms, the total emission may be 100 000 times higher than normal. Strong, sudden increases in radio emission are called *radio bursts*, and are classified into five types.

Type I bursts are common at metre wavelengths, and are the only bursts not related to flares. The type I bursts are connected with sunspots. They are often observed as storms lasting for hours or days, during which there are frequent, sharper $0.1-10$ s bursts.

Type II bursts are observed after flares as radio noise at metre wavelengths, lasting for about half an hour. These bursts are apparently produced when the flare particles propagate out through the solar atmosphere. As the particles lose energy, the radio emission frequency decreases.

Type III bursts are sharper, shorter spikes produced by clouds of relativistic electrons.

Type IV bursts last for about an hour and are observed at all wavelengths. They are due to synchrotron emission by the flare electrons.

Type V bursts last for a few minutes and are observed in a broad wavelength band.

X-ray and UV Radiation. The x-ray emission of the Sun is also related to active regions. Signs of activity are bright *x-ray regions* and smaller *x-ray bright points*, which last for around ten hours. The inner solar corona also emits x-rays. Near the solar poles, there are *coronal holes*, where the x-ray emission is weak.

Ultraviolet pictures of the solar surface show it as much more irregular than it appears in visible light. Most of the surface does not emit much UV radiation, but there are large active regions that are very bright in the ultraviolet.

Several satellites have made observations of the Sun at UV and x-ray wavelengths, for example the Skylab space station (1973–1974) and the Solar Maximum Mission satellite (1980). These observations have made possible detailed studies of the outer layers of the Sun. Observations of other stars have revealed coronae, chromospheres and magnetic variations similar to those in the Sun. Thus the new observational techniques have brought the physics of the Sun and the stars nearer to each other.

13.4 Examples

13.4.1   Assume that the Sun converts 0.8% of its mass into energy. Find an upper limit for the age of the Sun, assuming that its luminosity has remained constant.

The total amount of energy released is

$$E = mc^2 = 0.008\, M_\odot c^2 = 0.008 \times 2 \times 10^{30}\,\text{kg} \times (3 \times 10^8\,\text{ms}^{-1})^2 = 1.4 \times 10^{45}\,\text{J} \ .$$

The time needed to radiate this energy is

$$t = \frac{E}{L_\odot} = \frac{1.4 \times 10^{45}\,\text{J}}{3.9 \times 10^{26}\,\text{W}} = 3.6 \times 10^{18}\,\text{s} \approx 10^{11} \ \text{years} \ .$$

13.5 Exercises

13.5.1   The solar constant, i.e. the flux density of the solar radiation at the distance of the Earth is $1390\,\text{W m}^{-2}$.

a) Find the flux density on the surface of the Sun, when the apparent diameter of the Sun is 32′.

b) How many square metres of solar surface is needed to produce 1000 megawatts?

13.5.2   Some theories have assumed that the effective temperature of the Sun 4.5 billion years ago was 5000 K and radius 1.02 times the current radius. What was the solar constant then? Assume that the orbit of the Earth has not changed.

Chapter 14 Variable Stars

Stars with changing magnitudes are called *variabies* (Fig. 14.1). Variations in the brightness of stars were first noted in Europe at the end of the 16th century, when *Tycho Brahe's supernova* lit up (1572) and the regular light variation of the star o Ceti (Mira) was observed (1596). The number of known variables has grown steadily as observational precision has improved (Fig. 14.2). The most recent catalogues contain about 40 000 stars known or suspected to be variable.

Strictly speaking, all stars are variable. As was seen in Chap. 12, the structure and brightness of a star change as it evolves. Although these changes are usually slow, some evolutionary phases may be extremely rapid. In certain evolutionary stages, there will also be periodic variations, for example pulsations of the outer layers of a star.

Small variations in stellar brightness are also caused by hot and cool spots on a star's surface, appearing and disappearing as it rotates about its axis. The luminosity of

Fig. 14.1. The variables are stars changing in brightness. Two variables in Scorpio, R and S Sco (Photograph Yerkes Observatory)

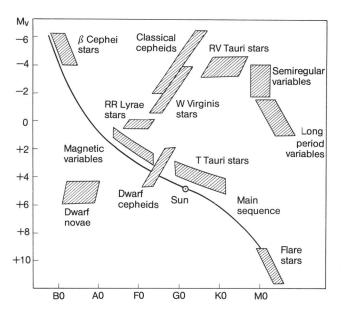

Fig. 14.3. The light curve of a variable is determined from continuous photometric measurement of the brightness of the star

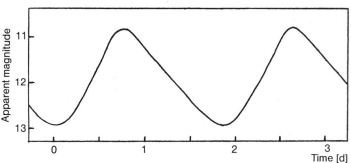

the Sun changes slightly because of the sunspots. Probably there are similar spots on almost all stars.

Initially, stellar brightnesses were determined visually by comparing stars near each other. Later on, comparisons were made on photographic plates. At present, the most accurate observations are made photoelectrically or using a CCD camera. The magnitude variation as a function of time is called the *light curve* of a star (Fig. 14.3). From it, one obtains the *amplitude* of the magnitude variation and its *period*, if the variation is periodic.

The basic reference catalogue of variable stars is the General Catalogue of Variable Stars by the Soviet astronomer *Boris Vasilyevich Kukarkin*. New, supplemented editions appear at times; the fourth edition from 1985 – 1987, edited by P. N. Kholopov, contains about 29 000 variables.

14.1 Classification

When a new variable is discovered, it is given a name according to the constellation in which it is located. The name of the first variable in a given constellation is R, followed by the name of the constellation (in the genitive case). The symbol for the second variable is S, and so on, to Z. After these, the two-letter symbols RR, RS, ... to ZZ are used, and then, AA to QZ (omitting J). This is only enough for 334 variables, a number that has long been exceeded in most constellations. The numbering therefore continues: V335, V336, etc. (V stands for variable). For some stars, the established Greek letter symbol has been kept, although they have later been found to be variable (e.g. δ Cephei).

The classification of variables is based on the shape of the light curve, and on the spectral class and observed radial motions. The spectrum may also contain dark absorption lines from material around the star. Observations can be made outside the optical region as well. Thus the radio emission of some variables (e.g. flare stars) increases strongly, simultaneously with their optical brightness. Examples of radio- and x-ray variables are the radio and x-ray pulsars, and the x-ray bursters.

Variables are usually divided into three main types: *pulsating, eruptive and eclipsing variables*. The eclipsing variables are binary systems in which the components periodically pass in front of each other. In these variables, the light variations do not correspond to any physical change in the stars. They have been treated in connection with the binary stars. In the other variables, the brightness variations are intrinsic to the stars. In the pulsating variables, the variations are due to the expansion and contraction of the outer layers. These variables are giants and supergiants that have reached an unstable stage in their evolution. The eruptive variables are usually faint stars ejecting mass. They are mostly members of close binary systems in which mass is transferred from one component to the other.

In addition, a few *rotating variables* are known, where the brightness variations are due to an uneven temperature distribution on the surface, starspots coming into sight when the star rotates. Such stars may be quite common − after all, our Sun is a weak rotating variable. The most prominent group of rotating variables are the magnetic A stars (e.g. the α^2 Canum Venaticorum stars). These stars have strong magnetic fields that may be giving rise to starspots. The periods of rotating variables range from about 1 day to 25 d, and the amplitudes are less than 0.1 mag.

14.2 Pulsating Variables

The wavelengths of the spectral lines of the pulsating variables change along with the brightness variations (Table 14.1). These changes are due to the Doppler effect, showing that the outer layers of the star are indeed pulsating. The observed gas velocities are in the range of $40 - 200 \, \text{km/s}$.

The diameter of the star may double during the pulsation, but usually the changes in size are minor. The main cause of the light variation is the periodic variation of the surface temperature. We have seen in Sect. 5.6 that the luminosity of a star depends sensitively on its effective temperature, $L \propto T_e^4$. Thus a small change in effective temperature leads to a large brightness variation.

The period of pulsation corresponds to a *proper frequency* of the star. Just like a tuning fork vibrates with a characteristic frequency when hit, a star has a fundamental frequency of vibration. In addition to the fundamental frequency, other frequencies,

Table 14.1. The main properties of pulsating variables (*N*, number of stars of the given type in Kukarkin's catalogue. *P*, pulsation period in days. Δm, pulsation amplitude in magnitudes)

Variable	N	P	Spectrum	Δm
Classical cepheids (δ Cep, W Vir)	700	$1 - 50$	F − K I	<2
RR Lyrae	4400	<1	B8 − F2 III	<0.7
Dwarf cepheids (δ Scuti)	20	$0.05 - 0.2$	F III	<1
β Cephei	20	$0.1 - 0.25$	B1 − B3 III	<0.1
Mira variables	4600	$80 - 1000$	M III	>2.5
RV Tauri	100	$30 - 150$	G − K I	<3
Semiregular	2200	$30 - 1000$	K − M I	<2.5
Irregular	1700	−	K − M I	<2

"overtones", are possible. The observed brightness variation can be understood as a superposition of all these modes of vibration. Around 1920, the English astrophysicist *Sir Arthur Eddington* showed that the period of pulsation P is inversely proportional to the square root of the mean density,

$$P \propto \rho^{-1/2}.$$

Normally a star is in stable hydrostatic equilibrium. If its outer layers expand, the density and temperature decrease. The pressure then becomes smaller and the force of gravity compresses the gas again. However, unless energy can be transferred to the gas motions, these oscillations will be damped.

The flux of radiative energy from the stellar interior could provide a source of energy for the stellar oscillations, if it were preferentially absorbed in regions of higher gas density. Usually this is not the case but in the *ionization zones*, where hydrogen and helium are partially ionized, the opacity in fact becomes larger when the gas is compressed. If the ionization zones are at a suitable depth in the atmosphere, the energy absorbed during compression and released during expansion of an ionization zone can drive an oscillation. Stars with surface temperatures of 6000 – 9000 K are liable to this instability. The corresponding section of the HR diagram is called the cepheid instability strip.

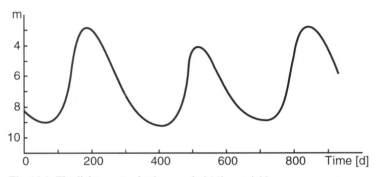

Fig. 14.4. The light curve of a long-period Mira variable

Mira Variables (Fig. 14.4). The Mira variables (named after Mira Ceti) are M supergiants, usually with emission lines in their spectrum. They are losing gas in a steady stellar wind. Their periods are normally 100 – 500 days, and for this reason, they are also sometimes called long-period variables. The amplitude of the light variations is typically about 6 magnitudes in the visual region. The period of Mira itself is about 330 days and its diameter is about 2 AU. At its brightest, Mira has the magnitude 2 – 4, but at light minimum, it may be down to 12. The effective temperature of the Mira variables is only about 2000 K. Thus 95% of their radiation is in the infrared, which means that a very small change in temperature can cause a very large change in visual brightness.

Cepheids. Among the most important pulsating variables are the cepheids, named after δ Cephei (Fig. 14.5). They are population I supergiants (stellar populations are dis-

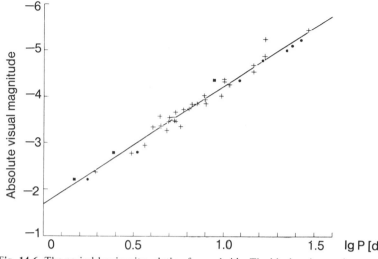

Fig. 14.5. The variation of brightness, colour and size of a cepheid during its pulsation

Fig. 14.6. The period-luminosity relation for cepheids. The black points and squares are theoretically calculated values, the crosses and the straight line represent the observed relation. [Drawing from Novotny, E. (1973): *Introduction to Stellar Atmospheres and Interiors* (Oxford University Press, New York) p. 359]

cussed in Sect. 18.2) of spectral class F-K. Their periods are 1 – 50 days and their amplitudes, 0.3 – 2.5 magnitudes. The shape of the light curve is regular, showing a fairly rapid brightening, followed by a slower fall off. There is a relationship between the period of a cepheid and its absolute magnitude (i.e. luminosity), discovered in 1912 by *Henrietta Leavitt* from cepheids in the Small Magellanic Cloud. This *period-luminosity relation* (Fig. 14.6) can be used to measure distances of stars and nearby galaxies.

We have already noted that the pulsation period is related to the mean density. On the other hand, the size of a star, and hence its mean density, is related to its total luminosity. Thus one can understand why there should be a relation between the period and the luminosity of a pulsating star.

The magnitudes M and periods P of classical cepheids are shown in Fig. 14.6. The relation between M and $\log P$ is linear. However, to some extent, the cepheid luminosities also depend on colour: bluer stars are brighter. For accurate distance determinations, this effect needs to be taken into consideration.

W Virginis Stars. In 1952, *Walter Baade* noted that there are in fact two types of cepheids: the classical cepheids and the *W Virginis stars*. Both types obey a period-luminosity relation, but the W Vir stars of a given period are 1.5 magnitudes fainter than the corresponding classical cepheids. This difference is due to the fact that the classical cepheids are young population I objects, whereas the W Vir stars are old stars of population II. Otherwise, the two classes of variables are similar.

Earlier, the W Vir period-luminosity relation had been used for both types of cepheids. Consequently, the calculated distances to classical cepheids were too small. For example, the distance to the Andromeda Galaxy had been based on classical cepheids, since only these were bright enough to be visible at that distance. When the correct period-luminosity relation was used, all extragalactic distances had to be doubled. Distances within the Milky Way did not have to be changed, since their measurements were based on other methods.

RR Lyrae Stars. The third important class of pulsating variables are the *RR Lyrae stars*. Their brightness variations are smaller than those of the cepheids, usually less than a magnitude. Their periods are also short, less than a day. Like the W Vir stars, the RR Lyrae stars are old population II stars. They are very common in the globular star clusters and were therefore previously called cluster variables.

The absolute magnitudes of the RR Lyrae stars are about $M_V = 0.6 \pm 0.3$. They are all of roughly the same age and mass, and thus represent the same evolutionary phase, where helium is just beginning to burn in the core. Since the absolute magnitudes of the RR Lyrae variables are known, they can be used to determine distances to the globular clusters.

Other Pulsating Variables. One additional large group of pulsating stars are the *semiregular* and *irregular variables*. They are supergiants, often very massive young stars with unsteady pulsations in their extended outer layers. If there is some periodicity in the pulsations, these variables are called semiregular; otherwise they are irregular. An example of a semiregular variable is Betelgeuze (α Orionis). The pulsation mechanism of these stars is not well understood, since their outer layers are convective, and the theory of stellar convection is still poorly developed.

In addition to the main types of pulsating variables, there are some smaller separate classes.

The *dwarf cepheids*, and the *δ Scuti stars*, which are sometimes counted as a separate type, are located below the RR Lyrae stars in the cepheid instability strip in the HR diagram. The dwarf cepheids are fainter and more rapidly varying than the classical cepheids. Their light curves often show a beating due to interference between the fundamental frequency and the first overtone.

The *β Cephei stars* are located in a different part of the HR diagram than the other variables. They are hot massive stars, radiating mainly in the ultraviolet. The variations are rapid and of small amplitude. The pulsation mechanism of the β Cephei stars is unknown.

The *RV Tauri stars* lie between the cepheids and the Mira variables in the HR diagram. Their period depends slightly on the luminosity. There are some unexplained features in the light curves of the RV Tauri stars, e.g. the minima are alternately deep and shallow.

14.3 Eruptive Variables

In the eruptive variables, there are no regular pulsations. Instead, sudden outbursts occur in which material is ejected into space. The scale of the outbursts goes from small local eruptions (flare stars) to the explosion of the entire star (supernovae) (Table 14.2).

Flare Stars. The *flare* or *UV Ceti stars* are dwarf stars of spectral class M. They are young stars, mostly found in young star clusters and associations. At irregular intervals, there are flare outbursts on the surface of the stars, similar to those on the Sun. The flares are related to disturbances in the surface magnetic fields.

Table 14.2. Main properties of eruptive variables (N, number of stars of the given type in Kukarkin's catalogue. Δm, change in brightness in magnitudes. The velocity is the expansion velocity in km/s, based on the Doppler shifts of the spectral lines)

Variable	N	Δm	Velocity
Supernovae	7	>20	4000 – 10000
Ordinary novae,		7 – 18	200 – 3500
recurrent novae	210	<10	600
Nova-like	80	<2	30 – 100
(P Cygni, symbiotic)			
Dwarf novae	330	2 – 6	(700)
(SS Cyg = U Gem, ZZ Cam)			
R Coronae Borealis	40	1 – 9	–
Irregular	1450	<4	(300)
(nebular variables,			
T Tau, RW Aur)			
Flare stars	750	<6	2000
(UV Ceti)			

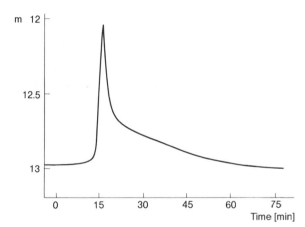

Fig. 14.7. The outbursts of typical flare stars are of short duration

The energy of the outbursts of the flare stars is apparently about the same as in solar flares, but because the stars are much fainter than the Sun, a flare can cause a brightening by up to 4 – 5 magnitudes. A flare lights up in a few seconds and then fades away in a few minutes (Fig. 14.7). The same star may flare several times in one day. The optical flare is accompanied by a radio outburst, like in the Sun. The flare stars were the first stars to be detected as radio sources.

T Tauri Stars. The *T Tauri stars* are so-called *nebular variables*, occurring in connection with bright or dark interstellar clouds. These stars are newly formed or just contracting towards the main sequence. The brightness variations of the T Tauri stars are irregular (Fig. 14.8). Their spectra contain bright emission lines, formed in the stellar chromosphere, and forbidden lines, which can only be formed at extremely low densities. The spectral lines also show that matter is streaming out from the stars.

Since the T Tauri stars are situated inside dense gas clouds, they are difficult to observe. However, this situation has improved with the development of radio and infrared techniques.

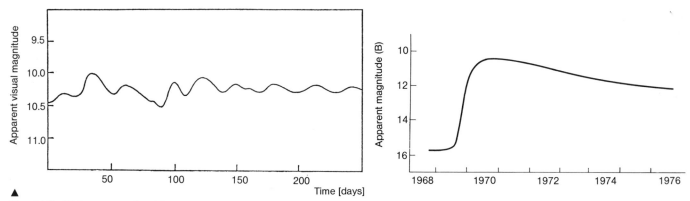

Fig. 14.8. Light curve of a T Tauri variable

Fig. 14.9. In 1969–1970, the star V1057 Cygni brightened by almost 6 magnitudes

Stars in the process of formation may change in brightness very rapidly. For example, in 1937, FU Orionis brightened by 6 magnitudes. This star is a strong source of infrared radiation, which shows that it is still enveloped by large quantities of interstellar dust and gas. A similar brightening by six magnitudes was observed in 1969 in V1057 Cygni (Fig. 14.9). Before its brightening, it was an irregular T Tauri variable; since then, it has remained a fairly constant tenth-magnitude AB star.

Novae. One of the best known types of eruptive variable are the *novae.* They are classified into several subtypes: *ordinary novae, recurrent novae, dwarf novae* (Fig. 14.10) and *nova-like variables.* The outbursts of all novae are rapid. Within a day or two, the brightness rises to a maximum, which may be 7–16 magnitudes brighter than the normal luminosity. This is followed by a gradual decline, which may go on for months or years. The light curve of a typical nova is shown in Fig. 14.11. This light curve of Nova Cygni 1975 has been composed from hundreds of observations, mostly by amateurs.

In recurrent novae, the brightening is somewhat less than 10 magnitudes and in dwarf novae, 2–6 magnitudes. The dwarf novae are also known as *U Geminorum* or *SS Cygni stars.* In both types, there are repeated outbursts. For recurrent novae, the time between outbursts is a few decades and for the dwarf novae, 20–600 days. The interval depends on the strength of the outburst: the stronger the outburst, the longer the time until the next one. The brightening in magnitudes is roughly proportional to the logarithm of the recharging interval. It is possible that ordinary novae obey the same relationship. However, their amplitude is so much larger that the time between outbursts should be thousands or millions of years.

Observations have shown all novae to be members of close binary systems. One component of the system is a normal star and the other is a white dwarf surrounded by a gas ring. (The evolution of close binary systems was considered in Sect. 12.6, where it was seen how this kind of system might have been formed.) The normal star fills its Roche surface, and material from it streams over to the white dwarf. When enough mass has collected on the surface of the white dwarf, the hydrogen is explosively ignited

Fig. 14.10. The light curve of the dwarf nova SS Cygni in the beginning of 1966

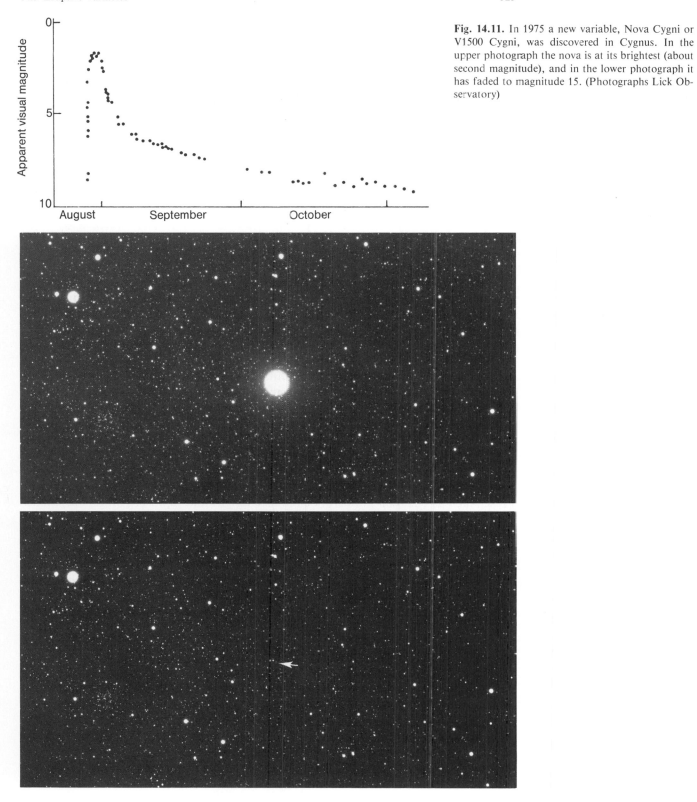

Fig. 14.11. In 1975 a new variable, Nova Cygni or V1500 Cygni, was discovered in Cygnus. In the upper photograph the nova is at its brightest (about second magnitude), and in the lower photograph it has faded to magnitude 15. (Photographs Lick Observatory)

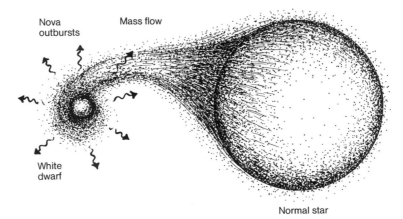

Nova outbursts

Mass flow

White dwarf

Normal star

Fig. 14.12. The novae are thought to be white dwarfs accreting matter from a nearby companion star. At times, nuclear reactions burning the accreted hydrogen are ignited, and this is seen as the flare-up of a nova

and the outer shell is ejected. The brightness of the star grows rapidly. As the ejected shell expands, the temperature of the star drops and the luminosity gradually decreases. However, the outburst does not stop the mass transfer from the companion star, and gradually the white dwarf accretes new material for the next explosion (Fig. 14.12).

The emission and absorption lines from the expanding gas shell can be observed in the spectrum of a nova. The Doppler shifts correspond to an expansion velocity of about 1000 km/s. As the gas shell disperses, its spectrum becomes that of a typical diffuse emission nebula. The expanding shell around a nova can also sometimes be directly seen in photographs.

A considerable fraction of the novae in our Galaxy are hidden by interstellar clouds and their number is therefore difficult to estimate. In the Andromeda Galaxy, observations indicate 25 – 30 nova explosions per year. The number of dwarf novae is much larger. In addition, there are nova-like variables, which share many of the properties of novae, such as emission lines from circumstellar gas and rapid brightness variations. These variables, some of which are *symbiotic stars*, are close binaries with mass transfer. Gas streaming from the primary hits a gas disc around the secondary in a hot spot, but there are no nova outbursts.

One very interesting variable is *η Carinae* (Fig. 14.13). At present, it is a sixth-magnitude star surrounded by a thick, extensive envelope of dust and gas. In the early 19th century, *η* Carinae was the second brightest star in the sky after Sirius. Around the middle of the century, it rapidly dimmed to magnitude 8, but during the 20th century, it has brightened somewhat. The circumstellar dust cloud is the brightest infrared source in the sky outside the solar system. The energy radiated by *η* Carinae is absorbed by the nebula and re-radiated at infrared wavelengths. It is not known whether *η* Carinae is related to the novae or whether it is a very young star that cannot evolve in the normal way because of the very thick cloud surrounding it.

Stars of the *R Coronae Borealis* type have "inverse nova" light curves. Their brightness may drop by almost ten magnitudes and stay low for years, before the star brightens to its normal luminosity. For example, R CrB itself is of magnitude 5.8, but may fade to magnitude 14.8. Figure 14.14 shows its most recent decline, based on observations by Finnish and French amateurs. The R CrB stars are rich in carbon and the decline is produced when the carbon condenses into a circumstellar dust shell.

Fig. 14.13. In the 19th century, η Carinae was one of the brightest stars in the sky; since then it has dimmed considerably. In an outburst in 1843 the star ejected an expanding nebula, which has been called "Homunculus". (Photograph NASA/HST)

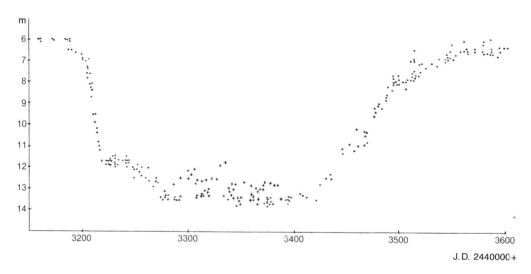

Fig. 14.14. The decline of R Coronae Borealis in 1977–1978; observations by Finnish and French amateur astronomers. (Kellomäki, Tähdet ja Avaruus 5/1978)

J.D. 2440000+

Supernovae. The stars with the largest brightness variations of all are the *supernovae*. In a few days, their brightness may rise more than 20 magnitudes, i.e. the luminosity may increase by a factor of a hundred million. After the maximum, there is a slow decline lasting several years.

The supernovae are exploding stars. In the explosion, a gas shell expanding with a velocity around 10 000 km/s is ejected. The expanding gas shell remains visible for

Fig. 14.15. The light curves of the two types of supernova. Above, SN 1972e, type I, and below, SN 1970g, type II. [Kirshner, R.P. (1976): Sci. Am. **235**, No. 6, 88]

thousands of years. A few tens of such supernova remnants have been discovered in the Milky Way. The remnant of the actual star may be a neutron star or a black hole.

The supernovae are classified as type I and II on the basis of their light curves (Fig. 14.15). *Type I* supernovae fade away in a regular manner, almost exponentially. The decline of a *type II* supernova is less regular and its maximum luminosity is smaller. The reason for this difference is that the exploding stars are of quite different character. Type I supernovae are produced by old, low-mass stars; those of type II, by young, massive stars.

In Chap. 12 it was mentioned that there are several possible ways in which a star may come to explode at the end of its evolution. The type II supernovae are the natural endpoint of the evolution of single stars. The type I supernovae, on the other hand, have about the mass of the Sun and should end up as white dwarfs. However, if a star is accreting mass from a binary companion, it will undergo repeated nova outbursts. Some of the accreted material will then be turned into helium or carbon and oxygen, and will collect on the star and increase its mass. Finally the mass may exceed the Chandrasekhar limit. The star will then collapse and explode as a supernova.

At least six supernova explosions have been observed in the Milky Way. Best known are the "guest star" seen in China in 1054 (whose remnant is the Crab Nebula), Tycho Brahe's supernova in 1572 and Kepler's supernova in 1604. On the basis of observations of other Sb − Sc-type spiral galaxies, the interval between supernova explosions in the Milky Way is predicted to be about 50 years. Some will be hidden by obscuring material, but the 380 years' interval since the last observed supernova explosion is unusually long.

On February 23, 1987, the first burst of light from a supernova in the Large Magellanic Cloud, the small companion galaxy of the Milky Way, reached the Earth (Fig. 14.16). This supernova, SN 1987 A, was of type II, and was the brightest supernova for 383 years. After its first detection, SN 1987 A was studied in great detail by all available means. Although the general ideas of Sects. 12.4 and 12.5 on the final stages of stellar evolution have been confirmed, there are complications. Thus e.g. the progenitor star was a blue rather than a red giant as expected, perhaps because of the lower abundance of heavy elements in the Large Magellanic Cloud compared to that in the Sun. The collapse of its core released vast amount of energy as a pulse of

Fig. 14.16. Supernova 1987 A in the Large Magellanic Cloud before and after the explosion. (Photographs ESO)

neutrinos, which was detected in Japan and the USA. The amount of energy released indicates that the remnant is a neutron star. As of February 1996, the remnant has however not been seen yet.

14.4 Examples

14.4.1 The observed period of a cepheid is 20 days and its mean apparent magnitude $m = 20$. From Fig. 14.6, its absolute magnitude is approximately $M = -5$. According to (4.12), the distance of the cepheid is

$$r = 10 \times 10^{(m-M)/5} = 10 \times 10^{(20+5)/5} = 10^6 \,\mathrm{pc} = 1 \,\mathrm{Mpc} \;.$$

14.4.2 The brightness of a cepheid varies 2 mag. If the effective temperature is 6000 K at the maximum and 5000 K at the minimum, how much does the radius change?

The luminosity varies between

$$L_{\mathrm{max}} = 4\pi R_{\mathrm{max}}^2 \sigma T_{\mathrm{max}}^4 \quad \text{and} \quad L_{\mathrm{min}} = 4\pi R_{\mathrm{min}}^2 \sigma T_{\mathrm{min}}^4 \;.$$

In magnitudes the difference is

$$\Delta m = -2.5 \lg \frac{L_{min}}{L_{max}} = -5 \lg \frac{R_{min}}{R_{max}} - 10 \lg \frac{T_{min}}{T_{max}} \ .$$

This gives

$$\lg \frac{R_{min}}{R_{max}} = -0.2 \Delta m - 2 \lg \frac{T_{min}}{T_{max}} = -0.4 - 2 \lg \frac{5000}{6000} = -0.24 \ ,$$

whence

$$\frac{R_{min}}{R_{max}} = 0.57 \ .$$

14.5 Exercises

14.5.1 The absolute visual magnitude of RR Lyrae variables is 0.6 ± 0.3. What is the relative error of distances due to the deviation in the magnitude?

14.5.2 The bolometric magnitude of a long period variable varies by one magnitude. The effective temperature at the maximum is 4500 K.

a) What is the temperature at the minimum, if the variation is due to temperature change only?
b) If the temperature remains constant, what is the relative variation in the radius?

14.5.3 In 1983 the radius of the Crab nebula was about $3'$. It is expanding $0.21''$ a year. Radial velocities of 1300 km s^{-1} with respect to the central star have been observed in the nebula.

a) What is the distance of the nebula, assuming its expansion is symmetric?
b) A supernova explosion has been observed in the direction of the nebula. Estimate, how long time ago?
c) What was the apparent magnitude of the supernova, if the absolute magnitude was typical -18?

Chapter 15 Compact Stars

In astrophysics, those stars in which the density of matter is much larger than in ordinary stars are known as compact objects. These include white dwarfs and neutron stars, and also black holes. In addition to a very high density, the compact objects are characterized by the fact that nuclear reactions have completely ceased in their interiors. Consequently they cannot support themselves against gravity by thermal gas pressure. In the white dwarfs and neutron stars, gravity is resisted by the pressure of a degenerate gas. In the black holes, the force of gravity is completely dominant and compresses the stellar material to infinite density.

15.1 White Dwarfs

The first white dwarf to be discovered was Sirius B, the companion of Sirius (Fig. 15.1). Its exceptional nature was realized in 1915, when it was discovered that its effective temperature was very high. Since it is faint, this meant that its radius had to be very small, slightly smaller than that of the Earth. The mass of Sirius B was known to be about equal to that of the Sun, so its density had to be extremely large.

Fig. 15.1. One of the best-known white dwarfs is Sirius B, the small companion of Sirius. In this picture taken with a 3-m telescope, it can be seen as a faint spot to the left of the primary star. (Photograph Lick Observatory)

The high density of Sirius B was confirmed in 1925, when the gravitational redshift of its spectral lines was measured. This measurement also provided early observational support to Einstein's general theory of relativity.

As explained on p. 336, the radius of a degenerate star is inversely proportional to the cube root of the mass. Unlike in a normal star, the radius decreases as the mass increases.

White dwarfs occur both as single stars and in binary systems. Their spectral lines are broadened by the strong gravitational field at the surface. In some white dwarfs, the spectral lines are further broadened by rapid rotation. Strong magnetic fields have also been observed.

Since a white dwarf has no internal sources of energy, it will slowly cool, changing in colour from white to red and finally to black. There ought to be a large number of invisible black dwarfs in the Milky Way.

Many of the important properties of white dwarfs have already been discussed in Chaps. 9–12 and 14 on ordinary and variable stars.

15.2 Neutron Stars

If the mass of a star is large enough, the density of matter may grow even larger than in normal white dwarfs. The equation of state of a classical degenerate electron gas then has to be replaced with the corresponding relativistic formula. In this case, decreasing the radius of the star no longer helps in resisting the gravitational attraction. Equilibrium is possible only for one particular value of the mass, the Chandrasekhar mass M_{Ch}, already introduced in Sect. 12.5. The value of M_{Ch} is about $1.4 M_{\odot}$, which is thus the upper limit to the mass of a white dwarf. If the mass of the star is larger than M_{Ch}, gravity overwhelms the pressure and the star will rapidly contract towards higher densities. The final stable state reached after this collapse will be a *neutron star* (Fig. 15.2). On the other hand, if the mass is smaller than M_{Ch}, the pressure dominates. The star will then expand until the density is small enough to allow an equilibrium state with a less relativistic equation of state.

When a massive star reaches the end of its evolution and explodes as a supernova, the simultaneous collapse of its core will not necessarily stop at the density of a white dwarf. If the mass of the collapsing core is larger than the Chandrasekhar mass, the collapse continues to a neutron star.

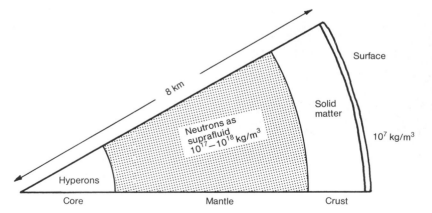

Fig. 15.2. The structure of a neutron star. The crust is rigid solid material, and the mantle a freely streaming superfluid

An important particle reaction during the final stages of stellar evolution is the *URCA process*, which was put forward by Schönberg and Gamow in the 1940's and which produces a large neutrino emission without otherwise affecting the composition of matter. (The URCA process was invented in Rio de Janeiro and named after the local casino. Apparently, money disappeared at URCA just as energy disappeared from inside stars in the form of neutrinos. It is claimed that the casino was closed by the authorities when this similarity became known.) The URCA process consists of the reactions

$$(Z, A) + e^- \rightarrow (Z-1, A) + \nu_e$$
$$(Z-1, A) \rightarrow (Z, A) + e^- + \bar{\nu}_e,$$

where Z is the number of protons in a nucleus; A, the mass number; e^-, an electron; and ν_e and $\bar{\nu}_e$, the electron neutrino and antineutrino. When the electron gas is degenerate, the latter reaction is suppressed by the Pauli exclusion principle. In consequence, the protons in the nuclei are transformed into neutrons. As the number of neutrons in the nuclei grows, their binding energies decrease. At densities of about 4×10^{14} kg/m^3, the neutrons begin to leak out of the nucleus, and at 10^{17} kg/m^3, the nuclei disappear altogether. Matter then consists of a neutron "porridge", mixed with about 0.5% electrons and protons.

Neutron stars are supported against gravity by the pressure of the degenerate neutron gas, just as white dwarfs are supported by electron pressure. The equation of state is the same, except that the electron mass is replaced by the neutron mass, and that the mean molecular weight is defined with respect to the number of free neutrons. Since the gas consists almost entirely of neutrons, the mean molecular weight is approximately one.

The typical diameters of neutron stars are about 10 km. Unlike ordinary stars, they have a well-defined solid surface. The atmosphere above it is a few centimetres thick. The upper crust is a metallic solid with a density growing rapidly inwards. Most of the star is a neutron superfluid, and in the centre, where the density exceeds 10^{18} kg/m^3, there may be a solid nucleus of heavier particles (hyperons).

The theory of neutron stars was developed in the 1930's, but the first observations were not made until the 1960's. At that time, a new type of rapidly pulsating radio source, the *pulsars*, was discovered and identified as neutron stars. In the 1970's, neutron stars were also seen as *x-ray pulsars* and *x-ray bursters* (Fig. 15.3).

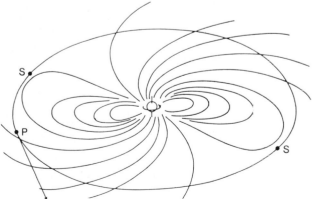

Fig. 15.3. The magnetic field around a rotating neutron star carries plasma with it. At a certain distance the speed of the plasma approaches the speed of light. At this distance, the radiating regions S emit radiation in a narrow forward beam. Radiation from the point P hits the observer located in the direction of the arrow. [Drawing from Smith, F. G. (1977): *Pulsars* (Cambridge University Press, Cambridge) p. 189]

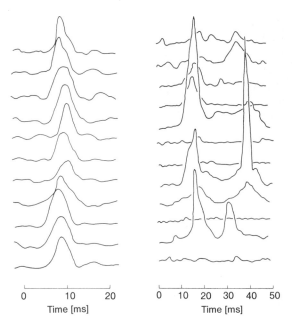

Fig. 15.4. Consecutive radio pulses at 408 MHz from two pulsars. To the left PSR 1642-03 and to the right PSR 1133+16. Observations made at Jodrell Bank. [Picture from Smith, F. G. (1977): *Pulsars* (Cambridge University Press, Cambridge) pp. 93, 95]

Pulsars. The pulsars were discovered in 1967, when *Anthony Hewish* and *Jocelyn Bell* in Cambridge, England, detected sharp, regular radio pulses coming from the sky. Since then, about four hundred pulsars have been discovered (Fig. 15.4). Their periods range from 0.0016 s (for the pulsar 1937+214) to several seconds.

The best-known pulsar is located in the Crab Nebula (Fig. 15.5). This small nebula in the constellation Taurus was noted by the French astronomer Charles Messier in the middle of the 18th century and became the first object in the Messier catalogue, M1. The Crab Nebula was found to be a strong radio source in 1948, and an x-ray source in

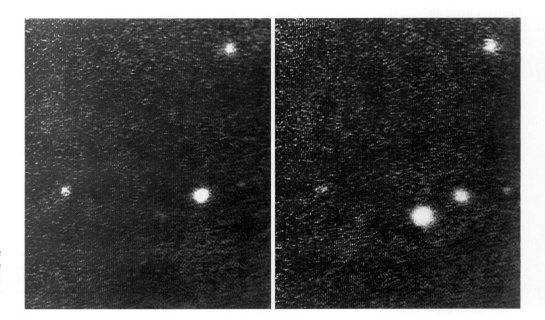

Fig. 15.5a. The pulsation of the Crab pulsar in visible light. The pulsar is visible in the right-hand picture. (Photograph Lick Observatory)

Fig. 15.5 b. Some pulsars shine brightly in gamma-rays. In the center the Crab pulsar, and on the upper left the gamma source Geminga, which was identified in 1992 to be the nearest pulsar with a distance of about 100 pc from the Sun. (Photo by Compton Gamma Ray Observatory)

1964. The pulsar was discovered in 1968. In the following year, it was observed optically, and was also found to be an x-ray emitter.

Neutron stars are difficult to study optically, since their luminosity in the visible region is very small (typically about $10^{-6} L_\odot$). For instance the Vela pulsar has been observed at a visual magnitude of about 25. This makes it one of the faintest objects ever observed. In the radio region, it is a very strong pulsating source.

A neutron star formed in the explosion and collapse of a supernova will initially rotate rapidly, because its angular momentum is unchanged while its radius is much smaller than before. In a few hours, the star will settle in a flattened equilibrium, rotating several hundred times per second.

The initial magnetic field of the neutron star will also be compressed in the collapse, so that there will be a strong field coupling the star to the surrounding material. The origin of the radio pulses can be understood if the magnetic field is tilted at an angle of $45° - 90°$ with respect to the rotation axis. There will then be a magnetosphere around the star, where the particles are tied to the magnetic field and rotate with it (Fig. 15.3). At a certain distance from the star, the speed of rotation approaches the speed of light. The rapidly moving particles will there emit radiation in a narrow cone in their direction of motion. As the star rotates, this cone sweeps around like a lighthouse beam, and is seen as rapid pulses. At the same time, relativistic particles stream out from the neutron star.

The angular momentum of the neutron star is steadily decreased by the emission of electromagnetic radiation, neutrinos, cosmic ray particles and possibly gravitational radiation. In consequence, the period of the pulsars will steadily increase. In addition, sometimes small sudden jumps in the period are observed. These might be a sign of rapid mass movements in the neutron star crust ("star quakes") or in its surroundings.

A few pulsars have been discovered in binary systems; the first one, PSR 1913 + 16, in 1974. The pulsar orbits about a companion, presumably another neutron star, with the orbital eccentricity 0.6 and the period 8 h. The observed period of the pulses is altered by the Doppler effect, and this allows one to determine the velocity curve of the pulsar. These observations can be made very accurately, and it has therefore been possible to follow the changes in the orbital elements of the system over a period of several years. For example, the periastron direction of the binary pulsar has been found to rotate about 4° per year. This phenomenon can be explained by means of the general theory of relativity; in the solar system, the corresponding rotation of the perihelion of Mercury is 43 arc sec per century.

The binary pulsar PSR 1913 + 16 has also provided the first strong evidence for the existence of gravitational waves. During the time of observation, the orbital period of the system has steadily decreased. This shows that the system is losing orbital energy at a rate that agrees exactly with that predicted by the general theory of relativity. The lost energy is radiated as gravitational waves.

X-ray Pulsars. Many kinds of variable x-ray sources were discovered in the 1970's. Among these, the x-ray pulsars and the x-ray bursters are related to neutron stars. In contrast to radio pulsars, the period of the pulsed emission of x-ray pulsars decreases with time. Since the origin of the x-ray pulses is thought to be similar to that of radio pulses, this means that the rotation rate of the neutron stars must increase. The pulse periods of x-ray pulsars are significantly longer than those of radio pulsars, from a few seconds to tens of minutes.

X-ray pulsars always belong to binary systems. Typically, the companion is an OB supergiant. The characteristic properties of x-ray pulsars can be understood from their binary nature. A neutron star formed in a binary system is first seen as a normal radio pulsar. Initially, the strong radiation of the pulsar prevents gas from falling onto it. However, as it slows down, its energy decreases, and eventually the stellar wind from the companion can reach its surface. The incoming gas is channelled to the magnetic polar caps of the neutron star, where it emits strong x-ray radiation as it hits the surface. This produces the observed pulsed emission. At the same time, the angular momentum of the incoming gas speeds up the rotation of the pulsar.

The emission curve of a typical fast x-ray pulsar, Hercules X-1, is shown in Fig. 15.6. The period of the pulses is 1.24 s. This neutron star is part of an eclipsing binary system, known from optical observations as HZ Herculis. The orbital properties of the system can therefore be determined. Thus e.g. the mass of the pulsar is about one solar mass, reasonable for a neutron star.

X-ray Bursters. X-ray bursters are irregular variables, showing random outbursts like dwarf novae (Fig. 15.7). The typical interval between outbursts is a few hours or days, but more rapid bursters are also kown. The strength of the outburst seems to be related to the recharging time.

The source of radiation in x-ray bursters cannot be the ignition of hydrogen, since the maximum emission is in the x-ray region. One explanation is that the bursters are binary systems like the dwarf novae, but containing a neutron star rather than a white dwarf. Gas from the companion settles on the surface of the neutron star, where hydrogen burns to helium. When the growing shell of helium reaches a critical temperature, it burns to carbon in a rapid helium flash. Since, in this case, there are no thick damping outer layers, the flash appears as a burst of x-ray radiation.

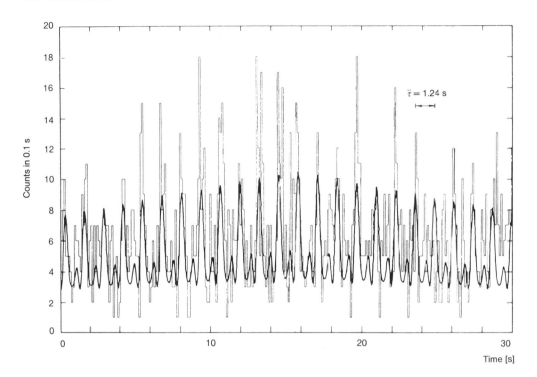

Fig. 15.6. The pulses of the x-ray pulsar Hercules X-1 have the period 1.24 s. The best-fitting curve has been superimposed on the observations. [Tananbaum, H. et al. (1972): Astrophys. J. (Lett.) **174**, L143]

It used to be believed that also the *gamma ray bursts*, very short and sharp gamma ray pulses, come from neutron stars. Satellite observations with the Compton Gamma Ray Observatory have now shown that the gamma ray bursts are almost uniformly distributed in the sky, unlike the known neutron stars. The question of their origin is therefore completely open.

Fig. 15.7. The variations of the rapid x-ray burster MXB 1730-335. An 100s interval is marked in the diagram. [Lewin, W. H. G. (1977): Ann N. Y. Acad. Sci **302**, 310]

* The Radius of White Dwarfs and Neutron Stars

The mass of a white dwarf or a neutron star determines its radius. This follows from the equation of hydrostatic equilibrium and from the pressure-density relation for a degenerate gas. Using the hydrostatic equilibrium equation (11.1)

$$\frac{dP}{dr} = - \frac{GM_r\rho}{r^2},$$

one can estimate the average pressure P:

$$\frac{dP}{dR} \approx \frac{P}{R} \propto \frac{M \cdot M/R^3}{R^2} = \frac{M^2}{R^5}.$$

Here we have used $\rho \propto M/R^3$. Thus the pressure obeys

$$P \propto \frac{M^2}{R^4}. \tag{1}$$

In the nonrelativistic case, the pressure of a degenerate electron gas is given by (11.12):

$$P \approx (h^2/m_e)(\mu_e m_H)^{-5/3} \rho^{5/3}, \quad \text{and hence}$$

$$P \propto \frac{\rho^{5/3}}{m_e \mu_e^{5/3}}. \tag{2}$$

By combining (1) and (2) we obtain

$$\frac{M^2}{R^4} \propto \frac{M^{5/3}}{R^5 m_e \mu_e^{5/3}}; \quad \text{i.e.} \quad R \propto \frac{1}{M^{1/3} m_e \mu_e^{5/3}} \propto M^{-1/3}.$$

Thus the smaller the radius of a white dwarf is, the larger its mass will be. If the density becomes so large that the relativistic equation of state (11.13) has to be used, the expression for the pressure is

$$P \propto \rho^{4/3} \propto \frac{M^{4/3}}{R^4}.$$

As the star contracts, the pressure grows at the same rate as demanded by the condition for hydrostatic support (1). Once contraction has begun, it can only stop when the state of matter changes: the electrons and protons combine into neutrons. Only a star that is massive enough can give rise to a relativistic degenerate pressure.

The neutrons are fermions, just like the electrons. They obey the Pauli exclusion principle, and the degenerate neutron gas pressure is obtained from an expression analogous to (2):

$$P_n \propto \frac{\rho^{5/3}}{m_n \mu_n^{5/3}},$$

where m_n is the neutron mass and μ_n, the molecular weight per free neutron. Correspondingly, the radius of a neutron star is given by

$$R_{ns} \propto \frac{1}{M^{1/3} m_n \mu_n^{5/3}}.$$

If a white dwarf consists purely of helium, $\mu_e = 2$; for a neutron star, $\mu_n \approx 1$. If a white dwarf and a neutron star have the same mass, the ratio of their radii is

$$\frac{R_{wd}}{R_{ns}} = \left(\frac{M_{ns}}{M_{wd}}\right)^{1/3} \left(\frac{\mu_n}{\mu_e}\right)^{5/3} \frac{m_n}{m_e} \approx 1 \times \left(\frac{1}{2}\right)^{5/3} \times 1840 \approx 600.$$

Thus the radius of a neutron star is about 1/1000 of that of a white dwarf. Typically, R_{ns} is about 10 km.

15.3 Black Holes

It was seen in Sect. 12.5 that there is an upper limit to the mass of a neutron star, the Oppenheimer-Volkov mass, above which not even the degenerate neutron gas pressure can resist the gravitational force. The precise value of this upper mass limit is difficult to calculate theoretically, since it depends on imperfectly known particle interaction probabilities, but it is estimated to be in the range $1.5 - 2\,M_\odot$.

If the mass of a star exceeds M_{OV}, and if it does not lose mass during its evolution, it can no longer reach any stable final state. The force of gravity will dominate over all other forces, and the star will collapse to a black hole.

A black hole is black because not even light can escape from it. Already at the end of the 18th century, Laplace showed that a sufficiently massive body would prevent the escape of light from its surface. According to classical mechanics, the escape velocity from a body of radius R and mass M is

$$v_e = \sqrt{\frac{2\,GM}{R}}\,.$$

This is greater than the speed of light, if the radius is smaller than the critical radius

$$R_S = 2\,GM/c^2\,.$$

The same value for the critical radius, the *Schwarzschild radius*, is obtained from the general theory of relativity. For example, for the Sun, R_S is about 3 km; however, the Sun's mass is so small that it cannot become a black hole by normal stellar evolution. Because the mass of a black hole formed by stellar collapse has to be larger than M_{OV}, the radius of the smallest black holes formed in this way is about $5 - 10$ km.

The properties of black holes have to be studied on the basis of the general theory of relativity. In the theory of relativity, each observer carries with him his own local measure of time. If two observers are at rest with respect to each other at the same point, their clocks go at the same rate. Otherwise, their clock rates are different, and the apparent course of events differs, too.

An *event horizon* is a surface through which no information can be sent out, even in principle. A black hole is surrounded by an event horizon at the Schwarzschild radius (Fig. 15.8). Near the event horizon, the different time definitions become significant. An observer falling into a black hole reaches the centre in a finite time, according to his own clock, and does not notice anything special as he passes through the event horizon. However, to a distant observer, he never seems to reach the event horizon; his velocity of fall seems to decrease towards zero as he approaches the horizon.

The slowing down of time also appears as a decrease in the frequency of light signals. The formula for the gravitational redshift can be written in terms of the Schwarzschild radius as

$$v_\infty = v\{1 - 2\,GM/(rc^2)\}^{1/2} = v(1 - R_S/r)^{1/2}\,.$$

Here, v is the frequency of radiation emitted at a distance r from the black hole and v_∞, the frequency observed by an infinitely distant observer. It can be seen that the frequency at infinity approaches zero for radiation emitted near the event horizon.

Fig. 15.8. A black hole is surrounded by a spherical event
horizon. In addition to this, a
rotating black hole is surrounded
by a flattened surface inside
which no matter can remain stationary. This region is called the
ergosphere

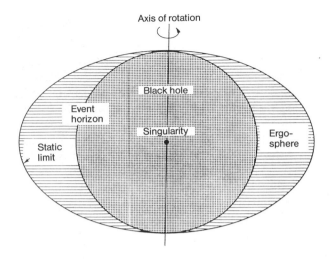

The tidal forces become extremely large near a black hole so that any material falling into the hole will be torn apart. All atoms and elementary particles are destroyed
near the central point, and the final state of matter is unknown to present-day physics.
The observable properties of a black hole do not depend on how it was made.

Not only all information on the material composition disappears as a star collapses
into a black hole; any magnetic field, for example, also disappears behind the event
horizon. A black hole can only have three observable properties: mass, angular momentum and electric charge.

At present, the only known way in which a black hole could be directly observed is
by means of the radiation from gas falling into it. For example, if a black hole is part of
a binary system, gas streaming from the companion will settle into a disc around the
hole. Matter at the inner edge of the disc will fall into the hole. The accreting gas will
lose a considerable part of its energy (up to 40% of the rest mass) as radiation, which
should be observable in the x-ray region.

Fig. 15.9. No periodicity is evident in the x-ray radiation of the
black hole candidate Cygnus X-1,
and thus it is probably not a
neutron star. [Schreier, E. et al.
(1971): Astrophys. J. (Lett.) 170,
L24]

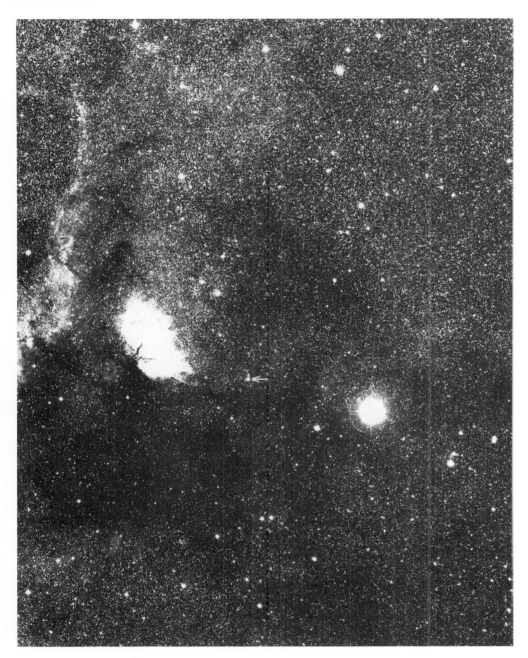

Fig. 15.10. The arrow shows the variable star V1357 Cyg. Its companion is the suspected black hole Cygnus X-1. The bright star to the lower right of V1357 Cyg is η Cygni, one of the brightest stars in the constellation Cygnus

Some rapidly and irregularly varying x-ray sources of the right kind have been discovered in the sky. The most promising black hole candidate is probably Cygnus X-1 (Fig. 15.9). Its luminosity varies on the timescale of 0.001 s, which means that the emitting region must be only 0.001 light-seconds or a few hundred kilometres in size. Only neutron stars and black holes are small and dense enough to give rise to such high-energy processes. Cygnus X-1 is the smaller component of the double system HDE 226868 (Fig. 15.10). The larger component is an optically visible supergiant with a mass $20 - 25\,M_\odot$. The mass of the unseen component has been calculated to be $10 - 15\,M_\odot$. If this is correct, the mass of the secondary component is much larger than the upper limit for a neutron star, and thus it has to be a black hole.

So far we have discussed only black holes with masses in the range of stellar masses. There is however no upper limit to the mass of a black hole. Many active phenomena in the nuclei of galaxies can be explained with black holes with masses of millions or thousands of millions solar masses (see Sect. 18.4 and 19.9).

Many frightening stories about black holes have been invented. It should therefore be stressed that they obey the same dynamical laws as other stars − they are not lurking in the darkness of space to attack innocent passers-by. If the Sun became a black hole, the planets would continue in their orbits as if nothing had happened.

15.4 Examples

15.4.1   Assume that the Sun collapses into a neutron star with a radius of 20 km.

a) What will be the mean density of the neutron star?
b) What will be its rotation period?

a) The mean density is

$$\rho = \frac{M_\odot}{(4/3)\,\pi R^3} = \frac{2 \times 10^{30}\,\text{kg}}{(4/3)\,\pi\,(20 \times 10^3)^3\,\text{m}^3} \approx 6 \times 10^{16}\,\text{kg m}^{-3} \ .$$

One cubic millimetre of this substance would weigh 60 million kilos.

b) To obtain an exact value, we should take into account the mass distributions of the Sun and the resulting neutron star. Very rough estimates can be found assuming that both are homogeneous. Then the moment of inertia is $I = (2/5)MR^2$, and the angular momentum is $L = I\omega$. The conservation of the angular momentum gives

$$\frac{2}{5}M_\odot R_\odot^2 \frac{2\pi}{P_\odot} = \frac{2}{5}M_\odot R^2 \frac{2\pi}{P} \ ,$$

whence

$$P = P_\odot \left(\frac{R}{R_\odot}\right)^2 = 25\,\text{d} \left(\frac{20 \times 10^3\,\text{m}}{6.96 \times 10^8\,\text{m}}\right)^2 = 2.064 \times 10^{-8}\,\text{d} \approx 0.0018\,\text{s} \ .$$

The Sun would make over 550 revolutions per second.

15.4.2 What should be the radius of the Sun if the escape velocity from the surface were to exceed the speed of light?

The escape velocity exceeds the speed of light if

$$\sqrt{\frac{2GM}{R}} > c \ ,$$

or

$$R < \frac{2GM}{c^2} = R_S \ .$$

For the Sun we have

$$R_S = \frac{2 \times 6.67 \times 10^{-11} \, \text{m}^3 \, \text{s}^{-2} \, \text{kg}^{-1} \times 1.989 \times 10^{30} \, \text{kg}}{(2.998 \times 10^8 \, \text{m s}^{-1})^2} = 2950 \, \text{m} \ .$$

15.5 Exercises

15.5.1 The mass of a pulsar is $1.5 \, M_\odot$, radius 10 km, and rotation period 0.033 s. What is the angular momentum of the pulsar? Variations of 0.0003 s are observed in the period. If they are due to radial oscillations ("starquakes"), how large are these oscillations?

15.5.2 In *Dragon's Egg* by Robert L. Forward a spaceship orbits a neutron star at a distance of 406 km from the centre of the star. The orbital period is the same as the rotation period of the star, 0.1993 s.

a) Find the mass of the star and the gravitational acceleration felt by the spaceship.
b) What is the effect of the gravitation on a 175 cm tall astronaut, if (s)he stands with her/his feet pointing towards the star? And if (s)he is lying tangential to the orbit?

15.5.3 A photon leaves the surface of star at a frequency v_e. An infinitely distant observer finds that its frequency is v. If the difference is due to gravitation only, the change in the enery of the photon, $h \Delta v$, equals the change in its potential energy. Find the relation between v and v_e, assuming the mass and radius of the star are M and R. How much will the solar radiation redshift on its way to the Earth?

Chapter 16   The Interstellar Medium

Although most of the mass of the Milky Way Galaxy is condensed into stars, interstellar space is not completely empty. It contains *gas* and *dust*, in the form both of individual clouds and of a diffuse medium. Interstellar space typically contains about one gas atom per cubic centimetre and 100 dust particles per cubic kilometre.

Altogether, about 10% of the mass of the Milky Way consists of interstellar gas. Since the gas is strongly concentrated in the galactic plane and the spiral arms, in these regions there are many places where the quantities of stars and interstellar matter are about equal. The dust (a better name would be "smoke", since the particle sizes are much smaller than in terrestrial dust) constitutes about one percent of the gas. High-energy cosmic ray particles are mixed with the gas and dust. There is also a weak, but still very important, galactic magnetic field.

At present, the most important observations of the interstellar medium are made at radio and infrared wavelengths, since the peak of the emission often lies at these wavelengths. But many forms of interstellar matter (such as solid bodies with diameters larger than 1 mm) would be almost impossible to detect on the basis of their emission or absorption. In principle, the mass of these forms of matter might be larger than the observed mass of all other forms put together. However, an upper limit on the total mass of interstellar matter, regardless of its form, can be derived on the basis of its gravitational effects. This is the so-called *Oort limit*. The galactic gravitational field is determined by the distribution of matter. By observing the motions of stars perpendicular to the galactic plane, the vertical gravitational force and hence the amount of mass in the galactic plane can be determined. The result is that the local density within 1 kpc of the Sun is $(7.3 - 10.0) \times 10^{-21}$ kg m^{-3}. The density of known stars is $(5.9 - 6.7) \times 10^{-21}$ kg m^{-3} and that of known interstellar matter about 1.7×10^{-21} kg m^{-3}. Thus there is very little room for unknown forms of mass in the solar neighbourhood.

16.1 Interstellar Dust

The first clear evidence for the existence of interstellar dust was obtained around 1930. Before that, it had been generally thought that space is completely transparent and that light can propagate indefinitely without extinction.

In 1930, *Robert Trumpler* published his study of the space distribution of the open clusters. The absolute magnitudes of the brightest stars could be estimated on the basis of the spectral type. Thus the distance r to the clusters could be calculated from the observed apparent magnitude m of the bright stars:

$$m - M = 5 \lg \frac{r}{10 \text{ pc}}. \tag{16.1}$$

Trumpler also studied the diameters of the clusters. The linear diameter D is obtained from the apparent angular diameter d by means of the formula

$$D = dr, \qquad (16.2)$$

where r is the distance of the cluster.

It caught Trumpler's attention that the more distant clusters appeared to be systematically larger than the nearer ones (Fig. 16.1). Since this could hardly be true, the distances of the more distant clusters must have been overestimated. Trumpler concluded that space is not completely transparent, but that the light of a star is dimmed by some intervening material. To take this into account, (16.1) has to be replaced with (4.18),

$$m - M = 5 \lg \frac{r}{10 \text{ pc}} + A , \qquad (16.3)$$

where $A > 0$ is the extinction in magnitudes due to the intervening medium. If the opacity of the medium is assumed to be the same at all distances and in all directions, A can be written

$$A = ar, \qquad (16.4)$$

where a is a constant. Trumpler obtained for the average value of a in the galactic plane, $a_{pg} = 0.79$ mag/kpc, in photographic magnitudes. At present, a value of 2 mag/kpc is used for the average extinction. Thus the extinction over a 5 kpc path is already 10 magnitudes.

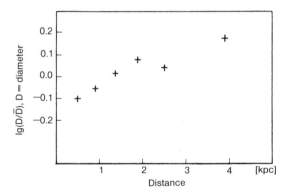

Fig. **16.1.** The diameters of open star clusters calculated with the distance given by formula (16.1) according to Trumpler (1930). The increase of the diameter with distance is not a real phenomenon, but an effect of interstellar extinction, which was discovered in this way

Extinction due to dust varies strongly with direction. For example, visible light from the galactic centre (distance 10 kpc) is dimmed by 30 magnitudes. Therefore the galactic centre cannot be observed at optical wavelengths.

Extinction is due to dust grains that have diameters near the wavelength of the light. Such particles scatter light extremely efficiently. Gas can also cause extinction by scattering, but its scattering efficiency per unit mass is much smaller. The total amount of gas allowed by the Oort limit is so small that scattering by gas is negligible in interstellar space. (This is in contrast with the Earth's atmosphere, where air molecules make a significant contribution to the total extinction).

Extinction by interstellar particles can be due to either absorption or scattering. In absorption, the radiant energy is transformed into heat, which is then re-radiated at in-

Fig. 16.2. Extinction by a distribution of particles. In the volume element with length dl and cross section dA, there are $n\,dA\,dl$ particles, where n is the particle density in the medium. If the extinction cross section of one particle is C_{ext}, the total area covered by the particles is $n\,dA\,dl\,C_{ext}$. Thus the fractional decrease in intensity over the distance dl is $dI/I = -n\,dA\,dl\,C_{ext}/dA = -n\,C_{ext}\,dl$

frared wavelengths corresponding to the temperature of the dust particles. In scattering, the direction of light propagation is changed, leading to a reduced intensity in the original direction of propagation.

An expression for interstellar extinction will now be derived. The size, index of refraction and number density of the particles are assumed to be known. For simplicity we shall assume that all particles are spheres with the same radius a and the geometrical cross section πa^2. The true extinction cross section of the particles C_{ext} will be

$$C_{ext} = Q_{ext}\,\pi\,a^2 \tag{16.5}$$

where Q_{ext} is the extinction efficiency factor.

Let us consider a volume element with length dl and cross section dA, normal to the direction of propagation (Fig. 16.2). It is assumed that the particles inside the element do not shadow each other. If the particle density is n, there are $n\,dl\,dA$ particles in the volume element and they will cover the fraction $d\tau$ of the area dA, where

$$d\tau = \frac{n\,dA\,dl\,C_{ext}}{dA} = n\,C_{ext}\,dl\ .$$

In the length dl, the intensity is thus reduced by

$$dI = -I\,d\tau\ . \tag{16.6}$$

On the basis of (16.6), $d\tau$ can be identified as the optical depth.

The total optical depth between the star and the Earth is

$$\tau(r) = \int_0^r d\tau = \int_0^r n\,C_{ext}\,dl = C_{ext}\,\bar{n}r\ ,$$

where $\bar{n}$ is the average particle density along the given path. According to (4.18), the extinction in magnitudes is

$$A = (2.5\lg e)\,\tau\ , \quad \text{and hence}$$

$$A(r) = (2.5\lg e)\,C_{ext}\,\bar{n}r\ . \tag{16.7}$$

This formula can also be inverted to calculate n, if the other quantities are known.

The extinction efficiency factor Q_{ext} can be calculated exactly for spherical particles with given radius and refractive index m. In general,

$$Q_{ext} = Q_{abs} + Q_{sca}\ , \quad \text{where}$$

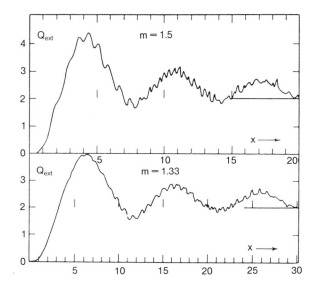

Fig. 16.3. Mie scattering: the extinction efficiency factor for spherical particles for the refractive indices $m = 1.5$ and $m = 1.33$ (refractive index of water). The horizontal axis is related to the size of the particle according to $x = 2\pi a/\lambda$, where a is the particle radius and λ, the wavelength of the radiation. [van de Hulst, H. C. (1957): *Light Scattering by Small Particles* (Wiley, New York) p. 177]

Q_{abs} = absorption efficiency factor, Q_{sca} = scattering efficiency factor.

If we define

$$x = 2\,\pi a/\lambda\,,\tag{16.8}$$

where λ is the wavelength of the radiation, then

$$Q_{ext} = Q_{ext}(x, m)\,.\tag{16.9}$$

The exact expression for Q_{ext} is a series expansion in x that converges more slowly for larger values of x. When $x \ll 1$, the process is called *Rayleigh scattering*; otherwise it is known as *Mie scattering*. Figure 16.3 shows Q_{ext} as a function of x for $m = 1.5$ and $m = 1.33$. For very large particles, $(x \gg 1)$ $Q_{ext} = 2$, as appears from Fig. 16.3. Purely geometrically, one would have expected $Q_{ext} = 1$; the two times larger scattering efficiency is due to the diffraction of light at the edges of the particle.

Other observable phenomena, apart from extinction, are also caused by interstellar dust. One of these is the *reddening* of the light of stars. (This should not be confused with the redshift of spectral lines.) Reddening is due to the fact that the amount of extinction becomes larger for shorter wavelengths. Going from red to ultraviolet, the extinction is roughly inversely proportional to wavelength. For this reason, the light of distant stars is redder than would be expected on the basis of their spectral class. The spectral class is defined on the basis of the relative strengths of the spectral lines which are not affected by extinction.

According to (4.21), the observed colour index B-V of a star is

$$\begin{aligned}
\text{B-V} &= M_B - M_V + A_B - A_V \\
&= (\text{B-V})_0 + E_{B\text{-}V}\,,
\end{aligned}\tag{16.10}$$

where $(\text{B-V})_0$ is the *intrinsic colour* of the star and $E_{B\text{-}V}$ the *colour excess*. As noted in Sect. 4.5 the ratio between the visual extinction A_V and the colour excess is approximately constant:

$$R = \frac{A_V}{E_{B-V}} = \frac{A_V}{A_B - A_V} \approx 3.0 . \tag{16.11}$$

R does not depend on either the properties of the star or the amount of extinction. This is particularly important in photometric distance determinations because of the fact that the colour excess E_{B-V} can be directly determined from the difference between the observed colour index B-V and the intrinsic colour (B-V)$_0$ known from the spectral class. One can then calculate the extinction

$$A_V \approx 3.0 \, E_{B-V} , \tag{16.12}$$

and finally, the distance. Since the interstellar medium is far from homogeneous, the colour excess method gives a much more reliable value than using some average value for the extinction in (4.19).

The wavelength dependence of the extinction, $A(\lambda)$, can be studied by comparing the magnitudes of stars of the same spectral class in different colours. These measurements have shown that $A(\lambda)$ approaches zero as λ becomes very large. In practice, $A(\lambda)$ can be measured up to a wavelength of about two micrometres. The extrapolation to zero inverse wavelength is then fairly reliable. Figure 16.4a shows $A(\lambda)$ as a function of inverse wavelength. It also illustrates how the quantities A_V and E_{B-V}, which are needed in order to calculate the value of R, are obtained from this *extinction* or *reddening curve*. Figure 16.4b shows the observed extinction curve. The points in the ultraviolet ($\lambda \leq 0.3$ μm) are based on rocket measurements.

It is clear from Fig. 16.4b that interstellar extinction is largest at short wavelengths in the ultraviolet, and decreases for longer wavelengths. In the infrared, it is only about ten percent of the optical extinction and in the radio region, it is vanishingly small. Objects that are invisible in the optical region can therefore be studied at infrared and radio wavelengths.

Another observed phenomenon caused by dust is the *polarization* of the light of the stars. Since spherical particles cannot produce any polarization, the interstellar dust

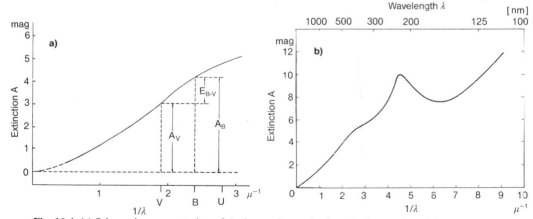

Fig. 16.4. (a) Schematic representation of the interstellar extinction. As the wavelength increases, the extinction approaches zero. [Greenberg, J. M. (1968): "Interstellar Grains", in *Nebulae and Interstellar Matter*, ed. by Middlehurst, B. M., Aller, L. H., Stars and Stellar Systems, Vol. VII (The University of Chicago Press, Chicago) p.224]. **(b)** Measured extinction curve, normalized to make $E_{B-V} = 1$. [Hoyle, F., Narlikar, J. (1980): *The Physics-Astronomy Frontier* (W. H. Freeman and Company, San Francisco) p. 156. Used by permission]

Fig. 16.5. In a homogeneous medium the extinction in magnitudes is proportional to the pathlength traversed. If the extinction in the direction of the galactic pole is Δm, then the extinction at the galactic latitude b will be $\Delta m/\sin b$

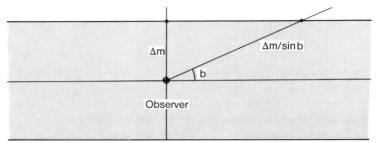

particles have to be nonspherical in shape. If the particles in a cloud are aligned by the interstellar magnetic field, they will polarize the radiation passing through the cloud. The degree of polarization and its wavelength dependence give information on the properties of the dust particles. By studying the direction of polarization in various directions, one can map the structure of the galactic magnetic field.

In the Milky Way, interstellar dust is essentially confined to a very thin, about 100 pc, layer in the galactic plane. The dust in other spiral galaxies has a similar distribution and is directly visible as a dark band in the disc of the galaxy (Fig. 19.10b).

The Sun is located near the central plane of the galactic dust layer, and thus the extinction in the direction of the galactic plane is very large, whereas the total extinction towards the galactic poles may be less than 0.1 magnitudes. This is apparent in the distribution of galaxies in the sky: at high galactic latitudes, there are many galaxies, while near the galactic plane, there is a $20°$ zone where hardly any galaxies are seen. This empty region is called the *zone of avoidance*.

If a homogeneous dust layer gives rise to a total extinction of Δm magnitudes in the vertical direction, then according to Fig. 16.5, the total extinction at galactic latitude b will be

$$\Delta m(b) = \Delta m/\sin b \,. \tag{16.13}$$

If the galaxies are uniformly distributed in space, then in the absence of extinction, the number of galaxies per square degree brighter than the magnitude m would be

$$\lg N_0(m) = 0.6\,m + C \,. \tag{16.14}$$

where C is a constant (see Exercise 18.5.1). However, due to extinction, a galaxy that would otherwise have the apparent magnitude m_0 will have the magnitude

$$m(b) = m_0 + \Delta m(b) = m_0 + \Delta m/\sin b \,, \tag{16.15}$$

where b is the galactic latitude. Thus the observable number of galaxies at latitude b will be

$$\lg N(m, b) = \lg N_0\,[m - \Delta m(b)] = 0.6\,[m - \Delta m(b)] + C$$
$$= \lg N_0(m) - 0.6\,\Delta m(b) \,; \quad \text{i.e.}$$

$$\lg N(m, b) = C' - 0.6\,\Delta m/\sin b \,, \tag{16.16}$$

where $C' = \lg N_0(m)$ does not depend on the galactic latitude. By making galaxy counts at various latitudes b, the extinction Δm can be determined. The value obtained from galaxy counts made at Lick Observatory is $\Delta m_{\mathrm{pg}} = 0.51$ mag.

The total vertical extinction of the Milky Way has also been determined from the colour excesses of stars. These investigations have yielded much smaller extinction values, about 0.1 mag. In the direction of the north pole, extinction is only 0.03 mag. The disagreement between the two extinction values is probably largely due to the fact that the dust layer is not really homogeneous. If the Sun is located in a local region of low dust content, the view towards the galactic poles might be almost unobstructed by dust.

Dark Nebulae. Observations of other galaxies show that the dust is concentrated in the spiral arms, in particular, at their inner edge. In addition, dust is concentrated in individual clouds, which appear as star-poor regions or *dark nebulae* against the background of the Milky Way. Examples of dark nebulae are the Coalsack in the southern sky (Fig. 16.6) and the Horsehead Nebula in Orion. Sometimes the dark nebulae form extended winding bands, and sometimes small, almost spherical, objects. Objects of the latter type are most easy to see against a bright background, e.g. a gas nebula (see Fig. 16.19). These objectes have been named *globules* by *Bart J. Bok*, who put forward the hypothesis that they are clouds that are just beginning to contract into stars.

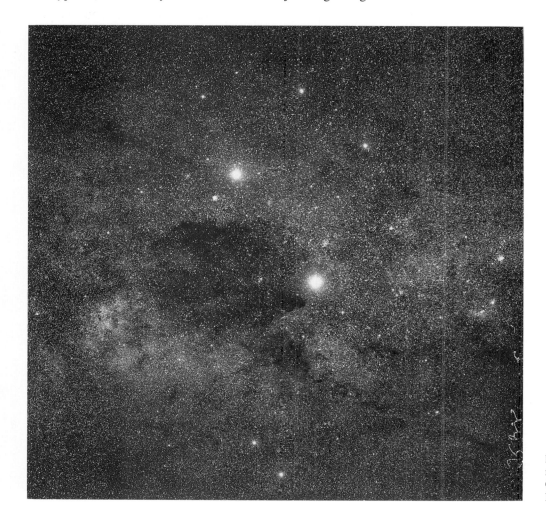

Fig. 16.6. The Coalsack is a dark nebula next to the Southern Cross. (Photograph K. Mattila, Helsinki University)

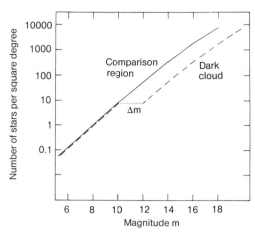

Fig. 16.7. Wolf diagram. The horizontal coordinate is the magnitude and the vertical coordinate is the number of stars per square degree in the sky brighter than that magnitude. A dark nebula diminishes the brightness of stars lying behind it by the amount Δm

The extinction by a dark nebula can be illustrated and studied by means of a *Wolf diagram*, shown schematically in Fig. 16.7. The diagram is constructed on the basis of star counts. The number of stars per square degree in some magnitude interval (e.g. between magnitudes 14 and 15) in the cloud is counted and compared with the number outside the nebula. In the comparison area, the number of stars increases monotonically towards fainter magnitudes. In the dark nebula, the numbers first increase in the same way, but beyond some limiting magnitude (10 in the figure), the number of stars falls below that outside the cloud. The reason for this is that the fainter stars are predominantly behind the nebula, and their brightness is reduced by some constant amount, Δm (2 magnitudes in the figure). The brighter stars are mostly in front of the nebula and suffer no extinction.

Reflection Nebulae. If a dust cloud is near a bright star, it will scatter, i.e. reflect the light of the star. Thus individual dust clouds can sometimes be observed as bright *reflection nebulae*. Some 500 reflection nebulae are known.

The regions in the sky richest in reflection nebulae are the areas around the Pleiades and around the giant star Antares. Antares itself is surrounded by a large red reflection nebula. This region is shown in Fig. 16.8.

Figure 16.9a shows the reflection nebula NGC 2068, which is located near a large, thick dust cloud a few degrees northwest of Orion's belt. It is one of the brightest reflection nebulae and the only one included in the Messier catalogue (M78). Figure 16.9b shows the reflection nebula NGC 1435 around Merope in the Pleiades.

Another bright and much-studied reflection nebula is NGC 7023 in Cepheus. It, too, is connected with a dark nebula. The illuminating star has emission lines in its spectrum (spectral type Be). Infrared stars have also been discovered in the area of the nebula, probably a region of star formation.

In 1922 *Edwin Hubble* published a fundamental investigation of bright nebulae in the Milky Way. On the basis of extensive photographic and spectroscopic observations, he was able to establish two interesting relationships. First, he found that emission nebulae only occur near stars with spectral class earlier than B0, whereas reflection nebulae may be found near stars of spectral class B1 and later. Secondly, Hubble discovered a relationship between the angular size R of the nebula and the apparent magnitude m of the illuminating star:

$$5 \lg R = -m + \text{const.} \tag{16.17}$$

a

b

c

Fig. 16.8 a – c. Bright and dark nebulae in Scorpio and Ophiuchus. Photograph (**a**) was taken in the blue colour region, $\lambda = 350-500$ nm, and (**b**) in the red colour region, $\lambda = 600-680$ nm. (The sharp rings in (**b**) are reflections of Antares in the correction lens of the Schmidt camera.) The nebulae located in the area are identified in drawing (**c**). B44 and H4 are dark nebulae. There is a large reflection nebula around Antares, which is faintly visible in the blue (**a**), but bright in the red (**b**) regions. Antares is very red (spectral class M1) and therefore the reflection nebula is also red. In contrast, the reflection nebulae around the blue stars ρ Ophiuchi (B2), CD-24° 12684 (B3), 22 Scorpii (B2) and σ Scorpii (B1) are blue and are visible only in (**a**). In (**b**) there is an elongated nebula to the right of σ Scorpii, which is invisible in (**a**). This is an emission nebula, which is very bright in the red hydrogen Hα line (656 nm). In this way, reflection and emission nebulae can be distinguished by means of pictures taken in different wavelength regions. [Photograph (**a**), E. Barnard, and (**b**), K. Mattila]

Thus the angular diameter of a reflection nebula is larger for a brighter illuminating star. Since the measured size of a nebula generally increases for longer exposures, i.e. fainter limiting surface brightness, the value of R should be defined to correspond to a fixed limiting surface brightness. The value of the constant in the Hubble relation depends on this limiting surface brightness. The Hubble relation for reflection nebulae is shown in Fig. 16.10, based on measurements by *Sidney van den Bergh* from Palomar Sky Atlas plates. Each point corresponds to a reflection nebula and the straight line represents the relation (16.17), where the value of the constant is 12.0 (R is given in arc minutes).

Fig. 16.9a. The reflection nebula NGC 2068 (M78) in Orion. In the middle of the nebula there are two stars of about magnitude 11. The northern one (at the top) is the illuminating star, while the other one probably lies in the foreground. (Photography Lunar and Planetary Laboratory, Catalina Observatory)

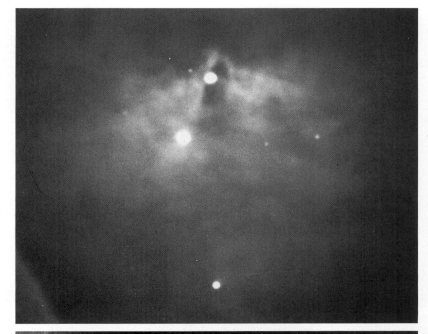

Fig. 16.9b. The reflection nebula NGC 1435 around Merope (23 Tau, spectral class B6) in the Pleiades. This figure should be compared with Fig. 17.1b, where Merope is visible as the lowest of the bright stars in the Pleiades. (National Optical Astronomy Observatories, Kitt Peak National Observatory)

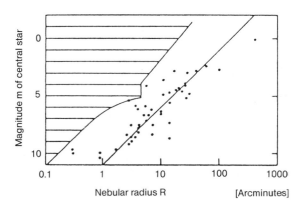

Fig. 16.10. The Hubble relation for reflection nebulae. The horizontal axis shows the radius R of the nebulae in arc minutes and the vertical axis, the (blue) apparent magnitude m of the central star. No measurements were made in the hatched region. [van den Bergh, S. (1966): Astron. J. **71**, 990]

The Hubble relation can be derived theoretically, if it is assumed that the illumination of a dust cloud is inversely proportional to the square of the distance to the illuminating star, and that the dust clouds are uniformly distributed in space. The theoretical Hubble relation also gives an expression for the constant on the right-hand side, which involves the albedo and the phase function of the grains.

The observations of reflection nebulae show that the albedo of interstellar grains must be quite high. It has not yet been possible to obtain its precise numerical value in this way, since the distances between the nebulae and their illuminating stars are not known well enough.

One may also consider the surface brightness of dark nebulae that are not close enough to a star to be visible as reflection nebulae. These nebulae will still reflect the diffuse galactic light from all the stars in the Milky Way. Calculations show that if the dust grains have a large albedo, then the reflected diffuse light should be bright enough to be observable, and it has indeed been observed. Thus the dark nebulae are not totally dark. The diffuse galactic light constitutes about $20-30\%$ of the total brightness of the Milky Way.

Dust Temperature. In addition to scattering, the interstellar grains also absorb radiation. The absorbed energy is re-radiated by the grains at infrared wavelengths corresponding to their temperatures. The temperature of dust in interstellar space (including dark nebulae) is about $10-20$ K. The corresponding wavelength according to Wien's displacement law (5.19) is $300-150$ μm. Near a hot star, the temperature of the dust may be $100-600$ K and the maximum emission is then at $30-5$ μm. In H II regions, the dust temperature is about $70-100$ K.

The rapid development of infrared astronomy in the 1970's has brought all the above-mentioned dust sources within the reach of observations (Fig. 16.11 a, b). In addition, infrared radiation from the nuclei of normal and active galaxies is largely thermal radiation from dust. Thermal dust emission is one of the most important sources of infrared radiation in astronomy.

One of the strongest infrared sources in the sky is the nebula around the star η Carinae (Fig. 16.12). The nebula consists of ionized gas, but infrared radiation from dust is also clearly visible in its spectrum. In even more extreme cases, the central star may be completely obscured, but revealed by the infrared emission from hot dust.

Composition and Origin of the Dust (Table 16.1). From the peaks in the extinction curve, it may be concluded that interstellar dust contains water ice and silicates, and

Fig. 16.11a, b. Interstellar dust is best seen in infrared wavelengths. Two examples of the images by the IRAS satellite. **(a)** In a view towards the Galactic Center, the dust is seen to be concentrated in a narrow layer in the galactic plane. Several separate clouds are also seen. **(b)** Most of the constellation Orion is covered by a complex area of interstellar matter. The brightest concentrations of dust below the center of the image, are in the region of the Horsehead Nebula and the Orion Nebula. (Photos NASA)

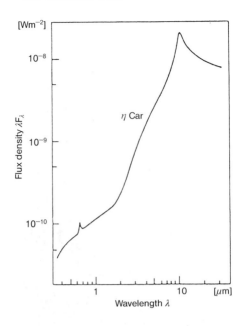

Fig. 16.12. More than 99% of the radiation from the η Carinae nebula (Fig. 14.13) is in the infrared. The peak in the visual region is from the hydrogen Hα line (0.66 μm). In the infrared, the silicate emission from dust is evident at 10 μm. [Allen, D. A. (1975): *Infrared, the New Astronomy* (Keith Reid Ltd., Shaldon) p. 103]

Table 16.1. Main properties of interstellar gas and dust

Property	Gas	Dust
Mass fraction	10%	0.1%
Composition	H I, H II, H$_2$ (70%)	Solid particles
	He (28%)	$d = 0.1 - 1$ μm
	C, N, O, Ne, Na, Mg,	H$_2$O (ice), silicates,
	Al, Si, S... (2%)	graphite + impurities
Particle density	1 cm^{-3}	10^{-13} cm^{-3} = 100 km^{-3}
Mass density	10^{-21} kg m^{-3}	10^{-23} kg m^{-3}
Temperature	100 K, 10^4 K, 50 K	10 – 20 K
	(H I, H II, H$_2$)	
Method of study	Absorption lines in stellar spectra.	Absorption and scattering of starlight.
	Optical:	Interstellar reddening
	Ca I, Ca II, Na I, K I,	Interstellar polarization
	Ti II, Fe I, CN, CH, CH$^+$	Thermal infrared emission
	Ultraviolet: H$_2$, CO, HD	
	Radio lines: hydrogen 21-cm emission and absorption;	
	H II, He II, C II recombination lines; molecular emission and absorption lines	

probably graphite as well. The sizes of the grains can be deduced from their scattering properties; usually they are smaller than one micrometre. The strongest scattering is due to grains of about 0.3 μm but smaller particles must also be present.

Dust grains are formed in the atmospheres of stars of late spectral types (K, M). Gas condenses into grains just as water in the Earth's atmosphere may condense into snow and ice. The grains are then expelled into interstellar space by the radiation pressure. Grains may also form in connection with star formation and possibly directly from atoms and molecules in interstellar clouds as well.

Table 16.2. Phenomena caused by the interstellar medium

Observable phenomenon	Cause
Interstellar extinction and polarization	Non-spherical dust grains aligned by magnetic field.
Dark nebulae, uneven distribution of stars and galaxies	Dust clouds
Interstellar absorption lines in stellar spectra	Atoms and molecules in the interstellar gas
Reflection nebulae	Interstellar dust clouds illuminated by nearby stars
Emission nebulae or H II regions (optical, infrared and radio emission)	Interstellar gas and dust cloud, where a nearby hot star ionizes the gas and heats the dust to 50 – 100 K
Optical galactic background (diffuse galactic light)	Interstellar dust illuminated by the integrated light of all stars
Galactic background radiation:	
a) short wavelength (<1 m)	Free-free emission from hot interstellar gas
b) long wavelength (>1 m)	Synchrotron radiation from cosmic ray electrons in the magnetic field
Galactic 21-cm emission	Cold (100 K) interstellar neutral hydrogen clouds (H I regions)
Molecular line emission (extended)	Giant molecular clouds, dark nebulae
Point-like OH, H_2O and SiO sources	Maser sources near protostars and long-period variables

16.2 Interstellar Gas

The mass of gas in interstellar space is a hundred times larger than that of dust. Although there is more gas, it is less easily observed, since the gas does not cause a general extinction of light. In the optical region, it can only be observed on the basis of a small number of spectral lines.

The existence of interstellar gas began to be suspected in the first decade of the 20th century, when in 1904 *Johannes Hartmann* observed that some absorption lines in the spectra of certain binary stars were not Doppler shifted by the motions of the stars like the other lines. It was concluded that these absorption lines were formed in gas clouds in the space between the Earth and the stars. In some stars, there were several lines, apparently formed in clouds moving with different velocities. The strongest lines in the visible region are those of neutral sodium and singly ionized calcium (Fig. 16.13). In the ultraviolet region, the lines are more numerous. The strongest one is the hydrogen Lyman α line (121.6 nm).

On the basis of the optical and uv lines, it has been found that many atoms are ionized in interstellar space. This ionization is mainly due to ultraviolet radiation from stars and, to some extent, to ionization by cosmic rays. Since the density of interstellar matter is very low, the free electrons only rarely encounter ions, and the gas remains ionized.

About thirty elements have been discovered by absorption line observations in the visible und ultraviolet region. With a few exceptions, all elements from hydrogen to zinc (atomic number 30) and a few additional heavier elements have been detected (Table 16.3). Like in the stars, most of the mass is hydrogen (about 70%) and helium (almost 30%). On the other hand, heavy elements are significantly less abundant than

Fig. 16.13. (a) The *D* lines D_1 and D_2 of interstellar sodium (rest wavelengths 589.89 and 589.00 nm) in the spectrum of the star HD 14134. Both lines consist of two components formed in the gas clouds of two spiral arms. The radial velocity difference of the arms is about 30 km/s. (Mt. Wilson Observatory). **(b)** The interstellar absorption lines of ionized calcium Ca II and ionized methylidyne CH$^+$ in the spectra of several stars. The emission spectrum of iron is shown for comparison in (**a**) and (**b**). (Lick Observatory)

in the Sun and other population I stars. It is thought that they have been incorporated into dust grains, where they do not produce any absorption lines. The element abundances in the interstellar medium (gas – dust) would then be normal, although the interstellar gas is depleted in heavy elements. This interpretation is supported by the observation that in regions where the amount of dust is smaller than usual, the element abundances in the gas are closer to normal.

Atomic Hydrogen. Ultraviolet observations have provided an excellent way of studying interstellar *neutral hydrogen*. The strongest interstellar absorption line, as has already been mentioned, is the hydrogen Lyman α line (Fig. 16.14). This line corresponds to the transition of the electron in the hydrogen atom from a state with principal quantum number $n = 1$ to one with $n = 2$. The conditions in interstellar space are such that almost all hydrogen atoms are in the ground state with $n = 1$. Therefore the Lyman α line is a strong absorption line, whereas the Balmer absorption lines, which arise from the excited initial state $n = 2$, are unobservable. (The Balmer lines are strong in stellar atmospheres with temperatures of about 10 000 K, where a large number of atoms are in the first excited state.)

The first observations of the interstellar Lyman α line were made from a rocket already in 1967. More extensive observations comprising 95 stars were obtained by the OAO 2 satellite. The distances of the observed stars are between 100 and 1000 parsec.

Table 16.3. Element abundances in the interstellar medium towards ζ Ophiuchi and in the Sun. The abundances are given relative to that of hydrogen, which has been defined to be 1 000 000. The last column gives the ratio of the abundances in the interstellar medium and in the Sun

Atomic number	Name	Chemical symbol	Interstellar abundance	Solar abundance	Abundance ratio
1	Hydrogen	H	1 000 000	1 000 000	1.00
2	Helium	He	85 000	85 000	1
3	Lithium	Li	0.000051	0.0015[a]	0.034
4	Beryllium	Be	<0.000070	0.000012	<5.8
5	Boron	B	0.000074	0.0046[a]	0.016
6	Carbon	C	74	370	0.20
7	Nitrogen	N	21	110	0.19
8	Oxygen	O	172	660	0.26
9	Fluorine	F	–	0.040	–
10	Neon	Ne	–	83	–
11	Sodium	Na	0.22	1.7	0.13
12	Magnesium	Mg	1.05	35	0.030
13	Aluminium	Al	0.0013	2.5	0.00052
14	Silicon	Si	0.81	35	0.023
15	Phosphorus	P	0.021	0.27	0.079
16	Sulfur	S	8.2	16	0.51
17	Chlorine	Cl	0.099	0.45	0.22
18	Argon	Ar	0.86	4.5	0.19
19	Potassium	K	0.010	0.11	0.094
20	Calcium	Ca	0.00046	2.1	0.00022
21	Scandium	Sc	–	0.0017	–
22	Titanium	Ti	0.00018	0.055	0.0032
23	Vanadium	V	<0.0032	0.013	<0.25
24	Chromium	Cr	<0.002	0.50	<0.004
25	Manganese	Mn	0.014	0.26	0.055
26	Iron	Fe	0.28	25	0.011
27	Cobalt	Co	<0.19	0.032	<5.8
28	Nickel	Ni	0.0065	1.3	0.0050
29	Copper	Cu	0.00064	0.028	0.023
30	Zinc	Zn	0.014	0.026	0.53

[a] Meteoritic abundance

Comparison of the Lyman α observations with observations of the 21-cm neutral hydrogen line have been especially useful. The distribution of neutral hydrogen over the whole sky has been mapped by means of the 21-cm line. However, the distances to nearby hydrogen clouds are difficult to determine from these observations. In the Lyman α observations one usually knows the distance to the star in front of which the absorbing clouds must lie.

Fig. 16.14. Interstellar absorption lines in the ultraviolet spectrum of ζ Ophiuchi. The strongest line is the hydrogen Lyman α line (equivalent width, more than 1 nm). The observations were made with the Copernicus satellite. [Morton, D. C. (1975): Astrophys. J. **197**, 85]

The average gas density within about 1 kpc of the Sun derived from the Lyman α observations is 0.7 atoms/cm³. Because the interstellar Lyman α line is so strong, it can be observed even in the spectra of very nearby stars. For example, it has been detected by the Copernicus satellite in the spectrum of Arcturus, whose distance is only 11 parsec. The deduced density of neutral hydrogen between the Sun and Arcturus is 0.02 – 0.1 atoms/cm³. Thus the Sun is situated in a clearing in the interstellar medium, where the density is less than one tenth of the average density.

If a hydrogen atom in its ground state absorbs radiation with a wavelength smaller than 91.2 nm, it will be ionized. Knowing the density of neutral hydrogen, one can calculate the expected distance a 91.2-nm photon can propagate before being absorbed in the ionization of a hydrogen atom. Even in the close neighbourhood of the Sun, where the density is exceptionally low, the mean free path of a 91.2-nm photon is only about a parsec and that of a 10-nm photon a few hundred parsec. Thus only the closest neighbourhood of the Sun can be studied in the extreme ultraviolet (xuv) spectral region.

The Hydrogen 21-cm Line. The spins of the electron and proton in the neutral hydrogen atom in the ground state may be either parallel or opposite. The energy difference between these two states corresponds to the frequency of 1420.4 MHz. Thus transitions between these two *hyperfine structure* energy levels will give rise to a spectral line at the wavelength of 21.049 cm (Fig. 16.15). The existence of the line was theoretically predicted by *Hendrick van de Hulst* in 1944, and was first observed by *Harold Ewen* and *Edward Purcell* in 1951. Studies of this line have revealed more about the properties of the interstellar medium than any other method – one might even speak of a special branch of 21-cm astronomy. The spiral structure and rotation of the Milky Way and other galaxies can also be studied by means of the 21-cm line.

Fig. **16.15.** The origin of the hydrogen 21-cm line. The spins of the electron and the proton may be either parallel or opposite. The energy of the former state is slightly larger. The wavelength of a photon corresponding to a transition between these states is 21 cm

Usually the hydrogen 21-cm line occurs in emission. Because of the large abundance of hydrogen, it can be observed in all directions in the sky. Some observed 21-cm line profiles are shown in Fig. 16.16. Rather than frequency or wavelength, the radial velocity calculated from the Doppler formula is plotted on the horizontal axis. This is because the broadening of the 21-cm spectral line is always due to gas motions either within the cloud (turbulence) or of the cloud as a whole. The vertical axis is mostly plotted in terms of the antenna temperature T_A (see Chap. 6), the usual radio astronomical measure of intensity. The brightness temperature of an extended source is then $T_b = T_A/\eta_B$, where η_B is the antenna loss.

Fig. 16.16. Hydrogen 21-cm emission line profiles in the galactic plane at longitudes 180°, 90° and 1° (in the direction $l = 0°$ there is strong absorption). The horizontal axis gives the radial velocity according to the Doppler formula, the vertical axis gives the brightness temperature. [Burton, W.B. (1974): "The Large Scale Distribution of Neutral Hydrogen in the Galaxy", in *Galactic and Extra-Galactic Radio Astronomy*, ed. by Verschuur, G.L., Kellermann, K.I. (Springer, Berlin, Heidelberg, New York) p. 91]

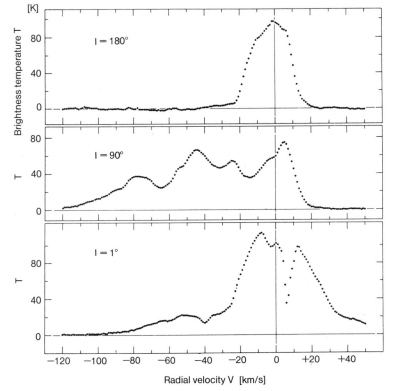

For the 21-cm line $h\nu/k = 0.07\,K$, and thus $h\nu/kT \ll 1$ for all relevant temperatures. One may therefore use the Rayleigh-Jeans approximation (6.4)

$$I_\nu = \frac{2\,\nu^2 kT}{c^2}\,. \tag{16.18}$$

In the solution of the equation of radiative transfer (5.29) the intensity can thus be directly related to a corresponding temperature. By definition, I_ν is related to the brightness temperatue T_b, and the source function S_ν is related to the excitation temperature T_{exc}, i.e.

$$T_b = T_{exc}(1 - e^{-\tau_\nu})\,. \tag{16.19}$$

In certain directions in the Milky Way there is so much hydrogen along the line of sight that the 21-cm line is optically thick, $\tau_\nu \gg 1$. In that case

$$T_b = T_{exc}\,, \tag{16.20}$$

i.e. the brightness temperature immediately yields the excitation temperature of the cloud. This is often referred to as the *spin temperature T_s*.

The excitation temperature need not always agree with the kinetic temperature of the gas. However, in the present case, the time between collisions is 400 years on the average, whereas the time for radiative transitions is 11 million years; thus the excitation temperature will be the same as the kinetic temperature. The observed temperature is T = 125 K.

The distance to a source cannot be obtained directly from the observed emission. Thus one can only study the number of hydrogen atoms in a cylinder with a 1-cm² base area extending from the observer to outside the Milky Way along the line of sight. This is called the *projected* or *column density* and is denoted by N. One may also consider the column density $N(v)\,dv$ of atoms with velocities in the interval $(v, v+dv)$.

It can be shown that if the gas is optically thin, the brightness temperature in a spectral line is directly proportional to the column density N of atoms with the corresponding radial velocity. Hence, if the diameter L of a cloud along the line of sight is known, the gas density can be determined from the observed line profile:

$$n = N/L .$$

The diameter L can be obtained from the apparent diameter, if the distance and shape of the cloud are assumed known.

The distances of clouds can be determined from their radial velocities by making use of the rotation of the Milky Way (Sect. 18.3). Thus if the observed peaks in the 21-cm line profiles (Fig. 16.16) are due to individual clouds, their distances and densities can be obtained. Since radio observations are not affected by extinction, it has been possible in this way to map the density distribution of neutral hydrogen in the whole galactic plane. The resulting distribution, based on observations at Leiden and Parkes, is shown in Fig. 16.17. It appears that the Milky Way is a spiral galaxy and that the interstellar hydrogen is concentrated in the spiral arms. The average density of interstellar hydrogen is 1 atom/cm³, but the distribution is very inhomogeneous. Typi-

Fig. 16.17. The distribution of neutral hydrogen in the galaxy from the Leiden and Parkes surveys. The density is given in atoms/cm³. [Oort, J.H., Kerr, F.T., Westerhout, G.L. (1958): Mon. Not. R. Astron. Soc. **118**, 379]

cally, the hydrogen forms denser regions, a few parsecs in size, where the densities may be $10-100$ atoms/cm^3. Regions where the hydrogen is predominantly neutral are known as *H I regions* (in contrast to H II regions of ionized hydrogen).

The hydrogen 21-cm line may also occur in absorption, when the light from a bright radio source, e.g. a quasar, passes through an intervening cloud. The same cloud may give rise to both an absorption and an emission spectrum. In that case, the temperature, optical thickness and hydrogen content of the cloud can all be derived.

Like interstellar dust, hydrogen is concentrated in a thin disc in the galactic plane. The thickness of the hydrogen layer is about twice that of the dust, or about 200 pc.

H II Regions. In many parts of space, hydrogen does not occur as neutral atoms, but is ionized. This is true in particular around hot O stars, which radiate strongly in the ultraviolet. If there is enough hydrogen around such a star, it will be visible as an emission nebula of ionized hydrogen. Such nebulae are known as *H II regions* (Figs. 16.18, 16.19).

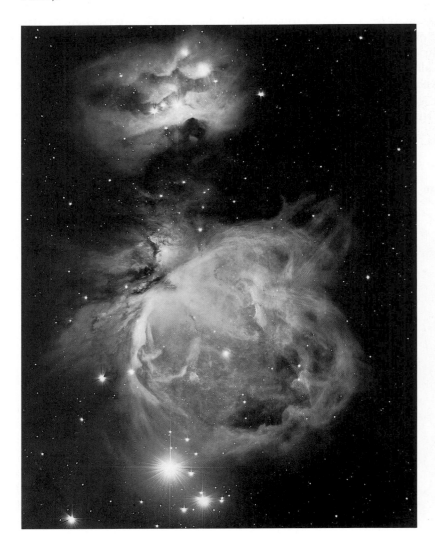

Fig. 16.18. The great nebula in Orion (M42, NGC 1976). The nebula gets its energy from newly formed hot stars. The dark regions are opaque dust clouds in front of the nebula. Radio and infrared observations have revealed a rich molecular cloud behind the nebula (Fig. 16.20). In the upper part of this picture is the gas nebula NGC 1977; in the lower part the bright star ι Orionis. (Lick Observatory)

Fig. 16.19. The Lagoon Nebula (M8, NGC 6523) in Sagittarius. This H II region contains many stars of early spectral types and stars that are still contracting towards the main sequence. Small, round dark nebulae, globules, are also visible against the bright background. These are presumably gas clouds in the process of condensation into stars. (National Optical Astronomy Observatories, Kitt Peak National Observatory)

A typical emission nebula is the great nebula in Orion, M42. It is visible even to the unaided eye, and is a beautiful sight when seen through a telescope. In the middle of the nebula is a group of four hot stars known as the Trapezium, which can be distinguished inside the bright nebula, even with a small telescope. The Trapezium stars emit strong ultraviolet radiation, which keeps the gas nebula ionized.

Unlike a star, a cloud of ionized gas has a spectrum dominated by a few narrow emission lines. The continuous spectrum of H II regions is weak. In the visible region the hydrogen Balmer emission lines are particularly strong. These are formed when a hydrogen atom recombines into an excited state and subsequently returns to the ground state via a sequence of radiative transitions. Typically, a hydrogen atom in a H II region remains ionized for several hundred years. Upon recombination it stays neutral for some months, before being ionized again by a photon from a nearby star.

The number of recombinations per unit time and volume is proportional to the product of the densities of electrons and ions,

$$n_{\rm rec} \propto n_{\rm e}\, n_{\rm i}\,. \tag{16.21}$$

In completely ionized hydrogen, $n_{\rm e} = n_{\rm i}$, and hence

$$n_{\rm rec} \propto n_{\rm e}^2\,. \tag{16.22}$$

Most recombinations will include the transition $n = 3 \rightarrow 2$, i.e. will lead to the emission of a H α photon. Thus the surface brightness of a nebula in the H α line will be proportional to the *emission measure*,

$$EM = \int n_e^2 \, dl, \tag{16.23}$$

where the integral is along the line of sight through the nebula.

The ionization of a helium atom requires more energy than that of a hydrogen atom, and thus regions of ionized helium are formed only around the hottest stars. In these cases, a large H II region will surround a smaller central He$^+$ or He^{2+} region. The helium lines will then be strong in the spectrum of the nebula.

Although hydrogen and helium are the main constituents of clouds, their emission lines are not always strongest in the spectrum. At the beginning of this century, it was even suggested that some strong unidentified lines in the spectra of nebulae were due to the new element nebulium. However, in 1927, *Ira S. Bowen* showed that they were *forbidden lines* of ionized oxygen and nitrogen, O$^+$, O^{2+} and N$^+$. Forbidden lines are extremely difficult to observe in the laboratory, because their transition probabilities are so small that at laboratory densities the ions are de-excited by collisions before they have had time to radiate. In the extremely diffuse interstellar gas, collisions are much less frequent, and thus there is a chance that an excited ion will make the transition to a lower state by emitting a photon.

Because of interstellar extinction, only the nearest H II regions can be studied in visible light. At infrared and radio wavelengths, much more distant regions can be studied. The most important lines at radio wavelengths are recombination lines of hydrogen and helium; thus the hydrogen transition between energy levels 110 and 109 at 5.01 GHz has been much studied. These lines are also important because with their help, radial velocities, and hence (using the galactic rotation law), distances of H II regions can be determined, just as for neutral hydrogen.

The physical properties of H II regions can also be studied by means of their continuum radio emission. The radiation is due to bremsstrahlung or free-free emission from the electrons. The intensity of the radiation is proportional to the emission measure EM defined in (16.23). H II regions also have a strong infrared continuum emission. This is thermal radiation from dust inside the nebula.

H II regions are formed when a hot O or B star begins to ionize its surrounding gas. The ionization steadily propagates away from the star. Because neutral hydrogen absorbs ultraviolet radiation so efficiently, the boundary between the H II region and the neutral gas is very sharp. In a homogeneous medium the H II region around a single star will be spherical, forming a so-called *Strömgren sphere*. For a B0 V star the radius of the Strömgren sphere is 50 pc and for an A0 V star only 1 pc.

The temperature of a H II region is higher than that of the surrounding gas, and it therefore tends to expand. After millions of years, it will have become extremely diffuse and will eventually merge with the general interstellar medium.

16.3 Interstellar Molecules

The first *interstellar molecules* were discovered in 1937–1938, when molecular absorption lines were found in the spectra of some stars. Three simple diatomic molecules were detected: *methylidyne* CH, its positive ion CH$^+$ and *cyanogen* CN. A few other molecules were later discovered by the same method in the ultraviolet. Thus *molecular*

Table 16.4. Some molecules observed in the interstellar medium

Molecule	Name	Year of discovery
Discovered in the optical and ultraviolet region:		
CH	methylidyne	1937
CH^+	methylidyne ion	1937
CN	cyanogen	1938
H_2	hydrogen molecule	1970
CO	carbon monoxide	1971
Discovered in the radio region:		
OH	hydroxyl radical	1963
CO	carbon monoxide	1970
CS	carbon monosulfide	1971
SiO	silicon monoxide	1971
SO	sulfur monoxide	1973
H_2O	water	1969
HCN	hydrogen cyanide	1970
NH_3	ammonia	1968
H_2CO	formaldehyde	1969
HCOOH	formic acid	1975
HCCNC	isocyanoacetylene	1991
H_2CCCC	cumulene carbene	1991
$(CH_3)_2O$	dimethyl ether	1974
C_2H_5OH	ethanol	1975
$HC_{11}N$	cyanopentacetylene	1981

hydrogen H_2 was discovered in the early 1970's, and *carbon monoxide*, which had been discovered by radio observations, was also detected in the ultraviolet. Molecular hydrogen is the most abundant interstellar molecule, followed by carbon monoxide.

Molecular Hydrogen. The detection and study of molecular hydrogen has been one of the most important achievements of uv astronomy. Molecular hydrogen has a strong absorption band at 105 nm, which was first observed in a rocket experiment in 1970 by *George R. Carruthers*, but more extensive observations could only be made with the Copernicus satellite. The observations showed that a significant fraction of interstellar hydrogen is molecular, and that this fraction increases strongly for denser clouds with higher extinction. In clouds with visual extinction larger than one magnitude essentially all the hydrogen is molecular.

Hydrogen molecules are formed on the surface of interstellar grains, which thus act as a chemical catalyst. Dust is also needed to shield the molecules from the stellar uv radiation, which would otherwise destroy them. Molecular hydrogen is thus found where dust is abundant. It is of interest to know whether gas and dust are well mixed or whether they form separate clouds and condensations.

UV observations have provided a reliable way of comparing the distribution of interstellar gas and dust. The amount of dust between the observer and a star is obtained from the extinction of the stellar light. Furthermore, the absorption lines of atomic and molecular hydrogen in the ultraviolet spectrum of the same star can be observed. Thus the total amount of hydrogen (atomic + molecular) between the observer and the star can also be determined.

Observations indicate that the gas and dust are well mixed. The amount of dust giving rise to one magnitude visual extinction corresponds to 1.9×10^{21} hydrogen atoms (one molecule is counted as two atoms). The mass ratio of gas and dust obtained in this way is 100.

Radio Spectroscopy. Absorption lines can only be observed if there is a bright star behind the molecular cloud. Because of the large dust extinction, no observations of molecules in the densest clouds can be made in the optical and ultraviolet spectral regions. Thus only radio observations are possible for these objects, where molecules are especially abundant.

Radio spectroscopy signifies an immense step forward in the study of interstellar molecules. In the early 1960's, it was still not believed that there might be more complicated molecules than diatomic ones in interstellar space. It was thought that the gas was too diffuse for molecules to form and that any that formed would be destroyed by ultraviolet radiation. The first molecular radio line, the hydroxyl radical OH, was discovered in 1963. Many other molecules have been discovered since then. In 1993, over 80 molecules had been detected, the heaviest one being the 13-atom molecule $HC_{11}N$.

Molecular lines in the radio region may be observed either in absorption or in emission. Radiation from diatomic molecules like CO (see Fig. 16.20) may correspond to three kinds of transitions. *Electronic transitions* correspond to changes in the electron cloud of the molecule. These are like the transitions in single atoms, and their wavelengths lie in the optical or ultraviolet region. *Vibrational transitions* correspond to changes in the vibrational energy of the molecule. Their energies are usually in the infrared region. Most important for radio spectroscopy are the *rotational transitions*, which are changes in the rotational energy of the molecule. Molecules in their ground

Fig. 16.20. Radio map of the distribution of carbon monoxide $^{13}C^{16}O$ in the molecular cloud near the Orion Nebula. The curves are lines of constant intensity. [Kutner, M. L., Evans II, N. J., Tucker K. D. (1976): Astrophys. J. **209**, 452]

state do not rotate, i.e. their angular momentum is zero, but they may be excited and start rotating in collisions with other molecules. For example, carbon sulfide CS returns to its ground state in a few hours by emitting a millimetre region photon.

A number of interstellar molecules are listed in Table 16.4. Many of them have only been detected in the densest clouds (mainly the Sagittarius B2 cloud at the galactic centre), but others are very common. The most abundant molecule H_2 cannot be observed at radio wavelengths, because it has no suitable spectral lines. The next most abundant molecules are carbon monoxide CO the hydroxyl radical OH and ammonia NH_3, although their abundance is only a small fraction of that of hydrogen. However, the masses of interstellar clouds are so large that the number of molecules is still considerable. (The Sagittarius B2 cloud contains enough ethanol, C_2H_5OH, for 10^{28} bottles of vodka.)

Most of the molecules in Table 16.4 have only been detected in dense molecular clouds occurring in connection with H II regions. Almost every molecule yet discovered has been detected in Sagittarius B2 near the galactic centre. Another very rich molecular cloud has been observed near the H II region Orion A, i.e. the well-known Orion Nebula M42 (Fig. 16.18).

Inside the actual H II regions there are no molecules, since they would be rapidly dissociated by the high temperature and strong ultraviolet radiation. Three types of molecular sources have been found near H II regions (Fig. 16.21):

a) large gas and dust envelopes around the H II region;
b) small dense clouds inside these envelopes;
c) very compact OH and H_2O maser sources.

The large envelopes have been discovered primarily by CO observations. OH and H_2CO have also been detected. Like in the dark nebulae, the gas in these clouds is

Fig. 16.21. Infrared map of the central part of the Orion Nebula. In the lower part are the four Trapezium stars. Above is an infrared source of about 0.5' diameter, the so-called Kleinmann-Low nebula (KL). BN is an infrared point source, the Becklin-Neugebauer object. Other infrared sources are denoted IRS. The large crosses indicate OH masers and the small crosses H_2O masers. On the scale of Fig. 16.18 this region would only be a few millimetres in size. [Goudis, C. (1982): *The Orion Complex: A Case Study of Interstellar Matter* (Reidel, Dordrecht) p. 176]

probably mainly molecular hydrogen. Because of the large size and density ($n = 10^3 - 10^4$ molecules/cm^3) of these clouds, their masses are very large, 10^5 or even 10^6 solar masses (Sgr B2). They are among the most massive objects in the Milky Way.

The dust in molecular clouds can be observed on the basis of its thermal radiation. Its peak falls at wavelengths of $10 - 100$ μm, corresponding to a dust temperature of $30 - 100$ K.

Some interstellar clouds contain very small *maser sources*. In these, the emission lines of OH, H_2O and SiO may be many million times stronger than elsewhere. The diameter of the radiating regions is only about $5 - 10$ AU. The conditions in these clouds are such that radiation in some spectral lines is amplified by stimulated emission as it propagates through the cloud. Hydroxyl and water masers occur in connection with dense H II regions and infrared sources, and appear to be related to the formation of protostars. In addition, maser emission (OH, H_2O and SiO) occurs in connection with Mira variables and some red supergiant stars. This maser emission comes from a molecule and dust envelope around the star, which also gives rise to an observable infrared excess.

16.4 The Formation of Protostars

The mass of the Milky Way is about 10^{11} solar masses. Since its age is about 10^{10} years, stars have been forming at the average rate of $10 M_\odot$ per year. This estimate is only an upper limit for the present rate, because earlier, the rate of star formation must have been much higher. Since the lifetime of O stars is only about a million years, a better estimate of the star formation rate can be made, based on the observed number of O stars. Accordingly, it has been concluded that at present, new stars are forming in the Milky Way at a rate of about three solar masses per year.

Stars are now believed to form inside large dense interstellar clouds mostly located in the spiral arms of the Galaxy. Under its own gravity, a cloud begins to contract and fragment into parts that will become protostars. The observations seem to indicate that stars are not formed individually, but in larger groups. Young stars are found in open clusters and in loose associations, typically containing a few hundred stars which must have formed simultaneously.

Theoretical calculations confirm that the formation of single stars is almost impossible. An interstellar cloud can contract only if its mass is large enough for gravity to overwhelm the pressure. The limiting mass is the Jeans mass (Sect. 7.10), given by

$$M_J \approx 3 \times 10^4 \sqrt{\frac{T^3}{n}} M_\odot \, ,$$

where n is the density in atoms/m^3.

In a typical interstellar neutral hydrogen cloud, $n = 10^6$ and $T = 100$ K, giving the Jeans mass, $30000 M_\odot$. In the densest dark clouds, $n = 10^{12}$ and $T = 10$ K and hence, $M_J = 1 M_\odot$.

It is thought that star formation begins in clouds of a few thousand solar masses and diameters of about 10 pc. The cloud begins to contract, but does not heat up because the liberated energy is carried away by radiation. As the density increases, the

Jeans mass thus decreases. Because of this, separate condensation nuclei are formed in the cloud, which go on contracting independently: the cloud *fragments*.

This contraction and fragmentation continues until the density becomes so large that the individual fragments become optically thick. The energy liberated by the contraction can then no longer escape, and the temperature will begin to rise. In consequence, the Jeans mass begins to increase, further fragmentation ceases and the rising pressure in existing fragments stops their contraction. Some of the protostars formed in this way may still be rotating too rapidly. These may split into two, thus forming double systems. The further evolution of protostars has been described in Sect. 12.2.

Although the view that stars are formed by the collapse of interstellar clouds is generally accepted, many details of the fragmentation process are still highly conjectural. Thus the effects of rotation, magnetic fields and energy input are very imperfectly known. Why a cloud begins to contract is also not certain; one theory is that passage through a spiral arm compresses clouds and triggers contraction (see Sect. 18.4). This would explain why young stars are predominantly found in the spiral arms of the Milky Way and other galaxies. The contraction of an interstellar cloud might also be initiated by a nearby expanding H II region or supernova explosion.

Table 16.5 The five phases of interstellar gas

	T [K]	n [cm^{-3}]
1. Very cool molecular gas clouds (mostly hydrogen H_2)	20	$> 10^3$
2. Cool gas clouds (mostly atomic neutral hydrogen)	100	20
3. Warm neutral gas enveloping the cooler clouds	6000	$0.05 - 0.3$
4. Hot ionized gas (mainly H II regions around hot stars)	8000	> 0.5
5. Very hot and diffuse ionized coronal gas, ionized and heated by supernova explosions	10^6	10^{-3}

16.5 Planetary Nebulae

Bright regions of ionized gas do not occur only in connection with newly formed stars, but also around stars in late stages of their evolution. The *planetary nebulae* are gas shells around small hot blue stars. As we have seen in connection with stellar evolution, instabilities may develop at the stage of helium burning. Some stars begin to pulsate, while in others, the whole outer atmosphere may be violently ejected into space. In the latter case, a gas shell expanding at $20-30$ km/s will be formed around a small and hot ($50\,000-100\,000$ K) star, the core of the original star.

The planetary nebulae were given their name in the 19th century, because certain small nebulae visually look quite like planets such as Uranus. The apparent diameter of the smallest known planetary nebulae is only a few arc seconds, whereas the largest ones (like the Helix Nebula) may be one degree in diameter (Fig. 16.22).

The expanding gas in planetary nebulae is ionized by ultraviolet radiation from the central star, and its spectrum contains many of the same bright emission lines as that of

Fig. 16.22. The Helix Nebula (NGC 7293). The planetary nebulae are formed during the final stages of evolution of solar-type stars. The centrally visible star has ejected its outer layers into space. (National Optical Astronomy Observatories, Kitt Peak National Observatory)

an H II region. The brightest emission lines are often due to forbidden transitions. For example, the green colour of the central parts of the Ring Nebula in Lyra are due to the forbidden lines of doubly ionized oxygen at 495.9 and 500.7 nm. The red colour of the outer parts is due to the hydrogen Balmer α line (656.3 nm) and the forbidden lines of ionized nitrogen (654.8 nm, 658.3 nm).

Planetary nebulae are generally much more symmetrical in shape than most H II regions, and they expand more rapidly. For example, the well-known Ring Nebula in Lyra (M57) has expanded visibly in photographs taken at 50-year intervals. In a few ten thousand years, the planetary nebulae disappear in the general interstellar medium and their central stars cool to become white dwarfs.

The total number of planetary nebulae in the Milky Way has been estimated to be 50000. About 1000 have actually been observed.

16.6 Supernova Remnants

In Chap. 12 we have seen that massive stars end their evolution in a supernova explosion. The collapse of the stellar core leads to the violent ejection of the outer layers, which then remain as an expanding gas cloud.

About 120 *supernova remnants* (SNR's) have been discovered in the Milky Way. Some of them are optically visible as a ring or an irregular nebula (e.g. the Crab Nebula; see Fig. 16.23), but most are detectable only in the radio region (because radio emission suffers no extinction).

Fig. 16.23. The Crab Nebula (M1, NGC 1952) in Taurus is the remnant of a supernova explosion observed in 1054. The photograph was taken at red wavelengths. The nebula is also a strong radio source. Its energy source is the central rapidly rotating neutron star, pulsar, which is the collapsed core of the original star. (Palomar Observatory)

In the radio region, the SNR's are extended sources similar to H II regions. However, unlike H II regions, the radiation from SNR's is often polarized. Another characteristic difference between these two kinds of sources is that whereas the radio brightness of H II regions grows or remains constant as the frequency increases, that of SNR's falls off almost linearly with increasing frequency (Fig. 16.24).

These differences are due to the different emission processes in H II regions and in SNR's. In an H II region, the radio emission is free-free radiation from the hot plasma. In a SNR it is *synchrotron radiation* from relativistic electrons moving in spiral orbits around the magnetic field lines. The synchrotron process gives rise to a continuous

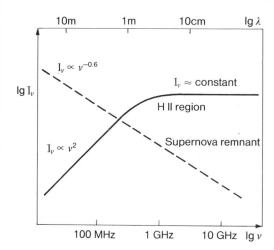

Fig. 16.24. The radio spectra of typical HII regions and supernova remnants. The radiation of H II regions is thermal and obeys the Rayleigh-Jeans law, $I_\nu \propto \nu^2$ at wavelengths larger than 1 m. In supernova remnants the intensity decreases with increasing frequency. [After Scheffler, H., Elsässer, H. (1987): *Physics of the Galaxy and the Interstellar Matter* (Springer, Berlin, Heidelberg, New York)]

Fig. 16.25. The Veil Nebula (NGC 6960 at the right, NGC 6992 at the left) in Cygnus is the remnant of a supernova explosion which occurred several ten thousand years ago. (Mt. Wilson Observatory)

spectrum extending over all wavelength regions. For example, the Crab Nebula looks blue or green in colour photographs because of optical synchrotron radiation.

In the Crab Nebula, red filaments are also visible against the bright background. Their emission is principally in the hydrogen Hα line. The hydrogen in a SNR is not ionized by a central star as in the H II regions, but by the ultraviolet synchrotron radiation.

The supernova remnants in the Milky Way fall into two classes. One type has a clearly ringlike structure (e.g. Cassiopeia A or the Veil Nebula in Cygnus; see Fig. 16.25); another is irregular and bright at the middle (like the Crab Nebula). In the remnants of the Crab Nebula type, there is always a rapidly rotating pulsar at the centre. This pulsar provides most of the energy of the remnant by continuously injecting relativistic electrons into the cloud. The evolution of this type of SNR reflects that of the pulsar and for this reason has a timescale of a few ten thousand years.

Ringlike SNR's do not contain an energetic pulsar; their energy comes from the actual supernova explosion. After the explosion, the cloud expands at a speed of 10000 – 20000 km/s. About 50 – 100 years after the explosion the remnant begins to form a spherical shell as the ejected gas starts to sweep up interstellar gas and to slow down in its outer parts. The swept-up shell expands with a decreasing velocity and cools until, after about 100000 years, it merges into the interstellar medium. The two types of supernova remnants may be related to the two types (I and II) of supernovae.

* Synchrotron Radiation

Synchrotron radiation was first observed in 1948 by *Frank Elder, Robert Langmuir* and *Herbert Pollack*, who were experimenting with an electron synchrotron, in which electrons were accelerated to relativistic energies in a magnetic field. It was observed that the electrons radiated visible light in a narrow cone along their momentary direction of motion. In astrophysics, synchrotron radiation was first invoked as an explanation of the radio emission of the Milky Way, discovered by *Karl Jansky* in 1931. This radiation had a spectrum and a large metre-wave brightness temperature (more than 10^5 K) which were inconsistent with ordinary thermal free-free emission from ionized gas. In 1950, *Hannes Alfven* and *Nicolai Herlofson* as well as *Karl-Otto Kiepenheuer* proposed that the galactic radio background was due to synchrotron radiation. According to Kiepenheuer the high-energy cosmic ray electrons would emit radio radiation in the weak galactic magnetic field. This explanation has turned out to be correct.

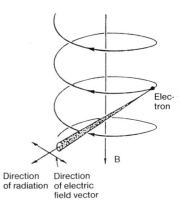

The emission of synchrotron radiation. A charged particle (electron) propagating in a magnetic field moves in a spiral. Because of the centripetal acceleration, the particle emits electromagnetic radiation

Synchrotron radiation is also an important emission process in supernova remnants, radio galaxies and quasars. It is a *non-thermal* radiation process, i.e. the energy of the radiating electrons is not due to thermal motions.

The origin of synchrotron radiation is schematically shown in the figure. The magnetic field forces the electron to move in a spiral orbit. The electron is thus constantly accelerated, and will emit electromagnetic radiation. According to the special theory of relativity, the emission of a relativistic electron will be concentrated in a narrow cone. Like the beam from a lighthouse, this cone sweeps across the observer's field of vision once for each revolution. Thus the observer sees a sequence of radiation flashes of very short duration compared with their interval. (In the total emission of a large number of electrons, separate flashes cannot be distinguished.) When this series of pulses is represented as a sum of different frequency components (Fourier transform), a broad spectrum is obtained with a maximum at

$$\nu_{max} = a B_\perp E^2 \ ,$$

where $B_\perp$ is the magnetic field component perpendicular to the velocity of the electron, and E its energy; a is a constant of proportionality.

The table gives the frequency and wavelength of the maximum as functions of the electron energy for the typical galactic field strength 0.5 nT:

λ_{max}	300 [nm]	30 [μm]	3 [mm]	30 [cm]	30 [m]	
ν_{max}	10^{15}	10^{13}	10^{11}	10^9	10^7	[Hz]
E	6.6×10^{12}	6.6×10^{11}	6.6×10^{10}	6.6×10^9	6.6×10^8	[eV]

To produce even radio synchrotron radiation, very energetic electrons are required, but these are known to be present in the cosmic radiation. In the optical galactic background radiation, the contribution from synchrotron radiation is negligible, but, for example, in the Crab Nebula, a significant part of the optical emission is due to this mechanism.

16.7 The Hot Corona of the Milky Way

As early as 1956, *Lyman Spitzer* showed that the Milky Way has to be surrounded by a large envelope of very hot gas. Almost two decades later the Copernicus satellite, whose scientific programme was directed by Spitzer, found evidence for this kind of gas, which began to be called *galactic coronal gas*, in analogy with the solar corona. The satellite observed emission lines of e.g. five times ionized oxygen (O VI), four times ionized nitrogen (N V) and triply ionized carbon (C IV). The formation of these lines requires a high temperature ($100\,000 - 1\,000\,000$ K), and a high temperature is also indicated by the broadening of the lines.

Galactic coronal gas is distributed through the whole Milky Way and extends several thousand parsecs from the galactic plane. Its density is only of the order of

10^{-3} atoms/cm^3 (recall that the mean density in the galactic plane is 1 atom/cm^3). Thus coronal gas forms a kind of background sea, from which the denser and cooler forms of interstellar matter, such as neutral hydrogen and molecular clouds, rise as islands. In the early 1980's the IUE satellite also detected similar coronae in the Large Magellanic Cloud and in the spiral galaxy M100. Coronal gas is probably quite a common and important form of matter in galaxies.

Supernova explosions are probably the source of both coronal gas and its energy. When a supernova explodes, it forms a hot bubble in the surrounding medium. The bubbles from neighbouring supernovae will expand and merge, forming a foamlike structure. In addition to supernovae, stellar winds from hot stars may provide some of the energy of the coronal gas.

16.8 Cosmic Rays and the Interstellar Magnetic Field

Cosmic Rays. Elementary particles and atomic nuclei reaching the Earth from space are called *cosmic rays*. They occur throughout interstellar space with an energy density of the same order of magnitude as that of the radiation from stars. Cosmic rays are therefore important for the ionization and heating of interstellar gas.

Since cosmic rays are charged, their direction of propagation in space is constantly changed by the magnetic field. Their direction of arrival therefore gives no information about their place of origin.

The most important properties of cosmic rays that can be observed from the Earth are their particle composition and energy distribution. As noted in Sect. 3.7, these observations have to be made in the upper atmosphere or from satellites, since cosmic ray particles are destroyed in the atmosphere.

The main constituent of the cosmic rays (about 90%), is hydrogen nuclei or protons. The second most important constituent (about 9%), is helium nuclei or α particles. The rest of the particles are electrons and nuclei more massive than helium.

Most cosmic rays have an energy smaller than 10^9 eV. The number of more energetic particles drops rapidly with increasing energy. The most energetic protons have an energy of 10^{20} eV, but such particles are very rare. (The largest particle accelerators reach only an energy of $5 \cdot 10^{11}$ eV.)

The distribution of low-energy (less than 10^8 eV) cosmic rays cannot be reliably determined from the Earth, since solar "cosmic rays", high-energy protons and electrons formed in solar flares fill the solar system and strongly affect the motion of low-energy cosmic rays.

The distribution of cosmic rays in the Milky Way can be directly inferred from gamma-ray and radio observations. The collisions of cosmic ray protons with interstellar hydrogen atoms gives rise to pions which then decay to form a gamma-ray background. The radio background is formed by cosmic ray electrons which emit synchrotron radiation in the interstellar magnetic field.

Both radio and gamma-ray emission are strongly concentrated in the galactic plane. From this it has been concluded that the sources of cosmic rays must also be located in the galactic plane. In addition, there are individual peaks in the backgrounds around known supernova remnants. In the gamma-ray region, such peaks are observed at e.g. the Crab Nebula and the Vela pulsar; in the radio region, the so-called North Polar Spur is a large, nearby ringlike region of enhanced emission.

Fig. 16.26. The polarization of starlight. The dashes give the direction and degree of the polarization. The thinner dashes correspond to stars with polarization smaller than 0.5%; the thicker dashes to stars with larger polarization. The scale is shown in the upper left-hand corner. Stars with polarization smaller than 0.08% are indicated by a small circle. [Mathewson, D. S., Ford, V. L. (1970): Mem. R.A.S. **74**, 139]

Apparently a large fraction of cosmic rays have their origin in supernovae. An actual supernova explosion will give rise to energetic particles. If a pulsar is formed, observations show that it will accelerate particles in its surroundings. Finally, the shock waves formed in the expanding supernova remnant will also give rise to relativistic particles.

On the basis of the relative abundances of various cosmic ray nuclei, it can be calculated how far they have travelled before reaching the Earth. It has been found that typical cosmic ray protons have travelled for a period of a few million years (and hence also a distance of a few million light-years) from their point of origin. Since the diameter of the Milky Way is about 100 000 light-years, the protons have crossed the Milky Way tens of times in the galactic field.

The Interstellar Magnetic Field. The strength and direction of the *interstellar magnetic field* are difficult to determine reliably. Direct measurements are impossible, since the magnetic fields of the Earth and the Sun are much stronger.

We have seen that interstellar grains give rise to interstellar polarization. In order to polarize light, the dust grains have to be similarly oriented; this can only be achieved by a general magnetic field. Figure 16.26 shows the distribution of interstellar polarization over the sky. Stars near each other generally have the same polarization. At low galactic latitudes, the polarization is almost parallel to the galactic plane, except where one is looking along a spiral arm.

More precise estimates of the strength of the magnetic field can be obtained from the rotation of the plane of polarization of the radio radiation from distant sources. This so-called Faraday rotation is proportional to the strength of the magnetic field and to the electron density. Another method is to measure the Zeeman splitting of the 21-cm radio line. These measurements have fairly consistently given a value of 0.1 − 1 nT for the strength of the interstellar magnetic field. This is about 1/1 000 000 of the interplanetary field in the solar system.

16.9 Examples

16.9.1   Estimate the dust grain size and number density in the galactic plane.

Let us compare the interstellar extinction curve in Fig. 16.4b with the Mie scattering curves in Fig. 16.3. We see that the leftmost parts of the curves may correspond to each others: the interval $0 < x < 5$ with $m = 1.5$ in Fig. 16.3 matches the interval $0 < 1/\lambda < 5\,\mu^{-1}$ in Fig. 16.4b. Remembering that $x = 2\pi a/\lambda$, this suggests a constant grain radius a, given by $2\pi a \approx 1\,\mu m$, or $a \approx 0.16\,\mu m$.

In the blue wavelength region ($\lambda = 0.44\,\mu m$), $x = 2.3$ and, according to the upper Fig. 16.3, $Q_{ext} \approx 2$. Using $A = 2$ mag for the interstellar extinction at $r = 1$ kpc, we get, substituting (16.5) into (16.7) $\bar{n} \approx 4 \times 10^{-7}\,m^{-3}$. This should give the order of magnitude of the interstellar dust density.

As a summary we could say that a considerable fraction of the interstellar extinction might be due to grains of diameter $0.3\,\mu m$ and particle density of the order of $10^{-7}\,m^{-3} = 100\,km^{-3}$.

16.9.2   Estimate the time interval between successive collisions of a hydrogen atom in interstellar gas.

Two atoms will collide if the separation between their centres becomes less than $2r$, where r is the radius of the atom. Thus, the microscopic cross section for the collision is $\sigma = \pi(2r)^2 = 4\pi r^2$. The macroscopic cross section, or the number of collisions of an H atom per unit length, is then $\Sigma = n\sigma$, where n is the number density of the H atoms. The mean free path l of an atom is the inverse of the macroscopic cross section, $l = 1/\Sigma$, and the time between two collisions is $t = l/v$, where v is the velocity of the atom.

Considering the numerical values, the Bohr radius of an H atom is $r = 5.3 \times 10^{-11}$ m. Taking $n = 1\,cm^{-3}$ we get $l = 2.8 \times 10^{13}\,m \approx 0.0009$ pc. The average velocity is not far from the root mean square velocity at $T = 125$ K, given by (6.10):

$$v = \sqrt{\frac{3kT}{m}} = 1760\,ms^{-1} \;.$$

These values of l and v give $t = l/v = 510$ y for the collision interval. Taking into account the velocity distribution in the gas, the mean free path appears to be shorter by a factor of $1/\sqrt{2}$, which reduces the time to about 400 years.

16.9.3   Consider the lowest rotational transition of the CO molecule. For ^{12}CO the frequency of this line is $v(^{12}CO) = 115.27$ GHz, and for ^{13}CO, $v(^{13}CO) = 110.20$ GHz. Estimate the optical thickness of each line in a molecular cloud, where the observed brightness temperatures of the lines are $T_b(^{12}CO) = 40$ K and $T_b(^{13}CO) = 9$ K.

For the ^{12}CO line, $hv/k = 5.5$ K. Thus, the Rayleigh-Jeans approximation is valid if the temperature is considerable higher than 5 K. This is not always the case, but the measured value of $T_b(^{12}CO)$ suggests that the approximation can be used.

Ignoring the background, (16.19) gives

$$T_b = T_{exc}(1 - e^{-\tau_\nu}) \ .$$

The optical thickness τ_ν is proportional to the opacity or the absorption coefficient α_ν (see (4.16)), and α_ν is evidently proportional to the number of CO molecules present. Other differencies between the lines are small, so we can write

$$\frac{\tau_\nu(^{12}CO)}{\tau_\nu(^{13}CO)} \approx \frac{n(^{12}CO)}{n(^{13}CO)} \ .$$

Adopting the terrestrial value $n(^{12}C)/n(^{13}C) = 89$, we set

$$\tau_\nu(^{12}CO) = 89 \tau_\nu(^{13}CO) \ .$$

Assuming the excitation temperatures equal, and denoting $\tau_\nu(^{12}CO)$ by τ, we get

$$T_{exc}(1 - e^{-\tau}) = 40 \ ,$$

$$T_{exc}(1 - e^{-\tau/89}) = 9 \ .$$

The solution of this pair of equations is

$$\tau_\nu(^{12}CO) = 23 \ , \quad \tau_\nu(^{13}CO) = 0.25 \ , \quad T_{exc} = 40 \ K \ .$$

Thus, the ^{12}CO line seems to be optically thick, and $T_{exc} = T_b(^{12}CO)$. If also the ^{13}CO line were optically thick, the brightness temperatures would be practically equal, and the optical thicknesses could not be determined.

16.10 Exercises

16.10.1 Two open clusters, which are seen near each other in the galactic plane, have angular diameters α and 3α, and distance moduli 16.0 and 11.0, respectively. Assuming their actual diameters are equal, find their distances and the interstellar extinction coefficient a in Eq. (16.4).

16.10.2 Estimate the free fall velocity on the surface of a spherical gas cloud contracting under the influence of its own gravity. Assume $n(H_2) = 10^3 \ cm^{-3}$ and $R = 5$ pc.

16.10.3 The force F exerted by a magnetic field B on a charge q moving with velocity v is $F = qv \times B$. If v is perpendicular to B, the path of the charge is circular. Find the radius of the path on an interstellar proton with a kinetic energy of 1 MeV. Use $B = 0.1$ nT for the galactic magnetic field.

Chapter 17 **Star Clusters and Associations**

Several collections of stars can be picked out in the sky, even with the naked eye. Closer study reveals that they really do form separate clusters in space. Such *open star clusters* are e.g. the Pleiades in Taurus and the Hyades around Aldebaran, the brightest star in Taurus. Almost the whole of the constellation Coma Berenices is also an open star cluster. Many objects appearing as nebulous patches to the unaided eye, when looked at with a telescope, turn out to be star clusters, like Praesepe in the constellation Cancer, or Misan, the double cluster in Perseus (Fig. 17.1). In addition to open clusters, some apparently nebulous objects are very dense *globular clusters*, such as those in Hercules and in Canes Venatici (Fig. 17.2).

a

b

c

Fig. 17.1a – c. Open clusters. **(a)** The Hyades slightly to the lower left in the photograph. Above them to the right the Pleiades (Photograph M. Korpi). **(b)** The Pleiades photographed with the Metsähovi Schmidt camera. The diameter of the cluster is about 1°. Reflection nebulae are visible around some of the stars (Photograph M. Poutanen and H. Virtanen, Helsinki University). **(c)** Misan or *h* and *χ* Persei, the double cluster in Perseus. The separation between the clusters is about 25′. Picture taken with the Metsähovi 60-cm Ritchey Chrétien telescope (Photograph T. Markkanen, Helsinki University)

Fig. 17.2. The globular cluster ω Centauri. Picture taken with the Danish 1.5-m telescope at La Silla, Chile. Thanks to the excellent seeing, one can see through the entire cluster in some places. (Photograph T. Korhonen, Turku University)

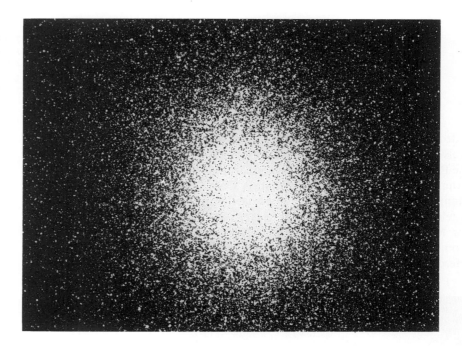

The first catalogue of star clusters was prepared by the French astronomer *Charles Messier* in 1784. Among his 103 objects, there were about 30 globular clusters and the same number of open clusters, in addition to gas nebulae and galaxies. A larger catalogue was the *New General Catalogue of Nebulae and Clusters of Stars* prepared by the Danish astronomer *John Louis Emil Dreyer*, which appeared in 1888. The catalogue numbers of objects in this list are preceded by the initials NGC. For example, the large globular cluster in Hercules is object M13 in the Messier catalogue and is also known as NGC 6205. The NGC catalogue was supplemented with the *Index Catalogue* in 1895 and 1910. The objects of this catalogue are given the initials IC.

17.1 Associations

In 1947, the Soviet astronomer *Viktor Amazaspovich Ambartsumyan* discovered that there are groups of young stars scattered over so large regions of the sky that they would be very difficult to identify merely on the basis of their appearance. These *associations* may have a few tens of members. One association is found around the star ζ Persei, and in the region of Orion, there are several associations.

Associations are groups of very young stars. They are usually identified on the basis either of absolutely bright main sequence stars or of T Tauri stars. According to the type, one speaks of OB associations and T Tauri associations. The most massive stars of spectral class O stay on the main sequence for only a few million years, and therefore associations containing them are necessarily young. The T Tauri stars are even younger stars that are in the process of contracting towards the main sequence.

Studies of the internal motions in associations show that they are rapidly dispersing. There are so few stars in an association that their gravity cannot hold them together for

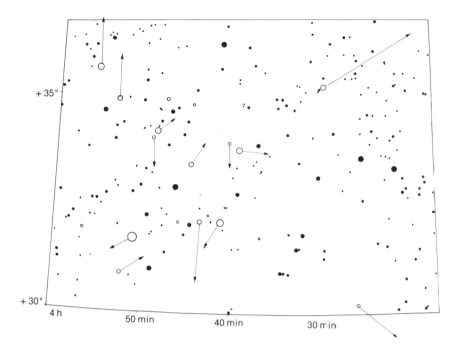

Fig. 17.3. ξ Persei association. O and B stars are shown as open circles. The proper motion vectors show the movements of the stars in the next 500000 years

any length of time. The observed motions have often confirmed that the stars in an association were very close together a few million years ago (Fig. 17.3).

Large amounts of interstellar matter, gas and dust nebulae, often occur in connection with associations, supplying information about the connection between star formation and the interstellar medium. Infrared observations have shown that stars are now forming or have recently formed in many dense interstellar clouds.

Associations are strongly concentrated in the spiral arms in the plane of the Milky Way. Both in the Orion region and in the direction of Cepheus, three generations of associations have been identified, the oldest ones being most extended and the youngest ones, most dense.

17.2 Open Star Clusters

The open clusters usually contain from a few tens to a few hundreds of stars. The kinetic energy of the cluster members, the differential rotation of the Milky Way (Sect. 18.3) and external gravitational disturbances tend to gradually disperse the open clusters. Still, many of them are fairly permanent; for example, the Pleiades is many hundreds of millions of years old, but nevertheless, quite a dense cluster.

The distances of star clusters – and also of associations – can be obtained from the photometric or spectroscopic distances of their brightest members. For the nearest clusters, in particular for the Hyades, one can use the method of *kinematic parallaxes*, which is based on the fact that the stars in a cluster all have the same average space velocity with respect to the Sun. The proper motions in the Hyades are shown in Fig. 17.4a. They all appear to be directed to the same point. Figure 17.4b explains how this convergence can be understood as an effect of perspective, if all cluster members have

a)

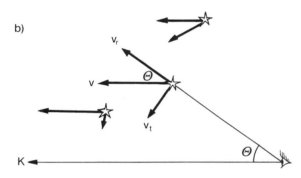

b)

Fig. 17.4. (a) Proper motions of the Hyades. The vectors show the movement of the stars in about 10000 years. [van Bueren, H. G. (1952): Bull. Astr. Inst. Neth. **11**]. **(b)** If all stars move in the same direction, their tangential velocity components appear to be directed towards the convergent point K

the same velocity vector with respect to the observer. Let Θ be the angular distance of a given star from the convergent point. The angle between the velocity of the star and the line of sight will then also be Θ. The velocity components along the line of sight and at right angles to it, v_r and v_t, are therefore given by

$$\begin{cases} v_r = v \cos \Theta \\ v_t = v \sin \Theta, \end{cases} \tag{17.1}$$

The radial velocity v_r can be measured from the Doppler shift of the stellar spectrum. The tangential velocity v_t is related to the proper motion μ and the distance r:

$$v_t = \mu r. \tag{17.2}$$

Thus the distance can be calculated:

$$r = v_t/\mu = \frac{v \sin \Theta}{\mu} = \frac{v_r}{\mu} \tan \Theta. \tag{17.3}$$

By means of this method, the distances of the individual stars can be determined from the motion of the cluster as a whole. Since the method of trigonometric parallaxes is reliable only out to a distance of 30 pc, the moving cluster method is an indispensable

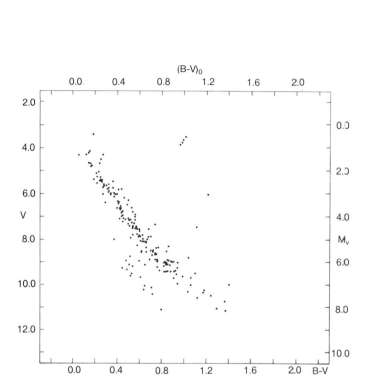

Fig. 17.5. Colour-magnitude diagram of the Hyades. Apparent visual magnitude on the left-hand vertical axis; absolute visual magnitude on the right-hand one

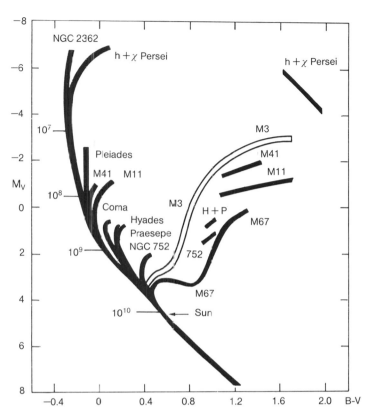

Fig. 17.6. Schematic colour-magnitude diagrams of star clusters. M3 is a globular cluster; the others are open clusters. Cluster ages are shown along the main sequence. The age of a cluster can be told from the point where its stars begin to turn off the main sequence. [Sandage, A. (1956): Publ. Astron. Soc. Pac. **68**, 498]

way of determining stellar distances. The distance of the Hyades obtained in this way is about 40 pc. The Hyades is the nearest open cluster.

The observed HR diagram or colour-magnitude diagram of the Hyades and other nearby star clusters show a very well-defined and narrow main sequence (Fig. 17.5). Most of the cluster members are main sequence stars; there are only a few giants. There are quite a few stars slightly less than one magnitude above the main sequence. These are apparently binary stars whose components have not been resolved. To see this, let us consider a binary, where both components have the same magnitude m and the same colour index. If this system is unresolved, the colour index will still be the same, but the observed magnitude will be $m - 0.75$, i.e. slightly less than one magnitude brighter.

The main sequences of open clusters are generally located in the same section of the HR or colour-magnitude diagram (Fig. 17.6). This is because the material from which the clusters formed has not varied much, i.e. their initial chemical composition has been fairly constant. In younger clusters, the main sequence extends to brighter and hotter stars and earlier spectral types. Usually one can clearly see the point in the diagram where the main sequence ends and bends over towards the giant branch. This point will depend very strongly on the age of the cluster. It can therefore be used in determining the ages of open clusters. Star clusters are of central importance in the study of stellar evolution.

The colour-magnitude diagrams of star clusters can also be used to determine their distances. The method is called *main sequence fitting*. By means of multicolour photometry, the reddening due to interstellar dust can be removed from the observed colours B-V of the stars, yielding the intrinsic colours $(B-V)_0$. Most star clusters are so far away from us that all cluster members can be taken to be at the same distance. The distance modulus

$$m_{V_0} - M_V = 5 \lg \frac{r}{10 \text{ pc}} , \tag{17.4}$$

will then be the same for all members. In (17.4), m_{V_0} is the apparent, M_V the absolute visual magnitude of a star, and r the distance. It has been assumed that the extinction due to interstellar dust A_V has been determined from multicolour photometry and its effect removed from the observed visual magnitude m_V:

$$m_{V_0} = m_V - A_V .$$

When the observed colour-magnitude diagram of the cluster is formed using the apparent magnitude m_{V_0} rather than the absolute magnitude M_V on the vertical axis, the only change will be that the position of the main sequence is shifted vertically by an amount corresponding to the distance modulus. The observed m_{V_0}-$(B-V)_0$ diagram may now be compared with the Hyades M_V-$(B-V)_0$ diagram used as a standard. By demanding that the main sequences of the two diagrams agree, the distance modulus and hence the distance can be determined. The method is very accurate and efficient. It can be used to determine cluster distances out to many kiloparsec.

17.3 Globular Star Clusters

Globular star clusters usually contain about 10^5 stars. The stars are spherically distributed, and the central densities are about ten times larger than in open clusters. The globular clusters are among the oldest stars in the Milky Way, and are therefore of great importance for studies of stellar evolution. There are about $150-200$ globular clusters in the Milky Way.

The colour-magnitude diagram of a typical globular cluster is shown in Fig. 17.7. The main sequence only contains faint red stars; there is a prominent giant branch, and the horizontal and asymptotic branches are clearly seen. The main sequence is lower than that of the open clusters, because the metal abundance is much lower in the globular clusters.

The horizontal branch stars have a known absolute magnitude, which has been calibrated using principally RR Lyrae type variables. Because the horizontal branch stars are bright, they can be observed even in distant clusters, and thus the distances of globular clusters can be well determined with their help.

Using the known distances, the linear sizes of globular clusters can be calculated. It is found that most of the mass is concentrated to a central core with radius about $1-10$ pc for different clusters. Outside this there is an extended envelope with a radius that may be $10-100$ times larger. At even larger radii stars will escape from the cluster because of the tidal force of the Galaxy.

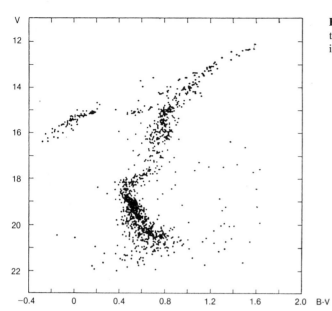

Fig. 17.7. Colour-magnitude diagram of the globular cluster M5. In addition to the main sequence, one can see the giant branch bending to the right and to its left the horizontal branch. [Arp, H. (1962): Astrophys. J. **135**, 311]

The masses of globular clusters can be roughly estimated from the virial theorem, if the stellar velocities in the cluster have been measured. More precise values are obtained by fitting theoretical models to the observed density and velocity distributions. In this way masses in the range $10^4 - 10^6 \, M_\odot$ have been obtained.

The globular clusters in the Milky Way fall into two classes. In the classification given in Table 18.1 these correspond to intermediate and halo population II. The disc globular clusters are concentrated towards the centre and the plane of the Milky Way and they form a system that is rotating with the general rotation of the Milky Way. In contrast, the halo clusters are almost spherically distributed in an extensive distribution reaching out to at least 35 kpc. The system of halo clusters does not rotate, but instead the velocities of individual clusters are uniformly distributed in all directions. The abundance of heavy elements is also different in the two classes of clusters. For disc clusters it is typically about 30% of the solar value, for halo clusters it is only about 1%. The smallest known heavy element abundances, about 10^{-3} times the solar value, have been detected in some halo globular clusters. They therefore give important information about the production of elements in the early Universe and during the formation of the Milky Way.

All globular clusters are old, and the halo clusters are among the oldest known astronomical objects. Determining a precise age is difficult, and requires both accurate observations of the turn-off point of the main sequence in the HR diagram, as well detailed theoretical stellar evolution models. The ages obtained have been about $(13-16) \times 10^9$ years. This age is close to the age of the Universe calculated from its rate of expansion (see Chap. 20).

17.4 Examples

17.4.1   Assume that a globular cluster has a diameter of 40 pc and contains 100 000 stars of one solar mass each.

a) Use the virial theorem to find the average velocity of the stars. You can assume that the average distance between stars equals the radius of the cluster.
b) Find the escape velocity.
c) Comparing these velocities, can you tell something about the stability of the cluster?

a) First, we have to estimate the potential energy. There are $n(n-1)/2 \approx n^2/2$ pairs of stars in the cluster, and the average distance of each pair is R. Thus the potential energy is about

$$U = -G\frac{m^2}{R}\frac{n^2}{2} \, ,$$

where $m = 1 \, M_\odot$. The kinetic energy is

$$T = \tfrac{1}{2} m v^2 n \, ,$$

where v is the root mean square velocity. According to the virial theorem we have $T = -1/2 \, U$, whence

$$\frac{1}{2}m v^2 n = \frac{1}{2}G\frac{m^2}{r}\frac{n^2}{2} \, .$$

Solving for the velocity we get

$$v^2 = \frac{Gmn}{2R}$$

$$= \frac{6.7 \times 10^{-11} \, \mathrm{m^3 \, kg^{-1} \, s^{-2}} \times 2.0 \times 10^{30} \, \mathrm{kg} \times 10^5}{40 \times 3.1 \times 10^{16} \, \mathrm{m}}$$

$$= 1.1 \times 10^7 \, \mathrm{m^2 \, s^{-2}} \, ,$$

which gives $v \approx 3 \, \mathrm{km \, s^{-1}}$.

b) The escape velocity from the edge of the cluster is

$$v_{\mathrm{e}} = \sqrt{\frac{2Gmn}{R}} = \sqrt{4v^2} = 2v = 6 \, \mathrm{km \, s^{-1}} \, .$$

c) No. The average velocity seems to be smaller than the escape velocity, but it was derived from the virial theorem assuming that the cluster is stable.

17.5 Exercises

17.5.1 A globular cluster consists of 100000 stars of the solar absolute magnitude. Calculate the total apparent magnitude of the cluster, if its distance is 10 kpc.

17.5.2 The Pleiades open cluster contains 230 stars within 4 pc. Estimate the velocities of the stars in the cluster using the virial theorem. For simplicity, let the mass of each star be replaced by $1\,M_\odot$.

Chapter 18   The Milky Way

On clear, moonless nights, a nebulous band of light can be seen stretching across the sky. This is the Milky Way (Fig. 18.1). The name is used both for the phenomenon in the sky and for the large stellar system causing it. The Milky Way system is also called the Galaxy − with a capital letter. The general term galaxy is used to refer to the countless stellar systems more or less like our Milky Way.

The band of the Milky Way extends round the whole celestial sphere. It is the reflection of a huge system consisting mostly of stars, among them the Sun. The stars of the Milky Way form a flattened disc-like system. In the direction of the plane of the disc, huge numbers of stars are visible, whereas relatively few are seen in the perpendicular direction. The faint light of distant stars merges into a uniform glow, and therefore the Milky Way appears as a nebulous band to the naked eye. A long-exposure photograph reveals hundreds of thousands of stars (Fig. 18.2).

In the early 17th century, *Galileo Galilei*, using his first telescope, discovered that the Milky Way consists of innumerable stars. In the late 18th century, *William*

Fig. 18.1. The nebulous band of the Milky Way stretches across the entire sky. (Photograph M. and T. Kesküla, Lund Observatory)

Fig. 18.2. About 40° section of the Milky Way between the Cygnus and Aquila constellations. The brightest star at the upper right is Vega (α Lyrae). (Photograph Palomar Observatory)

Herschel attempted to determine the size and shape of the Milky Way by means of star counts. Only early in the 20th century did the Dutch astronomer *Jacobus Kapteyn* obtain the first estimate of the size of the Milky Way. The true size of the Milky Way and the Sun's position in it became clear in the 1920's from *Harlow Shapley's* studies of the space distribution of globular clusters.

In studying the structure of the Milky Way, it is convenient to choose a spherical coordinate system so that the fundamental plane is the symmetry plane of the Milky Way. This is defined to be the symmetry plane of the distribution of neutral hydrogen, and it agrees quite closely with the symmetry plane defined by the distribution of stars in the solar neighbourhood (within a few kpc).

The basic direction in the fundamental plane has been chosen to be the direction of the centre of the Milky Way. This is located in the constellation Sagittarius ($\alpha = 17$ h 45.7 min, $\delta = -29°00'$, epoch 2000.0) at a distance of about 8.5 kpc. The galactic

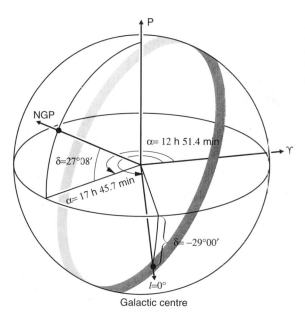

Fig. 18.3. The directions to the galactic centre and the North galactic pole (NGP) in equatorial coordinates. The galactic longitude is measured from the galactic centre along the galactic plane

latitude is counted from the plane of the Galaxy to its pole, going from $0°$ to $+90°$, and to the galactic south pole, from $0°$ to $-90°$. The galactic coordinate system is shown in Fig. 18.3.

18.1 Methods of Distance Measurement

In order to study the structure of the Milky Way, one needs to know how various kinds of objects, such as stars, star clusters and interstellar matter, are distributed in space. The most important ways of measuring the distances will first be considered (see Fig. 18.6a).

Trigonometric Parallaxes. The method of *trigonometric parallaxes* is based on the apparent yearly back-and-forth movement of stars in the sky, caused by the orbital motion of the Earth. Trigonometric parallaxes can be reliably measured out to a distance of about 30 pc; beyond 100 pc this method is no longer useful.

The Motion of the Sun with Respect to the Neighbouring Stars. The Local Standard of Rest. The motion of the Sun with respect to the neighbouring stars is reflected in their proper motions and radial velocities (Fig. 18.4). The point towards which the Sun's motion among the stars seems to be directed is called the *apex*. The opposite point is the *antapex*. The stars near the apex appear to be approaching; their (negative) radial velocities are smallest, on the average. In the direction of the antapex, the largest (positive) radial velocities are observed. On the great circle perpendicular to the apex-antapex direction, the radial velocities are zero on the average, but the proper motions are large. The average proper motions decrease towards the apex and the antapex, but always point from the apex towards the antapex.

In order to study the true motions of the stars, one has to define a coordinate system with respect to which the motions are to be defined. The most practical frame of

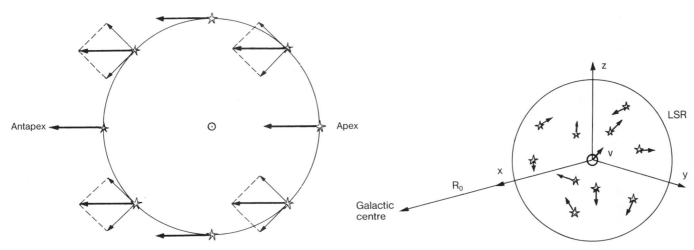

Fig. 18.4. Because of the motion of the Sun towards the apex, the average radial velocity of the nearby stars appears largest in the apex and antapex directions

Fig. 18.5. The local standard of rest, defined by the stars in the solar neighbourhood, moves with respect to the galactic centre. However, the average value of the stellar peculiar velocities with respect to the LSR is zero

reference is defined so that the stars in the solar neighbourhood are at rest, on the average. More precisely, this *local standard of rest* (LSR) is defined as follows:

Let us suppose the velocities of the stars being considered are distributed at random. Their velocities with respect to the Sun, i.e. their radial velocities, proper motions and distances, are assumed to be known. The local standard of rest is then defined so that the mean value of the velocity vectors is opposite to the velocity of the Sun with respect to the LSR. Clearly, the mean velocity of the relevant stars with respect to the LSR will then be zero.

The motion of the Sun with respect to the LSR is found to be:

Apex coordinates $\quad \alpha = 18\,\mathrm{h}\,00\,\mathrm{min}, \quad \delta = +30°; \quad l = 56°, \quad b = +23°$.

Solar velocity $\quad\quad v_0 = 19.7\,\mathrm{km/s}$.

The apex is located in the constellation of Hercules. When the sample of stars used to determine the LSR is restricted to a subset of all the stars in the solar neighbourhood, e.g. to stars of a given spectral class, the sample will usually have slightly different kinematical properties, and the coordinates of the solar apex will change correspondingly.

The velocity of an individual star with respect to the local standard of rest is called the *peculiar motion* of the star. The peculiar velocity of a star is obtained by adding the velocity of the Sun with respect to the LSR to the measured velocity. Naturally, the velocities should be treated as vectors.

The local standard of rest is at rest only with respect to a close neighbourhood of the Sun. The Sun and the nearby stars, and thus also the LSR, are moving round the centre of the Milky Way at a speed that is ten times greater than the typical peculiar velocities of stars in the solar neighbourhood (Fig. 18.5).

Statistical Parallaxes. The velocity of the Sun with respect to neighbouring stars is about 20 km/s. This means that in one year, the Sun moves about 4 AU with respect to the stars.

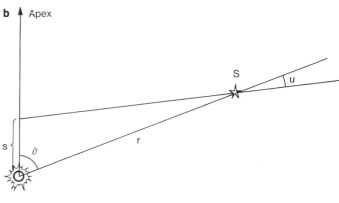

Fig. 18.6. (a) "The cosmic ladder." The main methods of astronomical distance determination with approximate maximum application distances. **(b)** When the Sun has moved the distance s towards the apex, the direction to the star S appears changed by the angle u

Let us consider a star S (Fig. 18.6), whose angular distance from the apex A is ϑ and which is at a distance r from the Sun $\odot$. The star will move away from the apex at the angular velocity $u = \mu_A$ because of solar motion. At the same time, the Sun will move the distance s. The sine theorem for triangles yields

$$r = s\,(\sin\vartheta)/\sin u \approx s \sin\vartheta/u\;, \tag{18.1}$$

because the angle u is small. In addition to the component μ_A due to solar motion, the observed proper motion has a component due to the peculiar velocity of the star. This can be removed by taking an average of (18.1) for a sample of stars, since the peculiar velocities of the stars in the solar neighbourhood can be assumed to be randomly distributed. By observing the average proper motion of objects known to be at the same distance one thus obtains their actual distance. A similar statistical method can be applied to radial velocities.

Objects that are at the same distance can be found as follows. We know that the distance modulus $m - M$ and the distance r are related according to:

$$m - M = 5 \log(r/10\,\mathrm{pc}) + A(r)\;, \tag{18.2}$$

where A is the interstellar extinction. Thus objects that have the same apparent and the same absolute magnitude will be at the same distance. It should be noted that we need not know the absolute magnitude as long as it is the same for all stars in the sample. Suitable classes of stars are main sequence A4 stars, RR Lyrae variables and classical cepheids with some given period. The stars in a cluster are also all at the same distance. This method has been used, for example, to determine the distance to the Hyades as explained in Sect. 17.2.

Parallaxes based on the peculiar or apex motion of the Sun are called statistical or secular parallaxes.

Photometric Parallaxes. The determination of the distance directly from (18.2) is called the photometric method of distance determination and the corresponding parallax, the *photometric parallax*. The most difficult task when using this method usually involves finding the absolute magnitude; there are many ways of doing this. For example, the two-dimensional MKK spectral classification allows one to determine the absolute magnitude from the spectrum. The absolute magnitudes of cepheids can be obtained from their periods. A specially useful method for star clusters is the procedure of main sequence fitting. A condition for the photometric method is that the absolute magnitude scale first be calibrated by some other method.

Trigonometric parallaxes do not reach very far. For example, not a single supergiant or cepheid distance has been measured by this method. The method of statistical parallaxes is indispensable for calibrating the absolute magnitudes of bright objects. When this has been done, the photometric method can be used to obtain distances of objects even further away.

Other examples of indicators of brightness, luminosity criteria, are characteristic spectral lines or the periods of cepheids. Again, their use requires that they first be calibrated by means of some other method. It is a characteristic feature of astronomical distance determinations that the measurement of large distances is based on knowledge of the distances to nearer objects.

18.2 Stellar Statistics

The Stellar Luminosity Function. By systematically observing all stars in the solar neighbourhood, one can find their distribution of absolute magnitudes. This is given by the luminosity function $\Phi(M)$, which gives the relative number of stars with absolute magnitudes in the range $(M - 1/2, M + 1/2)$. No stars appear to be forming at present in the region of space where the luminosity function has been determined. The age of the Milky Way is $10 - 20$ Gy, which means that all stars less massive than $0.9\ M_\odot$ will still be on the main sequence. On the other hand, more massive stars, formed early in the history of the Milky Way, will have completed their evolution and disappeared. Low-mass stars have accumulated in the luminosity function for many generations of star formation, whereas bright, high-mass stars are the result of recent star formation.

By taking into account the different lifetimes of stars of different masses and hence of different magnitudes, one can determine the initial luminosity function $\psi(M)$,

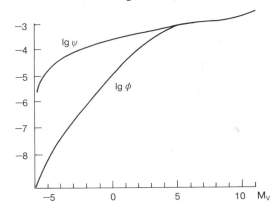

Fig. 18.7. The luminosity function $\Phi(M_V)$ and the initial luminosity function $\Psi(M_V)$ for main sequence stars in the solar neighbourhood. The functions give the number of stars per cubic parsec in the magnitude interval $(M_V - 1/2, M_V + 1/2)$

which gives the brightness distribution at the time of star formation, the zero age main sequence luminosity function. The relation between the function ψ and the observed luminosity function is

$$\psi(M) = \Phi(M) T_0/t_{\mathrm{E}}(M), \tag{18.3}$$

where T_0 is the age of the Milky Way and $t_{\mathrm{E}}(M)$ is the lifetime of stars of magnitude M. The initial luminosity function is shown in Fig. 18.7.

The Fundamental Equation of Stellar Statistics. The Stellar Density. A crucial problem for studies of the structure of the Milky Way is to find out how the density of stars varies in space. The number of stars per unit volume at a distance r in the direction (l, b) from the Sun is given by the stellar density $D = D(r, l, b)$.

The stellar density cannot be directly observed except in the immediate neighbourhood of the Sun. However, it can be calculated if one knows the luminosity function and the interstellar extinction as a function of distance in a given direction. In addition, the number of stars per unit solid angle (e.g. per square arc second) can be determined as a function of limiting apparent magnitude by means of star counts (Fig. 18.8).

Fig. 18.8. The stellar density is determined by means of star counts. In practice, the counting is done on photographic plates. (Cartoon S. Harris)

Let us consider the stars within the solid angle ω in the direction (l, b) and in the distance range $(r, r + dr)$. We let their luminosity function $\Phi(M)$ be the same as in the solar neighbourhood and their unknown stellar density D. The absolute magnitude M of the stars of apparent magnitude m is, as usual,

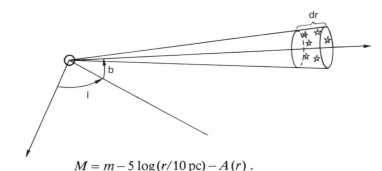

$$M = m - 5 \log(r/10 \, \text{pc}) - A(r) \, .$$

The number of stars of apparent magnitude m in the volume element $dV = \omega r^2 \, dr$ at distance r is (Fig. 18.9)

$$dN(m) = D(r, l, b) \, \Phi\left[m - 5 \lg \frac{r}{10 \, \text{pc}} - A(r) \right] dV \, . \tag{18.4}$$

The stars of apparent magnitude m in the given area of the sky will in reality be at many different distances. In order to obtain their total number $N(m)$, one has to integrate $dN(m)$ over all distances r:

$$N(m) = \int_0^\infty D(r, l, b) \, \Phi\left[m - 5 \lg \frac{r}{10 \, \text{pc}} - A(r) \right] \omega r^2 \, dr \, . \tag{18.5}$$

Equation (18.5) is called the *fundamental equation of stellar statistics*. Its left-hand side, the number of stars of apparent magnitude m in the solid angle ω, is obtained from the observations: one counts the stars of different magnitudes in a chosen area of

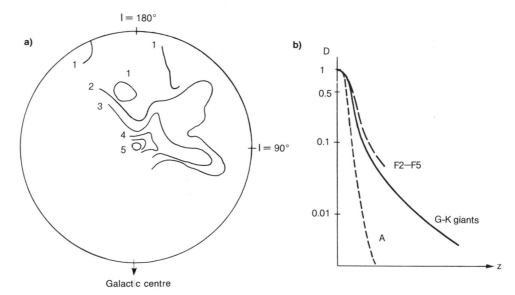

Fig. 18.10a, b. The stellar density near the Sun. **(a)** The stellar density of spectral classes A2 − A5 in the galactic plane, according to S. W. McCuskey. The numbers next to the isodensity curves give the number of stars in $10\,000 \, \text{pc}^3$. **(b)** The distribution of different spectral classes perpendicularly to the galactic plane according to T. Elvius. The density in the galactic plane has been normalized to one

a photographic plate. The luminosity function is known from the solar neighbourhood. The extinction $A(r)$ can be determined for the chosen areas, for instance, by means of multicolour photometry. In order to solve the integral equation (18.5) for $D(r, l, b)$, several methods have been developed, but we shall not go into them here.

Figure 18.10a shows the stellar density in the solar neighbourhood in the plane of the Milky Way, and Fig. 18.10b in the direction perpendicular to the plane. There are several individual concentrations, but e.g. spiral structure cannot be observed in such a limited region of space.

The Distribution of Bright Objects. Using stellar statistical methods, one can only study the close neighbourhood of the Sun, out to about 1 kpc at the most. Absolutely faint objects cannot be observed very far. Since the solar neighbourhood appears to be fairly representative of the general properties of the Milky Way, its study is naturally important, giving information e.g. on the distributions and luminosity functions of stars of various spectral types. However, in order to get an idea of the larger-scale structure of the Milky Way, one has to make use of objects that are as absolutely bright as possible, and which can be observed even at large distances.

Examples of suitable objects are stars of early spectral types, H II regions, OB associations, open star clusters, cepheids, RR Lyrae stars, supergiants and giants of late spectral types, and globular clusters. Some of these objects differ greatly in age, such as the young OB associations, on the one hand, the old globular clusters, on the other. Any differences in their space distribution tell us about changes in the general structure of the Milky Way.

The young optical objects, the H II regions, OB associations and open clusters, are strongly concentrated in the plane of the Milky Way (Table 18.1). Figure 18.11 shows that they also appear to be concentrated in three drawn-out bands, at least within the observed region. Since these types of objects in other galaxies are known to be part of a spiral structure, the observed bands in the Milky Way have been interpreted as portions of three spiral arms passing through the solar neighbourhood. Stars of later spectral types seem to be much more evenly distributed. Apart from a few special directions, interstellar dust limits observations in the galactic plane to within $3-4$ kpc.

Table 18.1. Populations of the Milky Way

	Halo population II	Intermediate population II	Disc population	Old population I	Young population I
Typical objects	Subdwarfs, globular clusters, RR Lyr ($P > 0.4$ d)	Long period variables	Planetary nebulae, novae, bright red giants	A stars, Me dwarfs, classical cepheids	Gas, dust, supergiants, T Tau stars
Average age [10^9 y]	$17-12$	$15-10$	$12-2$	$2-0.1$	0.1
Distance from galactic plane [pc]	2000	700	400	160	120
Vertical velocity [km/s]	75	25	18	10	8
Metal abundance	0.001	0.005	$0.01-0.02$	0.02	$0.03-0.04$

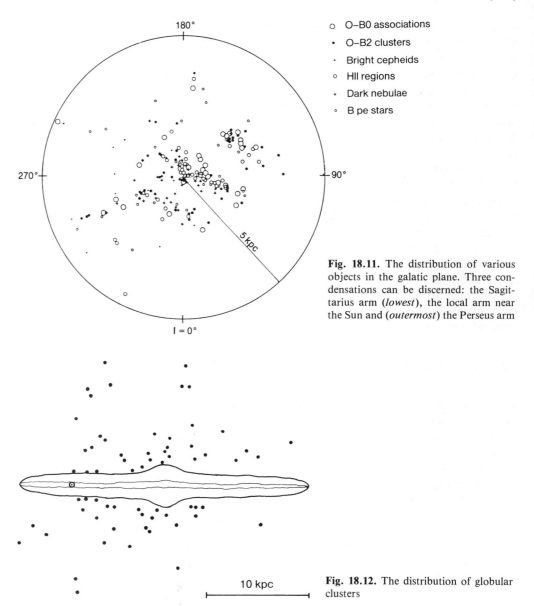

Fig. 18.11. The distribution of various objects in the galatic plane. Three condensations can be discerned: the Sagittarius arm (*lowest*), the local arm near the Sun and (*outermost*) the Perseus arm

Fig. 18.12. The distribution of globular clusters

Old objects, particularly the globular clusters, have an almost spherical distribution about the centre of the Milky Way (Fig. 18.12). The space density of old objects increases towards the galactic centre. They can be used to determine the distance of the Sun from the galactic centre; the value of this distance is about 8.5 kpc.

Stellar Populations. Studies of the motions of the stars in the Milky Way have revealed that the orbits of stars moving in the galactic plane are almost circular. These stars are also usually young, a few hundred million years at the most. They also contain a relatively large amount of heavy elements, about 2% − 4%. The interstellar material similarly moves in the galactic plane in almost circular orbits. On the basis of their motions and their chemical composition, the interstellar medium and the youngest stars are collectively referred to as *population I*.

Outside the plane of the Milky Way, an almost spherically symmetric halo extends out to 50 kpc and beyond. The stellar density is largest near the galactic centre and decreases outwards. The halo contains very little interstellar matter, and its stars are old, up to 15 Gy. These stars are also very metal-poor. Their orbits may be very eccentric and show no preference for the galactic plane. On the basis of these criteria, one defines stars of *population II*. Typical population II objects are the globular clusters, and the RR Lyrae and W Virginis stars.

The stars of population II have large velocities with respect to the local standard of rest, up to more than 300 km/s. In reality, their velocities at the solar distance from the galactic centre are quite small and may sometimes be opposite to the direction of motion of the LSR. The large relative velocities only reflect the motion of the LSR with a velocity of about 220 km/s round the galactic centre.

Between these two extremes, there is a sequence of intermediate populations. In addition to populations I and II, one generally also speaks of a disc population, including the Sun, for instance. The typical motions, chemical composition and age of the various populations (Table 18.1) contain information about the evolution of our Galaxy and about the formation of its stars.

18.3 The Rotation of the Milky Way

Differential Rotation. Oort's Formulae. The flatness of the Milky Way is already suggestive of a general rotation about an axis normal to the galactic plane. Observations of the motions both of stars and of interstellar gas have confirmed this rotation and shown it to be differential. This means that the angular velocity of rotation depends on the distance from the galactic centre (Fig. 18.13). Thus the Milky Way does not rotate like a rigid body. Near the Sun, the rotational velocity decreases with radius.

The observable effects of the galactic rotation were derived by the Dutch astronomer *Jan H. Oort*. Let us suppose the stars are moving in circular orbits about the galactic centre (Fig. 18.14). This approximation is acceptable for population I stars and gas. The star S, seen from the Sun $\odot$ at galactic longitude l at distance r, has circular velocity V at distance R from the centre. Similarly, for the Sun, the galactic radius and velocity are R_0 and V_0. The relative radial velocity v_r of the star with respect to the Sun is the difference between the projections of the circular velocities on the line of sight:

$$v_r = V \cos\alpha - V_0 \sin l \,, \tag{18.6}$$

where α is the angle between the velocity vector of the star and the line of sight. From Fig. 18.14, the angle $CS\odot = \alpha + 90°$. By applying the sine theorem to the triangle $CS\odot$ one obtains

$$\frac{\sin(\alpha + 90°)}{\sin l} = \frac{R_0}{R} \quad \text{or}$$

$$\cos\alpha = \frac{R_0}{R} \sin l \,. \tag{18.7}$$

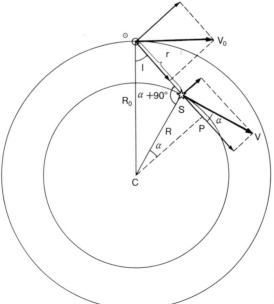

Fig. 18.13a – d. The effect of differential rotation on the radial velocities and proper motions of stars. **(a)** Near the Sun the orbital velocities of stars decrease outwards in the Galaxy. **(b)** The relative velocity with respect to the Sun is obtained by subtracting the solar velocity from the velocity vectors in **(a)**. **(c)** The radial components of the velocities with respect to the Sun. This component vanishes for stars on the same orbit as the Sun. **(d)** The tangential components of the velocities

Fig. 18.14. In order to derive Oort's formulae, the velocity vectors of the Sun and the star S are divided into components along the line ⊙S and normal to it

Denoting the angular velocity of the star by $\omega = V/R$ and that of the Sun by $\omega_0 = V_0/R_0$, one obtains the observable radial velocity in the form

$$v_r = R_0(\omega - \omega_0) \sin l \,. \tag{18.8}$$

The tangential component of the relative velocity of the Sun and the star is obtained as follows. From Fig. 18.14,

$$v_t = V \sin \alpha - V_0 \cos l = R \omega \sin \alpha - R_0 \omega_0 \cos l \,.$$

The triangle $\odot CP$ gives

$$R \sin \alpha = R_0 \cos l - r \,, \quad \text{and hence}$$

$$v_t = R_0(\omega - \omega_0) \cos l - \omega r \,. \tag{18.9}$$

Oort noted that in the close neighbourhood of the Sun $(r \ll R_0)$, the difference of the angular velocities will be very small. Therefore a good approximation for the exact equations (18.8) and (18.9) is obtained by keeping only the first term of the Taylor series of $\omega - \omega_0$ in the neighbourhood of $R = R_0$:

$$\omega - \omega_0 = \left(\frac{d\omega}{dR}\right)_{R=R_0}(R - R_0) + \ldots \,.$$

Using $\omega = V/R$ and $V(R_0) = V_0$, one finds

$$\omega - \omega_0 \approx \frac{1}{R_0^2}\left[R_0\left(\frac{dV}{dR}\right)_{R=R_0} - V_0\right](R - R_0) \,.$$

For $R \approx R_0 \gg r$, the difference $R - R_0 \approx -r \cos l$. One thus obtains an approximate form of (18.8):

$$v_r \approx \left[\frac{V_0}{R_0} - \left(\frac{dV}{dR}\right)_{R=R_0}\right] r \cos l \sin l \quad \text{or}$$

$$v_r \approx A r \sin 2l \,, \tag{18.10}$$

where A is a characteristic parameter of the solar neighbourhood of the Galaxy, the *first Oort constant*.

$$A = \frac{1}{2}\left[\frac{V_0}{R_0} - \left(\frac{dV}{dR}\right)_{R=R_0}\right] \,. \tag{18.11}$$

For the tangential relative velocity, one similarly obtains, since $\omega r \approx \omega_0 r$:

$$v_t \approx \left[\frac{V_0}{R_0} - \left(\frac{dV}{dR}\right)_{R=R_0}\right] r \cos^2 l - \omega_0 r \,.$$

Because $2 \cos^2 l = 1 + \cos 2l$, this may be written

$$v_t \approx A r \cos 2l + B r \,, \tag{18.12}$$

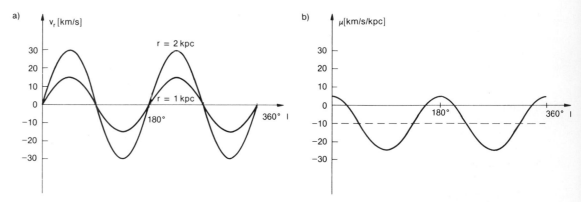

Fig. 18.15a,b. The velocity components due to differential rotation according to Oort's formulae as functions of galactic longitude. **(a)** Radial velocities for objects at a distance of 1 and 2 kpc. (Compare with Fig. 18.13.) Strictly, the longitude at which the radial velocity vanishes depends on the distance. Oort's formulae are valid only in the close vicinity of the Sun. **(b)** Proper motions

where A is the same as before and B, the *second Oort constant*, is

$$B = -\frac{1}{2}\left[\frac{V_0}{R_0} + \left(\frac{dV}{dR}\right)_{R=R_0}\right]. \tag{18.13}$$

The proper motion $\mu = v_t/r$ is then given by the expression

$$\mu \approx A\cos 2l + B. \tag{18.14}$$

Equation (18.10) says that the observed radial velocities of stars at the same distance should be a double sine curve as a function of galactic longitude. This has been confirmed by observations (Fig. 18.15a). If the distance to the stars involved is known, the amplitude of the curve determines the value of the Oort constant A.

Independently of distance, the proper motions of the stars form a double sine wave as a function of galactic longitude, as seen in Fig. 18.15b. The amplitude of the curve is A and its mean value, B.

In 1927, on the basis of this kind of analysis, Oort established that the observed motions of the stars indicated a diffential rotation of the Milky Way. Taking into account an extensive set of observational data, the International Astronomical Union IAU has confirmed the present recommended values for the Oort constants:

$$A = 15\,\text{km s}^{-1}\,\text{kpc}^{-1}, \quad B = -10\,\text{km s}^{-1}\,\text{kpc}^{-1}.$$

The Oort constants obey some interesting relations. By subtracting (18.13) from (18.11), one obtains

$$A - B = \frac{V_0}{R_0} = \omega_0. \tag{18.15}$$

Adding (18.13) and (18.11) gives

$$A + B = -\left(\frac{dV}{dR}\right)_{R=R_0}. \tag{18.16}$$

Knowing the values of A and B, one can calculate the angular velocity $\omega_0 = 0.0053''/$ year, which is the angular velocity of the local standard of rest around the galactic centre. If one could see the pointlike centre, this would be its proper motion.

The circular velocity of the Sun and the LSR can be measured in an independent way by using extragalactic objects as a reference. In this way, a value of about 220 km/s has been obtained for V_0. Using (18.15), one can now calculate the distance of the galactic centre R_0. The result is about 8.5 kpc, in good agreement with the distance to the centre of the globular cluster system. The direction to the galactic centre obtained from the distribution of radial velocities and proper motions by means of (18.10) and (18.14) also agrees with other measurements.

The orbital period of the Sun in the Galaxy according to these results is about 2.5×10^8 years. Since the Sun's age is nearly 5×10^9 years, it has made about 20 revolutions around the galactic centre. At the end of the last revolution, the Carboniferous period had ended on Earth and the first mammals would soon appear.

The Distribution of Interstellar Matter. Radio radiation from interstellar gas, in particular that of neutral hydrogen, is not strongly absorbed or scattered by interstellar dust. It can therefore be used to map the structure of the Milky Way on large scales. Radio signals can be detected even from the opposite edge of the Milky Way.

The position of a radio source, for example an H I cloud, in the Galaxy cannot be directly determined. However, an indirect method exists, based on the differential rotation of the Galaxy.

Figure 18.16 is a schematic view of a situation in which gas clouds on the circles P_1, P_2, ... are observed in the direction $l(-90° < l < 90°)$. The angular velocity increases inwards, and therefore the greatest angular velocity along the line of sight is obtained at the point P_k, where the line of sight is tangent to a circle. This means that the radial velocity of the clouds in a fixed direction grows with distance up to the maximum velocity at cloud P_k:

$$v_{r,\,max} = R_k(\omega - \omega_0)\,, \qquad\qquad\qquad (18.17)$$

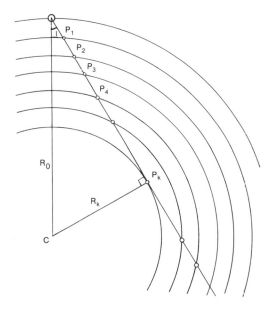

Fig. 18.16. Clouds P_1, P_2,... seen in the same direction at various distances

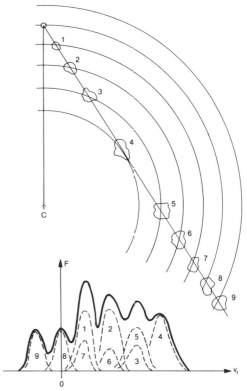

Fig. 18.17. The radial velocity as a function of distance (shown schematically)

Fig. 18.18. Clouds at different distances have different velocities and therefore give rise to emission lines with different Doppler shifts. The observed flux density profile (*continuous curve*) is the sum of the line profiles of all the individual line profiles (*dashed curves*). The numbers of the line profiles correspond to the clouds in the upper picture ▶

where $R_k = R_0 \sin l$. The distance of cloud P_k from the Sun is $r = R_0 \cos l$. When r increases further, v_r decreases monotonically. Figure 18.17 shows how the observed radial velocity in a given direction varies with distance r, if the gas moves in circular orbits and the angular velocity decreases outwards.

The neutral hydrogen 21-cm line has been particularly important for mapping the Milky Way. Figure 18.18 gives a schematic view of how the hydrogen spectral line is made up of the radiation of many individual concentrations of neutral hydrogen, clouds or spiral arms. The line component produced by each cloud has a wavelength which depends on the radial velocity of the cloud and a strength depending on its mass and density. The total emission is the sum of these contributions.

By making observations at various galactic longitudes and assuming that the clouds form at least partly continuous spiral arms, the distribution of neutral hydrogen in the galactic plane can be mapped. Figure 16.17 shows a map of the Milky Way obtained from 21-cm line observations of neutral hydrogen. It appears that the neutral hydrogen is concentrated in spiral arms. However, interpretation of the details is difficult because of the uncertainties of the map. In order to obtain the distances to the gas clouds, one has to know the *rotation curve*, the circular velocity as a function of the galactic radius. This is determined from the same radial velocity observations and involves assumptions concerning the density and rotation of the gas. The interpretation of the spiral structure obtained from radio observations is also still uncertain. For example, it is difficult to fit the radio spiral structure to the one obtained near the Sun from optical observations of young stars and associations.

The Rotation, Mass Distribution and Total Mass of the Milky Way. In (18.17), the galactic longitude l gives the the galactic radius R_k of the clouds with maximum radial velocity. By making observations at different longitudes, one can therefore use (18.17) to determine the angular velocity ω of the gas for various distances from the galactic centre. (Circular motions must be assumed.) In this way, the rotation curve $\omega = \omega(R)$ and the corresponding velocity curve $V = V(R)$ $(= \omega R)$ are obtained.

Figure 18.19 shows the rotation curve of the Milky Way. Its central part rotates like a rigid body, i.e. the angular velocity is independent of the radius. Outside this region, the velocity first drops and then begins to rise gradually. A maximum velocity is reached at about 8 kpc from the centre. Near the Sun, about 8.5 kpc from the centre, the rotational velocity is about 220 km/s. According to earlier opinions, the velocity continues to decrease outwards. This would mean that most of the mass is inside the solar radius. This mass could then be determined from Kepler's third law. According to (7.34),

$$M = R_0 V_0^2 / G .$$

Using the values $R_0 = 8.5$ kpc and $V_0 = 220$ km/s, one obtains

$$M = 1.9 \times 10^{41} \text{ kg} = 1.0 \times 10^{11} \, M_\odot .$$

The escape velocity at radius R is

$$V_e = \sqrt{\frac{2GM}{R}} = V\sqrt{2} . \tag{18.18}$$

This gives an escape velocity near the Sun $V_e = 310$ km/s. One therefore should not see many stars moving in the direction of galactic rotation, $l = 90°$, with velocities larger than 90 km/s with respect to the local standard of rest, since the velocity of such stars would exceed the escape velocity. This has been confirmed by observations.

The preceding considerations have been based on the assumption that near the Sun, the whole mass of the Galaxy can be taken to be concentrated in a central point. If this were true, the rotation curve should be of the Keplerian form, $V \propto R^{-1/2}$. That this is not the case can be established from the values of the Oort constants.

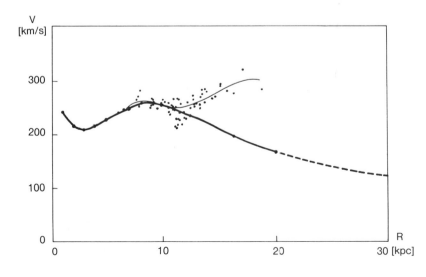

Fig. 18.19. Rotation curve of the Milky Way based on the motions of hydrogen clouds. Each point represents one cloud. The thick line represents the rotation curve determined by Maarten Schmidt in 1965. If all mass were concentrated within the radius 20 kpc, this curve would continue according to Kepler's third law (*broken line*). The rotation curve determined by Leo Blitz on the basis of more recent observations begins to rise again at 12 kpc

The derivative of the Keplerian relation

$$V = \sqrt{\frac{GM}{R}} = \sqrt{GM} R^{-1/2}$$

yields

$$\frac{dV}{dR} = -\frac{1}{2}\sqrt{GM} R^{-3/2} = -\frac{1}{2}\frac{V}{R} .$$

Using the properties (18.15) and (18.16) of the Oort constants, one finds

$$(A - B)/(A + B) = 2 \tag{18.19}$$

for a Keplerian rotation curve. This disagrees with the observed value and thus the assumed Keplerian law does not apply.

The mass distribution in the Milky Way can be studied on the basis of the rotation curve. One looks for a suitable mass distribution, such that the correct rotation curve is reproduced. Recently, distant globular clusters have been discovered, showing that the Milky Way is larger than expected. Also, observations of the rotation curve outside the solar circle suggest that the rotational velocity might begin to rise again. These results suggest that the mass of the Galaxy might be as much as ten times larger than had been thought.

18.4 The Structure and Evolution of the Milky Way

Spiral Structure. As mentioned earlier, the Milky Way appears to be a *spiral galaxy*, but there is no general agreement on the detailed form of the spiral pattern. For example, in 1976, *Y. M. Georgelin* and *Y. P. Georgelin* determined the distances of H II regions by radio and optical observations. In the optical region their method is independent of assumptions about the galactic rotation law. They then fitted four spiral arms through the H II regions (Fig. 18.20).

The cause of the spiral structure is a long-standing problem. A small perturbation in the disc will quickly by stretched into a spiral shape by differential rotation. However, such a spiral would disappear in a few galactic revolutions, a few hundred million years.

An important step forward in the study of the spiral structure was the *density wave theory* developed by *C. C. Lin* and *Frank H. Shu* in the 1960's. The spiral structure is taken to be a wavelike variation of the density of the disc. The spiral pattern rotates as a solid body with an angular velocity about half that of the galactic rotation, while the stars and gas in the disc pass through the wave.

The density wave theory explains in a natural way why young objects, like molecular clouds, H II regions and bright young stars are found in the spiral arms. As gas passes through the wave, it is strongly compressed. The internal gravity of the gas clouds then becomes more important and causes them to collapse and form stars.

It takes about 10^7 years for the material to pass through a spiral arm. By that time, the hot, bright stars have finished their evolution, their ultraviolet radiation has ceased and the H II regions have disappeared. The less massive stars formed in the spiral arms are spread out in the disc by their peculiar velocities.

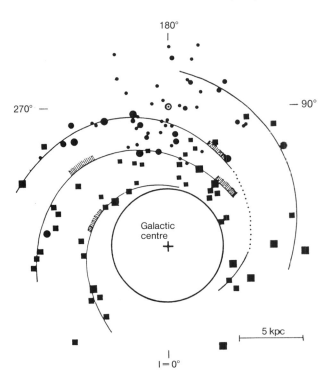

Fig. 18.20. The positions of H II regions according to optical (*circles*) and radio (*squares*) observations. Four spiral arms have been fitted to the data

It is not yet clear what gives rise to the spiral wave. For some further remarks, see Sect. 19.3.

The Galactic Centre. Our knowledge of the centre of the Milky Way is mostly based on radio and infrared observations. In the optical region the view to the centre is blocked by the dark clouds in the Sagittarius spiral arm about 2 kpc from us. The galactic centre is interesting because it may be a small-scale version of the much more violently active nuclei of some external galaxies (see Sect. 19.7). It therefore provides opportunities to study at close hand phenomena related to active galaxies. Since active galactic nuclei are throught to contain black holes with masses larger than $10^7 M_\odot$, there may also be a large black hole at the galactic centre.

As one approaches the galactic centre the stellar density continues to rise towards a sharp central peak. In contrast, the galactic gas disc has a central hole of radius about 3 kpc. According to one model the central bulge of the Milky Way is bar-shaped. The

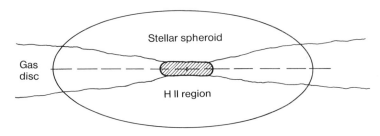

Fig. 18.21. The structure of the central parts of the Milky Way

effect of this bar is to channel gas into the galactic nucleus leaving a gas-free zone at larger radii.

Inside the central hole is a dense nuclear gas disc. Its radius is about 1.5 kpc in neutral hydrogen, but most of its mass is molecular and concentrated within 300 pc of the nucleus. In this region the mass of molecular gas is about $10^8 M_\odot$, or 5% of the total molecular mass of the Milky Way. The molecular clouds are probably confined by the pressure from surrounding very hot ($T \sim 10^8$ K) gas. This hot gas may then expand vertically, forming a galactic wind. Gas lost to a wind or to star formation is replenished with infalling gas from larger radii.

The central 10 pc are dominated by the radio continuum source Sgr A (see Colour Plate 31) and a dense star cluster observed in the infrared. There is also molecular gas with complex motions and signs of star formation activity. Within Sgr A is a unique point-like radio continuum source known as Sgr A*. The position of Sgr A* agrees to within 1″ with the centre of a cluster of stars that is much denser than anything observed in the galactic disc. If the galactic centre contains a large black hole, Sgr A* is the natural candidate.

The luminosity of the galactic centre could be provided by the central star cluster, but there are still reasons to expect that a large black hole may be present. The central mass distribution can be estimated by modelling the observed motions of stars and gas (cf. Sect. 19.2). The best fit with the observations is obtained with models with an extended stellar mass distribution together with a point-mass of a few times $10^6 M_\odot$. The size of Sgr A* measured by very long baseline interferometry is less than 10 AU. The most plausible explanation of this very compact structure is that Sgr A* is indeed a black hole with mass of a few million solar masses.

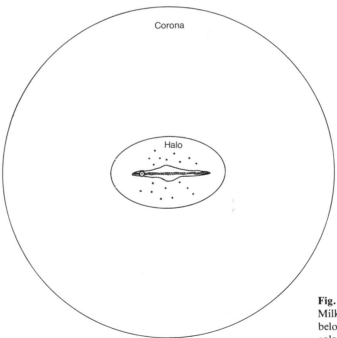

Fig. 18.22. General structure of the Milky Way. The globular clusters belong to the halo surrounding the galactic disc

18.5 Examples

18.5.1 Show that if the stars are uniformly distributed in space and there is no inter-stellar extinction, the number of stars brighter than apparent magnitude m is

$$N_0(m) = N_0(0) \times 10^{0.6m} .$$

Let us suppose first that all stars have the same absolute magnitude M. The distance in parsecs of those stars having the apparent magnitude m is

$$r = 10 \times 10^{0.2(m-M)} .$$

In order to appear brighter than m, a star has to be within a sphere of radius r. Because the stellar density is constant, the number of such stars is proportional to the volume:

$$N_0(m) \propto r^3 \propto 10^{0.6m} . \tag{1}$$

The result does not depend on the absolute magnitudes of the stars, so that the same result still applies when the magnitudes are not equal, as long as the luminosity function does not depend on distance. Thus Eq. (1) is generally valid under the stated conditions.

18.5.2 *The Estimation of Distances by Means of Oort's Formulae*
An object in the galactic plane at longitude $l = 45°$ has the radial velocity of 30 km/s with respect to the LSR. What is its distance?

According to (18.10),

$$v_{\rm r} = A\, r \sin 2l .$$

Thus

$$r = \frac{v_{\rm r}}{A \sin 2l} = \frac{30}{15} \frac{\text{km s}^{-1}}{\text{km s}^{-1}\text{kpc}^{-1}} = 2\,\text{kpc} .$$

In practice, the peculiar velocities are so large that this method cannot be used for distance determination. Oort's formulae are mostly suitable only for statistical studies.

18.5.3 *Discussion of the Gravitational Field of a Uniform Disk*

It can be shown that the gravitational field of a homogeneous infinite thin disk is constant and directed towards the plane of the disk. If the mass per unit area on the disk is σ, the gravitational field is

$$g = 2\pi G\sigma .$$

A test particle located outside the plane will therefore get a constant acceleration towards the plane. Taking a numerical example, assume a mass of $10^{11}\, M_\odot$ distributed uniformly on a circular disk 20 kpc in diameter. The mass per unit area will be

$$\sigma = \frac{10^{11} \times 2 \times 10^{30}\,\text{kg}}{\pi (10^4 \times 3.086 \times 10^{16}\,\text{m})^2} = 0.67\,\text{kg m}^{-2} .$$

The corresponding gravitational field is

$$g = 2.8 \times 10^{-10} \, \mathrm{m \, s^{-2}} \ .$$

Let a star be located at $d = 1$ kpc above the plane, initially at rest (not very near the edge of the disk, in order to keep our approximation valid). The disk will pull it towards the plane, and when the star crosses the plane, it has acquired a velocity given by

$$v = \sqrt{2gd} = 130 \, \mathrm{km \, s^{-1}} \ .$$

The time required to reach the plane is

$$t = v/g = 15 \times 10^6 \, \mathrm{y} \ .$$

18.6 Exercises

18.6.1 Assume that the Sun and a star move around the Galaxy at the same speed in the same circular orbit in the galactic plane. Show that the proper motion of the star is independent of its distance. How big is this proper motion?

18.6.2 a) A cepheid has a radial velocity of $80 \, \mathrm{km \, s^{-1}}$, and its galactic longitude is $145°$. What is the distance of the cepheid? b) The period of the cepheid is 3.16 d, and the apparent visual magnitude is 12.3. What is the distance derived from this information? Are the distances consistent?

18.6.3 a) How many of the nearest stars (Table E.15) are also among the brightest stars (Table E.16)? Explain. b) If the stellar density were constant, how many stars would there be within the distance of Canopus?

18.6.4 a) Assume that the Galaxy is a homogeneous disk and the Sun lies in the central plane of the disk. The absolute magnitude of a star is M, galactic latitude b, and distance from the central plane z. What is the apparent magnitude of the star, if the extinction inside the Galaxy is a mag kpc^{-1}? b) Assume that the thickness of the galactic disk is 200 pc. Find the apparent magnitude of a star with $M = 0.0$, $b = 30°$, distance $r = 1$ kpc, and $a = 1$ mag kpc^{-1}.

Chapter 19   Galaxies

The galaxies are the fundamental building blocks of the Universe. Some of them are very simple in structure, containing only normal stars and showing no particular individual features. There are also galaxies that are almost entirely made of neutral gas. On the other hand, others are complex systems built up from many separate components – stars, neutral and ionized gas, dust, molecular clouds, magnetic fields, cosmic rays.... The galaxies may form small groups or large clusters in space. At the centre of many galaxies, there is a compact nucleus that may sometimes be so bright that it overwhelms all the normal radiation of the galaxy.

The luminosities of the brightest normal galaxies may correspond to 10^{12} solar luminosities, but most of them are much fainter – the smallest ones that have been discovered are about $10^5 L_\odot$. Since galaxies do not have sharp outer edges, to some extent their masses and radii depend on how these quantities are defined. If only the bright central parts are included a giant galaxy may typically have a mass of about $10^{13} M_\odot$ and a radius of 30 kpc, and a dwarf, correspondingly, $10^7 M_\odot$ and 0.5 kpc. In addition, it seems that the outer parts of most galaxies contain large quantities of non-luminous matter that might increase galaxy masses by an order of magnitude.

The density of matter may be very different in different galaxies and in different parts of the same galaxy. Thus the evolution of a galaxy will be the result of processes occurring on vastly different time and energy scales, and no generally accepted comprehensive picture of it exists as yet. In the following, the most important observed properties of galaxies will be presented. Many of them still await an explanation in current theories of galaxy evolution.

19.1 The Classification of Galaxies

A useful first step towards an understanding of galaxies is a classification based on their various forms. Although such a morphological classification must always be to some extent subjective, it provides a framework within which the quantitative properties of galaxies can be discussed in a systematic fashion. However, it should always be remembered that the picture thus obtained will be limited to those galaxies that are large and bright enough to be easily visible in the sky. An idea of the consequent limitations can be obtained from Fig. 19.1, showing the radii and magnitudes of normal galaxies. One sees that only within a narrow region of this diagram can galaxies be easily found. If a galaxy has too large a radius for its magnitude (small surface brightness), it will disappear in the background light from the night sky. On the other hand, if its radius is too small it looks like a star and is not noticed on a photographic plate. In the following we shall mainly be concerned with bright galaxies that fit within these limits.

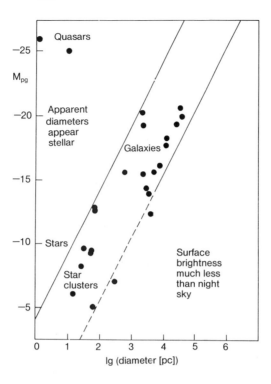

Fig. 19.1. Magnitudes and diameters of observable extragalactic objects. Objects to the upper left look like stars. The quasars in this region have been discovered on the basis of their spectra. Objects to the lower right have a surface brightness much smaller than that of the night sky. In recent years large numbers of low surface brightness galaxies have been discovered in this region. [Arp, H. (1965): Astrophys. J. **142**, 402]

If a classification is to be useful it should at least roughly correspond to important physical properties of the galaxies. Most classifications accord in their main features with the one put forward by *Edwin Hubble* in 1926. Hubble's own version of the *Hubble sequence* is shown in Fig. 19.2. The various types of galaxies are ordered in a sequence from early to late types. There are three main types: *elliptical*, *lenticular*, and *spiral* galaxies. The spirals are divided into two sequences, *normal* and *barred* spirals. In addition, Hubble included a class of *irregular galaxies*.

The elliptical galaxies appear in the sky as elliptical concentrations of stars, in which the density falls off in a regular fashion as one goes outwards. Usually there are no signs of interstellar matter (dark bands of dust, bright young stars). The ellipticals differ from

Fig. 19.2. The Hubble sequence in Hubble's 1936 version. At this stage the existence of type S0 was still doubtful. Photographs of the Hubble types are shown in Figs. 19.6 and 19.15 (E); 19.3 and 19.4 (S0 and S); 19.12 (S and Irr II); 19.5 (Irr I and dE). [Hubble, E. P. (1936): *The Realm of the Nebulae* (Yale University Press, New Haven)]

each other only in shape and on this basis they are classified as E0, E1, ..., E7. If the major and minor axes of an elliptical galaxy are a and b, its type is defined to be En, where

$$n = 10 \left(1 - \frac{b}{a}\right). \tag{19.1}$$

An E0 galaxy thus looks circular in the sky. The apparent shape of an E galaxy depends on the direction from which it is seen. In reality, an E0 galaxy may therefore be truly spherical or it may be a circular disc viewed directly from above.

A later addition to the Hubble sequence is a class of *giant elliptical galaxies* denoted cD. These are generally found in the middle of clusters of galaxies. They consist of a central part looking like a normal elliptical surrounded by an extended fainter halo of stars.

In the Hubble sequence the lenticulars or S0 galaxies are placed between the elliptical and the spiral types. Like the ellipticals they contain only little interstellar matter and show no signs of spiral structure. However, in addition to the usual elliptical stellar component they also contain a flat disc made up of stars. In this respect they are like spiral galaxies (Figs. 19.3, 19.4).

The characteristic feature of spiral galaxies is a more or less well-defined spiral pattern in the disc. Spiral galaxies consist of a central *bulge*, which is structurally similar to an E galaxy, and a stellar disc, as in an S0 galaxy. In addition to these there is a thin disc of gas and other interstellar matter, in which young stars are being born, which forms the spiral pattern. There are two sequences of spirals, normal Sa-Sb-Sc, and barred SBa − SBb − SBc spirals. In the barred spirals the spiral pattern ends at a central bar, whereas in the normal spirals the spiral pattern may end at an inner ring or continue all the way to the centre. The position of a galaxy within the spiral sequence is determined on the basis of three criteria (which are not always in agreement): later types have a smaller central bulge, more narrow spiral arms and a more open spiral pattern. The Milky Way Galaxy is thought to be of type SABbc (intermediate between Sb and Sc, and between normal and barred spirals).

The classical Hubble sequence is essentially based on bright galaxies; faint galaxies have been less easy to fit into it (Figs. 19.5, 19.6). For example, the irregular galaxies of the original Hubble sequence can be divided into the classes Irr I and Irr II. The Irr I galaxies form a continuation of the Hubble sequence towards later types beyond the Sc galaxies. They are rich in gas and contain many young stars. Type Irr II are dusty, somewhat irregular small ellipticals. Other types of dwarf galaxies are often introduced. One example is the *dwarf spheroidal* type dE, similar to the ellipticals but with a much less centrally concentrated star distribution. Another is the *blue compact galaxies* (also called extragalactic H II regions), in which essentially all the light comes from a small region of bright, newly formed stars.

19.2 Luminosities and Masses

Distances. In order to determine the absolute luminosities and linear dimensions of galaxies one needs to know their distances. Distances are also needed in order to estimate the masses of galaxies because these estimates depend on the absolute linear size. Dis-

Fig. 19.3. The classification of normal spiral and S0 galaxies. (Mt. Wilson Observatory)

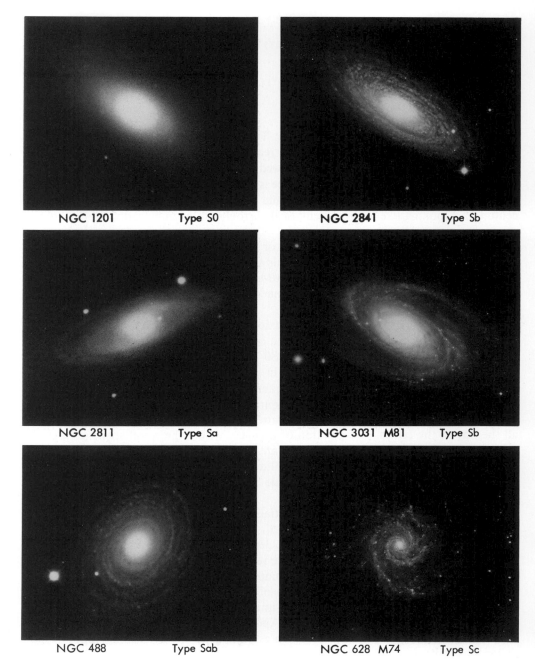

NGC 1201   Type S0

NGC 2841   Type Sb

NGC 2811   Type Sa

NGC 3031 M81   Type Sb

NGC 488   Type Sab

NGC 628 M74   Type Sc

tances within the Local Group can be measured with the same methods as used inside the Milky Way, most importantly by means of variable stars. On the very large scale (beyond 50 Mpc) the distances can be deduced on the basis of the expansion of the Universe (see Sect. 20.1). In order to connect these two regions one needs methods of distance determination based on the properties of individual galaxies.

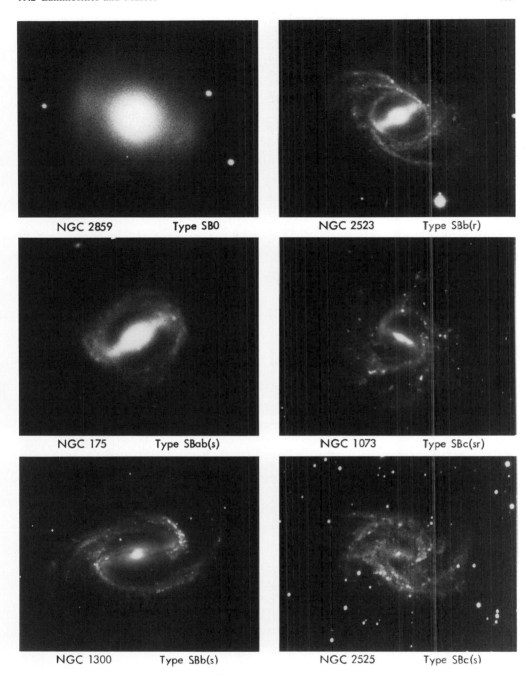

To some extent local distances can be determined using structural components of galaxies, such as the sizes of H II regions or the magnitudes of globular clusters. However, to measure distances of tens of megaparsecs one needs a distance-independent method to determine the absolute luminosities of entire galaxies. Several such methods have been proposed. For example, a luminosity classification has been introduced for

Fig. 19.5. **(a)** The Small Magellanic Cloud (Hubble type Irr I), a dwarf companion of the Milky Way. (Royal Observatory, Edinburgh). **(b)** The Sculptor Galaxy, a dE dwarf spheroidal. (ESO)

a

b

Fig. 19.6. M32 (type E2), a small elliptical companion of the Andromeda Galaxy. (NOAO/Kitt Peak National Observatory)

late spiral types by *Sidney van den Bergh*. This is based on a correlation between the luminosity of a galaxy and the prominence of its spiral pattern.

Other distance indicators are obtained if there is some intrinsic property of the galaxy that is correlated with its total luminosity, and which can be measured independently of the distance. Such properties are the colour, the surface brightness and the internal velocities in galaxies. All of these have been used to measure distances to both spiral and elliptical galaxies. For example, the absolute luminosity of a galaxy should depend on its mass. The mass, in turn, will be reflected in the velocities of stars and gas in the galaxy. Accordingly, there is a relationship between the absolute luminosity and the velocity dispersion (in ellipticals) and the rotational velocity (in spirals). Since rotational velocities can be measured very accurately from the width of the hydrogen 21-cm line, the latter relationship (known as the *Tully-Fisher relation*) is perhaps the best distance indicator currently available.

The luminosities of the brightest galaxies in clusters have been found to be reasonably constant. This fact can be used to measure even larger distances, providing a method which is important in cosmology.

Luminosities. The definition of the total luminosity of a galaxy is to some extent arbitrary, since galaxies do not have a sharp outer edge. The usual convention is to measure the luminosity of a galaxy out to a given value of the surface brightness, e.g. to 26.5 mag/sq. arc sec. For a given Hubble type the total luminosity L may vary widely.

As in the case of stars, the distribution of galaxy luminosities is described by the *luminosity function* $\Phi(L)$ (Fig. 19.7). This is defined so that the space density of galaxies with luminosities between L and $L+dL$ is $\Phi(L)\,dL$. It can be determined from the observed magnitudes of galaxies, once their distances have been estimated in some way. In practice, one assumes some suitable functional form for $\Phi(L)$, which is then fitted to the observations. One common form is *Schechter's luminosity function*,

$$\Phi(L)\,dL = \Phi^* \left(\frac{L}{L^*}\right)^\alpha e^{-L/L^*} d\left(\frac{L}{L^*}\right) . \tag{19.2}$$

The values of the parameters Φ^*, L^*, α are observationally determined for different types of objects; in general, they will be functions of position.

The shape of the luminosity function is described by the parameters α and L^*. The relative number of faint galaxies is described by α. Since its observed value is about -1.1, the density of galaxies grows monotonically as one goes towards fainter luminosities. The luminosity function falls off steeply above the luminosity L^*, which therefore represents a characteristic luminosity of bright galaxies. The observed L^* corresponds to an absolute magnitude $M^* = -21.0$ mag. The corresponding magnitude for the Milky Way Galaxy is probably -20.2 mag. The cD giant galaxies do not obey this brightness distribution; their magnitudes may be -24 mag and even brighter.

The parameter Φ^* is proportional to the space density of galaxies and is therefore a strong function of position. Since the total number density of galaxies predicted by relation (19.2) is infinite, we define $n^* =$ density of galaxies with luminosity $>L^*$. The observed average value of n^* over a large volume of space is $n^* = 3.5 \times 10^{-3}$ Mpc^{-3}.

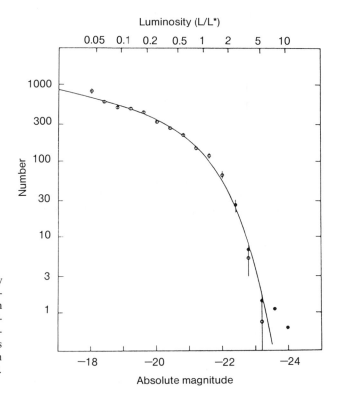

Fig. 19.7. Compound luminosity function of thirteen clusters of galaxies. The open symbols have been obtained by omitting the cD galaxies. The distribution is then well-described by (19.2). The cD galaxies (*filled symbols*) cause a deviation at the bright end. [Schechter, P. (1976): Astrophys. J. **203**, 297]

The mean separation between galaxies corresponding to this density is 4 Mpc. Since most galaxies are fainter than L^*, and since, in addition, they often belong to groups, we see that the distances between normal galaxies are generally not much larger than their diameters.

Masses. The distribution of mass in galaxies is a crucial quantity both for cosmology and for theories of the origin and evolution of galaxies. Observationally it is determined from the velocities of the stars and interstellar gas. Total masses of galaxies can also be derived from their motions in clusters of galaxies. The results are usually given in terms of the corresponding mass-luminosity ratio M/L, using the solar mass and luminosity as units. The value measured in the solar neighbourhood of the Milky Way is $M/L = 3$. If M/L were constant, the mass distribution could be determined from the observed luminosity distribution by multiplying by M/L.

The masses of elliptical galaxies may be obtained from the stellar velocity dispersion given by the broadening of spectral lines. The method is based on the virial theorem (see Sect. 7.9), which says that, in a system in equilibrium, the kinetic energy T and the potential energy U are related according to the equation

$$2T + U = 0 .\tag{19.3}$$

Since ellipticals rotate slowly, the kinetic energy of the stars may be written

$$T = Mv^2/2 ,\tag{19.4}$$

where M is the total mass of the galaxy and v the velocity width of the spectral lines. The potential energy is

$$U = -GM^2/2R ,\tag{19.5}$$

where R is a suitable average radius of the galaxy that can be estimated or calculated from the light distribution. Introducing (19.4) and (19.5) into (19.3), we obtain

$$M = 2v^2R/G .\tag{19.6}$$

From this formula, the mass of an elliptical galaxy can be calculated when v^2 and R are known. Some observations of velocities in elliptical galaxies are given in Fig. 19.8. These are further discussed in Sect. 19.4. The value of M/L derived from such observations is about 10 within a radius of 10 kpc. The mass of a bright elliptical might thus be up to $10^{13} M_\odot$.

The masses of spiral galaxies are obtained from their *rotation curve* $v(R)$, which gives the variation of their rotational velocity with radius. Assuming that most of the mass is in the almost spherical bulge, the mass within radius R, $M(R)$, can be estimated from Kepler's third law:

$$M(R) = Rv(R)^2/G .\tag{19.7}$$

Some typical rotation curves are shown in Fig. 19.9. In the outer parts of many spirals, $v(R)$ does not depend on R. This means that $M(R)$ is directly proportional to the radius – the further out one goes, the larger the interior mass is. Since the outer parts of spi-

Fig. 19.8. Velocity of rotation $V(R)$ [km/s] and velocity dispersion $\sigma(R)$ [km/s] as functions of radius for types E2 and E5. The latter galaxy is rotating, the former is not. [Davies, R. L. (1981): Mon. Not. R. Astron. Soc. **194**, 879]

rals are very faint, at large radii the value of M/L is directly proportional to the radius. For the disc one finds that $M/L = 8$ for early and $M/L = 4$ for late spiral types. The largest measured total mass is $2 \times 10^{12} M_\odot$.

In order to measure the mass at even larger radii where no emission can be detected, motions in systems of galaxies have to be used. One possibility is to use pairs of galaxies. In principle, the method is the same as for binary stars. However, because the orbital period of a binary galaxy is about 10^9 years, only statistical information can be obtained in this way. The results are still uncertain, but seem to indicate values of $M/L = 20-30$ at pair separations of about 50 kpc.

A fourth method for determining galaxy masses is to apply the virial theorem to clusters of galaxies, assuming that these are in equilibrium. The kinetic energy T in (19.4) can then be calculated from the observed redshifts, and the potential energy U

Fig. 19.9. Rotation curves for seven spiral galaxies. [Rubin, V. C., Ford, W. K., Thonnard, N. (1978): Astrophys. J. (Lett.) **225**, L107]

from the separations between cluster galaxies. If it is assumed that the masses of galaxies are proportional to their luminosities, it is found that M/L is about 200 within 1 Mpc of the cluster centre. However, there is a large variation from cluster to cluster.

Present results suggest that as one samples larger volumes of space, one obtains larger values for the mass-luminosity ratio. Thus a large fraction of the total mass of galaxies must be in an invisible and unknown form, mostly found in the outer parts. This is known as the *missing mass problem* and is one of the central unsolved questions of extragalactic astronomy.

19.3 Galactic Structures

Ellipticals and Bulges. In all galaxies the oldest stars have a more or less round distribution. In the Milky Way this component is represented by the population II stars. Its inner parts are called the bulge, and its outer parts are often referred to as the halo. There does not appear to be any physically significant difference between the bulge and the halo. The population of old stars can be best studied in ellipticals, which only contain this component. The bulges of spiral and S0 galaxies are very similar to ellipticals of the same size.

The surface brightness distribution in elliptical galaxies depends essentially only on the distance from the centre and the orientation of the major and minor axis. If r is the radius along the major axis, the surface brightness $I(r)$ is well described by *de Vaucouleurs' law*:

$$\log \frac{I(r)}{I_e} = -3.33 \left[\left(\frac{r}{r_e} \right)^{1/4} - 1 \right] . \tag{19.8}$$

The constants in (19.8) have been chosen so that half of the total light of the galaxy is radiated from within the radius r_e and the surface brightness at that radius is I_e. The parameters r_e and I_e are determined by fitting (19.8) to observed brightness profiles. Typical values for elliptical, normal spiral and S0 galaxies are in the ranges $r_e = 1-10$ kpc and $I_e = 20-23$ magnitudes per square arc second.

Although de Vaucouleurs' law is a purely empirical relation, it still gives a remarkably good representation of the observed light distribution. However, in the outer regions of elliptical galaxies departures may often occur: the surface brightness of dwarf spheroidals often falls off more rapidly than (19.8), perhaps because the outer parts of these galaxies have been torn off in tidal encounters with other galaxies. In the giant galaxies of type cD, the surface brightness falls off more slowly (see Fig. 19.10). It is thought that this is connected with their central position in clusters of galaxies.

Although the isophotes in elliptical galaxies are ellipses to a good approximation their ellipticities and the orientations of their major axes may vary as a function of radius. Different galaxies differ widely in this respect indicating that the structure of ellipticals is not as simple as it might appear. In particular, the fact that the direction of the major axis sometimes changes within a galaxy suggests that the ellipticals may not be axially symmetric in shape.

From the distribution of surface brightness, the three-dimensional structure of a galaxy may be inferred as explained in *Three-Dimensional Shape of Galaxies. The rela-

Fig. 19.10. The distribution of surface brightness in E and cD galaxies. Ordinate: surface magnitude, mag/sq. arc sec; abscissa: (radius [kpc])$^{1/4}$. Equation (19.8) corresponds to a straight line in this representation. It fits well with an E galaxy but for type cD the luminosity falls off more slowly in the outer regions. Comparison with Fig. 19.11 shows that the brightness distribution in S0 galaxies behaves in a similar fashion. cD galaxies have often been erroneously classified as S0. [Thuan, T. X., Romanishin, W. (1981): Astrophys. J. **248**, 439]

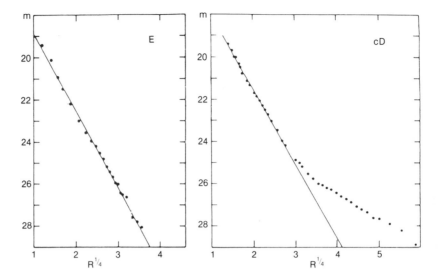

tion (19.8) gives a brightness profile that is very strongly peaked towards the centre. The real distribution of axial ratios for ellipticals can be statistically inferred from the observed one. On the (questionable) assumption that they are rotationally symmetric, one obtains a broad distribution with a maximum corresponding to types E3−E4. If the true shape is not axisymmetric, it cannot even statistically be uniquely determined from the observations.

Discs. A bright, massive stellar disc is characteristic for S0 and spiral galaxies, which are therefore called *disc galaxies*. According to some observations, in some ellipticals there is also a faint disc hidden behind the bright bulge. In the Milky Way the disc is formed by population I stars.

The distribution of surface brightness in the disc is described by the expression

$$I(r) = I_0 e^{-r/r_0} \ . \tag{19.9}$$

Figure 19.11 shows how the observed radial brightness distribution can be decomposed into a sum of two components: a centrally dominant bulge and a disc contributing sig-

Fig. 19.11. The distribution of surface brightness in types S0 and Sb. Ordinate: mag/sq. arc sec; abscissa: radius [arc sec]. The observed surface brightness has been decomposed into a sum of bulge and disc contributions. Note the larger disc component in type Sb. [Boroson, T. (1981): Astrophys. J. Suppl. **46**, 177]

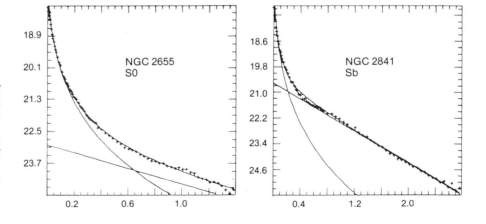

nificantly at larger radii. The central surface brightness I_0 is typically $21-22$ mag/sq. arc sec, and the radial scale length $r_0 = 1-5$ kpc. In Sc galaxies the total brightness of the bulge is generally slightly smaller than that of the disc, whereas in earlier Hubble types the bulge has a larger total brightness. The thickness of the disc, measured in galaxies that are seen edge-on, may typically be about 1.2 kpc. Sometimes the disc has a sharp outer edge at about $4 r_0$.

The Interstellar Medium. Elliptical and S0 galaxies contain very little interstellar gas. However, in some ellipticals neutral hydrogen amounting to about 0.1% of the total mass has been detected, and in the same galaxies there are also often signs of recent star formation. In some S0 galaxies much larger gas masses have been observed, but the relative amount of gas is very variable from one galaxy to another. The lack of gas in these galaxies is rather unexpected, since during their evolution the stars release much more gas than is observed.

The relative amount of neutral hydrogen in spiral galaxies is correlated with their Hubble type. Thus Sa spirals contain about 2%, Sc spirals 10%, and Irr I galaxies up to 30% or more.

The distribution of neutral atomic hydrogen has been mapped in detail in nearby galaxies by means of radio observations. In the inner parts of galaxies the gas forms a thin disc with a fairly constant thickness of about 200 pc, sometimes with a central hole of a few kpc diameter. The gas disc may continue far outside the optical disc, becoming thicker and often warped from the central disc plane.

Most of the interstellar gas in spiral galaxies is in the form of molecular hydrogen. The hydrogen molecule cannot be observed directly, but the distribution of carbon monoxide has been mapped by radio observations. The distribution of molecular hydrogen can then be derived by assuming that the ratio between the densities of CO and H_2 is everywhere the same, although this may not always be true. It is found that the distribution obeys a similar exponential law as the young stars and H II regions, although in some galaxies (such as the Milky Way) there is a central density minimum. The surface density of molecular gas may be five-times larger than that of H I, but because of its strong central concentration its total mass is only perhaps two-times larger.

The distribution of cosmic rays and magnetic fields in galaxies can be mapped by means of radio observations of the synchrotron radiation from relativistic electrons. The strength of the magnetic field deduced in this way is typically $0.5-1$ nT. The observed emission is polarised, showing that the magnetic field is fairly well-ordered on large scales. Since the plane of polarisation is perpendicular to the magnetic field, the large-scale structure of the magnetic field can be mapped. However, the plane of polarisation is changed by Faraday rotation and for this reason observations at several wavelengths are needed in order to determine the direction of the field. The results show that the field is generally strongest in the plane of the disc and is directed along the spiral arms in the plane. The field is thought to have been produced by the combined action of rising elements of gas, perhaps produced by supernova explosions, and the differential rotation, in principle in the same way as the production of solar magnetic fields was explained in Chap. 13.

*Three-Dimensional Shape of Galaxies

Equations (19.8) and (19.9) describe the distribution of galactic light projected on the plane of the sky. The actual three-dimensional luminosity distribution in a galaxy is obtained by inverting the projection. This is easiest for spherical galaxies.

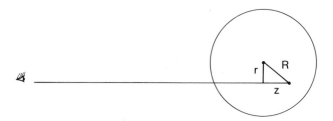

Let us suppose that a spherical galaxy has the projected luminosity distribution $I(r)$ [e.g. as in (19.8)]. With coordinates chosen according to the figure, $I(r)$ is given in terms of the three-dimensional luminosity distribution $\rho(R)$ by

$$I(r) = \int_{-\infty}^{\infty} \rho(R)\,dz \ .$$

Since $z^2 = R^2 - r^2$, a change of the variable of integration yields

$$I(r) = 2 \int_{r}^{\infty} \frac{\rho(R)\,R\,dR}{\sqrt{R^2 - r^2}} \ .$$

This is a so-called Abel integral equation for $\rho(R)$, which has a known solution:

$$\rho(R) = -\frac{1}{\pi R}\frac{d}{dR}\int_{R}^{\infty}\frac{I(r)\,r\,dr}{\sqrt{r^2 - R^2}} = -\frac{1}{\pi}\int_{R}^{\infty}\frac{(dI/dr)\,dr}{\sqrt{r^2 - R^2}} \ .$$

Introducing the observed $I(r)$ into this expression, one obtains the actual luminosity distribution $\rho(R)$. The figure shows the distribution obtained from de Vaucouleurs' law.

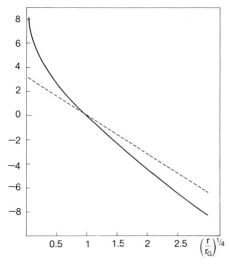

The three-dimensional luminosity distribution corresponding to de Vaucouleurs' law. Ordinate log (density), abscissa: (radius)$^{1/4}$. The projected distribution (19.8) is a straight line

If the galaxy is not spherical, its three-dimensional shape can only be determined if its inclination with respect to the line of sight is known. Since galactic discs are thin and of constant thickness, the inclination i of a disc galaxy is obtained directly from the axis ratio of its projected image:

$$\sin i = b/a \ .$$

When the inclination is known, the real axis ratio of the bulge q_0 can be determined from the projected value q. For a rotationally symmetric bulge, the relation between q and q_0 is

$$\cos^2 i = \frac{1-q^2}{1-q_0^2} \; .$$

The flattenings of disc galaxy bulges obtained from this relation lie in the range $q_0 = 0.3 - 0.5$.

Since the inclinations of ellipticals are generally unknown, only the statistical distribution of q can be determined from that of q_0.

19.4 Dynamics of Galaxies

We have seen how the masses of galaxies can be derived from observed velocities of stars and gas. The same observations can be used to study the internal distribution of mass in more detail.

Slowly Rotating Systems. The dynamics of elliptical galaxies and disc-galaxy bulges are studied by means of the Doppler shifts and broadenings of stellar absorption lines. Since a given absorption line is the sum of contributions from many individual stars, its Doppler shift gives their mean velocity while its broadening is increased by an amount depending on the dispersion of stellar velocities around the mean. By observing how the wavelengths and widths of spectral lines behave as functions of the radius, one can get some insight into the distribution of mass in the galaxy.

Examples of the observed radial dependence of the rotational velocity and velocity dispersion derived for some ellipticals were given in Fig. 19.8. The observed rotational velocities are often small (< 100 km/s), while the velocity dispersion may typically be about 200 km/s. If elliptical galaxies were in fact ellipsoids of revolution there should be a statistical relationship (when projection effects have been taken into account) between flatness, rotational velocity and velocity dispersion. Such a relationship has been observed for fainter ellipticals and for disc-galaxy bulges. However, some of the brightest ellipticals rotate very slowly. Therefore their flattening cannot be due to rotation.

The radial dependence of the velocity dispersion gives information on the distribution of mass within the galaxy. Since it also depends on how the shapes of stellar orbits in the galaxy are distributed, its interpretation requires detailed dynamical models.

Rotation Curves. In spiral galaxies the distribution of mass can be studied directly using the observed rotational velocities of the interstellar gas. This can be observed either at optical wavelengths from the emission lines of ionised gas in H II regions or at radio wavelengths from the hydrogen 21-cm line. Typical galactic rotation curves are shown in Fig. 19.9.

The qualitative behaviour of the rotation curve in all spiral galaxies is similar to the rotation curve of the Milky Way: there is a central portion, where the rotational velocity is directly proportional to the radius, corresponding to solid-body rotation. At a few kpc radius, the curve turns over and becomes flat, i.e. the rotational velocity does not depend on the radius. In early Hubble types the rotation curve rises more steeply near the centre and reaches larger velocities in the flat region (Sa about 300 km/s, Sc about 200 km/s). A higher rotational velocity indicates a larger mass according to (19.7), and thus Sa types must have a larger mass density near the centre. This is not unexpected since a more massive bulge is one of the defining properties of early type spirals.

A decrease of the rotational velocity at large radii would be an indication that most of the mass is inside that radius. In some galaxies such a decrease has been detected, in others the rotational velocity remains constant as far out as the observations can reach.

Spiral Structure. Spiral galaxies are relatively bright objects. Some have a well-defined, large-scale two-armed spiral pattern (Fig. 19.12), whereas in others the spiral structure is made up of a large number of short filamentary arms. From galaxies where the pattern is seen in front of the central bulge, it has been deduced that the sense of winding of the spiral is trailing with respect to the rotation of the galaxy.

The spiral structure is most clearly seen in the interstellar dust, H II regions, and the OB associations formed by young stars. The dust often forms thin lanes along the inner edge of the spiral arms with star-forming regions on their outside. Enhanced synchrotron radio emission associated with spiral arms has also been detected.

The spiral pattern is generally thought to be a wave in the density of the stellar disc, as discussed in Sect. 18.4. As the interstellar gas streams through the density wave, a shock, marked by the dust lanes, is formed as the interstellar gas is compressed, leading to the collapse·of molecular clouds and the formation of stars. The density wave theory predicts characteristic streaming motions within the arm, which have been detected in some galaxies by observations of the H I 21 cm line.

It is not known how the spiral wave was produced. In multi-armed galaxies the spiral arms may be short-lived, constantly forming and disappearing, but extensive, regular, two-armed patterns have to be more long-lived. In barred spirals, the bar can drive a spiral wave in the gas. Some normal spirals may have been produced by the tidal force from another galaxy passing nearby. Finally, in some galaxies a two-armed spiral may have been spontaneously generated by an instability of the disc.

19.5 Stellar Ages and Element Abundances in Galaxies

From the Milky Way we know that stars of populations I and II are different not only in respect to their spatial distribution but also in respect to their ages and heavy element abundances. This fact gives important evidence about the formation of the Milky Way, and it is therefore of interest if a similar connection can be found in other galaxies.

The indicators of composition most easily measured are the variations of colour indices inside galaxies and between different galaxies. Two regularities have been discovered in these variations: First, according to the *colour-luminosity relationship* for elliptical and S0 galaxies, brighter galaxies are redder. Secondly, there is a *colour-aperture effect*, so that the central parts of galaxies are redder. For spirals this relationship is due to the presence of young, massive stars in the disc, but it has also been observed for elliptical and S0 galaxies.

Galactic spectra are composed of the spectra of all their stars added together. Thus the colours depend both on the ages of the stars (young stars are bluer) and on the heavy element abundance Z (stars with larger Z are redder). The interpretation of the observational results thus has to be based on detailed modelling of the stellar composition of galaxies, or *population synthesis*.

Stars of different spectral classes contribute different characteristic absorption features to the galaxy spectrum. By observing the strength of various spectral features, one can find out about the masses, ages and chemical compositions of the stars that make up the galaxy. For this purpose a large number of characteristic properties of the spec-

Fig. 19.12. (**a**) A spiral galaxy from above: M51 (type Sc). The interacting companion is NGC 5195 (type Irr II) (Lick Observatory). (**b**) A spiral galaxy from the side: the Sb spiral NGC 4565. (NOAO/Kitt Peak National Observatory)

trum, strengths of absorption lines and broad-band colours are measured. One then attempts to reproduce these data using a representative collection of stellar spectra. If no satisfactory solution can be found, more stars have to be added to the model. The final result is a population model giving the stellar composition of the galaxy. If this is combined with theoretical stellar evolution calculations, the evolution of the light of the galaxy can also be computed.

Population synthesis of E galaxies show that practically all their stars were formed simultaneously about 15×10^9 years ago. Most of their light comes from red giants, whereas most of their mass resides in lower main sequence stars of less than one solar mass. Since all stars have roughly the same age, the colours of elliptical galaxies are directly related to their metallicities. Thus the colour-luminosity relation indicates that Z in giant ellipticals may be double that in the solar neighbourhood, whereas it may smaller by a factor 100 in dwarfs. Similarly, the radial dependence of the colours can be explained if the value of Z at the centre is an order of magnitude larger than it is at larger radii.

The stellar compositions of disc galaxy bulges are generally similar to those of ellipticals. The element abundances in the gases in spirals can be studied by means of the emission lines from H II regions ionised by newly formed stars. In this case too, the metallicity increases towards the centre but the size of the variation varies in different galaxies and is not yet well understood.

19.6 Systems of Galaxies

The galaxies are not smoothly distributed in space; rather they form systems of all sizes: galaxy pairs, small *groups*, large *clusters* and *superclusters* formed from several groups and clusters. The larger a given system, the less its density exceeds the mean density of the Universe. On average the density is twice the background density for systems of radius 5 Mpc, and 10% above the background at radius 20 Mpc.

Interactions of Galaxies. The frequencies of different types of galaxies vary in the various kinds of groups. This could either be because certain types of galaxies are formed preferentially in certain environments, or because interactions between galaxies have changed their shapes. There are many observed interacting systems where strong tidal forces have produced striking distortions, "bridges" and "tails", in the member galaxies.

The interactions between galaxies are not always dramatic. For example the Milky Way has two satellites, the Large and Small Magellanic Clouds (see Fig. 19.5), which are Irr-I-type dwarf galaxies at about 60 kpc distance. It is thought that approximately 5×10^8 years ago, these passed the Milky Way at a distance of about 10–15 kpc leaving behind the *Magellanic Stream*, a 180° long thin stream of neutral hydrogen clouds. Systems of this type, where a giant galaxy is surrounded by a few small companions, are quite common. Computations show that in many such cases the tidal interactions are so strong that the companions will merge with the parent galaxy at the next close approach. This is likely to happen to the Magellanic Clouds.

During earlier epochs in the Universe, when the density was larger, interactions between galaxies must have been much more common than at present. Thus it has been proposed that a large fraction of bright galaxies have undergone major mergers at some stage in their history. In particular, there are good reasons to believe that the slowly rotating, non-axisymmetric giant ellipticals may have formed by the merger of disc galaxies.

Groups. The most common types of galaxy systems are small, irregular groups of a few tens of galaxies. A typical example is the *Local Group*, which contains two larger galaxies in addition to the Milky Way − the Andromeda Galaxy M31, an Sb spiral of about the same size as the Milky Way with two dwarf companions, and the smaller Sc spiral M33. The rest of the about 30 members of the Local Group are dwarfs; about 12 are of type dE and 8 type Irr I. The diameter of the Local Group is about 1.5 Mpc.

Clusters. A system of galaxies may be defined to be a cluster if it contains a larger number (at least 50) of bright galaxies. The number of members and the size of a cluster depend on how they are defined. One way of doing this is to fit the observed distribution of galaxies within a cluster with an expression of the form (19.8). In this way a characteristic cluster radius of about $2-5$ Mpc is obtained. The number of members depends both on the cluster radius and on the limiting magnitude. A large cluster may contain several hundred galaxies that are less than two magnitudes fainter than the characteristic luminosity $L*$ of (19.2).

Clusters of galaxies can be ordered in a sequence from extended, low-density, irregular systems (sometimes called clouds of galaxies) to denser and more regular structures (Fig. 19.13). The galaxy type composition also varies along this sequence in the sense that in the loose irregular clusters the bright galaxies are predominantly spirals, whereas the members of dense clusters are almost exclusively type E and S0. The nearest cluster of galaxies is the Virgo Cluster at a distance of about 15 Mpc. It is a relatively irregular cluster, in which a denser central region containing early galaxy types is surrounded by a more extended distribution of mainly spiral galaxies. The nearest regular cluster is the Coma Cluster, roughly 90 Mpc away. In the Coma Cluster a central pair of giant ellipticals is surrounded by a flattened (axis ratio about 2:1) system of early-type galaxies.

X-ray emission from hot gas has been detected in many clusters. In irregular clusters the gas temperature is about 10^7 K and the emission is generally concentrated near individual galaxies; in the regular ones the gas is hotter, 10^8 K, and the emission tends to be more evenly distributed over the whole cluster area. By and large the x-ray emission follows the distribution of the galaxies. The amount of gas needed to explain this emission is about equal to the mass of the galaxies − thus it cannot solve the missing mass problem. X-ray emission lines from multiply ionised iron have also been observed. On basis of these lines it has been concluded that the metal abundance of intergalactic gas is roughly equal to that of the Sun. For this reason it is most likely that the gas has been ejected from the galaxies in the cluster.

Superclusters. Groups and clusters of galaxies may form even larger systems, superclusters. For example, the Local Group belongs to the *Local Supercluster*, a flattened system whose centre is the Virgo Cluster containing tens of smaller groups and clouds of galaxies. The Coma Cluster is part of another supercluster. The diameters of superclusters are $10-20$ Mpc. However, on this scale, it is no longer clear whether one can reasonably speak of individual systems. Perhaps it would be more accurate to think of the distribution of galaxies as a continuous network, in which the large clusters are connected by walls and strings formed by smaller systems. Between these there remain empty regions containing very few galaxies, which can be up to 50 Mpc in diameter (Fig. 19.14).

Fig. 19.13. (a) The irregular Virgo Cluster of galaxies. **(b)** The regular Coma Cluster (ESO and Karl-Schwarzschild-Observatory)

a

b

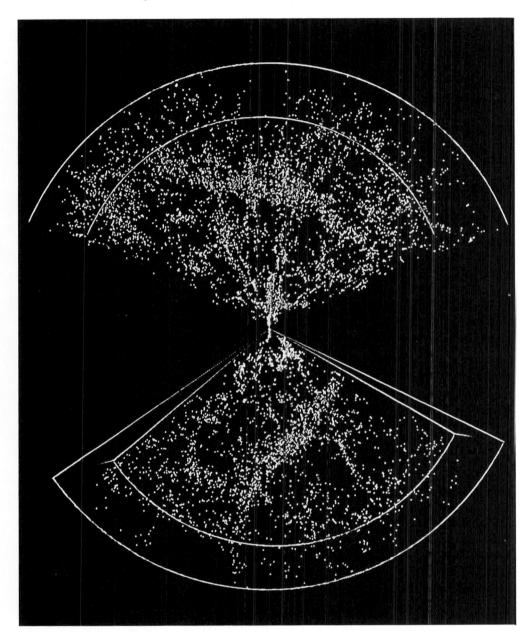

Fig. 19.14. Distribution of 9325 galaxies. The upper sector covers the region $8.5° \leq \delta \leq 55.5°$, $8\,h \leq \alpha \leq 17\,h$ in the Northern sky, the lower sector the region $-40° \leq \delta \leq -2.5°$, $20.8\,h \leq \alpha \leq 4\,h$ in the Southern sky. The radial coordinate is the measured velocity. If this were due only to the expansion of the Universe the velocity would be directly proportional to the distance (Sect. 20.1). In reality, clusters are stretched along the line of sight by the peculiar motions of galaxies. [da Costa et al. (1994): Astrophys. J. **424**, L1]

19.7 Active Galaxies and Quasars

So far in this chapter we have been concerned with the properties of normal galaxies. In some galaxies, however, the normal galaxy is overshadowed by violent activity. This activity is produced in the nucleus, which is then called an *active galactic nucleus (AGN)*.

The luminosities of active galactic nuclei may be extremely large, sometimes much larger than that of the rest of the galaxy. It seems unlikely that a galaxy could maintain such a large power output for long. For this reason it is thought that active galaxies do not form a separate class of galaxy but rather represent a passing stage in the evolution of normal galaxies.

Activity appears in many different forms. Some galaxies have an exceptionally bright nucleus similar to a large region of ionised hydrogen. These may be young galaxies in which, near the centre, large numbers of stars are forming and evolving into supernovae (starburst nuclei). In other nuclei the radiation cannot have been produced by stars, and the most plausible source of energy in these nuclei is the gravitational energy of a supermassive black hole (mass $> 10^8 \, M_\odot$). In some galaxies, the spectral lines are unusually broad indicating large internal velocities. These may be either rotational velocities near a black hole or due to explosive events in the nucleus. In some galaxies jets are seen coming out of the nucleus. Many active galaxies radiate a non-thermal spectrum, apparently synchrotron radiation produced by fast electrons in a magnetic field.

The classification of active galaxies has been developed rather unsystematically, since many of them have been discovered only recently and have not been completely studied. For example, the *Markarian galaxies* catalogued by *Benyamin Yerishevich Markarian* in the early 1970's are defined by strong ultraviolet emission. Many Markarian galaxies are Seyfert galaxies; others are galaxies undergoing a burst of star formation. The N galaxies form another class closely similar to the Seyfert galaxies.

Two natural basic classes of active galaxies are the *Seyfert galaxies* and the *radio galaxies*. The former are spirals, the latter are ellipticals. Some astronomers think that the Seyfert galaxies represent the active stage of normal spiral galaxies and the radio galaxies that of ellipticals.

Seyfert Galaxies. The Seyfert galaxies are named after *Carl Seyfert* who discovered them in 1943. Their most important characteristics are a bright, point-like central nucleus and a spectrum showing broad emission lines. The continuous spectrum has a non-thermal component, which is most prominent in the ultraviolet. The emission lines are thought to be produced in gas clouds moving close to the nucleus with large velocities.

On the basis of the spectrum Seyfert galaxies are classified as type 1 or 2. In a type-1 spectrum the allowed lines are broad (corresponding to a velocity of 10^4 km/s), much broader than the forbidden lines. In type 2 all lines are similar and narrower ($< 10^3$ km/s). Transitions between these types and intermediate cases have sometimes been observed. The reason for the difference is thought to be that the allowed lines are formed in denser gas near the nucleus and the forbidden lines are formed in more diffuse gas further out. In type-2 Seyfert galaxies the denser gas is missing or obscured.

Almost all Seyfert galaxies with known Hubble types are spirals; the possible exceptions are of type 2. They are strong infrared sources. Type 1 galaxies often show strong x-ray emission.

The true Seyfert galaxies are relatively weak radio sources. However, there are compact radio galaxies with an optical spectrum that is essentially the same as for Seyfert galaxies. These should probably be classified with the Seyfert galaxies. In general the stronger radio emission seems to come with a type-2 spectrum.

It is estimated that about 1% of all bright spiral galaxies are Seyfert galaxies. The luminosities of their nuclei are about $10^{36} - 10^{38}$ W, of the same order as all the rest of the galaxy. Brightness variations are common.

Radio Galaxies. By definition radio galaxies are galaxies that are powerful radio sources. The radio emission of a radio galaxy is non-thermal synchrotron radiation. The radio luminosity of radio galaxies is typically $10^{33} - 10^{38}$ W and may thus be as large as the total luminosity of a normal galaxy. The main problem in explaining radio emission is to understand how the electrons and magnetic fields are produced and, above all, where the electrons get their energy.

The forms and sizes of the radio-emitting regions of radio galaxies have been studied ever since the 1950's, when radio interferometers achieved the resolution of optical telescopes. The characteristic feature of a strong radio galaxy is a double structure: there are two large radio-emitting regions on opposite sides of the observed galaxy. The radio-emitting regions of some radio galaxies are as far apart as 6 Mpc, almost ten times the distance between the Milky Way and Andromeda galaxies. One of the smallest double radio sources is the galaxy M87 (Fig. 19.15) whose two components are only a few kpc distant from each other.

The double structures of radio galaxies appear to be produced by ejections from the nucleus. However, the electrons in the radio lobes cannot be coming from the centre of the galaxy because they would lose all their energy during such a long transit. Therefore electrons have to be continuously accelerated within the radio-emitting regions. Within the radio lobes there are almost point-like regions, hot spots. These are generally symmetrically placed with respect to the nucleus, and are apparently consequences of nuclear ejections.

"Tailed" radio sources also exist. Their radio emission mainly comes from one side of the galaxy, forming a curved tail, which is often tens of times longer than the diameter of the galaxy. The best examples are NGC 1265 in the Perseus cluster of galaxies, and 3C129, which appears to be in an elliptical orbit around a companion galaxy. The tail is interpreted as the trail left by the radio galaxy in intergalactic space.

Another special feature revealed by the radio maps is the presence of *jets*, narrow lines of radio emission, usually starting in the nucleus and stretching far outside the galaxy. The best known may be the M 87 jet, which has also been observed as an optical and x-ray jet. The optically observed jet is surrounded by a radio source. A similar radio source is seen on the opposite side of the nucleus where no optical jet is seen. Our nearest radio galaxy Centaurus A also has a jet extending from the nucleus to near the edge of the galaxy. VLBI observations of radio jets have also revealed *superluminal motions*: in many compact sources the components appear to be separating faster than the speed of light. Since such velocities are impossible according to the theory of relativity, the observed velocities can only be apparent, and several models have been proposed to account for them.

Quasars. The first quasar was discovered in 1963 when *Maarten Schmidt* interpreted the optical emission lines of the known radio source 3C273 as hydrogen Balmer lines redshifted by 16%. Such large redshifts are the most remarkable characteristics of the quasars. Properly speaking the word quasar is an abbreviation for quasistellar radio source, and some astronomers prefer to use the designation QSO (quasistellar object) since not all quasars emit radio radiation.

Optically, the quasars appear almost as point sources, although improved observational techniques have revealed an increasing number of quasars located inside more or less normal galaxies (Fig. 19.16). Although the first quasars were discovered by radio observations, only a small fraction of all optically identified quasars are bright radio

a

4885 MHz

b

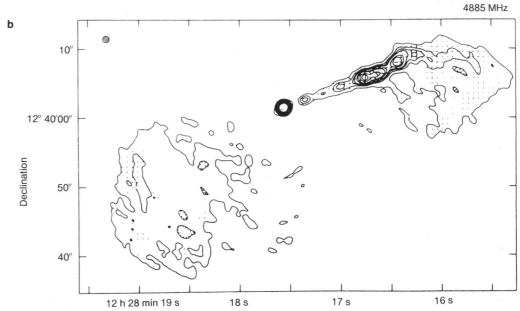

Fig. 19.15. (a) The active galaxy M87. A blue jet comes out of the nucleus of a normal E0 galaxy (NOAO/Kitt Peak National Observatory). **(b)** In the radio map made using the VLA the jet is observed to be two-sided. The area shown is much smaller than in the upper picture. [Owen, F. N., Hardee, P. E., Bignell, R. C. (1980): Astrophys. J. (Lett.) **239**, L11]

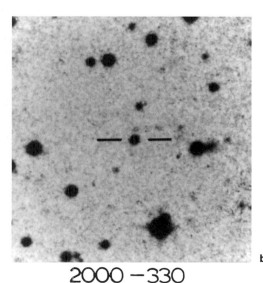

Fig. 19.16. (a) QSO 0351+026, a quasar-galaxy interacting pair. The components have the same redshift. (b) One of the most distant known quasars, PKS 2000−330 with the redshift of $z = 3.78$. (NOAO/Kitt Peak National Observatory; Jet Propulsion Laboratory)

sources. Most radio quasars are point sources, but some have a double structure like the radio galaxies. Satellite x-ray pictures also show the quasars to be point-like.

In the visible region the quasar spectra are dominated by spectral lines with rest wavelengths in the ultraviolet region. The first observed quasar redshifts were $z = 0.16$ and 0.37, and at this writing (early 1993) there are over a dozen quasars with redshifts greater than 4. The present record is 4.9. The light left the quasar when the Universe was less than one-tenth of its present age. The large inferred distances of the quasars mean that their luminosities have to be extremely large. Typical values lie in the range of $10^{38} - 10^{41}$ W. The brightness of quasars may vary rapidly, within a few days or less. Thus the emitting region can be no larger than a few light-days, i.e., about 100 AU.

The quasars often have both emission and absorption lines in their spectra. The emission lines are very broad and are probably produced in the quasar itself. The absorption spectrum consists of densely distributed narrow lines that are thought to be hydrogen Lyman α lines formed in gas clouds along the line of sight to the quasar. The clouds producing this *"Lyman α forest"* are young galaxies or protogalaxies, and they therefore provide important evidence about the formation of galaxies.

Some astronomers have questioned the cosmological interpretation of the redshift. Halton Arp has discovered small systems of quasars and galaxies in which some of the components have widely discrepant redshifts. For this reason Arp thinks that the quasar redshifts are produced by some unknown process. This claim is highly controversial.

Unified Models. Although the forms of galactic activity may at first sight appear diverse, they can be unified within a fairly widely accepted schematic model. According to this model most galaxies contain a compact central nucleus, usually supposed to be a black hole, with mass $10^7 - 10^9 M_\odot$, surrounded by a disc or ring of gas. The source of energy is the gravitational energy released as gas is accreted into the black hole. The disc may also give rise to a jet, where some of the energy is converted into perpendicular motions along the rotational axis. Thus active galactic nuclei are similar to the nucleus of the Milky Way, although the masses of both the black hole and the gas disc may be much larger.

A first characteristic parameter of the model described is obviously the total luminosity. For example, the only essential difference between Seyfert 1 galaxies and radio-quiet quasars is the larger luminosity of quasars. A second basic parameter is the radio brightness, which may be related to the strength of a jet. On the basis of their radio luminosity one can connect Seyfert galaxies and radio-quiet quasars on one hand, and radio galaxies and radio quasars on the other.

The third important parameter of unified models is the angle from which we happen to view the nuclear disc. For example, if the disc is seen edge-on the actual nucleus is obscured by the disc. This could explain the difference between Seyfert types 1 and 2: in type 2 we do not see the broad emission lines formed near the black hole, but only the narrower lines from the disc. Similarly, a galaxy that looks like a double radio source when seen edge-on, would look like a radio quasar if the disc were seen face-on. In the latter case there is a possibility that we may be seeing an object directly along the jet. It will then appear as a *blazar*, an object with rapid and violent variations in brightness and polarisation, and very weak or invisible emission lines. If the jet is almost relativistic its transverse velocity may appear larger than the speed of light, and thus superluminal motions can also be understood. One prediction of the unified model is that there should be a large number of quasars where the nucleus is obscured by the disc as in the Seyfert 2 galaxies. Some possible candidates have been found.

Gravitational Lenses. An interesting phenomenon first discovered in connection with quasars is that of gravitational lenses. Since light rays are bent by gravitational fields, a mass (e.g., a galaxy) placed between a distant quasar and the observer will distort the image of the quasar (Fig. 19.17). The first example of this effect was discovered in 1979 when it was found that two quasars, 5.7" apart in the sky, had essentially identical spectra. It was concluded that the "pair" was really a double image of a single quasar. Since then, several other gravitationally lensed quasars have been discovered.

Gravitational lenses have also been discovered in clusters of galaxies. Here the gravitational field of the cluster distorts the images of distant galaxies into arcs around the cluster centre. In addition, in 1993 microlensing was observed in the Milky Way, where the brightness of a star is momentarily increased by the lensing effect from a mass pass-

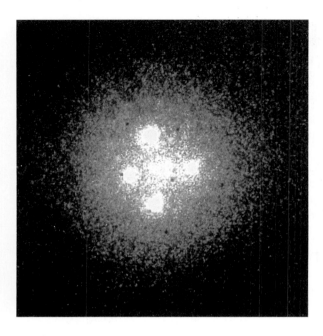

Fig. 19.17. The four components in the Einstein cross are gravitationally lensed images of the same quasar. (ESA/NASA)

ing in front of the star. Thus the study of gravitational lens effects offers a new promising method of obtaining information on the distribution of mass in the Universe.

19.8 The Origin and Evolution of Galaxies

Our present ideas about the origin and evolution of galaxies are still rather preliminary. However, there is a fairly wide consensus about the general framework within which more detailed questions can be investigated. Some aspects of the general picture have been touched on in connection with the Milky Way. Elliptical galaxies and disc galaxy bulges were formed in the collapse of a gas cloud about 15×10^9 years ago. In some galaxies some of the gas was left to settle into a disc before turning into stars. Within some of these galaxies (spirals) star formation still continues in the disc; in others (S0s) all the gas has been consumed.

By means of numerical simulations of the collapse of gas clouds and star formation in them, one can try to compute the evolution of the spectrum and the chemical abundances under various assumptions. The results of the models can be compared with observational data on galaxy spectra and the distribution of elements in galaxies.

There are many complicating factors. Gas may be expelled from the galaxy or there may be an influx of fresh gas. Interactions with the surroundings may radically alter the course of evolution − in dense systems they may lead to the complete merging of the individual galaxies into one giant elliptical. Little is known of how the formation of stars is affected by the general dynamical state of the galaxy and of how an active nucleus may influence the formation process.

19.9 Exercises

19.9.1   The galaxy NGC 772 is an Sb spiral, similar to M31. Its angular diameter is 7′ and apparent magnitude 12.0. The corresponding values of M31 are 3.0° and 5.0. Find the ratio of the distances of the galaxies

a) assuming their sizes are equal,
b) assuming they are equally bright.

19.9.2   The brightness of the quasar 3C279 has shown changes with the time scale of one week. Estimate the size of the region producing the radiation. The apparent magnitude is 18. If the distance of the quasar is 2000 Mpc, what is its absolute magnitude and luminosity? How much energy is produced per AU^3?

Chapter 20 **Cosmology**

After the demise of the Aristotelian world picture, it took hundreds of years of astronomical observations and physical theories to reach a level at which a satisfactory modern scientific picture of the physical universe could be formed. The decisive steps in this development were the clarification of the nature of the galaxies in the 1920's and the general theory of relativity developed by Einstein in the 1910's. Research in cosmology tries to answer questions such as: How large and how old is the Universe? How is matter distributed? How were the elements formed? What will be the future of the Universe? The central tenet of modern cosmology is the model of the expanding universe. On the basis of this model, it has been possible to approach these questions.

20.1 Cosmological Observations

The Olbers Paradox. The simplest cosmological observation may be that the sky is dark at night. This fact was first noted by *Johannes Kepler*, who, in 1610, used it as evidence for a finite universe. As the idea of an infinite space filled with stars like the Sun became widespread in consequence of the Copernican revolution, the question of the dark night sky remained a problem. In the 18th and 19th centuries, *Edmond Halley, Loys de Chéseaux* and *Heinrich Olbers* considered it in their writings. It has become known as the *Olbers paradox* (Fig. 20.1).

Fig. 20.1. The Olbers paradox. If the stars were uniformly distributed in an unending, unchanging space, the sky should be as bright as the surface of the Sun, since each line of sight would eventually meet the surface of a star. A two-dimensional analogy can be found in an optically thick pine forest where the line of sight meets a trunk wherever one looks. (Photo M. Poutanen and H. Karttunen)

The paradox is the following: Let us suppose the Universe is infinite and that the stars are uniformly distributed in space. No matter in what direction one looks, sooner or later the line of sight will encounter the surface of a star. Since the surface brightness does not depend on distance, each point in the sky should appear to be as bright as the surface of the Sun. This clearly is not true. The modern explanation of the paradox is that the stars have only existed for a finite time, so that the light from very distant stars has not yet reached us. Rather than proving the world to be finite in space, the Olbers paradox has shown it to be of a finite age.

Extragalactic Space. In 1923, *Edwin Hubble* showed that the Andromeda Galaxy M31 was far outside the Milky Way, thus settling a long-standing controversy concerning the relationship between the nebulae and the Milky Way. The numerous galaxies seen in photographs form an extragalactic space vastly larger than the dimensions of the Milky Way. It is important for cosmology that the distribution and motions of the basic components of extragalactic space, the galaxies and clusters of galaxies, should everywhere be the same as in our local part of the Universe.

Galaxies generally occur in various systems, ranging from small groups to clusters of galaxies and even larger superclusters. The largest structures observed are about 100 Mpc in size (see Sect. 19.6). They are thus significantly smaller than the volume of space (a few thousand Mpc in size) in which the distribution of galaxies has been investigated. One way of studying the large-scale homogeneity of the galaxy distribution is to count the number of galaxies brighter than some limiting magnitude m. If the galaxies are uniformly distributed in space, this number should be proportional to $10^{0.6m}$ (see Fig. 20.2 and Exercise 18.5.1). For example, the galaxy counts made by Hubble in 1934, which included 44 000 galaxies, were consistent with a galaxy distribution independent of position (homogeneity) and of direction (isotropy). Hubble found no "edge" of the Universe, nor have later galaxy counts found one.

Similar counts have been made for extragalactic radio sources. (Instead of magnitudes, flux densities are used. If F is the flux density, then because $m = -2.5 \lg(F/F_0)$, the number count will be proportional to $F^{-3/2}$.) These counts mainly involve very dis-

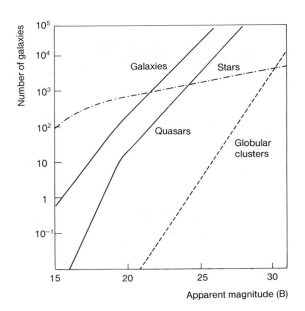

Fig. 20.2. The number of galaxies as a function of limiting magnitude obeys the $10^{0.6m}$ law of a uniform distribution down to the apparent magnitude $B = 20$. The flattening of the relation at fainter magnitudes can be explained by the curvature and expansion of the Universe

Fig. 20.3. The quasar 3C 295 and its spectrum. The quasars are among the most distant cosmological objects. (Photograph Palomar Observatory)

tant radio galaxies and quasars (Fig. 20.3). The results seem to indicate that the radio sources were either much brighter or much more common at earlier epochs than at present. This constitutes evidence in favour of an evolving, expanding universe.

The Hubble Law (Fig. 20.4). In the late 1920's, Hubble discovered that the spectral lines of galaxies were shifted towards the red by an amount proportional to their distances. If the redshift is due to the Doppler effect, this means that the galaxies move away from each other with velocities proportional to their separations, i.e. that the Universe is expanding as a whole.

In terms of the redshift $z = (\lambda - \lambda_0)/\lambda_0$, the Hubble law can be written as

$$z = (H/c)r \ , \tag{20.1}$$

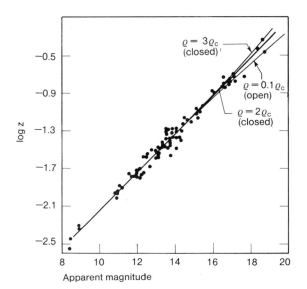

Fig. 20.4. The Hubble law for the brightest galaxies in clusters and the predictions of the Friedmann models. A choice of model cannot yet be made on the basis of observations

where c is the speed of light, H is the *Hubble constant* and r, the distance of the galaxy. For small velocities ($V \ll c$), the Doppler redshift $z = V/c$, and hence

$$V = Hr , \tag{20.2}$$

which is the most commonly used form of the Hubble law.

For a set of observed "standard candles", i.e. galaxies whose absolute magnitudes are close to some mean M_0, the Hubble law corresponds to a linear relationship between the apparent magnitude m and the logarithm of the redshift, $\lg z$. This is because a galaxy at distance r has an apparent magnitude $m = M_0 + 5\lg(r/10\,\mathrm{pc})$, and hence Hubble's law yields

$$m = M_0 + 5\lg(cz/H\,10\,\mathrm{pc}) = 5\lg z + C , \tag{20.3}$$

where the constant C depends on H and M_0. Suitable standard candles are e.g. the brightest galaxies in clusters and Sc galaxies of a known luminosity class.

Hubble's law has been confirmed for redshifts up to about $z = 1$. At small redshifts, it is masked by the redshifts due to the imperfectly known local peculiar velocities of galaxies. At large redshifts, the lack of suitable standard candles makes it difficult to check the exact form of the Hubble law.

If the Universe is expanding, the galaxies were once much nearer to each other. If the rate of expansion had been unchanging, the inverse of the Hubble constant, $T = H^{-1}$, would represent the age of the Universe. Since the expansion is thought to be gradually slowing down, the inverse Hubble constant gives an upper limit on the age of the Universe (Fig. 20.5). According to present estimates, $50\,\mathrm{km\ s^{-1}\ Mpc^{-1}} < H < 100\,\mathrm{km\,s^{-1}Mpc^{-1}}$, corresponding to $10\,\mathrm{Gy} < T < 20\,\mathrm{Gy}$.

The considerable uncertainty in the value of the Hubble constant is mainly due to the difficulty in determining extragalactic distances. (See p. 445 for a short survey of this problem.) A second problem is that the measured values of the velocity V, corrected to take into account the motion of the Sun within the Local Group, contain a significant component due to the peculiar motions of the galaxies. These peculiar velocities

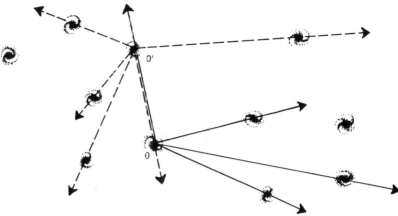

Fig. 20.5. Because the expansion of the Universe is slowing down, the inverse Hubble constant gives an upper limit of the age. The real age depends on the value of the deceleration parameter

Fig. 20.6. A regular expansion according with the Hubble law does not mean that the Milky Way (0) is the centre of the Universe. Observers at any other galaxy (0′) will see the same Hubble flow (*dotted lines*)

are caused by local mass concentrations like groups and clusters of galaxies. It is possible that the Local Group has a significant velocity towards the centre of the Local Supercluster (the Virgo Cluster). Because the Virgo Cluster is often used to determine the value of H, neglecting this peculiar velocity leads to a large error in H. The size of the peculiar velocity is not yet well known, but it is probably about 250 km/s.

The form of the Hubble law might give the impression that the Milky Way is the centre of the expansion, in apparent contradiction with the Copernican principle. Figure 20.6 shows that, in fact, the same Hubble law is valid at each point in a regularly expanding universe. There is no particular centre of expansion.

The Thermal Microwave Background Radiation. The most important cosmological discovery since Hubble's law was made in 1965. In that year, *Arno Penzias* and *Robert Wilson* discovered that there is a universal microwave radiation, with a spectrum corresponding to that of blackbody radiation (see Sect. 5.6) at a temperature of about 3 K (Fig. 20.7). For their discovery, they received the Nobel prize in physics in 1979.

The existence of a thermal cosmic radiation background had been predicted in the late 1940's by *George Gamow,* who was one of the first to study the initial phases of expansion of the Universe. According to Gamow, the Universe at that time was filled with extremely hot radiation. As it expanded, the radiation cooled, until at present, its temperature would be a few kelvins. After its discovery by Penzias and Wilson, the cosmic background radiation has been studied at wavelengths from 50 cm to 0.5 cm. The most recent measurements, made from the COBE (Cosmic Background Explorer) satellite show that it corresponds closely to a Planck spectrum at 2.73 ± 0.06 K.

The existence of a thermal background gives strong support to the belief that the Universe was extremely hot in its early stages. The background is very nearly isotropic, which supports the isotropic and homogeneous models of the Universe. The COBE satellite has also detected temperature variations of a relative amplitude 6×10^{-6} in the background. These fluctuations are interpreted as a gravitational redshift of the background produced by the mass concentrations that would later give rise to the ob-

Fig. 20.7. Observations of the cosmic microwave background radiation made by the COBE satellite in 1990 are in agreement with a blackbody law at 2.7 K

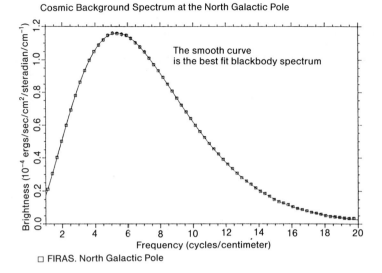

served structures in the Universe. They are the direct traces of initial irregularities in the big bang, and provide important constraints for theories of galaxy formation.

The Isotropy of Matter and Radiation. The microwave background radiation is strictly isotropic to an accuracy of 0.1%. Several other phenomena confirm the isotropy of the Universe. The distribution of radio sources, the x-ray background, and faint distant galaxies, as well as Hubble's law are all isotropic. The observed isotropy is also evidence that the Universe is homogeneous, since a large-scale inhomogeneity would be seen as an anisotropy.

The Age of the Universe. Important cosmological observations that do not depend on specific cosmological models are estimates of the ages of the Earth, the Sun and of star clusters. From the decay of radioactive isotopes, the age of the Earth is estimated to be 4600 million years. The age of the Sun is thought to be slightly larger than this. The ages of the oldest star clusters in the Milky Way are 10–15 Gy.

The values thus obtained give a lower limit to the age of the Universe. In an expanding universe, the inverse Hubble constant gives an upper limit to that age. It is most remarkable that the directly determined ages of cosmic objects are so close to the upper limit given by the Hubble constant. This is strong evidence that the Hubble law really is due to the expansion of the Universe. It also shows that the oldest star clusters formed very early in the history of the Universe.

The Relative Helium Abundance. A cosmological theory should also give an acceptable account of the origin and abundances of the elements. Even the abundance of the elementary particles and the lack of antimatter are cosmological problems that have begun to be investigated in the context of theories of the early Universe.

Observations show that the oldest objects contain about 25% by mass of helium, the most abundant element after hydrogen. The amount of helium produced is sensitive to the temperature of the Universe, which is related to that of the background radiation. The computations made for the standard models of the expanding Universe (the Friedmann models) yield a helium abundance of exactly the right size.

*The Hubble Constant

It is rather amusing that the name "Hubble constant" should be used for a quantity which, by its very definition, is not constant in time and whose measured value has changed within the range of 550km s^{-1} $\text{Mpc}^{-1} - 40 \text{ km s}^{-1} \text{Mpc}^{-1}$. In most cosmological models, the value of H changes continually. Thus it would be better to speak of the *Hubble parameter H*, and to denote its present value by H_0.

Within the distance range where Hubble's constant can be determined, one may use the approximate formula V/r for H_0. Here, r is the distance of the measured object and V is the expansion velocity at that distance. V need not be the velocity obtained from the redshift, since the measured redshift is also affected by the peculiar velocities of the Milky Way and of the object in question (see *Three Redshifts, p. 457). The uncertainty in the velocity V can be reduced either by first subtracting estimates of the local peculiar velocities or by using objects so distant that the relative importance of peculiar velocities becomes insignificant.

The Milky Way is near the edge of the Local Supercluster centred on the Virgo Cluster. This mass concentration is believed to have slowed down the expansion, so that the Milky Way has a peculiar velocity of about $200 - 400 \text{km/s}$ towards the Virgo Cluster. A possible rotation of the Supercluster might also affect the local velocities.

Distances are even more difficult to determine reliably than the local peculiar velocities. From Table 1 showing the history of measurements of H_0, it can be seen that its observed value has decreased by almost a factor of ten since the first measurements. This decrease is mainly a reflection of the improvement in distance determination. The range of accessible velocities has also increased from Hubble's few hundred km/s to about 30000 km/s in modern investigations.

Table 1. Measurements of the Hubble constant [$\text{km s}^{-1} \text{Mpc}^{-1}$]

1936	Hubble	536	(Brightest stars in galaxies used as distance indicators; calibrated within Local Group)
1950	Baade	200	(Local Group distances estimated too small because the period-luminosity relation for W Vir variables had been applied to classical cepheids)
1958	Sandage	100	(Some of Hubble's "brightest stars" were H II regions)
1975	Sandage	55	(The small value may be partly due to a still controversial increase of the distances of the nearest groups)
1993	Sandage	45	(In the early 1990's Sandage has published several papers favouring low value of H_0)

In the years $1975 - 1993$, many different methods have given results in the range $H_0 = 50 - 100 \text{ km s}^{-1}$ Mpc^{-1}. In order to reduce the 50% uncertainty still remaining in the value of H_0, several difficulties have to be overcome.

1) The distances are determined by comparing the apparent magnitude or angular diameter of an object with those of nearby objects with distances determined by other means. Thus one needs to find suitable objects with known standard properties for comparison. For example, if the diameters of the largest H II regions in a galaxy are used as standards, one should know whether these vary from one galaxy to another. In fact, there is a large variation, depending on the luminosity class of the galaxy and possibly on other factors. By taking this into account, the uncertainties can be reduced. Table 2 gives some examples of the kinds of objects and methods used for extragalactic distance determination. The estimated uncertainties may be optimistic.

Table 2. Extragalactic distance indicators

Object/method	Range [Mpc]	Typical uncertainty (no calibration error)
Cepheid/period-luminosity	0 – 4	20%
Largest H II regions/diameter	0 – 10	30%
Brightest stars	0 – 10	30%
H I line width/Tully-Fisher method	0 – 100	20%
Sc galaxies/luminosity classes	0 – 100	40%
Brightest cluster galaxies	20 – 2000	30%

The galaxies M81 (*left*) and M101 (*right*) in the constellation Ursa Major have played an important role in determining the distance scale of the universe. (Photos Palomar Observatory)

2) In distance determinations, it is of crucial importance that the dispersion of the standard properties does not introduce systematic errors depending on distance. There is always a risk of luminosity selection, meaning that at large distances, standard objects that are brighter than average are selected. Their distances will then be underestimated and the Hubble constant correspondingly overestimated. It is likely that the effects of luminosity selection have not yet been completely eliminated from the determinations of H_0.

3) Cosmic distances are determined step by step (see Fig. 18.6a). A distance error at an earlier step is directly reflected at all later steps. For example, when the distances within the Local Group doubled as a result of Baade's work, all distances outside the Local Group also doubled. And when the estimated distance of the Hyades star cluster was changed from 40 to 50 pc, all extragalactic distances changed correspondingly. The distances in the Local Group are essentially based on the cepheid period-luminosity relation. This relation is calibrated by means of cepheids in Milky Way star clusters, and the distances of the star clusters are obtained by fitting their main sequences to that of the Hyades.

20.2 The Cosmological Principle

One hopes that as ever larger volumes of the Universe are observed, its average properties will become simple and well defined. Figure 20.8 attempts to show this. It shows a distribution of galaxies in the plane. As the circle surrounding the observer O becomes

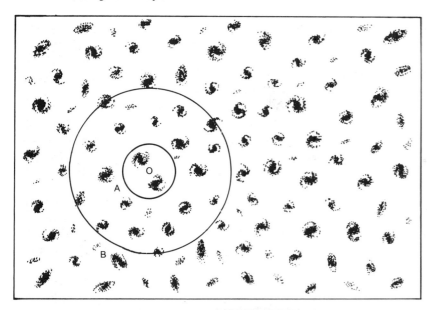

Fig. 20.8. The cosmological principle. In the small circle (A) about the observer (0) the distribution of galaxies does not yet represent the large-scale distribution. In the larger circle (B) the distribution is already uniform on the average

Fig. 20.9. The galaxies seem to be distributed in a "foamlike" way. Dense strings and shells are surrounded by relatively empty regions. [Seldner, M. et al. (1977): Astron. J. **82**, 249]

larger, the mean density inside the circle becomes practically independent of its size. The same behaviour occurs, regardless of the position of the centre of O: at close distances, the density varies randomly (Fig. 20.9), but in a large enough volume, the average density is constant. This is an example of the *cosmological principle*: apart from local irregularities, the Universe looks the same from all positions in space.

The cosmological principle is a far-reaching assumption, which has been invoked in order to put constraints on the large variety of possible cosmological theories. If in addition to the cosmological principle one also assumes that the Universe is isotropic, then the only possible cosmic flow is a global expansion. In that case, the local velocity difference V between two nearby points has to be directly proportional to their separation ($V = Hr$); i.e. Hubble's law must apply.

The plane universe of Fig. 20.8 is homogeneous and isotropic, apart from local irregularities. Isotropy at each point implies homogeneity, but homogeneity does not require isotropy. An example of an anisotropic, homogeneous universe would be a model containing a constant magnetic field: because the field has a fixed direction, space cannot be isotropic.

We have already seen that astronomical observations support the homogeneity and isotropy of our observable neighbourhood, the *metagalaxy*. On the grounds of the cosmological principle, these properties may be extended to the whole of the Universe.

The cosmological principle is closely related to the Copernican principle that our position in the Universe is in no way special. From this principle, it is only a short step to assume that on a large enough scale, the local properties of the metagalaxy are the same as the global properties of the Universe.

Homogeneity and isotropy are important simplifying assumptions when trying to construct *cosmological models* which can be compared with local observations. They may therefore reasonably be adopted, at least as a preliminary hypothesis.

20.3 Homogeneous and Isotropic Universes

Under general conditions, space and time coordinates in a universe may be chosen so that the values of the space coordinates of observers moving with the matter are constant. It can be shown that in a homogeneous and isotropic universe, the line element (Appendix C) then takes the form

$$ds^2 = -c^2 dt^2 + R^2(t) \left[\frac{dr^2}{1 - kr^2} + r^2(d\theta^2 + \cos^2\theta \, d\phi^2) \right] , \qquad (20.4)$$

known as the *Robertson-Walker line element*. (The radial coordinate r is defined to be dimensionless.) $R(t)$ is a time-dependent quantity representing the scale of the Universe. If R increases with time, all distances, including those between galaxies, will grow. The coefficient k may be $+1$, 0 or -1, corresponding to the three possible geometries of space, the *elliptic* or *closed*, the *parabolic* and the *hyperbolic* or *open* model.

The space described by these models need not be Euclidean, but can have positive or negative curvature. Depending on the curvature, the volume of the universe may be finite or infinite. In neither case does it have a visible edge.

The two-dimensional analogy to elliptical ($k = +1$) geometry is the surface of a sphere (Fig. 20.10): its surface area is finite, but has no edge. The scale factor $R(t)$ represents the size of the sphere. When R changes, the distances between points on the surface change in the same way. Similarly, a three-dimensional "spherical surface", or

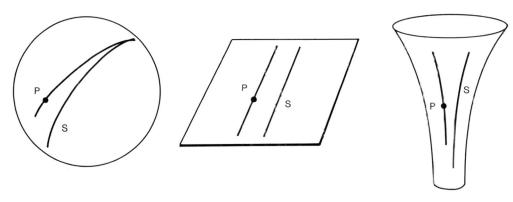

Fig. 20.10. The two-dimensional analogues of the Friedmann models: A spherical surface, a plane and a saddle surface

the space of elliptical geometry, has a finite volume, but no edge. Starting off in an arbitrary direction and going on for long enough, one always returns to the initial point.

When $k = 0$, space is *flat* or Euclidean, and the expression for the line element (20.4) is almost the same as in Minkowski space. The only difference is the scale factor $R(t)$. All distances in Euclidean space change with time. The two-dimensional analogue of this space is a plane.

The volume of space in the hyperbolic geometry ($k = -1$) is also infinite. A two-dimensional idea of the geometry in this case is given by a saddle surface.

In a homogeneous and isotropic universe, all physical quantities will depend on time through the scale factor $R(t)$. For example, from the form of the line element, it is evident that all distances will be proportional to R (Fig. 20.11). Thus, if the distance to a galaxy is r at time t, then at time t_0 it will be

$$[R(t_0)/R(t)]r \ . \tag{20.5}$$

Similarly, all volumes will be proportional to R^3. From this it follows that the density of any conserved quantity (e.g. mass) will behave as R^{-3}.

It can be shown that the wavelength of radiation in an expanding universe is proportional to R, like all other lengths. If the wavelength at the time of emission, corresponding to the scale factor R, is λ, then it will be λ_0 when the scale factor has increased to R_0:

$$\frac{\lambda_0}{\lambda} = \frac{R_0}{R} \ . \tag{20.6}$$

The redshift is $z = (\lambda_0 - \lambda)/\lambda$, and hence

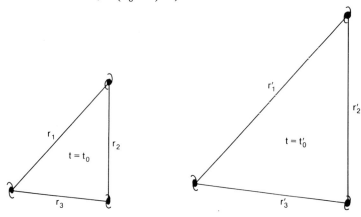

Fig. 20.11. When space expands, all galaxy separations grow with the scale factor R: $r' = [R(t_0')/R(t_0)]r$

$$1 + z = R_0/R \; ; \tag{20.7}$$

i.e. the redshift of a galaxy expresses how much the scale factor has changed since the light was emitted. For example, the light from a quasar with $z = 1$ was emitted at a time when all distances were half their present values.

For small values of the redshift, (20.7) approaches the usual form of Hubble's law. This can be seen as follows. When z is small, the change in R during the propagation of a light signal will also be small and proportional to the light travel time t. Because $t = r/c$ approximately, where r is the distance of the source, the redshift will be proportional to r. If the constant of proportionality is denoted by H/c, one has

$$z = Hr/c \; . \tag{20.8}$$

This is formally identical to Hubble's law (20.1). However, the redshift is now interpreted in the sense of (20.7).

As the universe expands, the photons in the background radiation will also be redshifted. The energy of each photon is inversely proportional to its wavelength, and will therefore behave as R^{-1}. It can be shown that the number of photons will be conserved, and thus their number density will behave as R^{-3}. Combining these two results, one finds that the energy density of the background radiation is proportional to R^{-4}. The energy density of blackbody radiation is proportional to T^4, where T is the temperature. Thus the temperature of cosmic background radiation will vary as R^{-1}.

20.4 The Friedmann Models

The results of the preceding section are valid in any homogeneous and isotropic universe. In order to determine the precise time-dependence of the scale factor $R(t)$ a theory of gravity is required.

In 1917, *Albert Einstein* presented a model of the Universe based on his general theory of relativity. It described a geometrically symmetric (spherical) space with finite volume but no boundary. In accordance with the cosmological principle, the model was homogeneous and isotropic. It was also *static*: the volume of space did not change.

In order to obtain a static model, Einstein had to introduce a new repulsive force, the *cosmological term,* in his equations. The size of this cosmological term is given by the *cosmological constant* Λ. Einstein presented his model before the redshifts of the galaxies were known, and taking the Universe to be static was then reasonable. When the expansion of the Universe was discovered, this argument in favour of the cosmological constant vanished. However, opinions are still divided about whether a cosmological term should be included in the equations of general relativity. Einstein himself later called it the biggest blunder of his life.

The St. Petersburg physicist *Alexander Friedmann* and later, independently, the Belgian *Georges Lemaître* studied the cosmological solutions of Einstein's equations for $\Lambda = 0$. In this case, only evolving, expanding or contracting models of the Universe are possible. From the Friedmann models, exact formulae for the redshift and the Hubble law may be derived.

The general relativistic derivation of the law of expansion for the Friedmann models will not be given here. It is interesting that the existence of three types of models and their law of expansion can be derived from purely Newtonian considerations, with results in complete agreement with the relativistic treatment. The detailed derivation is given on p.453, but the essential character of the motion can be obtained from a simple energy argument.

Let us consider a small expanding spherical region in the Universe. In a spherical distribution of matter, the gravitational force on a given spherical shell depends only on the mass inside that shell.

We can now consider the motion of a galaxy of mass m at the edge of our spherical region. According to the Hubble law, its velocity will be $V = Hr$ and the corresponding kinetic energy,

$$T = mV^2/2 \ . \tag{20.9}$$

The potential energy at the edge of a sphere of mass M is $U = -GMm/r$. Thus the total energy is

$$E = T + U = mV^2/2 - GMm/r \ , \tag{20.10}$$

which has to be constant. If the mean density of the Universe is ρ, the mass $M = (4\pi r^3/3)\rho$. The value of ρ corresponding to $E = 0$ is called the *critical density*, ρ_c. We have

$$E = \frac{1}{2}mH^2r^2 - \frac{GMm}{r} = \frac{1}{2}mH^2r^2 - Gm\frac{4\pi}{3}\frac{r^3\rho_c}{r}$$

$$= mr^2\left(\frac{1}{2}H^2 - \frac{4}{3}\pi G\rho_c\right) = 0 \ , \tag{20.11}$$

$$\Rightarrow \rho_c = \frac{3H^2}{8\pi G} \ . \tag{20.12}$$

The expansion of the Universe can be compared to the motion of a mass launched vertically from the surface of a celestial body. The form of the orbit depends on the initial energy. In order to compute the complete orbit, the mass M of the main body and the initial velocity have to be known. In cosmology, the corresponding parameters are the mean density and the Hubble constant.

The $E = 0$ model corresponds to the Euclidean Friedmann model, the so-called *Einstein-de Sitter* model. If the density exceeds the critical density, the expansion of any spherical region will turn to a contraction and it will collapse to a point. This corresponds to the closed Friedmann model. Finally, if $\rho < \rho_c$, the ever expanding hyperbolic model is obtained. The behaviour of the scale factor in these three cases is shown in Fig. 20.12.

These three models of the universe are called the *standard models*. They are the simplest relativistic cosmological models for $\Lambda = 0$. Other models based on Einstein's theory have been proposed, but no observational reason to abandon the simple standard models has yet been discovered.

The simple Newtonian treatment of the expansion problem is possible because Newtonian mechanics is approximately valid in small regions of the Universe. However, although the resulting equations are formally similar, the interpretation of the quantities involved (e.g. the parameter k) is not the same as in the relativistic context. The global

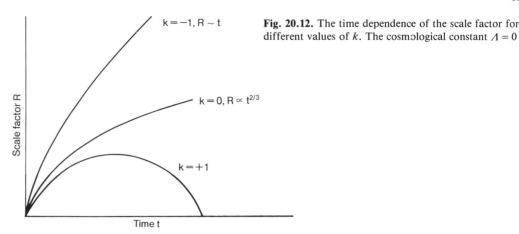

geometry of the Friedmann models can only be understood within the general theory of relativity.

A cosmological model is completely specified when the values of the mean density and the Hubble constant are known. Other quantities describing the expansion can then be expressed in terms of these.

Consider two points at a separation r. Let their relative velocity be V. Then

$$r = [R(t)/R(t_0)]\, r_0 \quad \text{and}$$
$$V = \dot{r} = [\dot{R}(t)/R(t_0)]\, r_0 \;, \tag{20.13}$$

and thus the Hubble constant is

$$H = V/r = \dot{R}(t)/R(t) \;. \tag{20.14}$$

The deceleration of the expansion is described by the *deceleration parameter q*, defined as

$$q = -R\ddot{R}/\dot{R}^2 \;. \tag{20.15}$$

The deceleration parameter describes the change of the rate of expansion R. The additional factors have been included in order to make it dimensionless, i.e. independent of the choice of units of length and time.

The value of the deceleration parameter can be expressed in terms of the mean density. The *density parameter Ω* is defined as $\Omega = \rho/\rho_c$, so that $\Omega = 1$ corresponds to the Einstein-de Sitter model. From the conservation of mass, it follows that $\rho_0 R_0^3 = \rho R^3$. Using expression (20.12) for the critical density, one then obtains

$$\Omega = \frac{8\pi G}{3}\, \frac{\rho_0 R_0^3}{R^3 H^2} \;. \tag{20.16}$$

On the other hand, using Eq. (4) on p. 453, q can be written

$$q = \frac{4\pi G}{3}\, \frac{\rho_0 R_0^3}{R^3 H^2} \;. \tag{20.17}$$

Thus there is a simple relation between Ω and q:

$$\Omega = 2q \ . \tag{20.18}$$

The value $q = 1/2$ of the deceleration parameter corresponds to the critical density $\Omega = 1$. Both quantities are in common use in cosmology. It should be noted that the density and the deceleration can be observed independently. The validity of (20.18) is thus a test for the correctness of general relativity.

*Newtonian Derivation of a Differential Equation for the Scale Factor $R(t)$

Let us consider a galaxy at the edge of a massive sphere (see figure). It will be affected by a central force

$$m\ddot{r} = -\frac{4\pi G}{3}\frac{r^3 \rho m}{r^2}$$

$$\ddot{r} = -\frac{4\pi}{3}G\rho r \ . \tag{1}$$

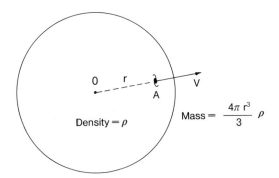

In these equations, the radius r and the density ρ are changing with time. They may be expressed in terms of the scale factor R:

$$r = (R/R_0)r_0 \ , \tag{2}$$

where R is defined to be R_0 when the radius $r = r_0$.

$$\rho = (R_0/R)^3 \rho_0 \ , \tag{3}$$

where the density $\rho = \rho_0$ when $R = R_0$. Introducing (2) and (3) in (1), one obtains

$$\ddot{R} = -\frac{a}{R^2} \ , \tag{4}$$

where $a = 4\pi GR_0^3\rho_0/3$. If (4) is multiplied on both sides by $\dot{R}$, the left-hand side yields

$$\dot{R}\ddot{R} = \frac{1}{2}\frac{d(\dot{R}^2)}{dt} \ ,$$

and thus (4) takes the form

$$d(\dot{R}^2) = -\frac{2a}{R^2}dR \ . \tag{5}$$

Let us define $R_0 = R(t_0)$. Integrating (5) from t_0 to t gives

$$\dot{R}^2 - \dot{R}_0^2 = 2a\left(\frac{1}{R} - \frac{1}{R_0}\right) . \tag{6}$$

The constants $\dot{R}_0$ and a can be eliminated in favour of the Hubble constant H_0 and the density parameter Ω_0. Because $\rho_c = 3H_0^2/8\pi G$,

$$2a = 8\pi G R_0^3 \rho_0/3 = H_0^2 R_0^3 \rho_0/\rho_c = H_0^2 R_0^3 \Omega_0 , \tag{7}$$

where $\Omega_0 = \rho_0/\rho_c$. Using expression (7) and $\dot{R}_0 = H_0 R_0$ in (6), one obtains

$$\dot{R}^2 = H_0^2 R_0^2\left(\Omega_0 \frac{R_0}{R} - \Omega_0 + 1\right) . \tag{8}$$

The time behaviour of the scale factor R depends on the value of the density parameter Ω_0. Because $\dot{R}^2 > 0$ always, according to (8)

$$\Omega_0 \frac{R_0}{R} - \Omega_0 + 1 \geq 0 \quad \text{or}$$

$$\frac{R_0}{R} \geq \frac{\Omega_0 - 1}{\Omega_0} . \tag{9}$$

If $\Omega_0 > 1$, this means that

$$R \leq R_0 \frac{\Omega_0}{\Omega_0 - 1} \equiv R_{\max} .$$

When the scale factor reaches its maximum value $R_{\max}$, then according to (8), $\dot{R} = 0$, and the expansion turns into a contraction. If $\Omega_0 < 1$, the right-hand side of (8) is always positive and the expansion continues forever.

The equation for the time dependence of the scale factor in the general theory of relativity contains the constant k which determines the geometry of space:

$$\dot{R}^2 = \frac{8\pi G R_0^3 \rho_0}{3R} - kc^2 . \tag{10}$$

Equations (10) and (6) [or (8)] can be made identical if one chooses

$$\frac{8\pi G R_0^2 \rho_0}{3} - \dot{R}_0^2 = H_0^2 R_0^2(\Omega_0 - 1) = kc^2 .$$

Thus complete agreement between the Newtonian and the relativistic equation for R is obtained. The values of the geometrical constant $k = +1, 0, -1$ correspond respectively to $\Omega_0 > 1, = 1$ and < 1.

When $k = 0$, the time dependence of the expansion is very simple. Setting $\Omega_0 = 1$ and using (10) and (7), one obtains

$$\dot{R}^2 = \frac{H_0^2 R_0^3}{R} .$$

The solution of this equation is

$$R = \left(\frac{3H_0 t}{2}\right)^{2/3} R_0 .$$

It is also easy to calculate the time from the beginning of the expansion: $R = R_0$ at the time

$$t_0 = \frac{2}{3}\frac{1}{H_0} .$$

This is the age of the Universe in the Einstein-de Sitter model.

20.5 Cosmological Tests

A central cosmological problem is the question of which Friedmann model best represents the real Universe. Different models make different observational predictions. However, present observations have not yet allowed an unambiguous determination of the geometry of the Universe. In the following, some possible tests will be considered.

The Critical Density. If the average density ρ is larger than the critical density ρ_c, the Universe is closed. For the Hubble constant $H = 100\,\mathrm{km\,s^{-1}Mpc^{-1}}$, the value of $\rho_c = 1.9 \times 10^{-26}\,\mathrm{kg\,m^{-3}}$, corresponding to roughly ten hydrogen atoms per cubic metre. Mass determinations for individual galaxies lead to smaller values for the density, favouring an open model. However, the density determined in this way is a lower limit, since there may be significant amounts of invisible mass outside the main parts of the galaxies.

   If most of the mass of clusters of galaxies is dark and invisible, it will increase the mean density nearer the critical value. Using the virial masses of clusters of galaxies (Sect. 19.2), one finds $\Omega_0 = 0.1$; and mass may still exist in the space between the clusters.

   It may be that the neutrino has a small mass (about 10^{-4} electron mass). A large neutrino background should have been produced in the big bang. In spite of the small suggested mass of the neutrino, it would still be large enough to make neutrinos the dominant form of mass in the Universe, and would probably make the density larger than critical. Laboratory measurement of the mass of the neutrino is very difficult, and the claims for a measured non-zero mass have not won general acceptance. Instead much recent work in cosmology has been based on the hypothesis of *Cold Dark Matter* (CDM), i.e. the idea that a significant part of the mass of the Universe is in the form of non-relativistic particles of unknown kind.

The Magnitude-Redshift Test. Although for small redshifts, all models predict the Hubble relationship $m = 5\lg z + C$ for standard candles, for larger redshifts there are differences, depending on the deceleration parameter q. This provides a way of measuring q.

   The models predict that galaxies at a given redshift look brighter in the closed models than in the open ones. In the redshift range $(z < 1)$ of the best standard candles, the brightest galaxies in clusters, these differences are unfortunately too small to have been detected. The difficulties are increased by possible brightness evolution, since one is looking back several thousand million years. Quasars can be observed at much higher redshifts, but no distance-independent way to estimate their brightnesses is known with certainty. Nevertheless, some studies of quasars have claimed evidence in favour of a closed model with $q = 1$.

The Angular Diameter-Redshift Test. Along with the magnitude-redshift test, the relation between angular diameter and redshift has been used as a cosmological test. Let us first consider how the angular diameter θ of a standard object varies with distance in static models with different geometries. In a Euclidean geometry, the angular diameter is inversely proportional to the distance. In an elliptical geometry, θ decreases more slowly with distance, and even begins to increase beyond a certain point. The reason for this can be understood by thinking of the surface of a sphere. For an observer at the pole, the angular diameter is the angle between two meridians marking the edges of his standard object. This angle is smallest when the object is at the equator, and grows without limit towards the opposite pole. In a hyperbolic geometry, the angle θ decreases more rapidly with distance than in the Euclidean case.

In an expanding closed universe, the angular diameter should begin to increase at a redshift of about one. This effect has been looked for in the diameters of radio galaxies and quasars. No turnover has been observed, but this may also be due to evolution of the radio sources or to the selection of the observational data. At smaller redshifts, the use of the diameters of clusters of galaxies has yielded equally inconclusive results.

Primordial Nucleosynthesis. The standard model predicts that 25% of the mass of the Universe turned into helium in the big "big bang". This amount is not sensitive to the density and thus does not provide a strong cosmological test. However, the amount of deuterium left over from helium synthesis does depend strongly on the density. Almost all deuterons formed in the big bang unite into helium nuclei. The larger the density, the more frequent the collisions were leading to the destruction of deuterium. Thus a small present deuterium abundance indicates a high cosmological density. Similar arguments apply to the amounts of ^{3}He and ^{7}Li produced in the big bang. The interpretation of the observed abundances is difficult, since they have been changed by later nuclear processes. Still, present results are consistent with a density corresponding to Ω_0 about 0.01. Note that this number only refers to the mass in the form of baryons, i.e. protons and neutrons. Since the virial masses of clusters of galaxies indicate that Ω_0 is at least $0.1-0.2$, this has stimulated models such as the CDM model, where most of the mass is not in the form of baryons.

Ages. The ages of different Friedmann models can be compared with known ages of various systems. For a Friedmann model with Hubble constant H and density parameter Ω, the age may be written

$$t_0 = f(\Omega)/H \; , \tag{20.19}$$

where the known function $f(\Omega) < 1$. For the critical density $f(\Omega = 1) = 2/3$. Thus if $H = 75$ km s^{-1} Mpc^{-1}, the age $t_0 = 9$ Gy. Larger values of Ω give smaller ages. If there are systems of the known age t_k, one has to demand that $t_0 > t_k$ or

$$f(\Omega) > t_k H \; . \tag{20.20}$$

In principle, this equation gives an upper limit for the allowed density in a model. However, the uncertainties in H and t_k are such that it has not yet led to interesting constraints on the density.

In summary, the density determinations and the deuterium abundance favour an open model of the Universe. The presence of massive neutrinos or other forms of invisible matter might raise the density above the critical value. Tests based on the relations between redshift and magnitude or angular diameter have not yet yielded reliable results. In the future, the use of quasars might change the situation.

*Three Redshifts

The redshift of a distant galaxy is the result of three different mechanisms acting together. The first one is the peculiar velocity of the observer with respect to the mean expansion: the Earth moves about the Sun, the Sun about the centre of the Milky Way, and the Milky Way and the Local Group of galaxies is falling towards the Virgo Cluster. The apparatus measuring the light from a distant galaxy is not at rest; the velocity of the instrument gives rise to a Doppler shift that has to be corrected for. Usually the velocities are much smaller than the speed of light. The Doppler shift is then

$$z_D = v/c \ .$$

For large velocities, the relativistic formula has to be used:

$$z_D = \sqrt{\frac{c+v}{c-v}} - 1 \ .$$

The redshift appearing in Hubble's law is the *cosmological redshift* z_c. It only depends on the values of the scale factor at the times of emission and detection of the radiation (R and R_0) according to

$$z_c = R_0/R - 1 \ .$$

The third type of redshift is the *gravitational redshift* z_g. According to general relativity, light will be redshifted by a gravitational field. For example, the redshift of radiation from the surface of a star of radius R and mass M will be

$$z_1 = \frac{\lambda_1 - \lambda_0}{\lambda_0} \quad \text{and} \quad z_2 = \frac{\lambda_2 - \lambda_1}{\lambda_1} \ ,$$

then the total redshift is

$$z_g = \frac{1}{\sqrt{1 - R_S/R}} - 1 \ ,$$

where $R_S = 2GM/c^2$ is the Schwarzschild radius of the star. The gravitational redshift of the radiation from galaxies is normally insignificant.

The combined effect of the redshifts can be calculated as follows. If the rest wavelength λ_0 is redshifted by the amounts z_1 and z_2 by two different processes, so that

$$z = \frac{\lambda_2 - \lambda_0}{\lambda_0} = \frac{\lambda_2}{\lambda_0} - 1 = \frac{\lambda_2}{\lambda_1} \frac{\lambda_1}{\lambda_0} - 1 \quad \text{or} \quad 1 + z = (1 + z_1)(1 + z_2) \ .$$

Similarly, the three redshifts z_D, z_c and z_g will combine to give an observed redshift z, according to

$$1 + z = (1 + z_D)(1 + z_c)(1 + z_g) \ .$$

20.6 History of the Universe

We have seen how the density of matter and of radiation energy and temperature can be computed as functions of the scale factor R. Since the scale factor is known as a function of time, these quantities can be calculated backwards in time.

At the earliest times, densities and temperatures were so immense that all theories about the physical processes taking place are highly conjectural. Nevertheless, first attempts have been made at understanding the most fundamental properties of the Universe on the basis of modern theories of particle physics. For example, no indications of significant amounts of antimatter in the Universe have been discovered. Thus, for some reason, the number of matter particles must have exceeded that of antimatter particles by a factor of 1.000000001. Because of this *symmetry breaking*, when 99.9999999

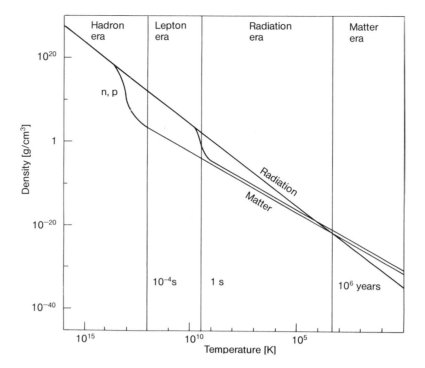

% of the hadrons were annihilated, 10^{-7} % was left later to form galaxies and everything else. It has been speculated that the broken symmetry originated in particle processes about 10^{-35} s after the initial time.

The breaking of fundamental symmetries in the early Universe may lead to what is known as *inflation* of the Universe. In consequence of symmetry breaking, the dominant energy density may be the zero-point energy of a quantum field. This energy density will lead to inflation, a strongly accelerated expansion, which will dilute irregularities and drive the density very close to the critical value. One may thus understand how the present homogeneity, isotropy and flatness of the Universe have come about. In the inflationary picture the mass density has to be very close to critical, $\Omega_0 = 1$. Since according to the theories of primordial nucleosynthesis the density of baryons has to be much smaller than critical, it has been assumed that much of the mass of the Universe may be in the form of unknown non-baryonic particles. If these particles are non-relativistic they are known as *cold dark matter*. Cold dark matter models have had some success in explaining the observed properties of galaxies and clusters of galaxies.

As the Universe expanded, the density decreased (Fig. 20.13) and conditions became such that known physical principles can be applied. During the hot early stages, photons and massive particles were continually changing into each other: high-energy photons collided to produce particle-antiparticle pairs, which then were annihilated and produced photons. As the Universe cooled, the photon energies became too small to maintain this equilibrium. There is a *threshold temperature* below which particles of a given type are no longer produced. For example, the threshold temperature for *hadrons* (protons, neutrons and mesons) is $T = 10^{12}$ K, reached at the time $t = 10^{-4}$ s. Thus the present building blocks of the atomic nuclei, protons and neutrons, are relics from the time $10^{-8} - 10^{-4}$ s, known as the *hadron era*.

The Lepton Era. In the time period $10^{-4} - 1\,\mathrm{s}$, the *lepton era*, the photon energies were large enough to produce light particles, such as electron-positron pairs. Because of matter-antimatter symmetry breaking, some of the electrons were left over to produce present astronomical bodies. During the lepton era, *neutrino decoupling* took place. Previously, the neutrinos had been kept in equilibrium with other particles by fast particle reactions. As the density and temperature decreased, so did the reaction rates, and finally they could no longer keep the neutrinos in equilibrium. The neutrinos decoupled from other matter and were left to propagate through space without any appreciable interactions. It has been calculated that there are at present 600 such cosmological neutrinos per cubic centimetre, but their negligible interactions make them extremely difficult to observe.

The Radiation Era. After the end of the lepton era, about 1 s after the initial time, the most important form of energy was electromagnetic radiation. This stage is called the *radiation era*. At its beginning, the temperature was about $10^{10}\mathrm{K}$ and at its end, about one million years later, when the radiation energy density had dropped to that of the particles, it had fallen to a few thousand degrees. At the very beginning of the radiation era, within a few hundred seconds, helium was produced.

Just before the epoch of helium synthesis, the number ratio of free protons and neutrons was changing because of the decay of the free neutrons. After about 100 s, the temperature had dropped to about $10^9\mathrm{K}$, which is low enough for deuterons to be formed. All remaining neutrons were then incorporated in deuterons; these, in turn, were almost entirely consumed to produce helium nuclei. Thus the amount of helium synthesized was determined by the number ratio of protons and neutrons at the time of deuterium production $t = 100\,\mathrm{s}$. Calculations show that this ratio was about $14:2$. Thus, out of 16 nucleons, 2 protons and 2 neutrons were incorporated in a helium nucleus. Consequently, $4/16 = 25\%$ of the mass turned into helium. This is remarkably close to the measured primordial helium abundance.

Only the isotopes $^2\mathrm{H}$, $^3\mathrm{He}$, $^4\mathrm{He}$ and $^7\mathrm{Li}$ were produced in appreciable numbers by nuclear processes in the big bang. The heavier elements have formed later in stellar interiors, in supernova explosions and perhaps in energetic events in galactic nuclei.

Radiation Decoupling. The Matter Era. As we have seen, the mass density of radiation (obtained from the formula $E = mc^2$) behaves as R^{-4}, whereas that of ordinary matter behaves as R^{-3}. Thus the radiation mass density decreases more rapidly. At the end of the radiation era, it became smaller than the ordinary mass density. The *matter era* began, bringing with it the formation of galaxies, stars, planets and human life. At present, the mass density of radiation is much smaller than that of matter. Therefore the dynamics of the Universe is completely determined by the density of massive particles.

Near the end of the radiation era, radiation decoupled from matter. This happened when the temperature had dropped to a few thousand degrees, and the protons and electrons combined to form hydrogen atoms. At present, light can propagate freely through space. The world is transparent to radiation: the light from distant galaxies is weakened only by the r^{-2} law and by the redshift. The redshift of decoupling is about 1000, which represents a fundamental farthest limit to observations by means of electromagnetic radiation.

It is an interesting coincidence that the epoch when the radiation energy density became larger than that of matter occurred so near the decoupling epoch. This fact accentuates the contrast between the radiation and the matter eras.

The Formation of Galaxies. As we go backward in time from the present, the distances between galaxies and clusters of galaxies become smaller. For example, the typical separation between galaxies is 100 times their diameter. At the redshift $z = 101$, most galaxies must have been practically in contact. For this reason, galaxies in their present form cannot have existed much earlier than about $z = 100$. Since the stars were presumably formed after the galaxies, all present astronomical systems must have formed later than this.

Galaxies are thought to have formed from the collapse of small density enhancements in the gas. As long as the gas remained strongly coupled to the radiation (while $R_0/R > 1000$), the radiation pressure prevented any overdensities in the ionized gas from growing. Any structures that were to give rise to galaxies and clusters of galaxies could begin to form only after the decoupling of the radiation.

The problem of the origin of galaxies is still open, and there is no agreement on what kind of perturbations in the smooth distribution of radiation and gas were responsible for present extragalactic systems. For example, it is not known whether the first structures to form were galaxies or clusters of galaxies. In the latter case, the galaxies would have formed by the fragmentation of a collapsing protocluster.

The minimum mass of a collapsing gas cloud is given by the *Jeans mass M_J*:

$$M_J \approx \frac{P^{3/2}}{G^{3/2}\rho^2} \, , \tag{20.24}$$

where ρ and P are the density and pressure in the cloud (see Sect. 7.10). The value of M_J before decoupling was

$$M_J = 10^{18} M_\odot \tag{20.25}$$

and after decoupling,

$$M_J = 10^5 M_\odot \, . \tag{20.26}$$

The reason for the large difference is that before decoupling, matter feels the large radiation pressure ($P = u/3$, see *Gas Pressure and Radiation Pressure, p. 267). After decoupling, this pressure no longer affects the gas.

The Jeans mass before decoupling is larger than that of any known systems, but after decoupling, it is interestingly close to the mass of globular clusters.

Clearly, astronomical systems may have been formed by the collapse of Jeans unstable gas clouds. A serious difficulty for this scenario is that the expansion of the Universe works against the collapse. The collapse therefore takes longer than in the static case. If present structures are to have had time to form, the density inhomogeneities present at the time of decoupling would have had to be large enough. One then has to ask where these initial perturbations came from. All that can be said at present is that they must have originated in the very early stages of the Universe. There is therefore hope that they will contain interesting information about the earliest times.

20.7 The Future of the Universe

The standard models allow two alternative prospects for the future development of the Universe. Expansion may either go on forever, or it may reverse to a contraction, where everything is eventually squeezed back to a point. In the final squeeze, the early history of the Universe would be repeated backwards: in turn, galaxies, stars, atoms and nucleons would be broken up. Finally, the state of the Universe could no longer be treated by present-day physics.

In the open models, the future is quite different. The evolution of the stars may lead to one of four end results: a white dwarf, a neutron star or a black hole may be formed, or the star may be completely disrupted. After about 10^{11} years, all present stars will have used up their nuclear fuel and reached one of these four final states.

Some of the stars will be ejected from their galaxies; others will form a dense cluster at the centre. In about 10^{27} years, the central star clusters will become so dense that a black hole is formed. Similarly, the galaxies in large clusters will collide and form very massive black holes.

Not even black holes last forever. By a quantum mechanical tunnelling process, mass can cross the event horizon and escape to infinity − the black hole is said to "evaporate". The rate of this phenomenon, known as the *Hawking process*, is inversely proportional to the mass of the hole. For a galactic-mass black hole, the evaporation time is roughly 10^{90} years. After this time, almost all black holes will have disappeared.

The ever expanding space now contains black dwarfs, neutron stars and planet-size bodies (unless the predictions of a finite proton lifetime of about 10^{31} years are confirmed; in that case all these systems will have been destroyed by proton decay). The temperature of the cosmic background radiation will have dropped to 10^{-20} K.

Even further in the future, other quantum phenomena come into play. By a tunnelling process, black dwarfs can change into neutron stars and these, in turn, into black holes. In this way, all stars are turned into black holes, which will then evaporate. The time required has been estimated to be $10^{10^{26}}$ years! At the end, only radiation cooling towards absolute zero will remain.

As noted previously, observations have not yet settled which alternative will come true. It is also highly doubtful whether our current cosmological theories really are secure enough to allow such far-reaching predictions. New theories and observations may completely change our present cosmological ideas.

20.8 Examples

20.8.1 Find the photon density of the 2.7 K background radiation.

The intensity of the radiation is

$$B_\nu = \frac{2h\nu^3}{c^2} \frac{1}{\exp{(h\nu/kT)} - 1}$$

and the energy density

$$u_\nu = \frac{4\pi}{c} B_\nu = \frac{8\pi h\nu^3}{c^3} \frac{1}{\exp{(h\nu/kT)} - 1} \ .$$

The number of photons per unit volume is found by dividing the energy density by the energy of a single photon, and integrating over all frequencies:

$$N = \int_0^\infty \frac{u_\nu d\nu}{h\nu} = \frac{8\pi}{c^3} \int_0^\infty \nu^2 \frac{d\nu}{\exp(h\nu/kT) - 1} .$$

We substitute $h\nu/kT = x$ and $d\nu = (kT/h)dx$:

$$N = 8\pi \left(\frac{kT}{hc}\right)^3 \int_0^\infty \frac{x^2 dx}{e^x - 1} .$$

The integral cannot be expressed in terms of elementary functions (however, it can be expressed as an infinite sum $2\sum_{n=0}^\infty (1/n^3)$), but it can be evaluated numerically. Its value is 2.4041. Thus the photon density at 2.7 K is

$$N = 8\pi \left(\frac{1.3805 \times 10^{-23} \,\mathrm{JK^{-1}} \times 2.7 \,\mathrm{K}}{6.6256 \times 10^{-34} \,\mathrm{Js} \times 2.9979 \times 10^8 \,\mathrm{ms^{-1}}}\right)^3 \times 2.4041$$

$$= 3.99 \times 10^8 \,\mathrm{m^{-3}} \approx 400 \,\mathrm{cm^{-3}} .$$

20.8.2   a) In a forest there are n trees per hectare, evenly spaced. The thickness of each trunk is D. What is the distance of the wood not seen for the trees? (Find the probability that the line of sight will hit a trunk within a distance x.) b) How is this related to Olbers' paradox?

a) Imaging a circle with radius x around the observer. A fraction $s(x)$, $0 \le s(x) \le 1$, is covered by trees. Then we'll move a distance dx outward, and draw another circle. There are $2\pi n x dx$ trees growing in the annulus limited by these two circles. They hide a distance $2\pi x n D dx$ or a fraction $nD dx$ of the perimeter of the circle. Since a fraction $s(x)$ was already hidden, the contribution is only $(1-s(x))nD dx$. We get

$$s(x+dx) = s(x) + (1-s(x))nD dx ,$$

which gives a differential equation for s:

$$\frac{ds(x)}{dx} = (1-s(x))nD .$$

This is a separable equation, which can be integrated:

$$\int_0^s \frac{ds}{1-s} = \int_0^x nD dx .$$

This yields the solution

$$s(x) = 1 - e^{-nDx} .$$

This is the probability that in a random direction we can see at most to a distance x. This function s is a cumulative probability distribution. The corresponding probability density is its derivative ds/dx. The mean free path λ is the expectation of this distribution:

$$\lambda = \int\limits_0^\infty x \left(\frac{ds(x)}{dx} \right) dx = \frac{1}{nD} \ .$$

For example, if there are 2000 trees per hectare, and each trunk is 10 cm thick, we can see to a distance of 50 m, on the average.

b) The result can easily be generalized into three dimensions. Assume there are n stars per unit volume, and each has a diameter D and surface $A = \pi D^2$ perpendicular to the line of sight. Then we have

$$s(x) = 1 - e^{-nAx} \ ,$$

$$\lambda = 1/nA \ .$$

For example, if there were one sun per cubic parsec, the mean free path would be 1.6×10^{14} parsecs. If the universe were infinitely old and infinite in size, the line of sight would eventually meet a stellar surface in any direction, although we could see very far indeed.

20.9 Exercises

20.9.1 The apparent diameter of the galaxy NGC 3159 is 1.3′, apparent magnitude 14.4, and radial velocity with respect to the Milky Way 6940 km s^{-1}. Find the distance, diameter and absolute magnitude of the galaxy. What potential sources of error can you think of?

20.9.2 The radial velocity of NGC 772 is 2562 km s^{-1}. Compare the distance obtained from this information with Exercise 19.11.1.

20.9.3 If the neutrinos have nonzero mass, the universe can be closed. What is the minimum mass needed for this? Assume that the density of neutrinos is 600 cm^{-3}, and the density of other matter is one tenth of the critical density.

Appendices

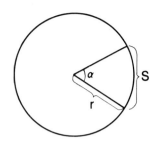

A. Mathematics

A.1 Geometry

Units of Angle and Solid Angle

Radian is the angular unit most suitable for theoretical studies. One radian is the angle subtended by a circular arc whose length equals the radius. If r is the radius of a circle and s the length of an arc, the arc subtends an angle

$$\alpha = s/r.$$

Since the circumference of the circle is $2\pi r$, we have

$$2\pi \text{ rad} = 360°, \quad \text{or} \quad 1 \text{ rad} = 180°/\pi.$$

In an analogous way, we can define a *steradian*, a unit of solid angle, as the solid angle subtended by a unit area on the surface of a unit sphere as seen from the centre. An area A on the surface of a sphere with radius r subtends a solid angle

$$\omega = A/r^2.$$

Since the area of the sphere is $4\pi r^2$, a full solid angle equals 4π steradians.

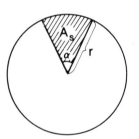

Circle

- Area $A = \pi r^2$
- Area of a sector $A_s = \frac{1}{2}\alpha r^2$

Sphere

- Area $A = 4\pi r^2$
- Volume $V = \frac{4}{3}\pi r^3$
- Volume of a sector $V_2 = \frac{2}{3}\pi r^2 h = \frac{2}{3}\pi r^3(1-\cos\alpha) = V_{\text{sphere}} \text{ hav } \alpha$
- Area of a segment $A_s = 2\pi rh = 2\pi r^2(1-\cos\alpha) = A_{\text{sphere}} \text{ hav } \alpha$

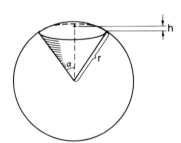

A.2 Taylor Series

Let us consider a differentiable real-valued function of one variable $f:R \to R$. The tangent to the graph of the function at x_0 is

$$y = f(x_0) + f'(x_0)(x - x_0),$$

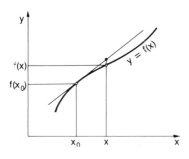

where $f'(x_0)$ is the derivative of f at x_0. Now, if x is close to x_0, the graph of the tangent at x will not be very far from the graph of the function itself. Thus, we can approximate the function by

$$f(x) \approx f(x_0) + f'(x_0)(x - x_0) \, .$$

The approximation becomes worse, the more the derivative f' varies in the interval $[x_0, x]$. The rate of change of f' is described by the second derivative f'', and so on. To improve accuracy, we have to also include higher derivatives. It can be shown that the value of the function f at x is (assuming that all the derivatives exist)

$$f(x) = f(x_0) + f'(x_0)(x - x_0) + \frac{1}{2} f''(x_0)(x - x_0)^2 + \dots$$
$$+ \frac{1}{n!} f^{(n)}(x_0)(x - x_0)^n + \dots$$

where $f^{(n)}(x_0)$ is the nth derivative at x_0 and $n!$ is the n-factorial, $n! = 1 \times 2 \times 3 \times \dots \times n$. This expansion is called the *Taylor series* of the function at x_0. The following list gives some useful Taylor series (in all these cases we have $x_0 = 0$):

$$\frac{1}{1+x} = 1 - x + x^2 - x^3 + \dots \qquad\qquad\qquad \text{converges if } |x| < 1$$

$$\sqrt{1+x} = 1 + \frac{1}{2}x - \frac{1}{8}x^2 + \frac{1}{16}x^3 - \dots$$

$$\sqrt{1-x} = 1 - \frac{1}{2}x - \frac{1}{8}x^2 - \frac{1}{16}x^3 \dots$$

$$\frac{1}{\sqrt{1+x}} = 1 - \frac{1}{2}x + \frac{3}{8}x^2 - \frac{5}{16}x^3 \dots$$

$$\frac{1}{\sqrt{1-x}} = 1 + \frac{1}{2}x + \frac{3}{8}x^2 + \frac{5}{16}x^3 \dots$$

$$e^x = 1 + x + \frac{1}{2!}x^2 + \frac{1}{3!}x^3 + \dots + \frac{1}{n!}x^n + \dots \qquad \text{converges for all } x$$

$$\ln(1+x) = x - \frac{1}{2}x^2 + \frac{1}{3}x^3 - \frac{1}{4}x^4 + \dots + (-1)^{n+1}\frac{1}{n}x^n + \dots \quad x \in (-1, 1]$$

$$\sin x = x - \frac{1}{3!}x^3 + \frac{1}{5!}x^5 - \dots \qquad\qquad\qquad \text{for all } x$$

$$\cos x = 1 - \frac{1}{2!}x^2 + \frac{1}{4!}x^4 - \dots \qquad\qquad\qquad \text{for all } x$$

$$\tan x = x + \frac{1}{3}x^3 + \frac{2}{15}x^5 + \dots \qquad\qquad\qquad |x| < \frac{\pi}{2} \, .$$

Many problems involve small perturbations, in which case it is usually possible to find expressions having very rapidly converging Taylor expansions. The great advantage of this is the reduction of complicated functions to simple polynomials. Particularly useful are linear approximations, such as

$$\sqrt{1+x} \approx 1 + \frac{1}{2}x, \qquad \frac{1}{\sqrt{1+x}} \approx 1 - \frac{1}{2}x, \qquad \text{etc.}$$

A.3 Vector Calculus

A *vector* is an entity with two essential properties: *magnitude* and *direction*. Vectors are usually denoted by boldface letters, *a, b, A, B*, etc. The *sum* of the vectors of *A* and *B* can be determined graphically by moving the origin of *B* to the tip of *A* and connecting the origin of *A* to the tip of *B*. The vector $-A$ has the same magnitude as *A*, is parallel to *A*, but points in the opposite direction. The *difference* $A - B$ is defined as $A + (-B)$.

Addition of vectors satisfies the ordinary rules of commutativity and associativity,

$$A + B = B + A ,$$

$$A + (B + C) = (A + B) + C .$$

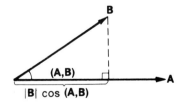

A point in a coordinate frame can be specified by giving its *position vector,* which extends from the origin of the frame to the point. The position vector *r* can be expressed in terms of *basis vectors*, which are usually *unit vectors*, i.e. have a length of one distance unit. In a rectangular xyz-frame, we denote the basis vectors parallel to the coordinate axes by $\hat{i}, \hat{j},$ and $\hat{k}$. The position vector corresponding to the point (x, y, z) is then

$$r = x\hat{i} + y\hat{j} + z\hat{k} .$$

The numbers x, y and z are the *components* of *r*. Vectors can be added by adding their components. For example, the sum of

$$A = a_x\hat{i} + a_y\hat{j} + a_z\hat{k} \quad \text{and} \quad B = b_x\hat{i} + b_y\hat{j} + b_z\hat{k}$$

is

$$A + B = (a_x + b_x)\hat{i} + (a_y + b_y)\hat{j} + (a_z + b_z)\hat{k} .$$

The magnitude of a vector *r* in terms of its components is

$$r = |r| = \sqrt{x^2 + y^2 + z^2} .$$

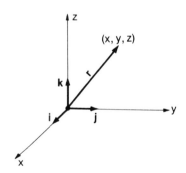

The *scalar product* of two vectors *A* and *B* is a real number (scalar),

$$A \cdot B = a_x b_x + a_y b_y + a_z b_z = |A||B| \cos(A, B) ,$$

where (A, B) is the angle between the vectors *A* and *B*. We can also think of the scalar product as the projection of, say, *A* in the direction of *B* multiplied by the length of *B*. If *A* and *B* are perpendicular, their scalar product vanishes. The magnitude of a vector expressed as a scalar product is $A = \sqrt{A \cdot A}$.

The *vector product* of the vectors A and B is a vector:

$$A \times B = (a_y b_z - a_z b_y)\hat{i} + (a_z b_x - a_x b_z)\hat{j} + (a_x b_y - a_y b_x)\hat{k} = \begin{vmatrix} \hat{i} & \hat{j} & \hat{k} \\ a_x & a_y & a_z \\ b_x & b_y & b_z \end{vmatrix}.$$

This is perpendicular to both A and B. Its length gives the area of the parallelogram spanned by A and B. The vector product of parallel vectors is a null vector. The vector product is anti-commutative:

$$A \times B = -B \times A.$$

Scalar and vector products satisfy the laws of distributivity:

$$A \cdot (B + C) = A \cdot B + A \cdot C \qquad A \times (B + C) = A \times B + A \times C$$

$$(A + B) \cdot C = A \cdot C + B \cdot C \qquad (A + B) \times C = A \times C + B \times C.$$

A *scalar triple product* is a scalar

$$A \times B \cdot C = \begin{vmatrix} a_x & a_y & a_z \\ b_x & b_y & b_z \\ c_x & c_y & c_z \end{vmatrix}.$$

Here, the cross and dot can be interchanged and the factors permuted cyclically without affecting the value of the product. For example, $A \times B \cdot C = B \times C \cdot A = B \cdot C \times A$, but $A \times B \cdot C = -B \times A \cdot C$.

A *vector triple product* is a vector, which can be evaluated using one of the expansions

$$A \times (B \times C) = B(A \cdot C) - C(A \cdot B).$$

$$(A \times B) \times C = B(A \cdot C) - A(B \cdot C).$$

In all these products, scalar factors can be moved around without affecting the product:

$$A \cdot kB = k(A \cdot B), \qquad A \times (B \times kC) = k(A \times (B \times C)) \qquad \text{etc.}$$

The position vector of a particle is usually a function of time: $r = r(t) = x(t)\hat{i} + y(t)\hat{j} + z(t)\hat{k}$. The velocity of the particle is a vector, tangent to the trajectory, obtained by taking the derivative of r with respect to time:

$$v = \frac{d}{dt} r(t) \equiv \dot{r} = \dot{x}\hat{i} + \dot{y}\hat{j} + \dot{z}\hat{k}.$$

The acceleration is the second derivative, $\ddot{r}$.

Derivatives of the various products obey the same rules as derivatives of products of real-valued functions:

$$\frac{d}{dt}(A \cdot B) = \dot{A} \cdot B + A \cdot \dot{B}, \qquad \frac{d}{dt}(A \times B) = \dot{A} \times B + A \times \dot{B}.$$

When computing a derivative of a vector product, one must be careful to retain the order of the factors, since the sign of the vector product changes if the factors are interchanged.

A.4 Conic Sections

As the name already says, conic sections are curves obtained by intersecting circular cones with planes.

Ellipse

- Equation in rectangular coordinates

$$\frac{x^2}{a^2} + \frac{y^2}{b^2} = 1 \quad \text{where}$$

    a: the semi-major axis
    b: the semi-minor axis $= a\sqrt{1-e^2}$
    e: the eccentricity $0 \le e < 1$.
- Distance of the foci from the centre $c = ea$
- Parameter $p = a(1-e^2)$
- Area $A = \pi a b$
- Equation in polar coordinates

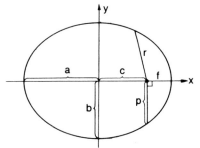

$$r = \frac{p}{1 + e \cos f}$$

    where the distance r is measured from one focus, not from the centre
- When $e \to 0$ the curve becomes a circle

Hyperbola

- Equations in rectangular and polar coordinates

$$\frac{x^2}{a^2} - \frac{y^2}{b^2} = 1, \quad r = \frac{p}{1 + e \cos f}.$$

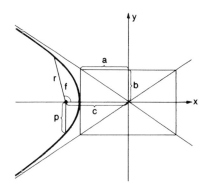

- Eccentricity $e > 1$
- Semi-minor axis $b = a\sqrt{e^2 - 1}$
- Parameter $p = a(e^2 - 1)$
- Asymptotes $y = \pm \dfrac{b}{a} x$

Parabola

Parabola is a limiting case between the previous ones; its eccentricity is $e = 1$.
- Equations

$$x = -ay^2, \quad r = \frac{p}{1 + \cos f}.$$

- Distance of the focus from the apex $h = 1/4a$
- Parameter $p = 1/2a$

A.5 Multiple Integrals

An integral of a function f over a surface A

$$I = \int_A f \, dA$$

can be evaluated as a double integral by expressing the surface element dA in terms of coordinate differentials. In rectangular coordinates,

$$dA = dx \, dy \,,$$

and in polar coordinates,

$$dA = r \, dr \, d\varphi \,.$$

The integration limits of the innermost integral may depend on the other integration variable. For example, the function $x e^y$ integrated over the shaded area is

$$I = \int_A x e^y \, dA = \int_{x=0}^{1} \int_{y=0}^{2x} x e^y \, dx \, dy$$

$$= \int_0^1 \left[\Big|_0^{2x} x e^y \right] dx = \int_0^1 (x e^{2x} - x) \, dx$$

$$= \Big|_0^1 \frac{1}{2} x e^{2x} - \frac{1}{4} e^{2x} - \frac{1}{2} x^2 = \frac{1}{4} \, (e^2 - 1) \,.$$

The surface need not be confined to a plane. For example, the area of a sphere is

$$A = \int_S dS \,,$$

where the integration is extended over the surface S of the sphere. In this case, the surface element is

$$dS = R^2 \cos \theta \, d\varphi \, d\theta \,,$$

and the area is

$$A = \int_{\varphi = 0}^{2\pi} \int_{\theta = -\pi/2}^{\pi/2} R^2 \cos \theta \, d\varphi \, d\theta$$

$$= \int_0^{2\pi} \left[\Big|_{-\pi/2}^{\pi/2} R^2 \sin \theta \right] d\varphi = \int_0^{2\pi} 2 R^2 \, d\varphi = 4 \pi R^2 \,.$$

Similarly, a volume integral

$$I = \int_V f \, dV$$

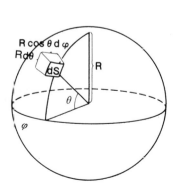

can be computed as a triple integral. In rectangular coordinates, the volume element dV

is

$$dV = dx\,dy\,dz \; ;$$

in cylindrical coordinates,

$$dV = r\,dr\,d\varphi\,dz \,,$$

and in spherical coordinates,

$$dV = r^2 \cos\theta\,dr\,d\varphi\,d\theta \qquad (\theta \text{ measured from the } xy\text{-plane})$$

or

$$dV = r^2 \sin\theta\,dr\,d\varphi\,d\theta \qquad (\theta \text{ measured from the } z\text{-axis}).$$

For example, the volume of a sphere with radius R is

$$V = \int_V dV$$

$$= \int_{r=0}^{R} \int_{\varphi=0}^{2\pi} \int_{\theta=-\pi/2}^{\pi/2} r^2 \cos\theta\,dr\,d\varphi\,d\theta$$

$$= \int_0^R \int_0^{2\pi} \left[\left.\vphantom{\int} \right|_{-\pi/2}^{\pi/2} r^2 \sin\theta \right] dr\,d\varphi = \int_0^R \int_0^{2\pi} 2r^2\,dr\,d\varphi$$

$$= \int_0^R 4\pi r^2\,dr = \left.\frac{4\pi r^3}{3}\right|_0^R = \frac{4}{3}\pi R^3 \,.$$

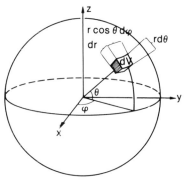

A.6 Numerical Solution of an Equation

We frequently meet equations defying analytical solution. Kepler's equation is a typical example. If we cannot do anything else, we can always apply some numerical method. Next, we shall present two very simple methods, the first of which is particularly suitable for calculators.

Method 1. We shall write the equation as $f(x) = x$. Next, we have to find an initial value x_0 for the solution. This can be done, for example, graphically, or by just guessing. Then we compute a succession of new iterates: $x_1 = f(x_0)$, $x_2 = f(x_1)$, ... and so on, until the difference of successive solutions becomes smaller than some preset limit. The last iterate x_i is the solution. After computing a few x_i's, it is easy to see if they are going to converge. If not, we rewrite the equation as $f^{-1}(x) = x$ and try again. (f^{-1} is the inverse function of f.)

As an example, let us solve the equation $x = -\ln x$. We guess $x_0 = 0.5$ and find

$$x_1 = -\ln 0.5 = 0.69 \,, \quad x_2 = 0.37 \,, \quad x_3 = 1.00 \,.$$

This already shows that something is wrong. Therefore, we change our equation to $x = e^{-x}$ and start again:

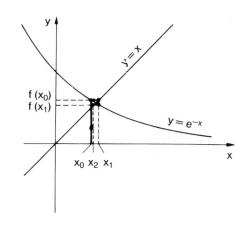

$$
\begin{array}{ll}
x_0 = 0.5 & x_3 = 0.58 \\
x_1 = e^{-0.5} & x_4 = 0.56 \\
\quad = 0.61 & x_5 = 0.57 \\
x_2 = 0.55 & x_6 = 0.57 \; .
\end{array}
$$

Thus the solution, accurate to two decimal places, is 0.57.

Method 2. In some pathological cases, the previous method may refuse to converge. In such situations, we can use the foolproof method of interval halving. If the function is continuous (as most functions of classical physics are) and we manage to find two points x_1 and x_2 such that $f(x_1) > 0$ and $f(x_2) < 0$, we know that somewhere between x_1 and x_2 there must be a point x in which $f(x) = 0$. Now, we find the sign of f in the midpoint of the interval, and select the half of the interval in which f changes sign. We repeat this procedure until the interval containing the solution is small enough.

We shall try also this method on our equation $x = -\ln x$, which is now written as $f(x) = 0$, where $f(x) = x + \ln x$. Because $f(x) \to -\infty$, when $x \to 0$, and $f(1) > 0$, the solution must be in the range $(0,1)$. Since $f(0.5) < 0$, we know that $x \in (0.5, 1)$. We continue in this way:

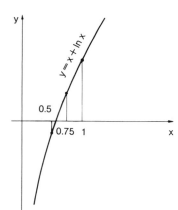

$$
\begin{array}{ll}
f(0.75) > 0 & x \in (0.5, 0.75) \\
f(0.625) > 0 & x \in (0.5, 0.625) \\
f(0.563) < 0 & x \in (0.563, 0.625) \\
f(0.594) > 0 & x \in (0.563, 0.594)
\end{array}
$$

The convergence is slow but certain. Each iteration restricts the solution to an interval which is half as large as the previous one, thus improving the solution by one binary digit.

B. Quantum Mechanics

B.1 Quantum Mechanical Model of Atoms. Quantum Numbers

Bohr's model can predict the coarse features of the hydrogen spectrum quite well. It cannot, however, explain all the finer details; nor is it applicable for atoms with several electrons. The reason for these failures is the classical concept of the electron as a particle with a well-defined orbit. *Quantum mechanics* describes the electron as a three-dimensional wave, which only gives the probability of finding the electron in a certain place. Quantum mechanics has accurately predicted all the energy levels of hydrogen atoms. Also, the energy levels of heavier atoms and molecules can be computed; however, such calculations are very complicated. The existence of *quantum numbers* can be understood from the quantum mechanical point of view as well.

The *principal quantum number n* describes the quantized energy levels of the electron. The classical interpretation of discrete energy levels allows only certain orbits given by (5.6). The orbital angular momentum of the electron is also quantized. This is described by the *angular momentum quantum number l*. The angular momentum corresponding to a quantum number l is

$$L = \sqrt{l(l+1)}\,\hbar\,.$$

The classical analogy would be to allow some elliptic orbits. The quantum number l can take only the values

$$l = 0, 1, \ldots n-1\,.$$

For historical reasons, these are often denoted by the letters s, p, d, f, g, h, i, j.

Although l determines the magnitude of the angular momentum, it does not give its direction. In a magnetic field, this direction is important, since the orbiting electron also generates a tiny magnetic field. In any experiment, only one component of the angular momentum can be measured at a time. In a given direction z (e.g. in the direction of the applied magnetic field), the projection of the angular momentum can have only the values

$$L_z = m_l \hbar\,,$$

where m_l is the *magnetic quantum number*

$$m_l = 0, \pm 1, \pm 2, \ldots, \pm l\,.$$

The magnetic quantum number is responsible for the splitting of spectral lines in strong magnetic fields, known as the *Zeeman effect*. For example, if $l = 1$, m_l can have $2l + 1 = 3$ different values. Thus, the line arising from the transition $l = 1 \rightarrow l = 0$ will split into three components in a magnetic field.

The fourth quantum number is the *spin* describing the intrinsic angular momentum of the electron. The spin of the electron is

$$S = \sqrt{s(s+1)}\,\hbar\,,$$

where the *spin quantum number* is $s = 1/2$. In a given direction z, the spin is

$$S_z = m_s \hbar \,,$$

where m_s can have one of two values:

$$m_s = \pm \tfrac{1}{2} \,.$$

All particles have a spin quantum number. Particles with an integral spin are called *bosons* (photon, mesons); particles with a half-intergral spin are *fermions* (proton, neutron, electron, neutrino etc.). Classically, spin can be interpreted as rotation of a particle; this analogy, however, should not be taken too literally.

The total angular momentum J of an electron is the sum of its orbital and spin angular momentum:

$$J = L + S \,.$$

Depending on the mutual orientation of the vectors L and S, the quantum number j of total angular momentum can have one of two possible values,

$$j = l + \tfrac{1}{2} \,, \quad l - \tfrac{1}{2}$$

(except if $l = 0$, when $j = 1/2$). The z-component of the total angular momentum can have the values

$$m_j = 0, \pm 1, \pm 2, \ldots \pm j \,.$$

Spin also gives rise to fine structure of spectral lines. Lines appear as close pairs (doublets).

B.2 Selection Rules and Transition Probabilities

The state of an electron in an atom is described by four quantum numbers: n, l, m_l, m_s; or, if we use the total angular momentum: n, l, j, m_j. Transitions between states are restricted by *selection rules*. Most favourable are *electric dipole transitions*, in which the dipole moment of the atom changes. The conservation laws require that in a transition we have

$$\Delta l = \pm 1, \quad \Delta m_l = 0, \pm 1 \quad \text{or,}$$

$$\Delta l = \pm 1, \quad \Delta j = 0, \pm 1, \quad \Delta m_j = 0, \pm 1 \,.$$

The probabilities of all other transitions are much smaller, and therefore they are called *forbidden transitions*. Magnetic dipole, quadrupole and all higher multipole transitions are such forbidden transitions.

B.3 Heisenberg's Uncertainty Principle

One of the oddities of quantum mechanics is that everything cannot be measured accurately at the same time. For example, even in principle, there is no way to determine the x-coordinate and the momentum p_x in the direction of x-axis with arbitrary precision simultaneously. These quantities have small uncertainties Δx and Δp_x, such that

$$\Delta x \, \Delta p_x \approx h \, .$$

A similar relation holds for other directions, too. Time and energy are also connected by an uncertainty relation,

$$\Delta E \, \Delta t \approx h \, .$$

This means that a particle can materialize out of nowhere if it disappears again within a time h/mc^2.

B.4 Exclusion Principle

Pauli's exclusion principle states that an atom with many electrons cannot have two electrons with exactly the same quantum numbers. This can be generalized to a gas of fermions (like electrons). The uncertainty principle allows us to find the size of a volume element (in phase space), which can hold at most two fermions; these must then have opposite spins. The phase space is a six-dimensional space, three coordinates giving the position of the particle and the other three, its momenta in the x, y and z-directions. Thus, a volume element is

$$\Delta V = \Delta x \, \Delta y \, \Delta z \, \Delta p_x \, \Delta p_y \, \Delta p_z.$$

From the uncertainty principle it follows that the smallest sensible volume element has a size of the order h^3. In such a volume, there can be at most two fermions of the same kind.

C. Theory of Relativity

Albert Einstein published his special theory of relativity in 1905 and the general theory of relativity ten years later. Especially the general theory, which is essentially a gravitation theory, has turned out to be very important for the theories of the evolution of the Universe. Therefore, it is appropriate to consider here some basic principles of relativity theory. A more detailed discussion would require sophisticated mathematics, and is beyond the scope of this very elementary book.

C.1 Basic Concepts

Everyone knows the famous Pythagorean theorem,

$$\Delta s^2 = \Delta x^2 + \Delta y^2 \, ,$$

where Δs is the length of the hypotenuse of a right-angle triangle, and Δx and Δy are the lengths of the other two sides. (For brevity, we have denotes $\Delta s^2 = (\Delta s)^2$.) This is easily generalized to three dimensions:

$$\Delta s^2 = \Delta x^2 + \Delta y^2 + \Delta z^2 \, .$$

This equation describes the *metric* of an ordinary rectangular frame in a Euclidean space, i.e. tells us how to measure distances.

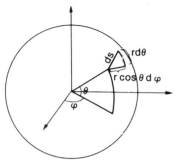

Generally, the expression for the distance between two points depends on the exact location of the points. In such a case, the metric must be expressed in terms of infinitesimal distances in order to correctly take into account the curvature of the coordinate curves. (A coordinate curve is a curve along which one coordinate changes while all the others remain constant.) An infinitesimal distance ds is called the *line element*. In a rectangular frame of a Euclidean space, it is

$$ds^2 = dx^2 + dy^2 + dz^2$$

and in spherical coordinates,

$$ds^2 = dr^2 + r^2(d\theta^2 + \cos^2\theta\, d\phi^2)\,.$$

Generally, ds^2 can be expressed as

$$ds^2 = \sum_i \sum_j g_{ij}\, dx^i\, dx^j\,,$$

where the x^i's are arbitrary coordinates, and the coefficients g_{ij} are components of the *metric tensor*. These can be functions of the coordinates, as in the case of the spherical coordinates. The metric tensor on an n-dimensional space can be expressed as an $n \times n$ matrix. Since $dx^i dx^j = dx^j dx^i$, the metric tensor is symmetric, i.e. $g_{ij} = g_{ji}$. If all coordinate curves intersect perpendicularly, the coordinate frame is *orthogonal*. In an orthogonal frame, $g_{ij} = 0$ for all $i \neq j$. For example, the spherical coordinates form an orthogonal frame, the metric tensor of which is

$$(g_{ij}) = \begin{pmatrix} 1 & 0 & 0 \\ 0 & r^2 & 0 \\ 0 & 0 & r^2\cos^2\theta \end{pmatrix}\,.$$

If it is possible to find a frame in which all the components of g are constant, the space is *flat*. In the rectangular frame of a Euclidean space, we have $g_{11} = g_{22} = g_{33} = 1$; hence the space is flat. The spherical coordinates show that even in a flat space, we can use frames in which the components of the metric tensor are not constant. The line element of a two-dimensional spherical surface is obtained from the metric of the spherical coordinate frame by assigning to r some fixed value R:

$$ds^2 = R^2(d\theta^2 + \cos^2\theta\, d\phi^2)\,.$$

The metric tensor is

$$(g_{ij}) = R^2 \begin{pmatrix} 1 & 0 \\ 0 & \cos^2\theta \end{pmatrix}\,.$$

This cannot be transformed to a constant tensor. Thus the surface of a sphere is a *curved space*.

If we know the metric tensor in some frame, we can compute a fourth-order tensor, the *curvature tensor* R_{ijkl}, which tells us whether the space is curved or flat. Unfortunately, the calculations are slightly too laborious to be presented here.

The metric tensor is needed for all computations involving distances, magnitudes of vectors, areas, and so on. Also, to evaluate a scalar product, we must know the metric. In fact, the components of the metric tensor can be expressed as scalar products of the basis vectors:

$$g_{ij} = e_i \cdot e_j .$$

If A and B are two arbitrary vectors,

$$A = \sum_i a^i e_i , \quad B = \sum_i b^i e_i ,$$

their scalar product is

$$A \cdot B = \sum_i \sum_j a^i b^j e_i \cdot e_j = \sum_i \sum_j g_{ij} a^i b^j .$$

C.2 Lorentz Transformation. Minkowski Space

The special theory of relativity abandoned the absolute Newtonian time, supposed to flow at the same rate for every observer. Instead, it required that the speed of light have the same value c in all coordinate frames. The constancy of the speed of light is the basic principle of the special theory of relativity.

Let us send a beam of light from the origin. It travels along a straight line at the speed c. Thus, at the moment t, its space and time coordinates satisfy the equation

$$x^2 + y^2 + z^2 = c^2 t^2 . \tag{C.1}$$

Next, we study what the situation looks like in another frame, $x'y'z'$, moving at a velocity v with respect to the xyz-frame. Let us select the new frame so that it coincides with the xyz-frame at $t = 0$. Also, let the time coordinate t' of the $x'y'z'$-frame be $t' = 0$ when $t = 0$. And finally, we assume that the $x'y'z'$-frame moves in the direction of the positive x-axis. Since the beam of light must also travel at the speed c in the $x'y'z'$-frame, we must have

$$x'^2 + y'^2 + z'^2 = c^2 t'^2 .$$

If we require that the new (dashed) coordinates be obtained from the old ones by a linear transformation, and also that the inverse transformation be obtained simply by replacing v by $-v$, we find that the transformation must be

$$x' = \frac{x - vt}{\sqrt{1 - v^2/c^2}}$$

$$y' = y$$

$$z' = z$$

$$t' = \frac{t - vx/c^2}{\sqrt{1 - v^2/c^2}} .$$

This transformation between frames moving at a constant speed with respect to each other is called the *Lorentz transformation*.

Becasue the Lorentz transformation is derived assuming the invariance of (C.1), it is obvious that the *interval*

$$\Delta s^2 = -c^2 \Delta t^2 + \Delta x^2 + \Delta y^2 + \Delta z^2$$

of any two events remains invariant in all Lorentz transformations. This interval defines a metric in four-dimensional spacetime. A space having such a metric is called the *Minkowski space* or *Lorentz space*. The components of the metric tensor g are

$$(g_{ij}) = \begin{pmatrix} -c^2 & 0 & 0 & 0 \\ 0 & 1 & 0 & 0 \\ 0 & 0 & 1 & 0 \\ 0 & 0 & 0 & 1 \end{pmatrix} .$$

Since this is constant, the space is flat. But it is no longer an ordinary Euclidean space, because the sign of the time component differs from the sign of the space components. In the older literature, a variable *ict* is often used instead of time, *i* being the imaginary unit. Then the metric looks Euclidean, which is misleading; the properties of a space cannot be changed just by changing notation.

In a Minkowski space position, velocity, momentum and other vector quantities are described by *four-vectors*, which have one time and three space components. The components of a four-vector obey the Lorentz transformation when we transform them from one frame to another, both moving along a straight line at a constant speed.

According to classical physics, the distance of two events depends on the motion of the observer, but the time interval between the events is the same for all observers. The world of special relativity is more anarchistic: even time intervals have different values for different observers.

C.3 General Relativity

The Equivalence Principle. Newton's laws relate the acceleration, a, of a particle and the applied force F by

$$F = m_i a ,$$

where m_i is the inertial mass of the particle, resisting the force trying to move the particle. The gravitational force felt by the particle is

$$F = m_g f ,$$

where m_g is the gravitational mass of the particle, and f is a factor depending only on other masses. The masses m_i and m_g appear as coefficients related to totally different phenomena. There is no physical reason to assume that the two masses should have anything in common. However, already the experiments made by Galilei showed that evidently $m_i = m_g$. This has been verified later with very high accuracy.

The *weak equivalence principle*, which states that $m_i = m_g$, can, therefore, be accepted as a physical axiom. The *strong equivalence principle* generalizes this: if we restrict our observations to a sufficiently small region of space-time, there is no way to tell whether we are subject to a gravitational field or are in uniformly accelerated motion. The strong equivalence principle is one of the fundamental postulates of general relativity.

Curvature of Space. General relativity describes gravitation as a geometric property of space-time. The equivalence principle is an obvious consequence of this idea. Particles moving through space-time follow the shortest possible paths, *geodesics*. The projection of a geodesic onto three-dimensional space need not be the shortest way between the points.

The geometry of space-time is determined by the mass and energy distribution. If this distribution is known, we can write the *field equations*, which are partial differential equations connecting the metric to the mass and energy distribution and describing the curvature of space-time.

In the case of a single-point mass, the field equations yield the *Schwarzschild metric*, the line element of which is

$$ds^2 = -\left(1 - \frac{2\,GM}{c^2 r}\right)c^2\,dt^2 + \frac{dr^2}{1 - \dfrac{2\,GM}{c^2 r}} + r^2(d\theta^2 + \cos^2\theta\,d\phi^2).$$

Here, M is the mass of the point; r, θ and ϕ are ordinary spherical coordinates. It can be shown that the components cannot be transformed to constants simultaneously: space-time must be curved due to the mass.

If we study a very small region of space-time, the curvature has little effect. Locally, the space is always a Minkowski space where special relativity can be used. Locality means not only a limited spatial volume but also a limited interval in time.

C.4 Tests of General Relativity

General relativity gives predictions different from classical physics. Although the differences are usually very small, there are some phenomena in which the deviation can be measured and used to test the validity of general relativity. At present, five different astronomical tests have verified the theory.

First of all, the orbit of a planet is no longer a closed Keplerian ellipse. The effect is strongest for the innermost planets, whose perihelia should turn little by little. Most of the motion of the perihelion of Mercury is predicted by Newtonian mechanics; only a small excess of 43 arc seconds per century remains unexplained. And it so happens that this is exactly the correction suggested by general relativity.

Secondly, a beam of light should bend when it travels close to the Sun. For a beam grazing the surface, the deviation should be about 1.75″. Such effects have been observed during total solar eclipses and also by observing pointlike radio sources, just before and after occultation by the Sun.

The third classical way of testing general relativity is to measure the redshift of a photon climbing in a gravitational field. We can understand the redshift as an energy loss while the photon does work against the gravitational potential. Or, we can think of the redshift as caused by the metric only: a distant observer finds that near a mass, time runs slower, and the frequency of the radiation is lower. The time dilation in the Schwarzschild metric is described by the coefficient of dt; thus it is not surprising that the radiation emitted at a frequency v from a distance r from a mass M has the frequency

$$v_\infty = v \sqrt{1 - \frac{2GM}{c^2 r}}\,, \tag{C.2}$$

if observed very far from the source. This gravitational redshift has been verified by laboratory experiments.

The fourth test employs the slowing down of the speed of light in a gravitational field near the Sun. This has been verified by radar experiments.

The previous tests concern the solar system. Outside the solar system, binary pulsars have been used to test general relativity. An asymmetric system in accelerated motion (like a binary star) loses energy as it radiates gravitational waves. It follows that the components approach each other, and the period decreases. Usually the gravitational waves play a minor role, but in the case of a compact source, the effect may be strong enough to be observed. The first source to show this predicted shortening of period was the binary pulsar PSR 1913 + 16.

D. Radio Astronomy Fundamentals

D.1 Antenna Definitions

The receiving (or transmitting) characteristics of an antenna are described by the *antenna power pattern* $P(\theta, \varphi)$, which, for an ideal antenna, is equivalent to the Fresnel diffraction pattern of an optical telescope. The power pattern $P(\theta, \varphi)$ is a measure of the antenna response to radiation as a function of the polar coordinates θ and φ. It is directly proportional to the radiation intensity $U(\theta, \varphi)$ [W sr^{-1}] and usually normalized so that the maximum intensity equals one, i.e.

$$P(\theta, \varphi) = \frac{U(\theta, \varphi)}{U_{\max}(\theta, \varphi)} \tag{D.1}$$

where $U_{\max}(\theta, \varphi)$ is the maximum radiation intensity.

The *beam solid angle* Ω_A is defined as the integral over the power pattern, i.e.

$$\Omega_A = \iint_{4\pi} P(\theta, \varphi)\, d\Omega\,. \tag{D.2}$$

This is the equivalent solid angle through which all the radiation would enter if the power were constant and equal to its maximum value within this angle.

If we integrate over the main beam only, i.e. to the limit where the power pattern reaches it first minimum, we obtain the *main beam solid angle* Ω_m

$$\Omega_m = \iint_{\text{mainlobe}} P(\theta, \varphi)\, d\Omega\,. \tag{D.3}$$

If the integration is carried out over all sidelobes, one gets the *sidelobe solid angle* Ω_s

$$\Omega_s = \iint_{\text{all sidelobes}} P(\theta, \varphi)\, d\Omega\,. \tag{D.4}$$

It is easy to see that

$$\Omega_A = \Omega_m + \Omega_s . \tag{D.5}$$

We now define the *beam efficiency*, η_B, as the ratio of the main beam solid angle Ω_m to the beam solid angle Ω_A, i.e.

$$\eta_B = \frac{\Omega_m}{\Omega_A} . \tag{D.6}$$

The beam efficiency thus measures the fraction of the total radiation that enters into the main lobe, and is one of the prime parameters that characterize a radio telescope.

Another important parameter is the *directivity*, D. The directivity is defined as the ratio of the maximum radiation intensity $U_{max}(\theta, \varphi)$ to the average radiation intensity $\langle U \rangle$, i.e.

$$D = \frac{U_{max}(\theta, \varphi)}{\langle U \rangle} . \tag{D.7}$$

Since the average radiation intensity is the total received power over 4π steradians

$$U = \frac{1}{4\pi} \iint_{4\pi} U(\theta, \varphi) \, d\Omega , \tag{D.8}$$

we can combine equations (D.1), (D.7) and (D.8) to obtain

$$D = \frac{4\pi}{\iint_{4\pi} P(\theta, \varphi) \, d\Omega} = \frac{4\pi}{\Omega_A} . \tag{D.9}$$

The directivity thus measures the amplification or *gain* of an antenna. If the antenna has no loss, the directivity is equal to the gain, G. However, all antennas have small resistive losses, which can be accounted for by defining a radiation efficiency factor, η_r (for most antennas very close to one). Then

$$G = \eta_r D . \tag{D.10}$$

Another basic antenna parameter is the *effective aperture* A_e, which, for all large antennas, is smaller than the geometrical antenna aperture, A_g. The ratio between the effective aperture and the geometrical aperture, η_A. i.e.

$$\eta_A = \frac{A_e}{A_g} , \tag{D.11}$$

is called the *aperture efficiency* and is another principal measure of the quality of a telescope.

From thermodynamical considerations, one can show that the effective aperture of a telescope is related to the beam solid angle by

$$\Omega_A A_e = \lambda^2 . \tag{D.12}$$

By combining Eqs. (D.9) and (D.12) we can therefore express the directivity as a function of the effective area

$$D = \frac{4 \pi A_e}{\lambda^2} \tag{D.13}$$

or, alternatively, by using the relation between directivity and gain, i.e.

$$G = \frac{4 \pi \eta_r A_e}{\lambda^2} . \tag{D.14}$$

Usually the radiation efficiency is included in the effective area or taken equal to one, in which case the expressions for directivity and gain appear equal.

D.2 Antenna Temperature and Flux Density

The power measured with a radio telescope can be related to an equivalent temperature, the *antenna temperature*, T_A. If we assume that the antenna is placed in an enclosure with a uniform temperature T_A, or equivalently, if the antenna is replaced by a matched load at the same temperature, it can be shown that the power P[W] at the antenna terminal is given by

$$P = w \Delta v = k T_A \Delta v , \tag{D.15}$$

where k is Boltzmann's constant, w is the spectral power (power per unit bandwidth) and Δv is the bandwidth of the radiometer. This relation provides a relation between the observed power P and the antenna temperature T_A. If the Rayleigh-Jeans approximation, $h v/k \ll T_A$, is not valid ($v \gtrsim 100$ GHz), then T_A has to be replaced by the *radiation temperature*

$$J(T_A) = \frac{h v/k}{\exp(h v/k T_A) - 1} .$$

Since the antenna is usually calibrated by comparing the incoming signal with a temperature-controlled matched load or a gas-discharge tube with a well-defined temperature, it is quite convenient to express the received signal as an antenna temperature. Using the antenna temperature has the additional advantage of providing a power measure that is independent of the radiometer bandwidth.

On the other hand, the *spectral power* for an antenna is defined as

$$w_v = \tfrac{1}{2} A_e \iint_{4\pi} I_v(\theta, \varphi) \, P(\theta, \varphi) \, d\Omega , \tag{D.16}$$

where $I_v(\theta, \varphi)$ is the brightness of the sky; the factor $\tfrac{1}{2}$ enters because a radio telescope receives only one polarization component. Using (D.15), we can now express the power in terms of the antenna temperature T_A, i.e.

$$T_A = \tfrac{1}{2} \frac{A_e}{k} \iint_{4\pi} I_v(\theta, \varphi) \, P(\theta, \varphi) \, d\Omega . \tag{D.17}$$

If the angular extent of the source is much smaller than the beamwidth of the antenna, i.e. for a "point source", the power pattern $P(\theta, \varphi) \sim 1$ and

$$T_A = \frac{S_v A_e}{2k}, \tag{D.18}$$

where the flux density S_v is given by

$$S_v = \iint\limits_{source} I_v(\theta, \varphi)\, d\Omega. \tag{D.19}$$

The unit for flux density is Jansky $[Jy] = [10^{-26}\,\mathrm{W\,m^{-2}\,Hz^{-1}}]$.

The sky or source brightness can also be given as a *brightness temperature* T_b, in which case

$$T_A = \tfrac{1}{2} A_e \iint\limits_{source} T_b(\theta, \varphi)\, P(\theta, \varphi)\, d\Omega. \tag{D.20}$$

Again, if the Rayleigh-Jeans approximation is not valid, all temperatures T_i should be replaced by the corresponding *radiation temperatures* $J(T_i)$.

E. Tables

Table E.1a. SI basic units

Quantity	Symbol	Unit	Abbr.	Definition
Length	$l, s, \ldots$	Metre	m	The length of the path travelled by light in vacuum during a time interval of $1/299\,792\,458$ of a second
Mass	m, M	Kilogram	kg	Equal to the mass of the international prototype of the kilogram
Time	t	Second	s	The duration of $9\,192\,631\,770$ periods of the radiation corresponding to the transition between the two hyperfine levels of the ground state of the caesium-133 atom
Electric current	I	Ampere	A	That constant current which, if maintained in two straight parallel conductors of infinite length, of negligible circular cross section, and placed 1 metre apart in a vacuum, would produce a force equal to 2×10^{-7} Newton per metre of length between these conductors
Temperature	T	Kelvin	K	The fraction $1/273.16$ of the thermodynamic temperature of the triple point of water
Amount of substance	n	Mole	mol	The amount of substance of a system which contains as many elementary entities as there are atoms in $0.012\,\mathrm{kg}$ of ^{12}C
Luminous intensity	I	Candela	cd	The luminous intensity in a given direction of a source which emits monochromatic radiation of frequency $540 \times 10^{12}\,\mathrm{Hz}$ and of which the radiant intensity in that direction is $1/683$ Watt per steradian

Table E.1b. Prefixes for orders of ten

Prefix	Symbol	Multiple
yocto	y	10^{-24}
zepto	z	10^{-21}
atto	a	10^{-18}
femto	f	10^{-15}
pico	p	10^{-12}
nano	n	10^{-9}
micro	μ	10^{-6}
milli	m	10^{-3}
centi	c	10^{-2}
deci	d	10^{-1}
deca	da	10^{1}
hecto	h	10^{2}
kilo	k	10^{3}
Mega	M	10^{6}
Giga	G	10^{9}
Tera	T	10^{12}
Peta	P	10^{15}
Exa	E	10^{18}
Zetta	Z	10^{21}
Yotta	Y	10^{24}

Table E.2. Constants and units. To avoid extensive changes of the less significant digits, older values of some constants of atomic physics are used in the text. More recent values are listed below

1 Radian	1 rad	$= 180°/\pi = 57.2957795° = 206264.8''$
1 Degree	$1°$	$= 0.01745329$ rad
1 Arc second	$1''$	$= 0.000004848$ rad
Velocity of light	c	$= 299792458$ m s^{-1}
Gravitational constant	G	$= 6.673\times10^{-11}$ m^3 kg^{-1} s^{-2}
		$= 4\pi^2$ AU^3 M^{-1} a$^{-2} = 3986005\times10^{8}$ m^3 $M_\oplus^{-1}$ s^{-2}
Planck constant	h	$= 6.6261\times10^{-34}$ Js
	$\hbar$	$= h/2\pi = 1.0546\times10^{-34}$ Js
Boltzmann constant	k	$= 1.3807\times10^{-23}$ JK^{-1}
Radiation density constant	a	$= 7.5659\times10^{-16}$ Jm^{-3} K^{-4}
Stefan-Boltzmann constant	σ	$= ac/4 = 5.6705\times10^{-8}$ Wm^{-2} K^{-4}
Atomic mass unit	amu	$= 1.6605\times10^{-27}$ kg
Electron volt	eV	$= 1.6022\times10^{-19}$ J
Electron charge	e	$= 1.6022\times10^{-19}$ C
Mass of electron	m_e	$= 9.1094\times10^{-31}$ kg $= 0.511$ MeV
Mass of proton	m_p	$= 1.6726\times10^{-27}$ kg $= 938.3$ MeV
Mass of neutron	m_n	$= 1.6749\times10^{-27}$ kg $= 939.6$ MeV
Mass of ^{1}H atom	m_H	$= 1.6735\times10^{-27}$ kg $= 1.0078$ amu
Mass of $^4_2$He atom	m_{He}	$= 6.6465\times10^{-27}$ kg $= 4.0026$ amu
Rydberg constant for ^{1}H	R_H	$= 1.0968\times10^{7}$ m^{-1}
Rydberg constant for ∞ mass	R_∞	$= 1.0974\times10^{7}$ m^{-1}
Gas constant	R	$= 8.3146$ JK^{-1} mol^{-1}
Normal atmospheric pressure	atm	$= 101325$ Pa $= 1013$ mbar $= 760$ mm Hg
Astronomical unit	AU	$= 1.49597870\times10^{11}$ m
Parsec	pc	$= 3.0857\times10^{16}$ m $= 206265$ AU $= 3.26$ ly
Light-year	ly	$= 0.9461\times10^{16}$ m $= 0.3066$ pc

Table E.3. Units of time

Unit	Equivalent to
Sidereal year	365.2564 d
Tropical year	365.2422 d (equinox to equinox)
Anomalistic year	365.2596 d (perihelion to perihelion)
Gregorian calendar year	365.2425 d
Julian year	365.2500 d
Julian century	36525 d
Eclipse year	346.6200 d (with respect to the ascending node of the Moon)
Lunar year	354.4306 d = 12 synodical months
Synodical month	29.5306 d
Sidereal month	27.3217 d
Anomalistic month	27.5546 d (perigee to perigee)
Draconitic month	27.2122 d (node to node)
Mean solar day	24 h mean solar time = 24 h 03 min 56.56 s (sidereal time)
Sidereal day	24 h sidereal time = 23 h 56 min 04.09 s (mean solar time)
Rotation period of the Earth (referred to fixed stars)	23 h 56 min 04.10 s
International atomic time on July 1, 1993	TAI = UTC + 28 s (exactly)
Terrestrial time	TT ≈ TAI + 32.184 s (SI seconds)

Table E.4. The Greek alphabet

$A\ \alpha$ alpha	$B\ \beta$ beta	$\Gamma\ \gamma$ gamma	$\Delta\ \delta$ delta	$E\ \varepsilon$ epsilon	$Z\ \zeta$ zeta	$H\ \eta$ eta	$\Theta\ \theta\ \vartheta$ theta	$I\ \iota$ iota	$K\ \kappa$ kappa
$\Lambda\ \lambda$ lambda	$M\ \mu$ mu	$N\ \nu$ nu	$\Xi\ \xi$ xi	$O\ o$ omicron	$\Pi\ \pi\ \tilde\omega$ pi	$P\ \rho$ rho	$\Sigma\ \sigma\varsigma$ sigma	$T\tau$ tau	$Y\ \upsilon$ upsilon
$\Phi\ \phi\ \varphi$ phi	$X\ \chi$ chi	$\Psi\ \psi$ psi	$\Omega\ \omega$ omega						

Table E.5. The Sun

Property	Symbol	Numerical value
Mass	$M_\odot$	1.989×10^{30} kg
Radius	$R_\odot$	6.96×10^{8} m
Effective temperature	T_e	5785 K
Luminosity	$L_\odot$	3.9×10^{26} W
Apparent visual magnitude	V	-26.78
Colour indices	$B - V$	0.62
	$U - B$	0.10
Absolute visual magnitude	M_V	4.79
Absolute bolometric magnitude	M_{bol}	4.72
Inclination of equator to ecliptic		$7°15'$
Equatorial horizontal parallax	$\pi_\odot$	$8.794''$
Motion: direction of apex		$\alpha = 270°,$ $\delta = 30°$
velocity in LSR		19.7 km/s
Distance from galactic centre		8.5 kpc

Table E.6. The Earth

	Symbol	Numerical value
Mass	$M_\oplus$	$M_\odot/332\,946 = 5.974 \times 10^{24}\,\text{kg}$
Mass, Earth + Moon	$M_\oplus + M_{\mathbb{C}}$	$M_\odot/328\,900.5 = 6.048 \times 10^{24}\,\text{kg}$
Equatorial radius	R_e	$6\,378\,140\,\text{m}$
Polar radius	R_p	$6\,356\,755\,\text{m}$
Flattening	f	$(R_e - R_p)/R_e = 1/298.257$
Surface gravity	g	$9.81\,\text{m/s}^2$

Table E.7. The Moon

	Symbol	Numerical value
Mass	$M_{\mathbb{C}}$	$M_\oplus/81.30 = 7.348 \times 10^{22}\,\text{kg}$
Radius	$R_{\mathbb{C}}$	$1738\,\text{km}$
Surface gravity	$g_{\mathbb{C}}$	$1.62\,\text{m/s}^2$
Mean equatorial horizontal parallax	$\pi_{\mathbb{C}}$	$57'$
Semi-major axis of orbit	a	$384\,400\,\text{km}$
Smallest distance from Earth	r_{min}	$356\,400\,\text{km}$
Greatest distance from Earth	r_{max}	$406\,700\,\text{km}$
Mean inclination of orbit to ecliptic	ι	$5°09'$

Table E.8. Planets

Name	Symbol	Equatorial radius R_e [km]	Mass planet m [kg]	Mass planet + satellites $m/M_\oplus$	Mass planet + satellites $m/M_\odot$	Density ρ [g/cm³]	Rotation period τ_{sid}
Mercury	☿	2439	3.30×10^{23}	0.0553	1/6023600	5.4	58.6 d
Venus	♀	6052	4.87×10^{24}	0.8150	1/408523.5	5.2	243.0 d
Earth	⊕	6378	5.97×10^{24}	1.0123	1/328900.5	5.5	23 h 56 min 04.1 s
Mars	♂	3397	6.42×10^{23}	0.1074	1/3098710	3.9	24 h 37 min 22.7 s
Jupiter	♃	71398	1.90×10^{27}	317.89	1/1047.355	1.3	9 h 55 min 30 s
Saturn	♄	60000	5.69×10^{26}	95.17	1/3498.5	0.7	10 h 30 min
Uranus	♅	26320	8.70×10^{25}	14.56	1/22869	1.1	17 h 14 min
Neptune	♆	24300	1.03×10^{26}	17.24	1/19314	1.7	16 h 07 min
Pluto	♇	1120	1×10^{22}	0.0023	1/150000000	2.1	6 d 09 h 17 min

Name	Symbol	Inclination of equator ε [°]	Flattening f	Temperature T [K]	Geometric albedo p	Absolute magnitude V (1,0)	Mean opposition magnitude V_0	Number of known satellites
Mercury	☿	0.0	0	615; 130	0.106	−0.42	−	0
Venus	♀	177.3	0	750	0.65	−4.40	−	0
Earth	⊕	23.44	0.003353	300	0.367	−3.86	−	1
Mars	♂	25.19	0.005186	220	0.150	−1.52	−2.01	2
Jupiter	♃	3.12	0.06481	140	0.52	−9.40	−2.70	16
Saturn	♄	26.73	0.10762	100	0.47	−8.88	+0.67	18
Uranus	♅	97.9	0.023	65	0.51	−7.19	+5.52	15
Neptune	♆	29.6	0.017	55	0.41	−6.87	+7.84	8
Pluto	♇	118?	0	45	0.3	−1.0	+15.12	1

Table E.9. Osculating elements of planetary orbits on Aug. 1, 1993 = JD 2449200.5

Planet	Semi-major axis of orbit a		Eccentricity e	Inclination ι [°]	Longitude of ascending node Ω [°]	Longitude of perihelion $\tilde{\omega}$ [°]	Mean anomaly M [°]	Sidereal period P_{sid}		Synodical period P_{syn} [d]
	[AU]	[10^6 km]						[a]	[d]	
Mercury	0.387	57.9	0.206	7.01	48.3	77.4	300.3	0.2408	87.97	115.9
Venus	0.723	108.2	0.0068	3.39	76.7	131.4	254.4	0.6152	224.70	583.9
Earth	1.000	149.6	0.0167	0.00	–	102.8	206.9	1.0000	365.26	–
Mars	1.524	227.9	0.093	1.85	49.6	336.0	230.8	1.8809	687.02	779.9
Jupiter	5.203	778.3	0.048	1.30	100.5	15.7	183.9	11.862	4333	398.9
Saturn	9.530	1425.7	0.053	2.49	113.7	93.2	238.3	29.42	10744	378.1
Uranus	19.24	2878	0.047	0.77	74.0	173.9	111.7	84.36	30810	369.6
Neptune	30.14	4509	0.006	1.77	131.8	33.5	257.2	165.5	60440	367.5
Pluto	39.81	5956	0.255	17.14	110.3	224.1	5.4	251.2	91750	366.7

Table E.10. Mean elements of planets (according to Le Verrier and Gaillot)
The mean elements do not contain periodic disturbances. The elements refer to the mean equinox and ecliptic of date, the variable T being the time in Julian centuries elapsed since Dec. 31, 1899, at 12 h UT (= 1900 Jan. 0.5 UT = JD 241 5020.0). One Julian century equals 36 525 days. Thus, for a Julian date JD, we have

$$T = \frac{JD - 241\,5020}{36\,525} \; .$$

The quantity L is the mean longitude, defined as $L = M + \tilde{\omega}$; $n = 360°/P_{sid}$ is the mean angular velocity of the orbital motion.

Mercury

L	$= 178°10'44.77'' + 538\,106\,655.62''\,T - 1.1289''\,T^2$
$\tilde{\omega}$	$= 75°53'49.67'' + 5592.49''\,T + 1.111''\,T^2$
Ω	$= 47°08'40.93'' + 4265.135''\,T + 0.835''\,T^2$
e	$= 0.205\,614\,94 + 0.000\,020\,30\,T - 0.000\,000\,04\,T^2$
ι	$= 7°00'10.85'' + 6.258''\,T - 0.056''\,T^2$
a	$= 0.387\,0984$
n	$= 14732.4197''/d$
P_{sid}	$= 87.969\,256\,d \qquad P_{syn} = 115.88\,d$

Venus

L	$= 342°46'03.24'' + 210\,669\,166.172''\,T + 1.1289''\,T^2$
$\tilde{\omega}$	$= 130°08'25.91'' + 4940.27''\,T - 5.93''\,T^2$
Ω	$= 75°47'17.04'' + 3290.50''\,T + 1.508''\,T^2$
e	$= 0.006\,816\,36 - 0.000\,053\,84\,T - 0.000\,000\,126\,T^2$
ι	$= 3°23'37.09'' + 4.508''\,T - 0.0156''\,T^2$
a	$= 0.723\,330\,15$
n	$= 5767.6698''/d$
P_{sid}	$= 224.700\,80\,d \qquad P_{syn} = 583.92\,d$

Earth

L	$= 99°41'48.72'' + 129\,602\,768.95''\,T + 1.1073''\,T^2$
$\tilde{\omega}$	$= 101°13'07.15'' + 6171.77''\,T + 1.823''\,T^2$
e	$= 0.016\,7498 - 0.000\,042\,58\,T - 0.000\,000\,137\,T^2$
a	$= 1.000\,001\,29$
n	$= 3548.192\,83''/d$
P_{sid}	$= 365.256\,361\,d$

Table E.10. (cont.)

Mars

L	$= 293°44'26.56'' + 68\,910\,106.509'' \, T + 1.1341'' \, T^2$
$\tilde{\omega}$	$= 334°13'05.81'' + 6625.42'' \, T + 1.2093'' \, T^2$
Ω	$= 48°47'12.04'' + 2797.0'' \, T - 2.17'' \, T^2$
e	$= 0.093\,308\,80 + 0.000\,095\,284 \, T - 0.000\,000\,122 \, T^2$
ι	$= 1°51'01.09'' - 2.3365'' \, T - 0.0945'' \, T^2$
a	$= 1.523\,6781$
n	$= 1886.5183'' / \mathrm{d}$
P_{sid}	$= 686.979\,82 \, \mathrm{d}$ $\qquad P_{\mathrm{syn}} = 779.94 \, \mathrm{d}$

Jupiter

L	$= 238°02'57.32'' + 10\,930\,687.148'' \, T + 1.204\,86'' \, T^2 - 0.005\,936'' \, T^3$
$\tilde{\omega}$	$= 12°43'15.34'' + 5795.862'' \, T + 3.802\,58'' \, T^2 - 0.012\,36'' \, T^3$
Ω	$= 99°26'36.19'' + 3637.908'' \, T + 1.2680'' \, T^2 - 0.030\,64'' \, T^3$
e	$= 0.048\,334\,75 + 0.000\,164\,180 \, T - 0.000\,000\,4676 \, T^2 - 0.000\,000\,0017 \, T^3$
ι	$= 1°18'31.45'' - 20.506'' \, T + 0.014'' \, T^2$
a	$= 5.202\,561$
n	$= 299.1283'' / \mathrm{d}$
P_{sid}	$= 4332.589 \, \mathrm{d}$ $\qquad P_{\mathrm{syn}} = 398.88 \, \mathrm{d}$

Saturn

L	$= 266°33'51.76'' + 4\,404\,635.5810'' \, T + 1.168\,35'' \, T^2 - 0.0021'' \, T^3$
$\tilde{\omega}$	$= 91°05'53.38'' + 7050.297'' \, T + 2.9749'' \, T^2 + 0.0166'' \, T^3$
Ω	$= 112°47'25.40'' + 3143.5025'' \, T - 0.547\,85'' \, T^2 - 0.0191'' \, T^3$
e	$= 0.055\,892\,32 - 0.000\,345\,50 \, T - 0.000\,000\,728 \, T^2 + 0.000\,000\,000\,74 \, T^3$
ι	$= 2°29'33.07'' - 14.108'' \, T - 0.055\,76'' \, T^2 + 0.000\,16'' \, T^3$
a	$= 9.554\,747$
n	$= 120.4547'' / \mathrm{d}$
P_{sid}	$= 10\,759.23 \, \mathrm{d}$ $\qquad P_{\mathrm{syn}} = 378.09 \, \mathrm{d}$

Uranus

L	$= 244°11'50.89'' + 1\,547\,508.765'' \, T + 1.137\,74'' \, T^2 - 0.002\,176'' \, T^3$
$\tilde{\omega}$	$= 171°32'55.14'' + 5343.958'' \, T + 0.8539'' \, T^2 - 0.002\,18'' \, T^3$
Ω	$= 73°28'37.55'' + 1795.204'' \, T + 4.722'' \, T^2$
e	$= 0.046\,3444 - 0.000\,026\,58 \, T + 0.000\,000\,077 \, T^2$
ι	$= 0°46'20.87'' + 2.251'' \, T + 0.1422'' \, T^2$
a	$= 19.218\,14$
n	$= 42.2309'' / \mathrm{d}$
P_{sid}	$= 30\,688.45 \, \mathrm{d}$ $\qquad P_{\mathrm{syn}} = 369.66 \, \mathrm{d}$

Neptune

L	$= 84°27'28.78'' + 791\,589.291'' \, T + 1.153\,74'' \, T^2 - 0.002\,176'' \, T^3$
$\tilde{\omega}$	$= 46°43'38.37'' + 5128.468'' \, T + 1.406\,94'' \, T^2 - 0.002\,176'' \, T^3$
Ω	$= 130°40'52.89'' + 3956.166'' \, T + 0.899\,52'' \, T^2 - 0.016\,984'' \, T^3$
e	$= 0.008\,997\,04 + 0.000\,006\,330 \, T - 0.000\,000\,002 \, T^2$
ι	$= 1°46'45.27'' - 34.357'' \, T - 0.0328'' \, T^2$
a	$= 30.109\,57$
n	$= 21.5349'' / \mathrm{d}$
P_{sid}	$= 60\,181.3 \, \mathrm{d}$ $\qquad P_{\mathrm{syn}} = 367.49 \, \mathrm{d}$

Table E.11. Planetary satellites

Planet / Satellite		Discoverer	Year of discovery	Semi-major axis of orbit a [10^3 km]	[in planetary radii]	Sidereal period P_{sid} [d]	Eccentricity e	Orbital inclination to planetary equator i [°]	Diameter d [km]	Mass [in planetary masses]	Density ρ [g/cm³]	Geometric albedo p	Mean opposition magnitude V_0
Earth													
	Moon	–	–	384.4	60.27	27.3217	0.055	18.28 – 28.58	3476	0.0123	3.34	0.12	– 12.74
Mars													
	Phobos	Hall	1877	9.38	2.76	0.3189	0.015	1.0	$27 \times 22 \times 19$	1.5×10^{-8}	1.6	0.06	11.3
	Deimos	Hall	1877	23.46	6.91	1.2624	0.0005	0.9 – 2.7	$15 \times 12 \times 11$	3×10^{-9}	1.9	0.07	12.4
Jupiter													
XVI	Metis	Synnott	1979	128	1.79	0.295	–	–	40	5×10^{-11}	–	0.05	17.5
XV	Adrastea	Jewitt Danielson Synnott	1979	129	1.80	0.298	–	–	$25 \times 20 \times 15$	1×10^{-11}	–	0.05	19.1
V	Amalthea	Barnard	1892	181	2.53	0.498	0.003	0.40	$270 \times 166 \times 150$	4×10^{-9}	2.2	0.05	14.1
XIV	Thebe	Synnott	1979	222	3.11	0.674	0.015	0.8	110×90	4×10^{-10}	1.3	0.05	15.7
I	Io	Galilei	1610	422	5.90	1.769	0.004	0.04	3630	4.7×10^{-5}	3.6	0.61	5.0
II	Europa	Galilei	1610	671	9.39	3.551	0.009	0.47	3138	2.5×10^{-5}	2.9	0.64	5.3
III	Ganymedes	Galilei	1610	1070	15.0	7.155	0.002	0.21	5262	7.8×10^{-5}	1.9	0.42	4.6
IV	Callisto	Galilei	1610	1883	26.3	16.689	0.007	0.51	4800	5.7×10^{-5}	1.9	0.20	5.6
XIII	Leda	Kowal	1974	11094	155.2	238.72	0.148	26.07	16	3×10^{-12}	3	–	20.2
VI	Himalia	Perrine	1905	11480	160.6	250.566	0.158	27.63	186	5×10^{-9}	3	0.03	14.8
X	Lysithea	Nicholson	1938	11720	163.9	259.22	0.107	29.02	36	4×10^{-11}	3	–	18.4
VII	Elara	Perrine	1905	11737	164.2	259.653	0.207	24.77	76	4×10^{-10}	3	0.03	16.8
XII	Ananke	Nicholson	1951	21200	297	631 R	0.169	147	30	2×10^{-11}	3	–	18.9
XI	Carme	Nicholson	1938	22600	316	692 R	0.207	164	40	5×10^{-11}	3	–	18.0
VIII	Pasiphae	Melotte	1908	23500	329	735 R	0.378	145	50	1×10^{-10}	3	–	17.0
IX	Sinope	Nicholson	1914	23700	332	758 R	0.275	153	36	4×10^{-11}	3	–	18.3
Saturn													
XVIII	Pan	Showalter	1990	133.58	2.22	0.57	–	–	$110 \times 90 \times 70$	–	–	0.5	–
XV	Atlas	Terrile	1980	137.67	2.28	0.60	0.000	0.3	40×20	–	–	0.9	18
XVI	Prometheus	Collins Carlson	1980	139.35	2.31	0.61	0.003	0.0	$140 \times 100 \times 80$	–	–	0.6	16
XVII	Pandora	Collins Carlson	1980	141.70	2.35	0.63	0.004	0	$110 \times 90 \times 70$	–	–	0.9	16
XI	Epimetheus	Cruikshank	1980	151.42	2.51	0.69	0.009	0.34	$140 \times 120 \times 100$	–	–	0.8	15
X	Janus	Pascu	1980	151.47	2.51	0.69	0.007	0.14	$220 \times 200 \times 160$	–	–	0.8	14
I	Mimas	W. Herschel	1789	185.52	3.08	0.94	0.020	1.53	392	8×10^{-8}	1.4	0.5	12.9
II	Enceladus	W. Herschel	1789	238.02	3.95	1.37	0.005	0.00	500	1.3×10^{-7}	1.1	1.0	11.7
XIII	Telesto	Smith Larson Reitsema	1980	294.66	4.89	1.89	–	–	$34 \times 28 \times 26$	–	–	0.5	18.5
III	Tethys	Cassini	1684	294.66	4.89	1.89	0.000	1.86	1060	1.3×10^{-6}	1.2	0.9	10.2
XIV	Calypso	Pascu Seidelmann Baum Currie	1980	294.66	4.89	1.89	–	–	$34 \times 22 \times 22$	–	–	0.6	18.7
IV	Dione	Cassini	1684	377.40	6.26	2.74	0.002	0.02	1120	1.9×10^{-6}	1.5	0.7	10.4
XII	Helene	Laques Lecacheux	1980	377.40	6.26	2.74	0.005	0.0	$36 \times 32 \times 30$	–	–	0.7	18
V	Rhea	Cassini	1672	527.04	8.74	4.52	0.001	0.35	1530	4.4×10^{-6}	1.3	0.7	9.7
VI	Titan	Huygens	1665	1221.83	20.3	15.95	0.03	0.33	5150	2.4×10^{-4}	1.9	0.21	8.3
VII	Hyperion	Bond	1848	1481.1	24.6	21.28	0.10	0.43	$410 \times 260 \times 220$	–	1.4	0.3	14.2
VIII	Iapetus	Cassini	1671	3561.3	59.1	79.33	0.03	14.7	1460	3.3×10^{-6}	1.2	0.5/0.05	11.1
IX	Phoebe	Pickering	1898	12952	215	550.48 R	0.16	177	220	–	–	0.06	16.4
Uranus													
VI	Cordelia	Voyager 2	1986	49.77	1.95	0.335	0.00	0.1	26	–	–	0.07	24.1
VII	Ophelia	Voyager 2	1986	53.79	2.10	0.376	0.01	0.1	30	–	–	0.07	23.8
VIII	Bianca	Voyager 2	1986	59.17	2.32	0.435	0.00	0.2	42	–	–	0.07	23.0
IX	Cressida	Voyager 2	1986	61.78	2.42	0.464	0.00	0.0	62	–	–	0.07	22.2
X	Desdemona	Voyager 2	1986	62.68	2.45	0.474	0.00	0.2	54	–	–	0.07	22.5
XI	Juliet	Voyager 2	1986	64.35	2.52	0.493	0.00	0.1	84	–	–	0.07	21.5
XII	Portia	Voyager 2	1986	66.09	2.59	0.513	0.00	0.1	108	–	–	0.07	21.0
XIII	Rosalind	Voyager 2	1986	69.94	2.74	0.558	0.00	0.3	54	–	–	0.07	22.5
XIV	Belinda	Voyager 2	1986	75.26	2.94	0.624	0.00	0.0	66	–	–	0.07	22.1
XV	Puck	Voyager 2	1985	86.01	3.37	0.762	0.00	0.31	154	–	–	0.07	20.2
V	Miranda	Kuiper	1948	129.39	5.06	1.413	0.003	4.2	480	8×10^{-7}	1.2	0.27	16.3
I	Ariel	Lassell	1851	191.02	7.47	2.520	0.003	0.3	1158	1.6×10^{-5}	1.7	0.34	14.2
II	Umbriel	Lassell	1851	266.30	10.42	4.144	0.005	0.36	1172	1.4×10^{-5}	1.4	0.18	14.8
III	Titania	W. Herschel	1787	435.91	17.06	8.706	0.002	0.14	1580	4.1×10^{-5}	1.7	0.27	13.7
IV	Oberon	W. Herschel	1787	583.52	22.83	13.463	0.001	0.10	1524	3.5×10^{-5}	1.6	0.24	13.9
Neptune													
III	Naiad	Voyager 2	1989	48.23	1.95	0.294	0.00	4.74	58	–	–	0.06	24.7
IV	Thalassa	Voyager 2	1989	50.07	2.02	0.311	0.00	0.21	80	–	–	0.06	23.8
V	Despina	Voyager 2	1989	52.53	2.12	0.335	0.00	0.07	144	–	–	0.06	22.6
VI	Galatea	Voyager 2	1989	61.95	2.50	0.429	0.00	0.05	158	–	–	0.06	22.3
VII	Larissa	Voyager 2	1989	73.55	2.97	0.555	0.01	0.20	208×178	–	–	0.06	22.0
VIII	Proteus	Voyager 2	1989	117.65	4.75	1.122	0.00	0.55	$436 \times 416 \times 402$	–	–	0.06	20.3
I	Triton	Lassell	1846	354.76	14.33	5.877 R	0.000	157.35	2706	2.1×10^{-4}	2.1	0.7	13.5
II	Nereid	Kuiper	1949	5513.4	222.6	360.136	0.751	27.6	340	2×10^{-7}	1.0	0.4	18.7
Pluto													
	Charon	Christy	1978	19.6	17	6.387	0.00	99	1186	0.22	–	0.5	16.8

Table E.12. Some well-known asteroids

Asteroid		Discoverer	Year of discovery	Semi-major axis of orbit a [AU]	Eccentricity e	Inclination ι [°]
1	Ceres	Piazzi	1801	2.77	0.08	10.6
2	Pallas	Olbers	1802	2.77	0.23	34.8
3	Juno	Harding	1804	2.67	0.26	13.0
4	Vesta	Olbers	1807	2.36	0.09	7.1
5	Astraea	Hencke	1845	2.58	0.19	5.3
6	Hebe	Hencke	1847	2.42	0.20	14.8
7	Iris	Hind	1847	2.39	0.23	5.5
8	Flora	Hind	1847	2.20	0.16	5.9
9	Metis	Graham	1848	2.39	0.12	5.6
10	Hygiea	DeGasparis	1849	3.14	0.12	3.8
433	Eros	Witt	1898	1.46	0.22	10.8
588	Achilles	Wolf	1906	5.18	0.15	10.3
624	Hektor	Kopff	1907	5.16	0.03	18.3
944	Hidalgo	Baade	1920	5.85	0.66	42.4
1221	Amor	Delporte	1932	1.92	0.43	11.9
1566	Icarus	Baade	1949	1.08	0.83	22.9
1862	Apollo	Reinmuth	1932	1.47	0.56	6.4
2060	Chiron	Kowal	1977	13.64	0.38	6.9
5145	Pholus	Rabinowitz	1992	20.46	0.58	24.7

Asteroid		Sidereal period of orbit P_{sid} [a]	Diameter d [km]	Rotation period τ_{sid} [h]	Geometric albedo p	Mean opposition magnitude B	Type
1	Ceres	4.6	946	9.08	0.07	7.9	C
2	Pallas	4.6	583	7.88	0.09	8.5	U
3	Juno	4.4	249	7.21	0.16	9.8	S
4	Vesta	3.6	555	5.34	0.26	6.8	U
5	Astraea	4.1	116	16.81	0.13	11.2	S
6	Hebe	3.8	206	7.27	0.16	9.7	S
7	Iris	3.7	222	7.14	0.20	9.4	S
8	Flora	3.3	160	13.60	0.13	9.8	S
9	Metis	3.7	168	5.06	0.12	10.4	S
10	Hygiea	5.6	443	18.00	0.05	10.6	C
433	Eros	1.8	20	5.27	0.18	11.5	S
588	Achilles	11.8	70	?	?	16.4	U
624	Hektor	11.7	230	6.92	0.03	15.3	U
944	Hidalgo	14.2	30	10.06	?	19.2	MEU
1221	Amor	2.7	5?	?	?	20.4	?
1566	Icarus	1.1	2	2.27	?	12.3	U
1862	Apollo	1.8	2.5	?	?	16.3	?
2060	Chiron	50.4	320?	?	?	17.3	?
5145	Pholus	92.6	190				

Table E.13. Periodic comets with several perihelion passages observed

	N_T	T	P	q	e	ω	Ω	ι	l	b	Q
Encke	56	1994 Feb. 9	3.28	0.331	0.850	186.3	334.0	11.9	160.2	−1.3	4.09
Grigg-Skjellerup	16	1992 Jul. 24	5.10	0.995	0.664	359.3	212.6	21.1	212.0	−0.3	4.93
Honda-Mrkos-Pajdušáková	8	1990 Sept. 12	5.30	0.541	0.822	325.8	88.6	4.2	54.5	−2.4	5.54
Tuttle-Giacobini-Kresák	8	1990 Feb. 8	5.46	1.068	0.656	61.6	140.9	9.2	202.1	8.1	5.14
Tempel 2	19	1994 Mar. 16	5.48	1.484	0.522	194.9	117.6	12.0	312.1	−3.1	4.73
Wirtanen	8	1991 Sept. 21	5.50	1.083	0.652	356.2	81.6	11.7	77.9	−0.8	5.15
Clark	8	1989 Nov. 28	5.51	1.556	0.501	208.9	59.1	9.5	267.7	−4.6	4.68
Forbes	8	1993 Mar. 15	6.14	1.450	0.568	310.6	333.6	7.2	284.5	−5.4	5.25
Pons-Winnecke	20	1989 Aug. 19	6.38	1.261	0.634	172.3	92.8	22.3	265.6	2.9	5.62
d'Arrest	15	1989 Feb. 4	6.39	1.292	0.625	177.1	138.8	19.4	316.0	1.0	5.59
Schwassmann-Wachmann 2	11	1994 Jan. 24	6.39	2.070	0.399	358.2	125.6	3.8	123.8	−0.1	4.82
Wolf-Harrington	8	1991 Apr. 4	6.51	1.608	0.539	187.0	254.2	18.5	80.8	−2.2	5.37
Giacobini-Zinner	12	1992 Apr. 14	6.61	1.034	0.707	172.5	194.7	31.8	8.3	3.9	6.01
Reinmuth 2	8	1994 Jun. 29	6.64	1.893	0.464	45.9	295.4	7.0	341.1	5.0	5.17
Perrine-Mrkos	8	1989 Mar. 1	6.78	1.298	0.638	166.6	239.4	17.8	46.6	4.1	5.87
Arend-Rigaux	7	1991 Oct. 3	6.82	1.438	0.600	329.1	121.4	17.9	91.7	−9.1	5.75
Borrelly	11	1987 Dec. 18	6.86	1.357	0.624	353.3	74.8	30.3	69.0	−3.4	5.86
Brooks 2	14	1994 Sept. 1	6.89	1.843	0.491	198.0	176.2	5.5	14.1	−1.7	5.40
Finlay	11	1988 Jun. 5	6.95	1.094	0.700	322.2	41.7	3.6	4.0	−2.2	6.19
Johnson	7	1990 Nov. 19	6.97	2.313	0.366	208.3	116.7	13.7	324.3	−6.4	4.98
Daniel	8	1992 Aug. 31	7.06	1.650	0.552	11.0	68.4	20.1	78.7	3.8	5.71
Holmes	8	1993 Apr. 10	7.09	2.177	0.410	23.2	327.3	19.2	349.4	7.4	5.21
Reinmuth 1	8	1988 May 10	7.29	1.869	0.503	13.0	119.2	8.1	132.0	1.8	5.65
Faye	19	1991 Nov. 15	7.34	1.593	0.578	204.0	198.9	9.1	42.6	−3.7	5.96
Ashbrook-Jackson	7	1993 Jul. 13	7.49	2.316	0.395	348.7	2.0	12.5	350.9	−2.4	5.34
Schaumasse	10	1993 Mar. 5	8.22	1.202	0.705	57.5	80.4	11.8	137.3	10.0	6.94
Wolf	14	1992 Aug. 28	8.25	2.428	0.406	162.3	203.4	27.5	7.6	8.1	5.74
Whipple	9	1994 Dec. 22	8.53	3.094	0.259	201.9	181.8	9.9	23.4	−3.7	5.25
Comas Solá	8	1987 Aug. 18	8.78	1.830	0.570	45.5	60.4	13.0	105.2	9.2	6.68
Väisälä 1	6	1993 Apr. 29	10.8	1.783	0.635	47.4	134.4	11.6	181.2	8.5	7.98
Tuttle	11	1994 Jun. 27	13.5	0.998	0.824	206.7	269.8	54.7	106.1	−21.5	10.3
Halley	30	1986 Feb. 9	76.0	0.587	0.967	111.8	58.1	162.2	305.3	16.4	35.3

N_T: number of perihelion passages observed
T: perihelion time
P: period of orbit [years]
q: perihelion distance [AU]

e: eccentricity
ω: argument of perihelion (1950.0)
Ω: longitude of ascending node (1950.0)
ι: inclination

l: longitude of perihelion, here defined as
      $l = \Omega + \arctan(\tan \omega \cos \iota)$
b: latitude of perihelion, $\sin b = \sin \omega \sin \iota$
Q: aphelion distance [AU]

Table E.14. Principal meteor showers

Shower	Period of visibility	Maximum	Radiant α	Radiant δ	Meteors per hour	Associated comet
Quadrantids	Jan. 1 − 5	Jan. 3 − 4	231°	+52°	30 − 40	−
Lyrids	Apr. 19 − 25	Apr. 22	275°	+35°	10	Thatcher
Eta Aquarids	May 1 − 12	May 5	336°	0°	5 − 10	Halley
Perseids	Jul. 20 − Aug. 18	Aug. 12	45°	+57°	40 − 50	Swift-Tuttle
Kappa Cygnids	Aug. 17 − 24	Aug. 20	290°	+55°	5	−
Orionids	Oct. 17 − 26	Oct. 21	95°	+15°	10 − 15	Halley
Taurids	Oct. 10 − Dec. 5	Nov. 1	55°	+15°	5	Encke
Leonids	Nov. 14 − 20	Nov. 17	151°	+22°	10	Tempel-Tuttle
Geminids	Dec. 7 − 15	Dec. 13 − 14	112°	+33°	40 − 50	−
Ursids	Dec. 17 − 24	Dec. 22	205°	+76°	5	−

Table E.15. Nearest stars

Name	α 2000.0 [h min]	δ [° ']	V	$B-V$	Spectrum	π [']	r [pc]	M_v	μ ['/a]	v_r [km/s]
Sun	−	−	− 26.78	0.62	G2 V	−		4.79		
α Cen C (Proxima)	14 30	− 62 41	11.05	1.97	M5e V	0.762	1.31	15.45	3.85	− 16
α Cen A	14 40	− 60 50	− 0.01	0.68	G2 V	0.745	1.34	4.35	3.68	− 22
α Cen B	14 40	− 60 50	1.33	0.88	K5 V	0.745	1.34	5.69	3.68	− 22
Barnard's star	17 58	4 34	9.54	1.74	M5 V	0.552	1.81	13.25	10.31	− 108
Wolf 359	10 56	7 01	13.53	2.01	M6e V	0.429	2.33	16.68	4.71	+ 13
BD + 36°2147	11 03	35 58	7.50	1.51	M2 V	0.401	2.49	10.49	4.78	− 84
α CMa (Sirius) A	6 45	− 16 43	− 1.46	0.00	A1 V	0.377	2.65	1.42	1.33	− 8
α CMa (Sirius) B	6 45	− 16 43	8.68	−	wd A	0.377	2.65	11.56	1.33	− 8
Luyten 726-8 A	1 39	− 17 57	12.45	−	M6e V	0.367	2.72	15.27	3.36	+ 29
Luyten 726-8 B (UV Cet)	1 39	− 17 57	12.95	−	M6e V	0.367	2.72	15.8	3.36	+ 32
Ross 154	18 50	− 23 50	10.6	−	M4e V	0.345	2.90	13.3	0.72	− 4
Ross 248	23 42	44 10	12.29	1.92	M6e V	0.317	3.15	14.80	1.59	− 81
ε Eri	3 33	− 9 28	3.73	0.88	K2 V	0.303	3.30	6.13	0.98	+ 16
Luyten 789-6	22 39	− 15 19	12.18	1.96	M6e V	0.303	3.30	14.60	3.26	− 60
Ross 128	11 48	0 48	11.10	1.76	M5 V	0.301	3.32	13.50	1.37	− 13
61 Cyg A	21 07	38 45	5.22	1.17	K5 V	0.294	3.40	7.58	5.21	− 64
61 Cyg B	21 07	38 45	6.03	1.37	K7 V	0.294	3.40	8.39	5.21	− 64
ε Ind	22 03	− 56 47	4.68	1.05	K5 V	0.291	3.44	7.00	4.69	− 40
α CMi (Procyon) A	7 39	5 13	0.37	0.42	F5 V	0.286	3.50	2.64	1.25	− 3
α CMi (Procyon) B	7 39	5 13	10.7	−	wd F	0.286	3.50	13.0	1.25	
BD + 59°1915 A	18 43	59 38	8.90	1.54	M4 V	0.283	3.53	11.15	2.30	0
BD + 59°1915 B	18 43	59 38	9.69	1.59	M5 V	0.283	3.53	11.94	2.28	+ 10
BD + 43°44 A	0 18	44 01	8.07	1.56	M1 V	0.282	3.55	10.32	2.90	+ 13
BD + 43°44 B	0 18	44 01	11.04	1.80	M6 V	0.282	3.55	13.29	2.90	+ 20
CD − 36°15693	23 06	− 35 52	7.36	1.46	M2 V	0.279	3.58	9.59	6.90	+ 10
τ Cet	1 44	− 15 56	3.50	0.72	G8 VI	0.276	3.62	5.72	1.91	− 16
BD + 5°1668	7 26	5 14	9.82	1.56	M5 V	0.268	3.73	11.98	3.74	+ 26
Luyten 725-32	1 12	− 17 00	11.6		M2e V	0.261	3.83	13.4	1.36	
CD − 39°14192	21 17	− 38 52	6.67	1.38	M0 V	0.260	3.85	8.75	3.46	+ 21
Kapteyn's star	5 12	− 45 01	8.81	1.56	M0 VI	0.256	3.91	10.85	8.81	+ 245
Krüger 60 A	22 28	57 42	9.85	1.62	M4 V	0.253	3.95	11.87	0.86	− 26
Krüger 60 B	22 28	57 42	11.3	1.8	M6e V	0.253	3.95	13.3	0.86	− 26
Ross 614 A	6 29	− 2 49	11.17	1.74	M7e V	0.250	4.00	13.16	0.99	+ 24
Ross 614 B	6 29	− 2 49	14.8		M	0.250	4.00	16.8	0.99	+ 24
BD − 12°4523	16 30	− 12 39	10.2	1.60	M4 V	0.249	4.02	12.10	1.18	− 13
van Maanen's star	0 49	5 23	12.37	0.56	wd G	0.236	4.24	14.26	2.97	+ 54
Wolf 424 A	12 33	9 01	13.16	1.80	M7e V	0.230	4.35	14.98	1.75	− 5
Wolf 424 B	12 33	9 01	13.4		M7e V	0.230	4.35	15.2	1.75	− 5
CD − 37°15492	0 06	− 37 21	8.63	1.45	M4 V	0.225	4.44	10.39	6.09	+ 23
BD + 50°1725	10 11	49 27	6.59	1.36	K7 V	0.219	4.57	8.32	1.45	− 26
CD − 46°11540	17 29	− 46 54	9.36	1.53	M4	0.216	4.63	11.03	1.10	
CD − 49°13515	21 34	− 49 00	8.67	1.46	M1 V	0.214	4.67	10.32	0.81	+ 8
CD − 44°11909	17 37	− 44 20	11.2		M5	0.213	4.69	12.8	1.16	
BD + 68°946	17 36	68 21	9.15	1.50	M4 V	0.209	4.78	10.79	1.32	− 22
BD − 15°6290	22 53	− 14 16	10.17	1.60	M5 V	0.207	4.83	11.77	1.15	+ 9
Luyten 145-141	11 46	− 64 50	11.44	0.19	wd A	0.206	4.85	13.01	2.68	
o^2 Eri A	4 15	− 7 39	4.43	0.82	K1 V	0.205	4.88	5.99	4.08	− 43
o^2 Eri B	4 15	− 7 39	9.53	0.03	wd A	0.205	4.88	11.09	4.11	− 21
o^2 Eri C	4 15	− 7 39	11.17	1.68	M4e V	0.205	4.88	12.73	4.11	− 45
BD + 15°2620	13 46	14 54	8.50	1.43	M4 V	0.205	4.88	10.02	2.30	+ 15
BD + 20°2465	10 20	19 52	9.43	1.54	M4e V	0.203	4.93	10.98	0.49	+ 11
α Aql (Altair)	19 51	8 52	0.76	0.22	A7 V	0.197	5.08	2.24	0.66	− 26
AC + 79°3888	11 48	78 42	10.94		M4 VI	0.195	5.13	12.38	0.89	− 117
70 Oph A	18 05	2 30	4.22	0.86	K0 V	0.195	5.13	5.67	1.12	− 7
70 Oph B	18 05	2 30	6.0		K5 V	0.195	5.13	7.45	1.12	− 10

Table E.16. Brightest stars ($V < 1.50$)

Name		α 2000.0	δ	V	$B-V$	Spectrum	M_v	r	μ	v_r	Remarks
		[h min]	[° ']					[pc]	['' /a]	[km/s]	
α CMa	Sirius	6 45.1	− 16 43	− 1.46	0.00	A1 V, wd A	+ 1.4	2.7	1.33	− 8	db $a = 7.5''$, $P = 50$ a
α Car	Canopus	6 24.0	− 52 42	− 0.73	0.16	F0Ib-II	− 4.6	60	0.02	+ 21	
α Cen	Rigil Kentaurus	14 39.6	− 60 50	− 0.29	0.72	G2V, K5 V	+ 4.1	1.3	3.68	− 22	db $a = 17.6''$, $P = 80$ a; Proxima 2.2° apart
α Boo	Arcturus	14 15.7	19 11	− 0.06	1.23	K2IIIp	− 0.3	11	2.28	− 5	
α Lyr	Vega	18 36.9	38 47	0.04	0.00	A0V	+ 0.5	8.1	0.34	− 14	
α Aur	Capella	5 16.7	46 00	0.08	0.79	G5III, G0III	− 0.6	14	0.44	+ 30	db $a = 0.054''$, $P = 0.285$ a
α Ori	Rigel	5 14.5	− 8 12	0.11	− 0.03	B8Ia	− 7.0	250	0.00	+ 21	db sep. 9''; spectr. db $P = 10$ d
α CMi	Procyon	7 39.3	5 13	0.37	0.42	F5V, wd F	+ 2.6	3.5	1.25	− 3	db $a = 4.5''$, $P = 41$ a
α Eri	Achernar	1 37.7	− 57 14	0.46	− 0.16	B3Vp	− 2.5	38	0.10	+ 19	
α Ori	Betelgeuse	5 55.2	7 24	0.50	1.85	M2I	− 6.0	200	0.03	+ 21	var. $V = 0.40 − 1.3$; spectr. db $P = 5.8$ a
β Cen	Hadar	14 03.8	− 60 22	0.60	− 0.23	B1II	− 5.0	120	0.04	− 11	db sep. 1.2''
α Aql	Altair	19 50.8	8 52	0.76	0.22	A7V	+ 2.2	5.1	0.66	− 26	
α Cru	Acrux	12 26.6	− 63 06	0.79	− 0.25	B0.5IV, B1 V	− 4.7	120	0.04	− 6	db sep. 5'', each spectr. db
α Tau	Aldebaran	4 35.9	16 31	0.85	1.53	K5III	− 0.8	21	0.20	+ 54	db sep. 31''
α Vir	Spica	13 25.2	− 11 10	0.96	− 0.23	B1V	− 3.6	80	0.05	+ 1	spectr. db $P = 4.0$ d; several components
α Sco	Antares	16 29.4	− 26 26	0.96	1.23	M1Ib, B2.5V	− 4.6	130	0.03	− 3	var. $V = 0.88 − 1.8$; db $a = 2.9''$, $P = 880$ a
β Gem	Pollux	7 45.3	28 01	1.15	1.00	K0III	+ 1.0	11	0.62	+ 3	
α PsA	Fomalhaut	22 57.6	− 29 37	1.16	0.09	A3V	+ 1.9	7.0	0.37	+ 7	
α Cyg	Deneb	20 41.4	45 17	1.25	0.09	A2Ia	− 7.2	500	0.00	− 5	
β Cru	Mimosa	12 47.7	− 59 41	1.26	− 0.24	B0III	− 4.6	150	0.05	+ 20	spectr. db $P = 0.16$ d and $7 − 8$ a
α Leo	Regulus	10 08.4	11 58	1.35	− 0.11	B7V	− 0.7	26	0 25	+ 4	db sep. 177''

Abbreviations: db: double star; *a*: semi-major axis of the orbit; *P*: period; sep.: separation; spectr.: spectroscopic; var.: variable; variation given in *V*-magnitude

Table E.17. Some double stars

Name		α 2000.0 [h min]	δ [° ']	m_{v1}	m_{v2}	sp$_1$	sp$_2$	Separation ['']	Parallax ['']
η Cas	Achird	0 49.1	57 49	3.7	7.5	G0 V	M0	12	0.176
γ Ari	Mesarthim	1 53.5	19 18	4.8	4.9	A1 p	B9 V	8	0.026
α Psc	Kaitain	2 02.0	2 46	4.3	5.3	A0 p	A3 m	2	0.017
γ And	Alamak	2 03.9	42 20	2.4	5.1	K3 IIb	B8 V + A0 V	10	0.010
δ Ori	Mintaka	5 32.0	−0 18	2.5	7.0	B0 III + O9 V	B2 V	52	0.014
λ Ori	Meissa	5 35.1	9 56	3.7	5.7	O8 e	B0.5 V	4	0.007
ζ Ori	Alnitak	5 40.8	−1 57	2.1	4.2	O9.5 Ibe	B0 III	2	0.024
α Gem	Castor	7 34.6	31 53	2.0	3.0	A1 V	A5 Vm	2	0.067
γ Leo	Algieba	10 20.0	19 51	2.6	3.8	K1 III	G7 III	4	0.013
ξ UMa	Alula Australis	11 18.2	31 32	4.4	4.9	G0 V	G0 V	3	0.137
α Cru	Acrux	12 26.6	−63 06	1.4	1.8	B0.5 IV	B1 V	4	0.008
γ Vir	Porrima	12 41.7	−1 27	3.7	3.7	F0 V	F0 V	4	0.099
α CVn	Cor Caroli	12 56.0	38 19	2.9	5.5	A0 p	F0 V	19	0.027
ζ UMa	Mizar	13 23.9	54 56	2.4	4.1	A1 Vp	A1 m	14	0.047
α Cen	Rigil Kentaurus	14 39.6	−60 50	0.0	1.2	G2 V	K5 V	21	0.745
ε Boo	Izar	14 45.0	27 04	2.7	5.3	K0 II − III	A2 V	3	0.016
δ Ser		15 34.8	10 32	4.2	5.3	F0 IV	F0 IV	4	0.021
β Sco	Grafias	16 05.4	−19 48	2.9	5.1	B1 V	B2 V	14	0.009
α Her	Rasalgethi	17 14.6	14 23	3.0 − 4.0	5.7	M5 Ib − II	G5 III + F2 V	5	0.008
ρ Her		17 23.7	37 09	4.5	5.5	B9.5 III	A0 V	4	−
70 Oph		18 05.5	2 30	4.3	6.1	K0 V	K5 V	2	0.195
$\varepsilon^1 \varepsilon^2$ Lyr		18 44.3	39 40	4.8	4.4	A4 V + F1 V	A8 V + F0 V	209	0.021
ε^1 Lyr		18 44.3	39 40	5.1	6.2	A4 V	F1 V	3	0.021
ε^2 Lyr		18 44.4	39 37	5.1	5.3	A8 V	F0 V	2	0.021
ζ Lyr		18 44.8	37 36	4.3	5.7	Am	F0 IV	44	0.031
θ Ser	Alya	18 56.2	4 12	4.5	4.9	A5 V	A5 V	22	0.030
γ Del		20 46.7	16 07	4.5	5.4	K1 IV	F7 V	10	0.026
ζ Aqr		22 28.8	−0 01	4.4	4.6	F3 V	F6 IV	2	0.039
δ Cep		22 29.2	58 25	3.5 − 4.3	7.5	F5 Ib − G2 Ib	B7 IV	41	0.011

Table E.18. Milky Way Galaxy

Property	Value
Mass	$> 2 \times 10^{11} M_\odot$
Disc diameter	30 kpc
Disc thickness (stars)	1 kpc
Disc thickness (gas and dust)	200 pc
Halo diameter	50 kpc
Sun's distance from galactic centre	8.5 kpc
Sun's rotational velocity	220 km/s
Sun's period of rotation	240×10^6 a
Direction of galactic centre (2000.0)	$\alpha = 17$ h 45.7 min $\delta = -29°00'$
Direction of galactic north pole (2000.0)	$\alpha = 12$ h 51.4 min $\delta = +27°08'$

Table E.19. Members of the Local Group of galaxies

Name	α 2000.0 [h min]	δ [° ']	Type	m_{pg}	M_{pg}	Distance [kpc]
Milky Way Galaxy	–	–	Sb or Sc	–	–	
NGC 224 = M 31	0 42.7	41 16	Sb	4.33	− 20.3	650
NGC 598 = M 33	1 33.9	30 39	Sc	6.19	− 18.5	740
LMC	5 23.7	− 69 45	Irr or SBc	0.86	− 17.8	50
SMC	0 52.7	− 72 49	Irr	2.86	− 16.2	60
NGC 205 = M 110	0 40.4	41 41	E6	8.89	− 15.8	650
NGC 221 = M 32	0 42.7	40 52	E2	9.06	− 15.6	650
NGC 6822	19 45.0	− 14 48	Irr	9.21	− 15.3	520
NGC 185	0 38.9	48 20	E0	10.29	− 14.4	650
IC 1613	1 04.8	2 07	Irr	10.00	− 14.4	740
NGC 147	0 33.2	48.30	dE4	10.57	− 14.1	650
Fornax	2 39.9	− 34 32	dE	9.1	− 12	190
And I	0 45.7	38 01	dE	13.5	− 11	650
And II	1 16.5	33 26	dE	13.5	− 11	650
And III	0 35.3	36 31	dE	13.5	− 11	650
And IV	0 42.4	40 35	dE	–	–	650
Leo I	10 08.5	12 19	dE	11.27	− 10	230
Sculptor	1 00.2	− 33 42	dE	10.5	− 9.2	90
Leo II	11 13.5	22 09	dE	12.85	− 9	230
Draco	17 20.1	57 55	dE	–	?	80
Ursa Minor	15 08.8	67 12	dE	–	?	80
Carina	6 41.6	− 50 58	dE	–	?	170
LGS 3	1 03.8	21 53	?	–	?	650
Sextans	10 13.0	− 1 36	dE	–	–	85

Table E.20. Optically brightest galaxies

Name	α 2000.0 [h min]	δ [° ']	Type	m_{pg}	Size ['']	Distance [Mpc]
NGC 55	0 15.1	− 39 13	Sc or Irr	7.9	30 × 5	2.3
NGC 205	0 40.4	41 41	E6	8.9	12 × 6	0.7
NGC 221 = M 32	0 42.7	40 52	E2	9.1	3.4 × 2.9	0.7
NGC 224 = M 31	0 42.8	41 16	Sb	4.3	163 × 42	0.7
NGC 247	0 47.2	− 20 46	S	9.5	21 × 8	2.3
NGC 253	0 47.6	− 25 17	Sc	7.0	22 × 5	2.3
SMC	0 52.6	− 72 48	Irr	2.9	216 × 216	0.06
NGC 300	0 54.9	− 37 41	Sc	8.7	22 × 16	2.3
NGC 598 = M 33	1 33.9	30 39	Sc	6.2	61 × 42	0.7
Fornax	2 39.9	− 34 32	dE	9.1	50 × 35	0.2
LMC	5 23.6	− 69 45	Irr or Sc	0.9	432 × 432	0.05
NGC 2403	7 36.9	65 36	Sc	8.8	22 × 12	2.0
NGC 2903	9 32.1	21 30	Sb	9.5	16 × 7	5.8
NGC 3031 = M 81	9 55.6	69 04	Sb	7.8	25 × 12	2.0
NGC 3034 = M 82	9 55.9	69 41	Sc	9.2	10 × 1.5	2.0
NGC 4258 = M 106	12 19.0	47 18	Sb	8.9	19 × 7	4.3
NGC 4472 = M 49	12 29.8	8 00	E4	9.3	10 × 7	11
NGC 4594 = M 104	12 40.0	− 11 37	Sb	9.2	8 × 5	11
NGC 4736 = M 94	12 50.9	41 07	Sb	8.9	13 × 12	4.3
NGC 4826 = M 64	12 56.8	21 41	?	9.3	10 × 4	3.7
NGC 4945	13 05.4	− 49 28	Sb	8.0	20 × 4	
NGC 5055 = M 63	13 15.8	42 02	Sb	9.3	8 × 3	4.3
NGC 5128 = Cen A	13 25.5	− 43 01	E0	7.9	23 × 20	
NGC 5194 = M 51	13 29.9	47 12	Sc	8.9	11 × 6	4.3
NGC 5236 = M 83	13 37.0	− 29 52	Sc	7.0	13 × 12	2.4
NGC 5457 = M 101	14 03.2	54 21	Sc	8.2	23 × 21	4.3
NGC 6822	19 45.0	− 14 48	Irr	9.2	20 × 10	0.7

Table E.21. Constellations

Latin name	Genitive	Abbreviation	English name
Andromeda	Andromedae	And	Andromeda
Antlia	Antliae	Ant	Air Pump
Apus	Apodis	Aps	Bird of Paradise
Aquarius	Aquarii	Aqr	Water-bearer
Aquila	Aquilae	Aql	Eagle
Ara	Arae	Ara	Altar
Aries	Arietis	Ari	Ram
Auriga	Aurigae	Aur	Charioteer
Bootes	Bootis	Boo	Herdsman
Caelum	Caeli	Cae	Chisel
Camelopardalis	Camelopardalis	Cam	Giraffe
Cancer	Cancri	Cnc	Crab
Canes Venatici	Canum Venaticorum	CVn	Hunting Dogs
Canis Major	Canis Majoris	CMa	Great Dog
Canis Minor	Canis Minoris	CMi	Little Dog
Capricornus	Capricorni	Cap	Shea-goat
Carina	Carinae	Car	Keel
Cassiopeia	Cassiopeiae	Cas	Cassiopeia
Centaurus	Centauri	Cen	Centaur

Cepheus	Cephei	Cep	Cepheus
Cetus	Ceti	Cet	Whale
Chamaeleon	Chamaeleontis	Cha	Chameleon
Circinus	Circini	Cir	Compasses
Columba	Columbae	Col	Dove
Coma Berenices	Comae Berenices	Com	Berenice's Hair
Corona Austrina	Coronae Austrinae	CrA	Southern Crown
Corona Borealis	Coronae Borealis	CrB	Northern Crown
Corvus	Corvi	Crv	Crow
Crater	Crateris	Crt	Cup
Crux	Crucis	Cru	Southern Cross
Cygnus	Cygni	Cyg	Swan
Delphinus	Delphini	Del	Dolphin
Dorado	Doradus	Dor	Swordfish
Draco	Draconis	Dra	Dragon
Equuleus	Equulei	Equ	Little Horse
Eridanus	Eridani	Eri	Eridanus
Fornax	Fornacis	For	Furnace
Gemini	Geminorum	Gem	Twins
Grus	Gruis	Gru	Crane
Hercules	Herculis	Her	Hercules
Horologium	Horologii	Hor	Clock
Hydra	Hydrae	Hya	Water Serpent
Hydrus	Hydri	Hyi	Water Snake
Indus	Indi	Ind	Indian
Lacerta	Lacertae	Lac	Lizard
Leo	Leonis	Leo	Lion
Leo Minor	Leonis Minoris	LMi	Little Lion
Lepus	Leporis	Lep	Hare
Libra	Librae	Lib	Scales
Lupus	Lupi	Lup	Wolf
Lynx	Lyncis	Lyn	Lynx
Lyra	Lyrae	Lyr	Lyre
Mensa	Mensae	Men	Table Mountain
Microscopium	Microscopii	Mic	Microscope
Monoceros	Monocerotis	Mon	Unicorn
Musca	Muscae	Mus	Fly
Norma	Normae	Nor	Square
Octans	Octantis	Oct	Octant
Ophiuchus	Ophiuchi	Oph	Serpent-bearer
Orion	Orionis	Ori	Orion
Pavo	Pavonis	Pav	Peacock
Pegasus	Pegasi	Peg	Pegasus
Perseus	Persei	Per	Perseus
Phoenix	Phoenicis	Phe	Phoenix
Pictor	Pictoris	Pic	Painter
Pisces	Piscium	Psc	Fishes
Piscis Austrinus	Piscis Austrini	PsA	Southern Fish
Puppis	Puppis	Pup	Poop
Pyxis	Pyxidis	Pyx	Compass
Reticulum	Reticuli	Ret	Net
Sagitta	Sagittae	Sge	Arrow
Sagittarius	Sagittarii	Sgr	Archer
Scorpius	Scorpii	Sco	Scorpion
Sculptor	Sculptoris	Scl	Sculptor
Scutum	Scuti	Sct	Sobieski's Shield
Serpens	Serpentis	Ser	Serpent
Sextans	Sextantis	Sex	Sextant
Taurus	Tauri	Tau	Bull

Table E.21. (cont.)

Latin name	Genitive	Abbreviation	English name
Telescopium	Telescopii	Tel	Telescope
Triangulum	Trianguli	Tri	Triangle
Triangulum Australe	Trianguli Australis	TrA	Southern Triangle
Tucana	Tucanae	Tuc	Toucan
Ursa Major	Ursae Majoris	UMa	Great Bear
Ursa Minor	Ursae Minoris	UMi	Little Bear
Vela	Velorum	Vel	Sails
Virgo	Virginis	Vir	Virgin
Volans	Volantis	Vol	Flying Fish
Vulpecula	Vulpeculae	Vul	Fox

Table E.22. Largest telescopes

Telescope	Location	Completion year	Mirror diameter [m]
W.M. Keck Telescope	Mauna Kea, Hawaii	1992	10
BTA Telescope	Zelenchukskaya, USSR	1975	6.0
Hale Telescope	Mt. Palomar, USA	1948	5.0
Multiple-Mirror Telescope (MMT)	Mt. Hopkins, USA	1979	4.5[a]
William Herschel Telescope	Canary Islands	1987	4.2
Cerro Tololo	Chile	1976	4.0
Anglo-Australian Telescope	Siding Spring, Australia	1974	3.9
Mayall Telescope	Kitt Peak, USA	1973	3.9
United Kingdom Infrared Telescope (UKIRT)	Mauna Kea, Hawaii	1978	3.8
Canada-France-Hawaii Telescope (CFHT)	Mauna Kea, Hawaii	1979	3.6
European Southern Observatory (ESO)	La Silla, Chile	1976	3.6
German-Spanish Astronomical Center	Calar Alto, Spain	1984	3.5
New Technology Telescope (ESO)	La Silla, Chile	1989	3.5
Lick Observatory	Mt. Hamilton, USA	1957	3.1
Nasa Infrared Telescope (IRTF)	Mauna Kea, Hawaii	1979	3.1

[a] Consists of six mirrors, each 1.8 m in diameter, the total area of which corresponds to one 4.5 m mirror

Table E.23. Largest parabolic radio telescopes

Telescope	Location	Completion year	Diameter [m]	Shortest wavelength [cm]	Remarks
Arecibo	Puerto Rico, USA	1963	305	5	Fixed disk, limited tracking
Effelsberg	Bonn, Federal Republic of Germany	1973	100	0.8	Largest completely steerable telescope
Jodrell Bank	Macclesfield, Great Britain	1957	76.2	10 – 20	First large paraboloid antenna
Yevpatoriya	Crimea, USSR	1979	70	1.5	
Parkes	Australia	1961	64	2.5	Innermost 17 m of dish can be used down to 3 mm-wavelengths
Goldstone	California, USA		64	1.5	Belongs to NASA

Table E.24. Millimetre and submillimetre wave radio telescopes and interferometers

Institute	Location	Altitude above sea level [m]	Diameter [m]	Shortest wave-length [mm]	Remarks; operational since
Max-Planck-Institut für Radioastronomie (Federal Republic of Germany) & University of Arizona (USA)	Mt. Graham, USA	3250	10	0.3	1994?
California Institute of Technology (USA)	Mauna Kea, Hawaii	4100	10.4	0.3	1986
Science Research Council (United Kingdom & Netherlands)	Mauna Kea, Hawaii	4100	15.0	0.5	The James Clerk Maxwell Telescope; 1986
California Institute of Technology (USA)	Owens Valley, USA	1220	10.4	0.5	3-antenna interferometer; 1980
University of Cologne (Federal Republic of Germany)	Gornergrat, Suisse	3120	3.0	0.5	1986
Sweden-ESO Southern Hemisphere Millimeter Antenna (SEST)	La Silla, Chile	2400	15.0	0.6	1987
Institut de Radioastronomie Millimetrique (IRAM; France & Federal Republic of Germany)	Plateau de Bure, France	2550	15.0	0.6	3-antenna interferometer; 1990 Fourth antenna 1993
University of Texas (USA)	McDonald Observatory, USA	2030	4.9	0.8	1967
IRAM	Pico Veleta, Spain	2850	30.0	0.9	1984
National Radio Astronomy Observatory (USA)	Kitt Peak, USA	1940	12.0	0.9	1983 (1969)
CSIRO (Australia)	Sydney, Australia	10	4.0	1.3	1977
Bell Telephone Laboratory (USA)	Holmdel, USA	115	7.0	1.5	1977
University of Massachusetts (USA)	New Salem, USA	300	13.7	1.9	Radom; 1978
University of California, Berkeley (USA)	Hat Creek Observatory	1040	6.1	2	3-antenna interferometer; 1968
Purple Mountain Observatory (China (P. R.))	Western China	3000	13.7	2	Radom; 1987
Daeduk Radio Astronomy Observatory (South Korea)	Seoul, South Korea	300	13.7	2	Radom; 1987
University of Tokyo (Japan)	Nobeyama, Japan	1350	45.0	2.6	1982
University of Tokyo (Japan)	Nobeyama, Japan	1350	10.0	2.6	5-antenna interferometer; 1984
Chalmers University of Technology (Sweden)	Onsala, Sweden	10	20.0	2.6	Radom; 1976

Answers to Exercises

Chapter 2

2.16.1 The distance is ≈ 7640 km, the northernmost point is 79°N, 45°W, in North Greenland, 1250 km from the North Pole.

2.16.2 The star can culminate South or North of zenith. In the former case we get $\delta = 65°$, $\phi = 70°$, and in the latter $\delta = 70°$, $\phi = 65°$.

2.16.3 a) $\phi > 58°4'$. If refraction is taken into account, the limit is $57°29'$.
b) $\phi = \delta = 7°24'$. c) $-59°14' \le \phi \le -0°46'$.

2.16.4 Pretty bad.

2.16.5 $\lambda_\odot = 70°22'$, $\beta_\odot = 0°0'$, $\lambda_\oplus = 250°22'$, $\beta_\oplus = 0°0'$.

2.16.6 c) $\Theta_0 = 18$ h.

2.16.7 a) $\alpha = 6$ h 45 min 9 s, $\delta = -16°43'$. b) $v_t = 16.7$ km s^{-1}, $v = 18.5$ km s^{-1}.
c) After $\approx 61\,000$ years. d) $\mu = 1.62''$ per year, parallax $0.42''$.

Chapter 3

3.9.1 a) 3.2 s. b) 1.35 cm and 1.80 cm. c) 60 and 80.

3.9.2 a) $0.001''$ (note that the aperture is a line rather than circular; therefore the coefficient 1.22 should not be used). b) 140 m.

Chapter 4

4.7.1 0.9.

4.7.2 The absolute magnitude will be -17.5 and apparent 6.7.

4.7.3 $N(m+1)/N(m) = 10^{3/5} = 3.98$.

4.7.4 $r = 2.1$ kpc, $E_{B-V} = 0.7$, and $(B-V)_0 = 0.9$.

4.7.5 a) 1.06 mg. b) $\tau = -\ln 0.85^6 = 0.98$.

Chapter 5

5.10.2 $n = 166$, which corresponds to $\lambda = 21.04$ cm. Such transitions would keep the population of the state $n = 166$ very high, resulting in downward transitions also from this state. Such transitions have not been detected. Hence, the line is produced by some other process.

5.10.3 If we express the intensity as something per unit wavelength, we get $\lambda_{max} = 1.1$ mm. If the intensity is given per unit frequency, we have $\lambda_{max} = 1.9$ mm. The total intensity is 9.6×10^{-7} W m^{-2} sterad^{-1}. At 550 nm the intensity is zero for any practical purposes.

5.10.4 a) $L = 1.35 \times 10^{29}$ W. Planck's law can be integrated numerically; a rather complicated expression can be obtained using the Wien approximation. Both ways, we find that 3.3% of the radiation goes to the visual band. Thus $L_V = 4.45 \times 10^{27}$ W. b) 10.3 km.

Chapter 6

6.2.1 $R = 2.0\,R_\odot$.

6.2.2 $T = 1380$ K. There are several strong absorption lines in this spectral region, reducing the brightness temperature.

6.2.4 $v_{rms} \approx 6700$ km s^{-1}.

Chapter 7

7.12.1 $v_a/v_p = (1-e)/(1+e)$. For the Earth this is 0.97.

7.12.2 $a = 1.4581$ AU, $v \approx 23.6$ km s^{-1}.

7.12.3 The period must equal the sidereal rotation period of the Earth. $r = 42\,339$ km $= 6.64\,R_\oplus$. Areas within 8.6° from the poles cannot be seen from geostationary satellites. The hidden area is 1.1% of the total surface area.

7.12.4 $\rho = 3\pi/(GP^2(\alpha/2)^3) \approx 1400$ kg m^{-3}.

7.12.5 $M = 90°$, $E = 90.96°$, $f = 91.91°$.

7.12.6 $a = 3.55 \times 10^7$ AU, $e = 1 + 3.97 \times 10^{-16}$, $r_p = 2.1$ km. The comet will hit the Sun.

Chapter 8

8.24.1 $P_{sid} = 11.9$a, $a = 5.20$ AU, $d = 144\,000$ km. Obviously, the planet is Jupiter.

8.24.2 b) 7.06 d.

8.24.3 a) If the radii of the orbits are a_1 and a_2, the angular velocity of the retrograde motion is

$$\frac{d\lambda}{dt} = \frac{\sqrt{GM}}{\sqrt{a_1 a_2}(\sqrt{a_1} + \sqrt{a_2})} \; .$$

b) In six days Pluto moves about $0.128°$, corresponding to 4 mm. For a main belt asteroid the displacement is almost 4 cm.

8.24.4 $p = 0.11$, $q = 2$, and $A = 0.2$.

8.24.5 The absolute magnitude is $V(1,0) = 23$. a) $m = 18.7$. b) $m = 14.2$. At least a 15 cm telescope is needed to detect the asteroid even one day before the collision.

8.24.6 Assuming the comet is a slow rotator we get a) 104 K, b) 328 K, c) 464 K.

Chapter 9

9.8.1 $c\,g\,a\,d\,f\,e\,b$; the actual spectral classes from top to bottom are A0, M5, O6, F2, K5, G2, B3.

Chapter 10

10.6.1 The period is $P = 1/\sqrt{2}$ years and the relative velocity $42\,100$ m s^{-1}. The maximum separation of the lines is 0.061 nm.

10.6.3 Substituting the values to the equation of Exercise 10.6.2 we get an equation for a. The solution is $a = 4.4$ AU. The mass of the planet is $0.0015\,M_\odot$.

Chapter 11

11.6.1 10.5.

11.6.2 a) 9.5×10^{37}. b) The neutrino production rate is 1.9×10^{38} s^{-1}, and each second 9×10^{28} neutrinos hit the Earth.

11.6.3 The mean free path is $1/\kappa\rho \approx 42\,000$ AU.

Chapter 12

12.10.1 $t_{ff} = 6.7 \times 10^5$ y. Stars are born at the rate $0.75\,M_\odot$ per year.

12.10.2 $t_t \approx 400\,000$ y, $t_n \approx 3 \times 10^8$ y.

12.10.3   About 900 million years.

Chapter 13

13.5.1   a) 6.4×10^7 W m^{-2}. b) $16\,\text{m}^2$.

13.5.2   807 W m^{-2}.

Chapter 14

14.5.1 $dr/r = -0.46\,dM = 0.14$.

14.5.2   a) $T = 3570$ K. b) $R_{\min}/R_{\max} = 0.63$.

14.5.3   a) 1300 pc. b) 860 years earlier; due to inaccuracies a safe estimate would be 860 ± 100 years. Actually, the explosion was observed in 1054. c) -7.4.

Chapter 15

15.5.1 $L = 2.3 \times 10^{40}$ kg m^2 s^{-1}. $dR = 45$ m.

15.5.2   a) $M = 0.5\,M_\odot$, $a = 0.49 \times 10^9$ m s$^{-2} \approx 4 \times 10^7\,g$. b) A standing astronaut will be subject to a stretching tidal force. The gravitational acceleration felt by the feet is 3479 m s$^{-2} \approx 355\,g$ larger than that felt by the head. If the astronaut lies tangentially, (s)he will experience a compressing force of $177\,g$.

15.5.3 $v = v_e(1 - GM/Rc^2)$. If $\Delta v/v$ is small, we have also $\Delta\lambda = (GM/Rc^2)\lambda_e$. A photon emitted from the Sun is reddened by $2.1 \times 10^{-6}\,\lambda_e$. In yellow light (550 nm) the change is 0.0012 nm.

Chapter 16

16.10.1   2.6 kpc and 0.9 kpc, $a = 1.5$ mag kpc^{-1}.

16.10.2   7 km s^{-1}.

16.10.3   0.01 AU.

Chapter 17

17.5.1   7.3.

17.5.2 0.5 km s^{-1}.

Chapter 18

18.6.1 $\mu = 0.0055'' \text{ y}^{-1}$.

18.6.2   a) 5.7 kpc. b) 11 kpc. Possible reasons for the discrepancy include: 1) The distance is so large that the approximations used for deriving Oort's formulae are not very good, 2) taking into account the interstellar extinction will reduce the distance in (b), 3) the peculier velocity of the star was neglected.

18.6.3   a) 3. b) ≈ 90000.

18.6.4   a) If the thickness of the disk is H, the light has to travel a distance $s = \min\{r, (H/2) \sec b\}$ in the interstellar medium. Thus the magnitude will be $m = M + 5 \lg(r/10 \text{ pc}) + as$. b) $s = 200$ pc, $m = 10.2$.

Chapter 19

19.12.1   a) 26, b) 25.

19.12.2   Diameter is about one light week ≈ 1200 AU. $M = -23.5$. Assuming this is roughly equal to the bolometric magnitude, we get $L \approx 2 \times 10^{11} L_\odot$, corresponding to $210 \, L_\odot \, \text{AU}^{-3}$.

Chapter 20

20.9.1 $r = v/H = 93$ Mpc (if $H = 75$), diameter is 35 kpc, $M = -20.4$. Potential sources of error include: 1) inaccuracy of the Hubble constant, 2) peculiar velocity of the galaxy, 3) intergalactic extinction, 4) only 2-dimensional projection is observed, and the edge depends on the limiting magnitude used.

20.9.2   If we use $H = 50 \text{ km s}^{-1} \text{Mpc}^{-1}$, we get $r = 51$ Mpc, or 74 times the distance of M31. Doubling H will halve the distance, which will still be higher than in Exercise 19.12.1. A possible explanation is the peculiar velocity of the galaxy.

20.9.3 $m_\nu = 1.5 \times 10^{-35}$ kg, or 0.00002 times the mass of the electron.

Further Reading

The following list of references is not intended as a complete bibliography. It gives a number of intermediate level or more advanced works, which can serve as starting points for those who wish to learn about some specific topic in more detail or depth.

General Reference Works

Allen, C. W. (1973): *Astrophysical Quantities*, 3rd ed. (Athlone, London)
Apparent Places of Fundamental Stars (annual) (Astronomisches Rechen-Institut, Heidelberg)
Burnham, S. (1978): *Celestial Handbook* (Dover, New York)
Lang, K. R. (1980): *Astrophysical Formulae*, 2nd ed. (Springer, Berlin, Heidelberg)
de Vaucouleurs, G., et al. (1964, 1976): *Reference Catalogue of Bright Galaxies* (University of Texas Press, Austin)

Chapter 1

Mitton, S. (ed.) (1977): *Cambridge Encyclopaedia of Astronomy* (Cape, London)
Shu, F. H. (1982): *The Physical Universe* (University Science Books, Mill Valley)
Unsöld, A. (1983): *The New Cosmos*, 3rd ed. (Springer, Berlin, Heidelberg)

Chapter 2

Explanatory Supplement to the Astronomical Ephemeris (1961) (HM Stationary Office, London)
Green, R. M. (1985): *Spherical Astronomy* (Cambridge University Press, Cambridge)

Chapter 3

Evans, D. S. (1968): *Observation in Modern Astronomy* (English Universities Press, London)
Kitchin, C. R. (1984): *Astrophysical Techniques* (Hilger, Bristol)
Wohlleben, R., Mattes, H., Krichbaum, T. (1991): *Interferometry in Radioastronomy and Radar Techniques* (Kluwer, Dordrecht)

Chapters 4 – 6

Harwit, M. (1973): *Astrophysical Concepts* (Wiley, New York)
Rybicki, G. B., Lightman, A. F. (1979): *Radiative Processes in Astrophysics* (Wiley, New York)

Chapter 7

Brouwer, D., Clemence, G. M. (1960): *Methods of Celestial Mechanics* (Academic, New York)
Danby, J. M. A. (1962): *Fundamentals of Celestial Mechanics* (MacMillan, New York)

Chapter 8

Beatty, J. K., O'Leary, B., Chaikin, A. (1981): *The New Solar System* (Cambridge University Press, Cambridge)

Brandt, J. C., Chapman, R. D. (1981): *Introduction to Comets* (Cambridge University Press, Cambridge)

Cole, G. H. A. (1978): *The Structure of Planets* (Wykeham, London)

Cook, A. H. (1980): *Interiors of the Planets* (Cambridge University Press, Cambridge)

Chapter 9

Böhm-Vitense, E. (1989): *Stellar Astrophysics*, Vols. 1 – 2 (Cambridge University Press, Cambridge)

Mihalas, D. (1978): *Stellar Atmospheres*, 2nd ed. (Freeman, San Francisco)

Novotny, E. (1973): *Introduction to Stellar Atmospheres and Interiors* (Oxford University Press, New York)

Chapter 10

Heinz, W. (1978): *Double Stars* (Reidel, Dordrecht)

Sahade, J., Wood, F. B. (1978): *Interacting Binary Stars* (Pergamon, Oxford)

Chapters 11 – 12

Clayton, D. D. (1968/1983): *Principles of Stellar Evolution and Nucleosynthesis* (McGraw-Hill, New York/University of Chicago Press, Chicago)

Novotny, E. (1973): *Introduction to Stellar Atmospheres and Interiors* (Oxford University Press, New York)

Schwarzschild, M. (1958/1965): *Structure and Evolution of the Stars* (Princeton University Press, Princeton/Dover, New York)

Chapter 13

Noyes, R. W. (1982): *The Sun, Our Star* (Harvard University Press, Cambridge, MA)

Priest, E. R. (1982): *Solar Magnetohydrodynamics* (Reidel, Dordrecht)

Stix, M. (1989): *The Sun – An Introduction* (Springer, Berlin, Heidelberg)

Chapter 14

Cox, J. P. (1980): *Theory of Stellar Pulsation* (Princeton University Press, Princeton)

Glasby, J. S. (1968): *Variable Stars* (Constable, London)

Chapter 15

Manchester, R. N., Taylor, J. H. (1977): *Pulsars* (Freeman, San Francisco)

Shapiro, S. L., Teukolsky, S. A. (1983): *Black Holes, White Dwarfs and Neutron Stars* (Wiley, New York)

Chapter 16

Dyson, J. E., Williams, D. A. (1980): *The Physics of the Interstellar Medium* (Manchester University Press, Manchester)

Longair, M. S. (1981): *High Energy Astrophysics* (Cambridge University Press, Cambridge)

Spitzer, L. (1978): *Physical Processes in the Interstellar Medium* (Wiley, New York)

Chapter 17

Hanes, D., Madore, B. (eds.) (1980): *Globular Clusters* (Cambridge University Press, Cambridge)

Spitzer, L. Jr. (1987): *Dynamical Evolution of Globular Clusters* (Princeton University Press, Princeton)

Chapter 18

Bok, B. & P. (1982): *The Milky Way*, 5th ed. (Harvard University Press, Cambridge, MA)

Mihalas, D., Binney, J. (1981): *Galactic Astronomy* (Freeman, San Francisco)

Scheffler, H., Elsässer, H. (1988): *Physics of the Galaxy and the Interstellar Matter* (Springer, Berlin, Heidelberg)

Chapter 19

Binney, J., Tremaine, S. (1987): *Galactic Dynamics* (Princeton University Press, Princeton)

Mihalas, D., Binney, J. (1981): *Galactic Astronomy* (Freeman, San Francisco)

Sandage, A. (1961): *The Hubble Atlas of Galaxies* (Carnegie Institution, Washington, D.C.)

Sérsic, J. L. (1982): *Extragalactic Astronomy* (Reidel, Dordrecht)

Weedman, D.W. (1986): *Quasar Astronomy* (Cambridge University Press, Cambridge)

Chapter 20

Peebles, P. J. E. (1971): *Physical Cosmology* (Princeton University Press, Princeton)

Raine, D.J. (1981): *The Isotropic Universe* (Adam Hilger, Bristol)

Sciama, D. W. (1971): *Modern Cosmology* (Cambridge University Press, Cambridge)

Weinberg, S. (1972): *Gravitation and Cosmology* (Wiley, New York)

Photograph Credits

We are grateful to the following institutions who kindly gave us permission to use illustrations (abbreviated references in the figure captions)

Anglo-Australian Observatory, photograph by David F. Malin
Arecibo Observatory, National Astronomy and Ionosphere Center, Cornell University
Arp, Halton C., Mount Wilson and Las Campanas Observatories (colour representation of plate by Jean Lorre)
Big Bear Solar Observatory, California Institute of Technology
Catalina Observatory, Lunar and Planetary Laboratory
CSIRO (Commonwealth Scientific and Industrial Research Organisation), Division of Radiophysics, Sydney, Australia
ESA (copyright Max-Planck-Institut für Aeronomie, Lindau, Harz, FRG)
European Southern Observatory (ESO)
Helsinki University Observatory
High Altitude Observatory, National Center for Atmospheric Research, National Science Foundation
Karl-Schwarzschild-Observatory Tautenburg of the Central Institute of Astrophysics of the Academy of Sciences of the GDR (Archives)
Lick Observatory, University of California at Santa Cruz
Lowell Observatory
Lund Observatory
Mount Wilson and Las Campanas Observatories, Carnegie Institution of Washington
NASA (National Aeronautics and Space Administration)
National Optical Astronomy Observatories, operated by the Association of Universities for Research in Astronomy (AURA), Inc., under contract to the National Science Foundation
National Radio Astronomy Observatory, operated by Associated Universities, Inc., under contract with the National Science Foundation (Plate 27: observers Richard J. Tuffs, Richard A. Perley, Martin T. Brown, Stephen F. Gull; Plate 31: observers Farhad Yusef-Zadeh, Mark R. Morris, Don R. Chance; Plate 33: observers Frazer N. Owen, Richard A. White; Plate 36: observers Peter A.G. Scheuer, Robert A. Laing, Richard A. Perley)
NRL-ROG (Space Research Laboratory, University of Groningen)
Palomar Observatory, California Institute of Technology
Yerkes Observatory, University of Chicago

Name and Subject Index

Colour Supplement

Photograph on previous page. NGC 6769-71 is a magnificent triple system, far south in the constellation of Pavo (the Peacock). The two left galaxies, NGC 6769 (uppermost) and NGC 6770, are of nearly the same magnitude, whereas NGC 6771 (to the right) is slightly fainter. All three show peculiarities indicating the presence of gravitational interaction. (Photograph European Southern Observatory)

Chapter 2. **Spherical Astronomy**

Plate 1. A long exposure shows the apparent motion of the celestial sphere. In photographs the colours of the stars are seen more clearly than with the naked eye. The brightest object is the planet Venus. (Photograph Pekka Parviainen)

Plate 2. Atmospheric refraction has distorted and discoloured the image of the setting Sun. A rare green segment can be seen above the Sun, and a mirage has formed below it. (Photograph Pekka Parviainen)

Chapter 3. **Observational Instruments**

Plate 3. Astronomical observatories are usually built in high, dry places with a clear sky. The picture shows the forest of domes at Kitt Peak in Arizona in the United States. At the right is the building of the 4 m telescope. (National Optical Astronomy Observatories)

Plate 4. The most powerful radio telescope in the world is the VLA or Very Large Array in New Mexico in the United States. It is composed of twenty-seven movable 25 m antennas. (Photograph National Radio Astronomy Observatory)

Chapter 8. **The Solar System**

Plate 5. Mercury can easily be seen with the naked eye in the morning or evening sky, if only one looks at the right time. The picture shows Venus (in the *upper part*) and Mercury (*lower*) in March 1985. (Photograph Hannu Heiskanen)

Colour pictures related to all chapters are collected in these pages

Plate 6. Eclipses are astronomical phenomena that tend to attract the attention of the general public. The picture shows a total eclipse of the Moon in January 1982. (Photograph Juhani Laurila)

Plate 7. The first colour pictures from the surface of Venus were obtained in 1982. Panoramic views from Venera 13 (*above*) and Venera 14 (*below*). At the edges, the camera is looking at the horizon and in the centre, at the ground around the base of the space probe. Parts of the base, the protective covers of the camera and a colour map can be seen

Plate 8. The surface of the Earth as seen from space. The picture shows a part of Egypt with the Nile in the middle and the Red Sea in the background. (Photograph Gemini 12/NASA)

Plate 9. The surface of the Moon was visited by twelve astronauts in 1969–1972. On the last visits, a lunar vehicle was used to extend the domain of exploration trips. (Photograph Apollo 15/NASA)

Plate 10. The red desert of Mars in the Chryse Plain. (Viking 1/NASA)

Plate 11. The surface of Jupiter shows many atmospheric structures. The picture was taken by Voyager 2 from a distance of 20 million kilometres. Io is seen over the Red Spot, and Europa on the right. (NASA)

Plate 12. The strangest moon of Jupiter is Io, where the Voyager space probes observed several volcanic eruptions. One of the largest active areas is the Pele region. (Voyager 1/NASA)

Plate 13. An approach picture of Saturn taken by Voyager 2 shows the planet and its rings from a distance of 40 million kilometres. The colours in the picture have been artificially enhanced in order to bring out the atmospheric details more clearly. (NASA)

Plate 14. Neptune was imaged by Voyager 2 in August 1989. The bright spots are high-altitude methane clouds, and the red hue is due to semitransparent haze covering the planet. (NASA)

Plate 15. The nucleus of Halley's comet was imaged by the Halley Multicolour Camera on board ESA's Giotto spacecraft in March 1986. The nucleus is seen as a dark elongated area in the center of the photograph. The bright areas are jets of dust emanating from the comet. (ESA/MPI für Aeronomie. Courtesy Dr. H.U. Keller)

Plate 16. Objective prism spectra of Regulus (*above*) and Saturn (*below*). The spectrum of Saturn is essentially the reflected spectrum of the Sun (spectral class G2V). The spectral class of Regulus is B7V

Plate 17. A part of the solar spectrum in the green light (about 520–540 nm). The dark absorption lines on the left are due to magnesium. (High Altitude Observatory)

Chapter 12. **Stellar Evolution**

Plate 18. M16 is a star cluster surrounded by interstellar matter. New stars are possibly born from the protruding condensates. (Photograph NASA/Hubble Space Telescope)

Chapter 13. **The Sun**

Plate 19. X-ray image of the Sun showing active regions (*bright areas*) and dark coronal holes. The picture was obtained by the Skylab x-ray telescope, which is sensitive to the wavelengths 0.2–3.2 nm and 4.4–5.4 nm. (NASA)

Plate 20. In 1980, images of the Sun at visual and ultraviolet wavelengths were obtained by the Solar Maximum Mission satellite. This false colour picture shows the outward density decrease in the corona. The streamer going to the upper right extends 1.5 million kilometres from the surface. (NASA)

Chapter 16. **The Interstellar Medium**

Plate 21. The region around ρ Ophiuchi as seen by the Infrared Astronomical Satellite (IRAS). The same region is seen in optical wavelengths in Fig. 16.8. The cool dust shows up in this far-infrared picture in emission. (NLR-ROG)

Plate 22. The interstellar dust appears in visible light as dark clouds or as reflection nebulae. The photograph shows the Trifid Nebula M20, which is a reddish gas (emission) nebula. Dark streaks of dust can be seen in front of it and a bluish reflection nebula below it. (National Optical Astronomy Observatories)

Plate 23. Different forms of interstellar matter are found in the Eagle Nebula M16 in Serpens: red emission nebulae and dark streaks of interstellar dust. (National Optical Astronomy Observatories)

Plate 24. The brightest infrared source in the sky is the nebula around η Carinae. The red emission is from hydrogen gas and the dark areas are dust. (European Southern Observatory)

Plate 25. The Dumbell Nebula M27 in Vulpecula. A hot star at the centre of the nebula ionizes the gas. The greenish colour is from twice ionized oxygen; the reddish hue of the outer edges, from ionized nitrogen and from the hydrogen Balmer lines. (European Southern Observatory)

Plate 26. Another beautiful example for a planetary nebula: the Ring Nebula M57 in Lyra. (European Southern Observatory)

Plate 27. The radio source Cassiopeia A is the remnant of a supernova that exploded approximately 300 years ago. The source was observed by the Very Large Array (VLA) in 1982–1983 at a wavelength of 6 cm. The field of view is 6×6 arc min. (National Radio Astronomy Observatory)

Plate 28. The Vela supernova remnant. The picture is a composite of three black-and-white photographs taken with the ESO 1 m Schmidt telescope. The filamentary gas in the supernova remnant is visible over a six-degree area. (European Southern Observatory)

Plate 29. The structure of the galactic magnetic field is apparent in this whole-sky radio map. The red areas are bright at radio wavelengths; the blue ones are faint. The interstellar medium is mainly confined to the galactic plane, but filaments along the magnetic field lines may extend quite a long way out of the plane. The map is based on measurements at 73 cm wavelength made at Bonn, Jodrell Bank and Parkes. (Max-Planck-Institut für Radioastronomie)

Chapter 18. **The Milky Way**

Plate 30. The image of the centre of our galaxy was produced from observations by the IRAS satellite. The yellow and green knots and blobs scattered along the band of the Milky Way are giant clouds of interstellar gas and dust heated by nearby stars. (NASA)

Plate 31. The bright radio source Sagittarius A is located at the galactic centre. This high-resolution map was made with the VLA in 1982–1984. The wavelength was 20 cm and the region shown covers approximately as large an area as the full moon. The yellow dot in the middle is a radio point source at the exact centre of the Milky Way. Its nature is unknown. In the upper part of the picture lies the "continuum arc", where narrow filamentary structures can be seen. The side of the picture corresponds to about 70 pc. (National Radio Astronomy Observatory)

Chapter 19. **The Galaxies**

Plate 32. The Andromeda Galaxy M 31 is our nearest large galaxy. Its central parts can be seen with the nak-

ed eye. A satellite trail is visible across the image. (National Optical Astronomy Observatories)

Plate 33. False-colour CCD image of the central galaxy (3 C 75) in the cluster of galaxies Abell 400. The two fuzzy images are the twin nuclei of the galaxy. The two objects with spikes are bright stars. Observations were made with the Kitt Peak 0.9 m telescope. (National Radio Astronomy Observatory)

Plate 34. Enhanced colour image of the galaxy M 81 in Ursa Major, showing the blue H II regions in the spiral arms and the red, old stars in the central bulge. (Halton C. Arp)

Plate 35. The Southern radio galaxy Centaurus A is a peculiar galaxy that radiates at all wavelengths. Visually it looks like an old, round elliptical galaxy crossed by a strong dust lane. (National Optical Astronomy Observatories)

Plate 36. Cygnus A (3 C 405), a typical radio galaxy. The actual galaxy is located at the bright point in the middle, between the two large radio lobes on both sides. The image was obtained with the VLA at 6–20 cm wavelength. (National Radio Astronomy Observatory)

Plate 1

Plate 2

5

Plate 3

Plate 5

Plate 6

Plate 8

Plate 9

Plate 10

Plate 11

9

Plate 12

Plate 13

Plate 14

Plate 15

Plate 16

Plate 17

11

Plate 18

Plate 19

Plate 20

Plate 21

13

Plate 22

Plate 23

Plate 24

Plate 25

Plate 26

15

Plate 27

Plate 28

Plate 29

Plate 30

Plate 31

Plate 32

Plate 33

Plate 34

Plate 35

Plate 36